Impunity:
Countering Illicit Power in War and Transition

Edited By
Michelle Hughes
and
Michael Miklaucic

This book is published in the United States as a joint effort of the Center for Complex Operations (CCO) and the Peacekeeping and Stability Operations Institute (PKSOI). The views expressed are those of the authors and do not necessarily reflect the official policy or position of CCO, PKSOI, National Defense University, the Department of the Army, the Department of Defense, or the U.S. Government. Authors of this publication enjoy full academic freedom, provided they do not disclose classified information, jeopardize operations security, or misrepresent official U.S. policy. Such academic freedom empowers them to offer new and sometimes controversial perspectives in the interest of furthering debate on key issues. This report is cleared for public release; distribution is unlimited.

*****

This publication is subject to Title 17, United States Code, Sections 101 and 105. It is in the public domain and may not be copyrighted.

*****

Comments pertaining to this report are invited and should be forwarded to: CCO Director of Research, Center for Complex Operations, Institute for National Strategic Studies, National Defense University, Abraham Lincoln Hall (Building 64), 260 Fifth Avenue, Fort Lesley J. McNair, Washington, DC 20319-5066

*****

All CCO publications are available on the CCO homepage for electronic dissemination; *http://cco.ndu.edu/Home.aspx*. All Peacekeeping and Stability Operations Institute (PKSOI) publications are available on the PKSOI homepage for electronic dissemination. Hard copies of this report also may be ordered while copies last from our homepage. PKSOI's homepage address is: *https://pksoi.army.mil*

ISBN: 978-0-9861865-7-8

Contents

FOREWORD v

       Lieutenant General H.R. McMaster

ACKNOWLEDGMENTS..ix

INTRODUCTION 1

       Michael Miklaucic

PART 1 — Case Studies from Conflict

1. *Criminal Patronage Networks and the Struggle to Rebuild the Afghan State*..........11

       Carl Forsberg and Tim Sullivan

2. *Jaish al-Mahdi in Iraq*..40

       Phil Williams and Dan Bisbee

3. *Haiti: The Gangs of Cité Soleil*..67

       D.C. (David) Beer

4. *Liberia: Durable Illicit Power Structures*..99

       William Reno

5. *Traffickers and Truckers: Illicit Afghan and Pakistani Power Structures with a Shadowy but Influential Role*..125

       Gretchen Peters

6. *Colombia and the FARC: From Military Victory to Ambivalent Political Reintegration?*...150

       Carlos Ospina

7. *The Philippines: The Moro Islamic Liberation Front - A Pragmatic Power Structure?*..170

       Joseph Franco

8. *Sierra Leone: The Revolutionary United Front*..190

       Ismail Rashid

9. *Sri Lanka: State Response to the Liberation Tigers of Tamil Eelam as an Illicit Power Structure*..217

       Thomas A. Marks and Lt. Gen. (Ret.) Tej Pratap Singh Brar

Part 2 — Confronting Illicit Power: Understanding Enablers, Ours and Theirs

10. *It Takes a Thief to Catch a Thief: Illicit Power and the Intelligence Challenge*..................241

    Michelle Hughes

11. *Weapons Trafficking and the Odessa Network: How One Small Think Tank was Able to Unpack One Very Big Problem, and the Lessons It Teaches Us*..................267

    David E. A. Johnson

12. *Financial Tools and Sanctions: Following the Money and the Joumaa Web*..................289

    Robert ("J. R.") McBrien

13. *Recruitment and Radicalization: The Role of Social Media and New Technology*..................313

    Maeghin Alarid

Part 3 — Licit Transitions: Building Institutions and Strengthening Capacity for Success

14. *Make It Matter: Ten Rules for Institutional Development that Works*..................331

    Mark Kroeker

15. *The Hitchhiker's Guide to Intelligence-Led Policing*..................341

    Clifford Aims

16. *Security Sector Reconstruction in Post-Conflict: The Lessons from Timor-Leste*..................347

    Deniz Kocak

17. *A Granular Approach to Combating Corruption and Illicit Power Structures*..................367

    Scott N. Carlson

CONCLUSION. *What Should We Have Learned by Now? Enduring Lessons from Thirty Years of Conflict and Transition*..................377

    Michelle Hughes

LIST OF CONTRIBUTORS..................400

FOREWORD

Strategies that weaken illicit power structures and strengthen legitimate state authority are vital to national and international security. As Dr. Henry Kissinger observed, we may be "facing a period in which forces beyond the restraints of any order determine the future." Because threats to security emanate from disorder in areas where governance and rule of law are weak, defeating terrorist, insurgent, and criminal organizations requires integrated efforts not only to attack enemy organizations, but also to strengthen institutions essential to sustainable security.

Successful outcomes in armed conflict require confronting illicit networks. A failure to do so effectively frustrated efforts to consolidate gains in Afghanistan and Iraq, and after more than a decade of war and development, the international community and the governments of those countries, continue to contend with the violence and instability that are the result. In Afghanistan, corruption and organized crime networks perpetuate state weakness and undermine the state's ability to cope with the regenerative capacity of the Taliban. The failure to counter militias and Iranian proxies that infiltrated the government and security forces in Iraq led to a return of large scale communal violence and set conditions (along with the Syrian Civil War) for the rise of a terrorist proto-state and a humanitarian catastrophe that has adversely impacted the entire Middle East. These and other cases illustrate how governments and international actors struggle to establish security and rule of law, and reveal incomplete plans and fragmented efforts that fail to address the causes of violence and state weakness.

While challenging, success in confronting illicit power structures is not impossible. While still works in progress, successful efforts, such as those in Colombia and Sierra Leone, are the result of integrated diplomatic, military, economic, development, informational, intelligence, and law enforcement efforts directed toward well-defined political outcomes. The case studies and analyses in this volume make clear that understanding the dynamics associated with illicit power and state weakness is essential to preventing or resolving armed conflict.

These case studies also point out that confronting illicit power requires coping with political and human dynamics in complex, uncertain environments. People fight today for the same fundamental reasons the Greek historian Thucydides identified nearly 2,500 years ago: fear, honor and interests. They further remind us that that illicit power structures often depend on the perpetuation of violence and the conflict economy. Crafting effective strategies to address the challenge of weak states must begin with an understanding of the factors that drive violence, weaken state authority, and strengthen illicit actors and power structures. Terrorist, insurgent, and criminal networks exploit fear and anger over injustice, portraying themselves as patrons or protectors of a community in competition with others for power, resources, or survival. Thus military and law enforcement capabilities provide only one component of what must be comprehensive, civilian and military approach to confronting illicit power.

Because of the political basis for illicit power, mediation between parties in conflict to reach an accommodation is often necessary to end violence and isolate illicit networks from popular support. However, attempts to mediate before security conditions improve and illicit network control over territory and populations is weakened are likely to fail. Political and security efforts must be aligned to consolidate gains and achieve sustainable outcomes. A comprehensive approach is necessary because consolidating security gains requires efforts both to strengthen the state and weaken a range of criminalized adversaries. *Ultimately, the state must possess the strength and legitimacy to control its territory, contend with the regenerative capacity of illicit power networks, and convince the vast majority of the population to advance their interests through politics rather than violence.*

Unfortunately, as many of the case studies in this volume reveal, plans are often driven by what international actors prefer to do or would like to avoid doing rather than what is necessary to address the causes of violence and state weakness. Because the greatest obstacle to confronting illicit power is oftentimes a lack of political will, international actors must help persuade leaders that it is in their interest to undertake necessary reforms. Those working on reform must understand that control of institutions is often the prize that drives competition between illicit organizations. Security sector and judicial reform require special emphasis to insulate investigative and adjudicative bodies from infiltration and subversion. Paradoxically, avoiding state building or sidestepping the political causes of state weakness in the hope of avoiding costly or protracted commitments often increases costs and extends efforts in time.

International organizations and indigenous leaders must also galvanize the will to defeat illicit organizations and undertake reforms to strengthen legitimate state authority, and demonstrate commitment through action. To isolate illicit networks from popular support, leaders must craft and build confidence in a long-term vision for the future in which communities believe their interests will be advanced and protected. Communications efforts must bolster the legitimacy of the state and discredit illicit organizations. Information campaigns must draw attention to the brutality and venality of illicit organizations and trace popular grievances back to their actions. Exposing the criminality of key actors and applying national and international law enforcement, financial actions, and other sanctions against illicit network leaders and facilitators is important to weaken illicit power structures psychologically, as well as physically.

Maintaining the will to defeat illicit power and consolidate security gains is difficult because the standard of success for illicit networks is low relative to the standard for those who confront the malignant effects of terrorism, insurgency, and organized crime. Illicit groups need only to impose pain rather than secure populations, incite fear rather than build confidence, foment hatred rather than foster tolerance, protract conflict rather than arbitrate peace, subvert institutions rather than build state effectiveness, and perpetuate destitution rather than rekindle hope. Licit actors are held to a much higher bar. But meeting the high standard for success, especially in traumatized and fragmented societies, requires that we, and our host nation partners, make a concerted effort to learn from recent and ongoing experiences and adapt efforts to changing local conditions. The case studies in this volume are an excellent starting point for learning and adapting to the problem of illicit power.

As readers study this book, however, they might be discouraged by the fact that our record of learning is poor. As the analysis reveals, failures to confront illicit power often stem from either an ignorance of history or the simplistic application of erroneous historical lessons. The neglect of history is often willful; wishful thinking makes the future appear easier and fundamentally different from the past. Budget pressures, the perception that the nation is fatigued by recent wars, and persistent fascination with technology encourage wishful thinking and a reductionist vision of future armed conflict divorced from its political and human nature.

From a military standpoint, a vision that uncouples the military instrument from policy as well as the causes of violence, portrays future conflict mainly as targeting exercises. Precision strikes, whether executed by drones or special operations forces, promise to deliver security from afar without requiring complex efforts associated with consolidating gains politically or confronting illicit power structures. While emerging technologies are essential for military effectiveness, employing these advanced capabilities without a strategy that addresses the causes of violence and state weakness confuses military activity with progress toward wartime goals. Paradoxically, the advanced military technologies that some argue will solve the problem of future armed conflict from standoff range are likely to drive future enemies further away from conventional battlegrounds. They tend to push illicit power structures closer to civilian populations and offer few solutions to the kinds of problems described in this volume.

A serious and effective effort to meet the challenges of illicit power in the 21st century will require technology, global partnership, and an integrated, comprehensive campaign driven by international commitment and broad political will. Of the many important lessons that emerge from these essays the most important is to be skeptical of concepts that divorce conflict from its political and human nature, particularly those that promise fast, cheap victories through technology while ignoring the need to confront illicit power in war and transition.

Lieutenant General H.R. McMaster (USA)
Director, Army Capabilities Integration Center and
Deputy Commanding General, Futures,
U.S. Army Training and Doctrine Command

ACKNOWLEDGMENTS

Behind every book is another book, and this is very true with "*Impunity: Countering Illicit Power in War and Transition.*" The book itself is the product of a process that began in 2005, with numerous conferences, workshops, seminars, and editorial sessions along the way. Among the many we must gratefully acknowledge is Michael Carr who served tirelessly as a technical editor for the book, and was a critical sounding board for us as we attempted to synthesize a broad range of research and ideas into a coherent whole. We cannot thank him enough. We also acknowledge the interns of the Center for Complex Operations at National Defense University (CCO) for their valuable contributions. Over the years 2012 – 2015 quite a few CCO interns played one role or another – some were organizers and note-takers at editorial workshops, others compiled expert lists, drafted memos and chapter abstracts, collected biographical information, and a variety of other critical tasks. They include Molly Jerome, Shannon Corson, Ava Cacciolfi, Thi Le, Hiram Reynolds, Samantha Fletcher, David Robinson, and Tamara Tanso. There may be others and if we have forgotten any, please forgive us. The book would not have been possible without support from the Office of the Secretary of Defense, with special thanks to Daniel Plafcan. We received additional critical support from the Department of Defense Minerva Initiative and particularly want to single out Dr. Erin Fitzgerald for supporting our effort. Dr. Fitzgerald encouraged us to use this book as a pilot effort to track the use of social science research in defense policy-making, and in professional military education. To do so we assembled a professional team consisting of Christopher Holshek and David Gordon whom we would like to thank. The book layout was done at the Peacekeeping and Stability Operations Institute of the U.S. Army (PKSOI) by the patient and tireless Jennifer Nevil; we thank PKSOI and specifically Dr. William Flavin and Dr. Karen Finkenbinder for their partnership in this undertaking. Back at NDU Judy Kim helped us with the cover art. Nadia Gerspacher of the United States Institute of Peace was one of the pioneers of the project leading to this publication, and we are grateful for her work in the early stages of the project. We also thank Larry Sampler for his early contribution, particularly the laconic but insightful observation that to counter illicit power structures "you have to either make them licit, take away their power, or dismantle their structure." Over the course of the decade since the original illicit power structures project was initiated it went into dormant periods, as project participants and investigators were called to different assignments. At times the project – and the book – were orphaned. We owe a great debt to Michael Dziedzic who was relentless in his determination that the intellectual capital created in the course of the project not be lost. He kept the project, and the book concept, going when others had dropped it. Finally and perhaps most importantly we thank all the authors who contributed to this book and to the larger effort. Their patience and professionalism are deeply appreciated.

INTRODUCTION

Michael Miklaucic

Much of the twentieth century was dominated by the competition between rival ideologies for structuring political life. Democratic capitalism ultimately triumphed over fascism in 1945, then over Communism in 1991. But this triumph did not presage "the end of history," as some optimistically wrote.[1] A more accurate portrayal of what lay in store was "the coming anarchy."[2] The end of the Cold War exposed the fragility underlying the Westphalian, rule-based system of sovereign states, particularly in those scores of states born during the decolonization era of the mid-twentieth century. Nearly 25 years after the demise of Communism, no fewer than 65 of the 193 United Nations member states are rated as "high warning," "alert," or "high alert" in the 2015 Fragile States Index, and only 53 are rated "very sustainable" through "stable."[3]

In his forward to the 2015 National Military Strategy of the United States, former Chairman of the Joint Chiefs of Staff General Martin Dempsey tells us, "Today's global security environment is the most unpredictable I have seen in 40 years of service.... We now face multiple, simultaneous security challenges from traditional state actors and transregional networks of sub-state groups—all taking advantage of rapid technological change." The Military Strategy "addresses the need to counter revisionist states that are challenging international norms as well as violent extremist organizations (VEOs) that are undermining transregional security," and central to our task "is strengthening our global network of allies and partners."[4] In the current post-"Big Footprint" era, when international and U.S. national security depend ever more on our partners throughout the world, the persistent fragility of so many states—and, indeed, of the state system itself—is troubling.

If our national and international security rests on a foundation of effective partnerships, then the epic challenge of the twenty-first century must be to sustain and strengthen effective sovereign partners to share this burden. But we find that the contemporary state is under duress, challenged by millenarian religious movements, divisive identity-based conflicts, pervasive anarchical violence, and what Dr. Brzezinski called the "global political awakening."[5] Some argue that the state is in decline, and even that we are entering a new Dark Ages.[6]

Among the most insidious challenges to the contemporary state is the corrosive capture and use of illicit power, which both weakens the state and creates among popula-

---

[1] Francis Fukuyama, "The End of History?" *National Interest*, Summer 1989.

[2] Robert D. Kaplan, "The Coming Anarchy: How Scarcity, Crime, Overpopulation, Tribalism, and Disease Are Rapidly Destroying the Social Fabric of Our Planet," *Atlantic*, Feb. 1994.

[3] Fund for Peace, "Fragile States Index 2015," http://library.fundforpeace.org/library/fragilestatesindex-2015.pdf.

[4] Joint Chiefs of Staff, "The National Military Strategy of the United States of America 2015," www.jcs.mil/Portals/36/Documents/Publications/National_Military_Strategy_2015.pdf.

[5] Zbigniew Brzezinski, "The Global Political Awakening," *New York Times*, Dec. 16, 2008.

[6] See, for example, Martin van Creveld, *The Rise and Decline of the State* (Cambridge, UK: Cambridge Univ. Press, 1999); Phil Williams, "From the New Middle Ages to a New Dark Age: The Decline of the State and U.S. Strategy," monograph, Strategic Studies Institute, U.S. Army War College, June 2008, www.strategicstudiesinstitute.army.mil/pubs/display.cfm?pubID=867.

tions the perception of inefficacy, incompetence, and, ultimately, illegitimacy. Part 1 of this book presents case studies of illicit power, and the domestic and international responses to it, in a range of states. These nine country case studies cumulatively alert analysts and practitioners to the gravity of the challenge posed by illicit power. Following these, the four chapters in Part 2 describe fundamental characteristics of illicit power structures, and the elements of the operational environment that enable illicit power to succeed, as well as those that enable the international community to counter it. The outcome of the struggle between illicit power and legitimate, accountable governance in any state hinges on the competition between them over these enablers. Part 3 gives four approaches to mitigating illicit power, which the international community must become adept at using and flexibly adapting to the specific conditions faced in particular environments. Part 4, the book's concluding section, offers insights that help clarify both the nature and the gravity of the challenge that is illicit power, as well as the most effective approaches to mitigating it.

Power captured and applied illicitly corrupts legitimate efforts to resolve conflict, make and maintain peace, and build effective states. Illicit power is no recent intruder in the affairs of states and men. But the unprecedented access of criminals, terrorists, and insurgents to lethal technology, military-grade weaponry, staggering sums of money, and world-class professional and technical services—including legal, financial, information and communications technology, and security—constitutes a game changer. In most of the cases examined in this book, the possessors of illicit power have been able to challenge the fundamental roles and legitimacy of the countries where they reside. Most notably and dangerously, the agents of illicit power have in some cases successfully challenged the state's monopoly of the legitimate use of coercive force. Note that we refer here not only to the state's monopoly of force, which is only in the rarest cases absolute, but to its monopoly of the *legitimate use* of that force. The legitimacy that the state alone should enjoy can be so compromised by illicit power that it migrates to any entity that can reliably provide rudimentary public services, such as basic security.

What do we mean by "illicit power"? The use of power, whether economic, political, or military, by unsanctioned structures, organizations, or networks, operating outside the parameters of law, national custom, or broad public acceptance, is illicit. Illicit power structures are entities that seek political or economic power through the use of violence, typically supported by criminal economic activity. The leadership of these structures may be within, or parallel to, the state, or they may constitute an armed opposition to it. Illicit power structures operate primarily outside the framework for establishing and maintaining the rule of law. Moreover, they erode that framework, although they frequently maintain activities and operations within the legal and licit sector as well. They are often referred to as "nonstate armed groups," but this term is insufficient since it would exclude illicit organizations such as the Janjaweed in western Sudan, the Interahamwe in Rwanda, and the criminal patronage networks in Afghanistan, none of which can fully be detached from the state.[7]

Indeed, illicit power structures may be embedded within the state or may be an armed opposition to the state. A perverse iteration is that both the state and its violent

---

[7] See chapter 5 of this volume: "Traffickers and Truckers" by Gretchen Peters.

opposition may be illicit power structures, colluding to profit from conflict and insecurity in war and transitions. This book uses "illicit power," "illicit power structures," and "illicit networks" to describe unsanctioned agents and organizations that pursue and use power to secure and protect parochial interests without regard for the larger public welfare. Included are terrorist, insurgent, and criminal organizations and agents. Importantly, unlike sanctioned agents or the state itself, illicit power is not accountable to the constituency of the territory where it operates or to the representatives of the rule-based system of states, as represented by the United Nations or other intergovernmental bodies. Illicit power acts with state-corrupting, system-eroding impunity.

The terms "illicit," "illegitimate," "illegal," and "informal" are sometimes used interchangeably and, thus, are often confused. In this book, we understand informal power structures to include a broad range of socially and culturally embedded hierarchies, which often existed long before the formal state institutions of modern government, or which are created as a way of consolidating power within specific factions or groups. They may be licit or illicit, and they may both complement and compete with formal governmental organizations. Illicit power structures are frequently but not always informal structures in that they are detached from the state, but "informal power structures" is a broader category that may include tribal, religious, clan, and local municipal structures.

We also distinguish illicit power from illegitimate power, since communities that benefit from illicit power will often regard its exercise as politically legitimate. Illicit power structures may even obtain power through legitimate electoral processes, or they may have their exercise of power legitimized by ill-advised elections organized by the international community (as in Bosnia). Even though illicit power structures may be regarded as legitimate by any identity group that benefits from their patronage, they will almost certainly be considered illegitimate by those that do not. Consider, for example, the Sinaloa drug trafficking cartel, which provides employment, security, and many rudimentary services to numerous Mexicans who may deem it a legitimate provider. Yet even within its area of operations, the majority views it as a criminal organization. Moreover, the cartel is not accountable to anyone or any authority other than its own membership. It acts with impunity; thus, it is an illicit power structure.

Finally, we distinguish illicit power structures from *illegal* power structures. The law in the form of constitutions, statutes, treaties, written regulations, and other formal instruments often lags far behind political, social, and economic developments. This is especially true in developing countries, where, for example, until very recently there were no laws prohibiting corruption, money laundering, or membership in prohibited organizations. Certain behaviors may undermine state performance and even state legitimacy without being illegal under the current laws of the given state. Yet these behaviors by illicit power structures must be countered even though the current law is inadequate—in some cases by legislating appropriate statutes, but more often through applying a broad range of transparency and accountability techniques, both public and private.

Diplomats, senior soldiers, policymakers, and development professionals have been aware of the presence and malignant impact of illicit power for decades. And yet, they have systematically chosen to sideline its careful examination, preferring instead to focus on more traditional forces within the international system. After being preoccupied throughout the Cold War with the Communist bloc and its proxies, Western powers next turned their attention to despotic, disruptive, and especially failed or failing states once the Communist world imploded in 1991. NATO enlargement, democratic expansion, and trade liberalization displaced the struggle against Communism. Ironically, liberalization of finance, labor markets, travel, communication, and rapid privatization in many ways enabled the growth and diversification of illicit power structures and networks.

The last vestige of the illusion of Westphalian universality on which the modern international system was optimistically built dissipated in the 1990s. The veneer of the Yugoslavian state dissolved, revealing strong and tangled currents of power in pursuit of brazenly parochial interests: Serb interests, Croatian interests, Kosovar interests. A world away from Europe, the first popularly elected president of Haiti in a generation was overthrown by a military coup representing conservative political forces unwilling to lose the perquisites of the status quo. Another world away, the clearly stated will of the Timorese people was subverted by armed militias in league with their former Indonesian overlords.

Even though illicit power has been one of the predominant obstacles to peace, stabilization, and state building, the international community has routinely overlooked this threat and, hence, been ill-prepared to confront it. In two cases in this volume, the failure to recognize this risk brought UN missions to the brink of failure. In Sierra Leone, the United Nations engaged in "best case" analysis, presuming that the Revolutionary United Front (RUF) intended to comply with the Lomé Peace Agreement. But when UN peacekeepers encroached on the RUF's control of diamond-mining districts in 2000, it took 500 peacekeepers hostage. The result, according to case study author Ismail Rashid, was that the UN Mission in Sierra Leone "tottered on the brink of collapse." In Haiti, President Jean-Bertrand Aristide exploited gangs in the Port-au-Prince ghetto of Cité Soleil as instruments of political warfare. In 2006, two years after the United Nations Stabilization Mission in Haiti (MINUSTAH) had deployed, these gangs carried out a campaign of kidnappings, murder, rape, and extortion that nearly derailed both the recently elected government of René Préval and MINUSTAH.

It is the failure to recognize the threat and take the challenge seriously that makes illicit power insidious. The new world order that so many hoped would succeed the long and frightening Cold War and replace violent ideological competition was devolving instead into the coming anarchy—a reality that many have willfully ignored. Humanitarian disaster after humanitarian disaster, some manmade, others natural, repeatedly drew in and continue to draw in the United States and the Western powers, who typically show up with no real understanding of the underlying power currents in the countries they are entering. Even development professionals, often with significant knowledge of the in-country environment and supported by host-country nationals, still lacked sufficient insight to understand the violent currents of illicit power.

Some prescient analysts, such as Stephen Stedman, did detect the pattern of this troubling trend across a wide geographical, cultural, and economic range. He recognized the malignant agents of illicit power as absolute, limited, or opportunistic spoilers in peace processes. According to Stedman, no compromise, agreement, or deal can be reached with an absolute spoiler, for whom the adversary's very existence is a form of illegitimacy. A limited spoiler, on the other hand, has specific finite demands, which, if met, may lead the spoiler to potentially successful peace negotiations. The opportunistic spoiler will continue to impede or subvert negotiations until a point is reached at which the costs to the spoiler of remaining outside the emerging agreement framework exceed the benefits of continued disruption.[8]

Other analysts, such as David Keen and William Reno, recognized the perverse yet rational incentives of conflict economies.[9] Spoilers evolved into conflict entrepreneurs, whose commitment to ending conflict was subordinate to their lucrative exploitation of the conflict. Meanwhile, world leaders' eyes remained fixed throughout the 1990s on tectonic events taking place in Russia and Eastern Europe. The ever-inflammatory potential of the Palestinian-Israeli dispute preoccupied the final year of the Bill Clinton presidency. The health of the international system as a whole remained a forest hidden behind several very large and problematic trees.

Then, on September 11, 2001, a profound discontinuity occurred, one of those historic singularities in time to which one can attribute much of what follows: the al-Qaeda attacks on the United States. The subsequent Global War on Terror was partly based on the realization that U.S. national security was threatened as much by state weakness and the illicit power structures that weaken states and take advantage of state weakness as by strong adversary states. The National Security Strategy of the United States, published in September 2002, gave central emphasis to the threat of illicit power: "Shadowy networks of individuals can bring great chaos and suffering to our shores for less than it costs to purchase a single tank. Terrorists are organized to penetrate open societies and to turn the power of modern technologies against us."[10]

When, on May 1, 2003, President George W. Bush proclaimed that "major combat operations in Iraq have ended," there was widespread expectation that this indicated the imminent end of Operation Iraqi Freedom, and an end of the second Iraq war. It was neither. The armed forces of the dictator Saddam Hussein had been decisively defeated, but powerful illicit power structures and networks arose to sustain—and, indeed, vastly intensify—the conflict that the U.S. invasion had ignited in Iraq. As case study authors Phil Williams and Dan Bisbee write, "The invasion itself, combined with the ineptitude of the occupation authorities, created several distinct but powerful strands of resistance, which erupted as a complex insurgency." By mid-2004, it was clear that major combat operations in Iraq had not ended, for a previously unrecognized adversary (or set of adversaries) had emerged. Meanwhile, Operation Enduring Freedom in Afghanistan

---

[8] Stephen John Stedman, "Spoiler Problems in Peace Processes," *International Security* 22, no. 2 (Fall 1997): 553.

[9] See for example, David Keen, "The Economic Functions of Violence in Civil Wars," ISS Adelphi Paper 320, 1998, 189; William Reno, *Warlord Politics and African States* (Boulder, CO: Lynne Rienner, 1998).

[10] White House, "The National Security Strategy of the United States of America," Sept. 2002, www.state.gov/documents/organization/63562.pdf.

was encountering a resurgent Taliban, supported by the Haqqani Network (described by Gretchen Peters in this volume) and by Pakistani military intelligence.

In toppling the Taliban and Saddam Hussein regimes, the U.S. armed forces performed admirably, both in Afghanistan and Iraq. The armed forces of both regimes were defeated within weeks, yet the wars raged on for years. The U.S. Joint Forces Command (USJFCOM) received a "war-fighter challenge" from the field, to provide insight into the nature of the emerging illicit power structures and networks that were proving so persistent and increasingly lethal. The U.S. Agency for International Development (USAID)—the official agency responsible for managing America's international humanitarian relief and economic development program—was struggling to provide effective relief in the so-called nonpermissive environments of Iraq and Afghanistan.

Development workers, often lacking relevant intelligence and analytical tools that would permit sophisticated network analysis, frequently found themselves caught in the crossfire between rival illicit power structures, or subverted in their efforts by illicit forces invisible to them. Even the United Nations, with its wealth of experience in peacekeeping and humanitarian work in the world's most conflicted locations, was caught in the vortex of violence that consumed Iraq.[11] Humanitarian agencies and organizations were likewise in uncharted territory: at the same time reluctant to associate openly with U.S. or international military forces, and, in many cases, dependent on them for security. Thus, the war-fighter challenge was compounded by the diplomatic, development, and humanitarian challenges.

That compound challenge, emerging from the defense, diplomacy, development, and humanitarian communities, led to the initiation of a project in late 2005, cosponsored by USJFCOM and USAID, to examine the nature, beliefs, motivations, and morphology of illicit power structures. This book is the legacy of the intellectual capital generated by that project. The project was robustly supported by experienced practitioners from across the U.S. interagency, the United States Institute of Peace, and the Pearson Peacekeeping Centre. It sought to peel away the veil of traditional institutional paradigms, reveal the underlying currents of power (both licit and illicit), and give practitioners clear guidance for countering illicit power. Only by developing a more sophisticated understanding of illicit power structures could one reduce their ability to subvert legitimate efforts at stabilization and reconstruction. Until then, illicit power would continue to capture development resources, use public resources in the service of parochial interests, and displace legitimate state functions and institutions. Although the effort ultimately timed out, foundering on the seemingly infinite permutations of illicit power, it was an attempt to develop a taxonomy or typology of illicit power structures, classifying them according to their worldview, motivations, operational modalities, and morphologies.[12]

Regardless of the idiosyncrasies of any particular illicit organization or network, the fundamental issue at stake is accountability. States are, or should be, accountable both

---

[11] On August 19, 2003, the UN headquarters at the Canal Hotel in Baghdad was bombed, resulting in over 100 wounded and 22 killed, including the UN special representative in Iraq, Sergio Vieira de Mello. Subsequently, the UN Iraq team relocated to Amman, Jordan, and worked remotely.

[12] Michael Miklaucic, "Contending with Illicit Power Structures: A Typology," in *Nonstate Actors as Standard Setters*, ed. Anne Peters, Lucy Koechlin, Till Förster, and Gretta Fenner Zinkernagel, Cambridge Univ. Press, 2009.

to their citizens and to the international community of states. Illicit power structures, organizations, and networks are accountable only to themselves. They have no commitment to the broader public good beyond their parochial interests. And to the extent that they succeed in carving out operating space within a polity, they erode that polity's legitimacy by creating accountability-free zones, or zones of impunity. When undermined by agents of illicit power that are unaccountable, the state cannot succeed and will eventually recede into irrelevance or near irrelevance, as is occurring in many countries in Central America, Africa, and the Middle East today.

To counter the subversive effects of illicit power structures, we must (a) convert them to licit structures, (b) deprive them of their power, or (c) dismantle them. In Afghanistan, attempts were made to co-opt warlord leaders and their networks into the state-building enterprise, in the hope that they would become constructive contributors to the undertaking rather than continue serving their parochial interests with impunity. As Carl Forsberg and Tim Sullivan write in this volume, "By co-opting leaders from the most powerful ethnic and mujahideen-era constituencies, the settlement provided a basic framework that has kept Afghanistan's ethnic fissures beneath the surface over the succeeding thirteen years." Ismail Rashid explains that in Sierra Leone, a plan was developed aimed at "severing RUF from its Liberian strategic center of gravity, particularly the planning and arms support provided by Taylor. It also aimed to restrict RUF profits from the illegal diamond trade through Liberia and Guinea," thus depriving RUF of its source of power. In the case of the Liberation Tigers of Tamil Eelam (LTTE), Lieutenant General (Ret.) Tej Pratap Singh Brar and Thomas Marks tell us "A last stand on a narrow stretch of northeastern beach ended in annihilation" of their structure.

Even after decades of experience with illicit power structures and networks, this volume shows that the international community rarely, if ever, approaches the challenge with a unanimously agreed strategic objective or a coherent plan to counter illicit power. Independent and autonomous agencies, national, international, and nongovernmental, primarily focus their resources on the objectives of their individual mandates and competences. Only on rare occasions does unity of effort, let alone unity of command, exist. Illicit power is reluctantly (and, usually, belatedly) recognized as a significant challenge, and even then it is typically addressed as a law enforcement matter rather than an international security challenge, as shown clearly in the Sierra Leone and Haiti cases. The resources assigned to address the problem of illicit power are frequently inadequate or inappropriate to the strategic challenge posed. Thus, when we ask, "What was the international community's plan?" the true answer is, there was none. The problem of illicit power has typically been addressed as an afterthought—often only when a much sought-after peace agreement or stabilization effort was already lost.

Efforts to categorize the manifestations of illicit power have been beneficial and instructive, even if only to demonstrate the enormous variety of the phenomenon. Stedman's typology remains valuable even though limited in descriptive detail. Scholars Richard Shultz and Andrea Dew, in their insightful analysis of contemporary warfare, describe the possessors of illicit power as insurgents, terrorists, and militias, but their purview did not extend to transnational criminal organizations.[13] Withholding value

---

[13] Richard H. Shultz and Andrea J. Dew, *Insurgents, Terrorists, and Militias: The Warriors of Contemporary Combat* (New York: Columbia Univ. Press, 2006).

judgments, John Arquilla characterizes the "masters of irregular warfare" as "insurgents, raiders, and bandits."[14] There are others, but all fall into the same trap: mirror imaging. We analysts, scholars, practitioners, and policymakers assume for illicit power structures and networks the same kinds of organizational logic and imperatives that motivate us in our own organizational behavior: loyalty, conventional notions of membership, and formal identification.

But sanctioned organizations are accountable and have a tangible formality (manifest in articles of incorporation, rules and regulations, reporting requirements, membership logs, and so on), which illicit power structures and networks do not have. There certainly may be codes of conduct or other informal rules governing behavior, but there is a fundamental difference in essence between structures that rely for their legitimacy on impunity and the successful protection of parochial interests, and those that operate within the parameters of public interest, accountability, and sanctioned legitimacy. So although analytic templates can be useful, we must avoid overreliance on taxonomies, typologies, and categories, for regardless of their complexity, they can never fully account for the sheer variety of illicit power. Nor must we permit ourselves to misanalyze the adversary or the operating threat environment because we assume attributes reflecting our own familiar paradigms for organizational behavior.

To effectively mitigate the disruptive and malignant effects of illicit power structures, organizations, and networks requires a substantial tolerance for compromise and trade-offs. A frequent mistake that often leads to ineffective policy is the assumption that all good things go—or even *can* go—together. We assume, for example, that policies to promote democratization and policies to promote stability will work in harmony. But this is not always the case, and efforts to accomplish both simultaneously may run aground due to inherent tensions between the two goals. To accomplish both peace and justice in or after violent conflict is, in some cases, too much to attempt at one time—and indeed, they may even be incompatible at any one time. Take, for example, the case of Colombia. Peace negotiations in Havana between the government of Colombia and the Revolutionary Armed Forces of Colombia (FARC), beginning in 2012, have struggled with the problem of justice. How should the many crimes committed by both sides in a multigenerational struggle be accounted for? The Colombian people, victimized their whole lives by the struggle, demand justice in the form of retribution—jail time for convicted "war criminals." But neither the Colombian armed forces nor the FARC cadres will willingly accept incarceration, and either can threaten to return to conflict. Some form of compromise will be required. In Sri Lanka, the armed forces accomplished decisive victory over a longtime tormentor, the LTTE, but only at the cost of massive collateral damage. Although no juridical judgment has been reached, the verdict of the international community has clearly been that measures taken by both adversaries defied international norms and acceptable behavior in war. Yet Sri Lankan leaders made a calculated judgment, based on experience, that nothing short of comprehensive military victory would prevent LTTE from carrying on the armed struggle. These are symptomatic of the inescapable moral ambiguities encountered when dealing with illicit power.

The temptation to intervene in conflict and transition situations is powerful and of-

---

[14] John Arquilla, *Insurgents, Raiders, and Bandits: How Masters of Irregular Warfare Have Shaped Our World* (Lanham, MD: Rowman & Littlefield, 2011).

ten justified. As international norms evolve and egregious abuse of power is no longer acceptable, such doctrines as the "responsibility to protect" (or R2P) have emerged, and the concept of "responsibility while protecting" suggests a set of principles and procedures to govern such interventions.[15] Would-be interveners must realize, however, that their interventions can catalyze local reactions, including the emergence of illicit power structures. The case study of the Jaish al-Mahdi, by Phil Williams and Dan Bisbee, offers a clear lesson: this organization arose in response to the U.S. intervention in Iraq, which many in Iraq saw as illicit. Also, the infusion of significant financial resources into a conflict environment can reinforce or facilitate the operations of illicit organizations and networks, providing them with critical resources that undermine the state.

The effort to better understand the use of illicit power and its agents must be sustained. The cost of failure is simply too high. The human cost of terrorism vastly exceeds the number of terrorism-caused fatalities—although that number alone tells a horrible tale. According to one source, "On average, there were 1,122 terrorist attacks, 2,727 deaths, and 2,899 injuries per month worldwide in 2014."[16] And the scourge is expanding geographically: "The number of countries experiencing more than 50 deaths rose to 24 in 2013; the previous high had been 19 in 2008."[17] Although this book is not about terrorists per se, their criminal violence is a brutal manifestation of the use of illicit power. The illicit economy was estimated in 2011 to have amounted to "some 3.6 percent of GDP (2.3-5.5 percent) or around US$2.1 trillion in 2009."[18] If we accept Paul Collier's estimate that civil wars cost at least $50 billion, and since "on average around two civil wars break out each year . . . the phenomenon is a $100 billion a year problem."[19] These rough estimates give a notion of the financial cost of activities frequently attributable to illicit power structures and networks. They indicate a substantial direct cost, as well as the obviously related opportunity cost. These figures exceed and clearly undermine total official development assistance, which amounted to $135 billion in 2013.[20]

The investment in development, democratization, stabilization, and reconstruction, not only by the United States but by the international community as a whole, has been substantial, yet all is at risk. The FARC in Colombia, as Carlos Ospina points out in this book, continues to threaten state legitimacy in areas under its control, jeopardizing years of state building and investment in Colombia. Unmitigated impunity and the exercise of illicit power compromise state legitimacy, as occurred in Haiti when the gangs of

---

[15] Gareth Evans, "The Limits of Sovereignty: The Case of Mass Atrocity Crimes," PRISM, keynote address to Research School of Asia and the Pacific Symposium, *Landscapes of Sovereignty in Asia and the Pacific,*" Australian National Univ., Canberra, Oct. 22, 2014, http://gevans.org/speeches/speech554.html.

[16] U.S. State Dept., "Annex of Statistical Information: Country Reports on Terrorism 2014," National Consortium for the Study of Terrorism and Responses to Terrorism, June 2015, www.state.gov/documents/organization/239628.pdf.

[17] Ewen MacAskill, "Fivefold Increase in Terrorism Fatalities Since 9/11," *Guardian*, Nov. 17, 2014, www.theguardian.com/uk-news/2014/nov/18/fivefold-increase-terrorism-fatalities-global-index).

[18] UNODC, "Estimating Illicit Financial Flows Resulting from Drug Trafficking and Other Transnational Organized Crime," report, Oct. 2011.

[19] Paul Collier, "Development and Conflict," Centre for the Study of African Economies, Oxford Univ., Oct. 1, 2004, www.un.org/esa/documents/Development.and.Conflict2.pdf.

[20] OECD, "Aid to Developing Countries Rebounds in 2013 to Reach an All-Time High," press release, Aug. 4, 2014, www.oecd.org/newsroom/aid-to-developing-countries-rebounds-in-2013-to-reach-an-all-time-high.htm.

Port-au-Prince terrorized the city. Ultimately, this brings us back to the self-interested problem of global security partnerships. The corrosive and debilitating effects of illicit power on struggling states mean that potential partner states cannot be relied on. Iraq and Afghanistan are only two examples of states in which literally billions of dollars of have been invested in the past decade, yet neither can be relied on to effectively support international security and our national security interests. The systemic risk posed by inaction or negligence regarding illicit power is that the domain of the rule-based system of states, within which the United States and many other countries have flourished, contracts, while the "coming anarchy" expands.

1. Criminal Patronage Networks and the Struggle to Rebuild the Afghan State

Carl Forsberg and Tim Sullivan

Throughout the international community's 2001-14 engagement in Afghanistan, few challenges proved as complex, pervasive, and threatening as corruption and organized crime.[1] Together, systemic corruption and organized crime undermined efforts to build Afghan institutions, consolidate security gains, achieve political progress, encourage economic growth, and create conditions for enduring stability. Corruption in Afghanistan reached crippling levels after the post-2001 political settlement, which was built on the distribution of political power between factions formed during the country's civil war. Benefiting from judicial impunity, these factions facilitated penetration of the Afghan state by what eventually became known as *criminal patronage networks* (CPNs). In the Afghan context, CPNs are a form of organized illicit power structure made particularly intractable by their integration into the government and by their access to international money.

Criminal patronage networks stymied reconstruction and hindered the 2009-12 U.S. and NATO troop surge leading up to Afghanistan's two major transitions in 2014: the country's first-ever transfer of power from one elected president to another, and the withdrawal of most international forces. While Afghanistan in 2014 was stronger than it had been in decades—possessing a sizable and increasingly capable national army and police—factionalism, corruption, and criminal subversion of state institutions left its politics fragile and security uncertain. The coalition government of President Ashraf Ghani and Chief Executive Officer Abdullah Abdullah, formed after the contentious 2014 presidential election, will be limited by factional fault lines, weak rule of law, and the continued influence of CPNs as Afghanistan adjusts to an era of reduced foreign support.

These challenges are not unique to Afghanistan. Conflicts elsewhere in recent decades have revealed that states emerging from insurgencies and civil wars—especially where rule of law is minimal and substantial international resources have been injected with inadequate oversight—are vulnerable to the rise of illicit power structures both inside and outside government. The Afghan experience is rich with lessons for the U.S. military and foreign policy establishment. In the years ahead, the United States and allies may again be compelled to assist or intervene in weak states experiencing protracted instability or rebuilding after years of violence. In such environments, there is a pressing requirement, not only for seamless integration of civilian and military efforts to establish security, enable law enforcement, and promote rule of law, but also for transparency and accountability within the state's critical institutions. All these efforts, meanwhile, must be grounded in a thorough understanding of that state's politics and tailored to generate among its key leaders the necessary will to undertake complementary reforms.

This chapter outlines the nature and origins of CPNs in Afghanistan, and the effect they have had on efforts by the United States and its allies to rebuild the country and

---

[1] The authors wish to thank Michael Miklaucic, Michelle Hughes, and Michael Dziedzic for their contributions in editing this chapter. We also acknowledge valuable feedback from a roundtable discussion with a panel of experts consisting of John Agoglia, William Byrd, Scott Carlson, Sasha Kishinchand, and James Schear. We also received written feedback from Michael Metrinko and Shahmahmood Miakhel.

insulate it against the Taliban and transnational terrorists. We then discuss the evolution of U.S. and International Security Assistance Force (ISAF) strategies to address these threats, examining both successes and failures. The chapter concludes with a review of lessons and implications of the Afghan experience for future armed conflict and stability operations.

Cause of the Conflict

The corruption prevalent in Afghanistan since 2001 does not reflect an intrinsic cultural phenomenon, as some have carelessly asserted. Rather, the subversion of the Afghan government by CPNs in recent decades can be understood only in the context of the country's turbulent post-1978 history. From the 1940s to the 1970s, the Afghan monarchy maintained peace and order throughout the country, overcoming the centuries-old problem of tribal revolts by developing mechanisms for addressing popular grievances. To maintain its legitimacy among the myriad tribes and ethnic groups, the monarchy vigilantly policed its own officials, and investigations and prosecutions of government functionaries were common. This formal justice system, which applied to Afghanistan's city dwellers and elites, functioned in tandem with an informal system of tribal and religious law that maintained order in rural areas. By the 1960s, most senior Afghan government officials won their positions on the basis of merit, without consideration of their ethnic or tribal origins, and Afghan officials strove to maintain a public image of being poor but honest.[2]

Afghanistan's golden decades came to a close in 1978, when Communist officers in the Afghan army assassinated President Sardar Mohammad Daoud Khan. The new Communist government launched an ambitious set of Marxist social, political, and economic reforms. Thousands of supporters of the old regime, particularly Pashtun aristocrats and radical Islamists, were arrested and imprisoned, and thousands more fled into exile abroad. These measures provoked a fierce popular backlash, forcing the Soviet Union to send over 100,000 troops to Afghanistan, beginning in 1979, to prop up the faltering regime.

Opposition to the Soviet occupation and its puppet government coalesced around the *mujahideen* – an Arab term for those waging jihad. Most of these resistance fighters were affiliated with one of the "Peshawar Seven" mujahideen commanders' networks, based in Peshawar, Pakistan. The Peshawar Seven depended to varying degrees on support from Pakistan's Inter-Services Intelligence Agency (ISI) and the CIA, which provided weapons, funding, and sanctuary. By 1989, the Soviet Union withdrew all troops from Afghanistan, and three years later, it ended its $35 billion annual subsidies to the Communist government, which collapsed in April 1992 as a result. Afghanistan's principal mujahideen commanders occupied Kabul but failed to agree on a power-sharing arrangement, initiating a second civil war (1992-96). The mujahideen's power squabbles and abuses against the Afghan people, and Pakistani disillusionment with its mujahi-

---

[2] On the Afghan government's effectiveness in maintaining law and order in the mid-twentieth century, see, for example, Shahmahmood Miakhel, "Myths and Impact of Bad Governance on Stability in Afghanistan," in *Afghanistan in Transition: Beyond 2014?* ed. Shanthie Mariet D'Souza (New Delhi: Pentagon Press, 2012).

deen allies, led to the emergence of the Taliban, a radical Islamist movement backed by Pakistan's ISI and led by ethnic Pashtuns. By 1998, the Taliban controlled most of Afghanistan. Some of the mujahideen leaders and their fighters defected to the Taliban, while others, particularly non-Pashtuns, formed a loose alliance and, with Iranian and Russian support, kept a small foothold in northern Afghanistan. They eventually coalesced into what would become known as the Northern Alliance.

In November 2001, the United States and its new ally, the Northern Alliance, toppled the Taliban regime following the Taliban's refusal to hand over al-Qaeda leadership responsible for the 9/11 attacks. By that time, two decades of civil war had destroyed the foundations of national unity and civil order. The country suffered from chronically weak governance, nonexistent rule of law, and fractured social structures. Conflict left deep divisions between Afghanistan's solidarity groups. While most mujahideen parties began as multiethnic movements, they took on distinct ethnic identities during the civil war as commanders began to affiliate along ethnic and tribal lines, giving rise to intense ethnic polarization by 2001.[3]

The upheaval of the 1990s had also destroyed Afghanistan's state and civil institutions. By the time the Taliban seized Kabul in 1996, state institutions existed in name only and had ceased performing their official functions. The Taliban attempt to replace Afghanistan's older hybrid system of civil and traditional law with Sharia law further undermined what remained of the legal system. War and the breakdown of any institutional check on illicit power drove many of Afghanistan's mujahideen and tribal leaders to embrace organized crime, including narcotics trafficking, arms smuggling, and arbitrary confiscation of land, to fund private local militias.[4]

The Post-2001 Political Settlement

During a UN-sponsored conference in Bonn in late 2001, international diplomats, led by the United States and the United Nations, tried to piece together the fragments of Afghanistan and form a post-Taliban government. The agreement achieved there between the primary Afghan and international participants—including the United States, Iran, Russia, representatives from various Afghan expatriate communities, and members of the network of commanders known as the Northern Alliance—laid out a three-year plan for creating a new constitution and electing a new government.[5] In the eyes of UN, U.S., and European officials, these agreements—later known as the Bonn Settlement—had

---

[3] Barnett Rubin, *The Fragmentation of Afghanistan: State Formation and Collapse in the International System*, 2nd ed. (New Haven, CT: Yale Univ. Press, 2002), 269-71; Antonio Giustozzi, "The Ethnicisation of an Afghan Faction: Junbesh-i-Milli from Its Origins to the Presidential Election," (working paper no. 67, Crisis States Programme, Development Research Centre, 2005), http://eprints.lse.ac.uk/13315/1/WP67.pdf.

[4] On illicit forms of revenue extraction by Northern Allianceaffiliated militias in the 1990s, see Antonio Giustozzi, *Empires of Mud: Wars and Warlords in Afghanistan* (New York: Columbia Univ. Press, 2009). On the evolution of tribally affiliated militias in Kandahar in the 1990s, see Carl Forsberg, "Politics and Power in Kandahar," Afghanistan Report no. 5, Institute for the Study of War, Apr. 2010, 1417, www.understandingwar.org/sites/default/files/Politics_and_Power_in_Kandahar.pdf.

[5] Ambassador James Dobbins, the U.S. envoy at Bonn (and formerly the Obama administration's special representative for Afghanistan and Pakistan) provides a thorough account of the conference in his 2008 memoir, *After the Taliban* (Dulles, VA: Potomac, 2008).

successfully established a process by which the Afghan people and the international community could build a democratic, sovereign Afghanistan. But Afghan factions understood the Bonn Settlement very differently. In their eyes, the deals struck at Bonn served as the origins of an ever-evolving grand bargain for distribution of power in the new state, between several of Afghanistan's ethnic and mujahideen factions. Afghan elites used the power they gained at Bonn to subvert UN hopes that the Bonn Settlement would lead to a more inclusive and accountable Afghan government. But instead, by co-opting leaders from the most powerful ethnic and mujahideen-era constituencies, the settlement provided a basic framework that has kept Afghanistan's ethnic fissures beneath the surface over the succeeding 13 years, as competition took the form of lobbying for positions within the government and for a larger share of international assistance.

CPNs' Relationship to the Political Settlement and Peace Process

One of the principal effects of the Bonn Settlement was to return mujahideen leaders to a dominant role in Afghanistan's internal politics. During 2002-4, in the new Afghan government, the predominantly Tajik[6] Jamiat-e Islami Party staked a claim to power that was disproportionately large compared to Afghanistan's actual Tajik population. Jamiat had been the party of legendary mujahideen resistance leader Ahmed Shah Massoud and former Afghan president Burhanuddin Rabbani. Following their assassinations in 2001 and 2011 respectively, Jamiat's paramount military commander, Mohammad Qasim Fahim, effectively assumed party leadership. Fahim had served as Afghanistan's minister of defense during 2001-4 and later returned to power as first vice president in 2009. By occupying Kabul with its militias and by gaining control over key ministries as a result of deals made during the Bonn Conference, Jamiat consolidated influence over the key organs of the central state, including the Ministry of Defense and Ministry of Interior. Its control over powerful state ministries enabled a system of patronage politics in which former mujahideen leaders distributed official appointments, businesses, and other opportunities to subordinates, who then proceeded to intimidate rivals, seize public and private land, and strengthen their own business interests. With no institutions to check their power, the ability of members of the new government and their clients to commit crimes with impunity in the years after 2001 established a precedent of violence and coercion as valid tools to be wielded by government elites.[7]

Across Afghanistan, mujahideen leaders who returned to influence after 2001 generally viewed the institutions of the Afghan state as personal patrimonies for factional interests and personal enrichment. This problem was exacerbated by a belief among senior mujahideen commanders that civil war would resume once international forces left the country. For example, when Fahim took Kabul in 2001, he told his commanders

---

[6] Tajiks, Afghanistan's second-largest ethnic group, hail principally from the northeast provinces. Tajiks represent 27 percent of the Afghan population, second to Pashtuns, at 42 percent. CIA, "Afghanistan," in *The World Factbook* (Langley, VA: CIA, 2013).

[7] Giustozzi, *Empires of Mud*, 289; Stephen Carter and Kate Clark, "No Shortcut to Stability: Justice, Politics, and Insurgency in Afghanistan," Chatham House, Dec. 2010, www.chathamhouse.org/sites/files/chathamhouse/public/Research/Asia/1210pr_afghanjustice.pdf; Sarah Chayes, *The Punishment of Virtue* (London: Penguin, 2006); Forsberg, "Politics and Power in Kandahar."

that the U.S. presence would be short-lived, and pushed Jamiat to continue consolidating power in expectation of a return to ethnic and factional conflict.[8] A sense of entitlement, arising from their long struggle against the Taliban and, before that, the Soviets, also drove the mujahideen elites. Fahim and his generation had joined the anti-Soviet Jihad while in their twenties and had spent two formative decades of their lives at war. As often happens when men who have never known peace seize power, the mujahideen felt entitled to use the state for personal and factional gain and to establish themselves as a new oligarchy. They ultimately lacked a vision for a more stable, unified Afghanistan after America's departure.

Hamid Karzai began his tenure as Afghanistan's interim president in 2002. A Pashtun from the southern province of Kandahar, he lacked a power base of his own and, thus, was at the mercy of the mujahideen, most notably Fahim. In the early years of his presidency, Karzai's only alternative to the warlords was U.S. backing. But for his first year and a half in office, the U.S. government itself was making accommodations with regional strongmen, whom it viewed as allies in counterterrorism operations, and advised Karzai to follow its lead.[9] Thus, Karzai's approach to governance was characterized by extreme caution and a conscious effort to avoid antagonizing the mujahideen parties. In particular, Karzai feared that enforcing the law would lead mujahideen elites to oppose the central government and challenge his leadership.[10] As he noted in a 2002 speech, "We must have peace, stabilize peace, make it certain, make it stand on its own feet and then go for justice . . . Justice becomes a luxury for now."[11]

In 2003, U.S. policy began to shift gradually toward strengthening Afghanistan's central government and helping Karzai co-opt his rivals. Karzai began to strengthen his hand vis-à-vis the mujahideen in 2003, thanks to changes in U.S. policy and his election as Afghanistan's first democratically elected president. The international community's initial inaction gave way to a series of state-building measures that gave Karzai greater leverage over regional mujahideen power brokers. With his 2004 election came enhanced presidential powers, including authority over all appointments, ranging from cabinet ministers to district police chiefs. The central government, with its tremendous powers of patronage, emerged as the center of gravity in Afghan politics.[12]

Despite the opportunities provided by his election, Karzai continued to govern cautiously. He adopted a strategy of balancing, dividing, and co-opting, rather than confronting, Afghanistan's fractious strongmen and their clients. Having assessed that they could cause less trouble inside the government than outside, Karzai parceled senior ministerial posts and provincial governorships to allies and competitors alike. Overt factionalism and violent conflict between armed militias was sublimated into competition

---

[8] Giustozzi, Empires of Mud, 290-93; Kai Eide, *Power Struggle over Afghanistan: An Inside Look at What Went Wrong – and What We Can Do to Repair the Damage* (New York: Skyhorse, 2010), 150-51.

[9] See Donald Rumsfeld, *Known and Unknown: A Memoir* (New York: Sentinel, 2011), 398-408; Ahmed Rashid, *Decent into Chaos* (New York: Penguin, 2008).

[10] International Crisis Group, "Afghanistan: Judicial Reform and Transitional Justice," Asia Report no. 45, Jan. 28, 2003, www.crisisgroup.org/en/regions/asia/south-asia/afghanistan/045-afghanistan-judicial-reform-and-transitional-justice.aspx.

[11] *BBC News*, "Karzai Sets Out Afghanistan Vision," June 14, 2002.

[12] Giustozzi, *Empires of Mud*; Carl Forsberg, "The Karzai-Fahim Alliance," (Washington, DC: Institute for the Study of War, 2012).

for state offices, patronage, and wealth. Although intimidation remained ubiquitous in both the public and private sectors, money replaced guns as the leading source of political influence. To carry on criminal activity and expand their financial and political power, regional networks had to infiltrate and manipulate, rather than openly defy, the Afghan government. Thus, Karzai and the patrons of illicit networks developed a symbiotic relationship, with each using the other to achieve their goals.

But co-option into the central government did little to reduce the leverage of strongmen and mujahideen-era networks over President Karzai. The 2009 presidential election, which gave Karzai a second term, demonstrated that time, economic growth, advances in education, and international donor assistance had not reduced his dependence on strongmen and their illicit networks. In fact, the 2009 election only strengthened the influence of Afghanistan's CPNs when Karzai selected Fahim, whose reputation for corruption and abuse of power was infamous even by the Karzai government's standards, to be his vice presidential running mate and made concessions to other corrupt power brokers to buy their support for his candidacy.

U.S. policy in 2009 only exacerbated Karzai's inclination to make concessions to strongmen. The newly elected Obama administration made it clear that it would not give Karzai the same privileged relationship that he had enjoyed with President Bush, and U.S. Special Representative for Afghanistan and Pakistan Richard Holbrooke emphasized that the United States wanted Karzai's rivals to enjoy a level playing field in the election. These changes made Karzai deeply insecure about U.S. support, and he compensated by embracing old power brokers.

Following the 2009 elections, Karzai made Fahim a key governing partner, devolving significant authority and the power of appointment to Fahim and his network. As a result, Jamiat reemerged as the most powerful political network within the Afghan government during the 2009-12 U.S. and NATO troop surge, with significant influence in the security forces and across Afghanistan's northern provinces. In exchange for autonomy to operate criminal patronage networks within the government, Fahim kept powerful Tajik networks in the security forces loyal to the president. He also prevented anti-Taliban Tajik leaders from undermining Karzai's attempts to broker peace with the Taliban. Even as Karzai co-opted much of Jamiat, Tajik leaders not included in the flow of patronage resources—such as former foreign minister and 2014 presidential candidate Abdullah Abdullah—together with ethnic-minority Uzbek and Hazara leaders, led a vocal opposition to Karzai.

Fahim's March 2014 death (from natural causes) and the 2014 presidential election marked a major transition for Afghanistan's Tajiks. After Abdullah Abdullah contested his loss in the second round of the 2014 election, on the grounds of massive voter fraud, U.S. and European diplomats negotiated a power-sharing arrangement by which Abdullah assumed the newly created post of chief executive officer under President Ghani. Abdullah's rise to power suggests that President Ghani will work in coalition with a new generation of Jamiat-affiliated Tajiks whose backgrounds are more civilian than military and who are less shaped by the country's pre-2001 conflicts. It remains to be seen whether, and how, these leaders will distance themselves and the Jamiat Party from the illicit activities that Fahim long protected.

Karzai's continued dependence on mujahideen networks and CPNs for political support ensured continued impunity for them through the end of his presidential term in September 2014. Even as Karzai modified his political alliances and shifted patronage from one group to another, he persistently refused to apply the rule of law to clients of his critical coalition partners. The result was a fragile political settlement wherein former mujahideen elites, integrated into senior positions in government, had free rein to extract for their personal enrichment (and that of their patronage networks) as many resources as their institution or post afforded.

Assessing the Nature and Operation of CPNs in Afghanistan

CPNs, enabled by Afghanistan's political settlement and a fragile war economy sustained by international aid, security assistance, and the narcotics trade, undermined U.S. and ISAF attempts to stabilize Afghanistan after 2001. They still pose a grave threat to the viability of the Afghan state. Corruption undermines the Afghan government's legitimacy, effectiveness, and cohesion, fuels popular discontent, generating active and passive support for the insurgency, and prevents the growth of a strong licit economy, thus perpetuating dependence on international assistance.

Along with weakening the country's critical institutions, systemic corruption and organized crime have helped foment instability and insurgent violence. Analysts consistently identified causal links between predatory governance and the growing insurgency, noting, for example, the connection between the Taliban's reemergence after 2003 and the abuse of power by government officials, security forces, and their affiliate networks.[13] More recently, by undermining popular confidence in the government's legitimacy, effectiveness, and long-term durability, corruption has discouraged the population from actively mobilizing against the insurgency, thus lending the Taliban passive support. Moreover, uncertainty about Afghanistan's future, and the anticipation, despite reassurances, of a large-scale international disengagement from the country, drives the short-term maximization-of-gains mentality among the country's CPNs. This has heightened ethnic and factional tensions, accelerated the networks' efforts to consolidate power, and created conditions for continued violence.

The most serious and destabilizing forms of corruption in Afghanistan were carried out systematically by CPNs with political protection from President Karzai's coalition partners within the Afghan government. In December 2010, Afghanistan's national security adviser, Rangin Dadfar Spanta, said of the corruption problem, "In this government we have mafia networks."[14] The networks, he continued, "begin with the financial banking system, with corruption networks, with reconstruction and security firms and also with drugs and the Taliban; they are in Parliament and they are in government." These CPNs engage in the capture and subversion of critical state functions, and they

---

[13] See, for example, Antonio Giustozzi, *Koran, Kalashnikov, and Laptop* (New York: Columbia Univ. Press, 2008); Anand Gopal, "The Battle for Afghanistan: Militancy and Conflict in Kandahar," policy paper, New America Foundation, Nov. 2010, http://newamerica.net/sites/newamerica.net/files/policydocs/kandahar_0.pdf; Tom Coglan, "The Taliban in Helmand: An Oral History," in *Decoding the New Taliban*, ed. Antonio Giustozzi (New York: Columbia Univ. Press, 2009).

[14] Matthew Rosenberg, "'Malign' Afghans Targeted," *Wall Street Journal*, Dec. 29, 2010.

have been stakeholders in the state's weakness since the continued fragility of institutions gives them impunity.

Those CPNs embedded in state institutions have generally upheld the status quo of Afghanistan's post-Bonn political settlement—understandable given the enormous benefits and wealth they have derived from it. That said, they were not above using targeted violence, coercion, and intimidation against Karzai, international forces, or state bureaucrats. CPNs' use of violence is not primarily a means of opposing either the Afghan government or the international intervention. Rather, CPNs use violence and coercion as a form of blackmail to manipulate the Afghan president, international donors, or government officials to direct a greater share of resources to their affiliates. Violence has thus served primarily as a negotiating tool selectively employed to protect CPN revenue streams and influence within the state.

Afghanistan's CPNs depend on flows of money to co-opt government officials and keep supporters' loyalty, so they vie with each other to capture critical revenue streams. They divert customs revenue at the international airports and border crossing points, steal international security and development assistance disbursed to the Afghan government, protect and facilitate the narcotics trade, and abuse public and private financial institutions. The revenues available to CPNs, made possible by a rapid influx of international assistance and ongoing judicial impunity, gave rise, during the decade after 2001, to the commoditization of political power in Afghanistan and the sale of government offices and official influence.

Borders, Airports, and Inland Customs Depots

From 2002 on, power brokers have sought to use their political influence to control police and customs officers at Afghanistan's border crossing points, international airports, and inland customs depots in order to collect bribes for licit commerce, protect smuggling, and siphon off customs revenue. Border police and Ministry of Finance officials responsible for collection of customs duties have been co-opted by regional strongmen, mujahideen networks, and narcotics traffickers, sharing the profits of lucrative cross-border trade. Several powerful provincial governors, including Muhammad Atta Noor in Balkh province and Gul Agha Sherzai in Nangarhar province, were accused of diverting tens of millions of dollars in customs revenues, owed to the central state, to build their own local patronage networks.

This criminal capture of state functions at borders, airports, and customs depots robs the Afghan state of revenue, inhibits economic growth, and leaves the country vulnerable to transnational threats. In 2012, the World Bank ranked Afghanistan's borders as the fourth hardest in the world to cross for the purposes of trade, creating a significant obstacle to the country's regional economic integration.[15] Corruption and organized crime at Afghanistan's critical ports of entry also directly undermine the state's security by enabling trafficking of narcotics, precursor chemicals, and weapons while facilitating insurgents' freedom of movement.

---

[15] World Bank, *Doing Business in a More Transparent World* (Washington, DC: World Bank, 2012).

Diversion of Foreign Assistance

CPNs also targeted a post-2001 influx of international assistance, by diverting resources intended for the Afghan state and by directly manipulating aid agencies and foreign donors. International aid channeled through Afghan ministries is susceptible to embezzlement by government officials due to the lack of internal accountability mechanisms, impunity from prosecution, and poor budgeting processes in most ministries. Black markets, meanwhile, have emerged throughout Afghanistan for the resale of equipment donated by the international community to the Afghan government or pilfered from international forces, in some cases leading to U.S.-provided military equipment being sold to insurgent groups.[16]

International donors' lack of oversight in disbursing assistance and awarding contracts enabled the diversion of resources administered directly by development agencies and international forces. Local officials have often colluded with CPNs to direct international contracts and assistance projects to the businesses—particularly construction and private security firms—of their allies in exchange for payments. In many cases, CPNs' influence directed international funds to development projects that went unused, such as empty schools and hospitals, and electrical plants that could not be connected to major power grids. The awarding of international contracts also suffered from limited accountability for much of the post-2001 period, enabling fraud and abuse by both Afghan and Western firms—including, in the most egregious cases, payoffs by contractors to local strongmen and the Taliban.[17]

The Narcotics Trade

Afghanistan's CPNs have also sought to profit from facilitating, protecting, and participating in the narcotics trade. According to the United Nations, Afghanistan accounted for 75 percent of the world's heroin exports in 2012, and the narcotics trade is estimated to account for up to 15 percent of the country's gross domestic product.[18] Afghan police, district governors, and provincial governors are particularly susceptible to narcotics-fueled corruption and often take protection payments from traffickers. In other cases, officials, including many border police commanders, use their state positions to engage directly in trafficking. The narcotics trade's profitability for local government appointments, in turn, has fueled the sale of government offices, with police appointments in the southern Afghanistan opium belt reportedly sold by Ministry of Interior officials for as much as one hundred thousand dollars.[19]

---

[16] See, for example, Matt Millham, "Afghan Market Flourishing with Coalition Goods," *Stars and Stripes*, May 25, 2012; Martin Kuz, "Afghan Commander Suspected of Acting as Crime Boss," *Stars and Stripes*, Sept. 3, 2012.

[17] "Warlord, Inc.: Extortion and Corruption Along the U.S. Supply Chain in Afghanistan," Report of the Majority Staff, Rep. John F. Tierney, Chair, Subcommittee on National Security and Foreign Affairs, Committee on Oversight and Government Reform, U.S. House of Representatives, June 2010; "Inquiry into the Role and Oversight of Private Security Contractors in Afghanistan," Report of the Committee on Armed Services, U.S. Senate, Sept. 28, 2010.

[18] Rod Nordland, "Production of Opium by Afghans Is Up Again," *New York Times*, Apr. 15, 2013.

[19] Barnett Rubin, "Still Ours to Lose: Afghanistan on the Brink," Testimony for the House Committee on International Relations and the Senate Committee on Foreign Relations, Sept. 21, 2006.

Financial System: The Case of Kabul Bank

CPNs also embedded themselves in Afghanistan's nascent financial system as owners of its fledgling banks. They built alliances with political elites, illicitly moving money, making payoffs, and distributing patronage in exchange for political protection for Ponzi schemes and unsustainably risky lending. More than any other institution, Kabul Bank illustrates the new political arrangement in Kabul, in which power became a commodity to be bought and sold. The commoditization of political influence explains how Afghanistan's largest bank gained political protection and hid its insolvency for years before its dramatic collapse.

Kabul Bank was founded in 2004 by two Afghan businessmen, Khalilullah Fruzi and Sherkhan Farnood, who were driven by profit, not politics. Both were quick to buy government protection and support and to build an alliance with Haseen Fahim, half brother of First Vice President Fahim, and, through him, with the Fahim family's business empire. In 2006, the bank also co-opted Mahmoud Karzai, the president's half brother, by lending Karzai the funds to purchase a seven percent share of the bank. Kabul Bank became a major financier of the Karzai-Fahim ticket in the 2009 presidential elections, and the bank subsequently paid millions in direct bribes to influence parliamentary votes in 2010. To further cement its influence within the government, Kabul Bank bribed senior government officials, with one estimate suggesting that it paid nearly $100 million in bribes and gifts by the end of 2010.[20]

Political influence paid off. The bank was awarded a number of key government contracts that covered its mushrooming liabilities. One of the most important contracts was for the pay of Afghan government employees, a sum of $75 million a month, much of it provided by international donors. The bank placed a two-week hold on withdrawal of assets, ensuring a ready source of cash and encouraging government employees to keep their money in Kabul Bank accounts, thus helping capitalize the bank. Other critical contracts awarded to Kabul Bank included collecting and managing $300 million in payments by pilgrims using the state-administered hajj service to travel to Mecca. By 2010, the bank claimed approximately $1 billion in total assets—34 percent of all assets in the Afghan banking system.[21]

Government backing enabled, but did not cover, the cost of Kabul Bank's rapidly mounting liabilities, which consisted principally of risky and illegal loans to its shareholders and political allies. In October 2010, when the Central Bank completed its investigation of Kabul Bank after the bank's collapse, it discovered $986 million in outstanding loans, many of which were paying no interest and had no collateral. Some $380 million of those loans had been made without documentation. After the bank was placed in conservatorship, government officials discovered that most of the loans were unrecoverable.[22]

---

[20] Adam B. Ellick and Dexter Filkins, "Political Ties Shielded Bank in Afghanistan," *New York Times*, Sept. 7, 2010; Dexter Filkins, "Letter from Kabul: The Great Afghan Bank Heist," New Yorker, Jan. 30, 2011.

[21] Alissa Rubin and James Risen, "Losses at Afghan Bank Could Be $900 million," *New York Times*, Jan. 30, 2011; Filkins, "Letter from Kabul; Gulkohi and Shakuri, "'$300m' Haj Account Is a Hot Potato," Killid Group, Aug. 5, 2010, http://tkg.af/english/reports/others/212-300m-haj-account-is-a-hot-potato.

[22] Independent Joint Anti-Corruption Monitoring and Evaluation Committee, "Report of the Public Inquiry into the Kabul Bank Crisis," Nov. 15, 2012; Alissa Rubin, "Afghan Elite Borrowed Freely from Kabul Bank," *New York Times*, Mar. 28, 2011.

The bank's loans had been used to finance the small clique of businessmen, close to the Fahim family, who tried to build monopolies in several sectors of the Afghan economy, including the airline industry. Under the Fahims' influence, Afghanistan's largest bank had, in essence, become an instrument of patronage employed by the ruling elites. The actions of Farnood, Fruzi, Fahim, and Mahmoud Karzai suggested designs for a state-sponsored oligarchy in which several ruling families would dominate the business sphere on behalf of the regime, dispensing access to capital and business opportunities as a form of patronage.

In the summer of 2010, U.S. Treasury officials discovered Kabul Bank's insolvency, leading to a run on the bank in September. The Afghan government stepped in to guarantee the bank and ultimately injected $800 million in bailout funds, a sum amounting to 56 percent of Afghanistan's GDP. Although Western donors adamantly insisted that they did not help bail out Kabul Bank, they continued to fund most of the Afghan government's budget and filled budget holes created by the bailout. Though a number of Afghan Central Bank officials were convicted in 2013 for failing to regulate the bank, shareholders connected to the Fahim and Karzai families escaped legal sanction for the duration of Karzai's presidency.[23]

CPNs, the Judicial System, and Impunity

Since 2001, CPNs have operated with impunity, consistently avoiding meaningful investigations and prosecution by exerting influence within law enforcement, investigative, and judicial institutions across the Afghan government. "Politics is constraining [our] ability to prosecute high-level corruption cases," noted Afghan Attorney General Mohammad Ishaq Aloko, whose office was particularly vulnerable to political interference and manipulation during his tenure. The internal accountability and oversight mechanisms of many critical Afghan institutions are similarly subject to intimidation and coercion. A lack of capacity within the judicial system, including poor legal education, low pay for judges and public prosecutors, and limited judicial security, exacerbates the challenge of political interference and corruption within the judiciary and the attorney general's office.

International Strategy and the Limits of Capacity Building

Since 2001, the international community has made numerous, often episodic attempts to combat the causes and symptoms of Afghanistan's crippling CPNs. These efforts, developed at headquarters and embassies in Afghanistan and by policymakers in national capitals, were often frustrated by a lack of coordination, continuity, and political will. Countercorruption efforts were also stymied by the international community's failure to develop a common understanding of the corruption problem as officials debated whether politics, limited capacity, or international spending was to blame. The international

---

[23] Independent Joint Anti-Corruption Monitoring and Evaluation Committee, "Report of the Public Inquiry into the Kabul Bank Crisis"; Quil Lawrence, "Another Bailout Looms, but This Time It is for Kabul," NPR, Feb. 28, 2011; "Kabul Bank: Pitiful," *Economist*, Dec. 1, 2012; Matthew Rosenberg, "Trail of Fraud and Vengeance Leads to Kabul Bank Convictions," *New York Times*, Mar. 5, 2013.

community's response was thus defined, not by any coherent strategy, but rather by a series of disparate overlapping strategies and approaches, the most common of which was capacity building.

During the first two years after 9/11, the U.S. government and its NATO allies failed to recognize the growing risk posed by organized crime, factionalism, and warlordism. Instead, U.S. policy often exacerbated the problem by using regional strongmen and their CPNs as allies in operations against al-Qaeda and Taliban fighters. The CIA and some military units gave their local proxies a degree of protection and impunity, as well as millions in cash, deeming the benefits that Afghan partners unrestrained by Afghan law provided for counterterrorism operations to outweigh the costs. By 2003, official U.S. government policy began to reverse course, even though the CIA continued its strategy of payoffs and cooperation with local proxies.[24] The U.S. government and its NATO partners grew concerned when armed clashes between regional militias of competing warlords, including Uzbek commander Abdul Rashid Dostum, Herat warlord Ismail Khan, and Balkh provincial governor Muhammad Atta Noor, left hundreds dead in 2002 and 2003. During 2003-5, the United States intervened to support Karzai and the central government in co-opting these strongmen. Zalmai Khalilzad, the U.S. ambassador to Kabul during 2003-5, helped implement a strategy by which the United States and Karzai sequentially confronted warlords, forced them to disarm and surrender their offices, and then co-opted them into the central government.

The United States also began to build a new Afghan National Army, which, it hoped, would lack the factional loyalties of the mujahideen-era militias and give the central government a monopoly on legitimate use of violence. Complementing U.S.-led efforts were those of the United Nations, which executed a comprehensive disarmament, demobilization, and reintegration (DDR) campaign to disband the militias of the regional warlords. By 2005, DDR had succeeded in disarming the country's most powerful warlords. But the program learned its limits, for militias were often reflagged as police units during 2001-6, and used the newfound state authority to pursue their factional and criminal interests. UN pressure on Karzai during this period led to removals of some corrupt officials, but only those not necessary to Karzai's political agenda. This pattern continued over the next decade.

Aside from these early attempts to reduce the influence of Afghanistan's warlords, however, most of the international community's efforts against corruption and organized crime since 2002 downplayed the political and factional causes of Afghan corruption, focusing instead on capacity building and technical assistance. These efforts, undertaken by the United States, ISAF, the United Nations, and donor nations, often sought to reduce opportunities for corruption and strengthen Afghan institutions. These efforts have included training and advisory programs spearheaded by the UK government, for Afghanistan's elite counternarcotics and investigative units. Various interna-

---

[24] Note, for example, the activities of the Kandahar Strike Force, including the 2009 murder of Kandahar police chief Matiullah Qati; reports of CIA payments to the presidential palace; and the 2013 charges made against U.S. special operations forces interpreter Zakaria Kandahari. Dexter Filkins, Mark Mazzetti, and James Risen, "Brother of Afghan Leader Said to Be Paid by C.I.A.," *New York Times*, Oct. 27, 2009; Yaroslav Trofimov, "After Afghan Raid, Focus on Captors," *Wall Street Journal*, Mar. 11, 2013; Matthew Rosenberg, "Karzai Says He Was Assured CIA Would Continue Delivering Bags of Cash," *New York Times*, May 4, 2013; Rod Nordland, "Interpreter Accused of Torturing and Killing Afghan Civilians Is Arrested, *New York Times*, July 7, 2013.

tional development agencies have sponsored technical assistance programs in ministries across the government, and considerable resources and attention have been dedicated to the mammoth task of recruiting, training, equipping, and professionalizing the Afghan National Security Forces.

Although some of these areas saw progress, corruption and organized crime remained common and even proliferated within Afghan institutions during Karzai's first term in office (2005-9). From the beginning, the international community underfunded and underemphasized the rebuilding of Afghanistan's rule of law institutions, particularly the police, the judiciary, and the prison system, as donors placed more attention on large-scale infrastructure development. At the 2001 Bonn Conference, the German government volunteered to raise and train a new Afghan police force, while the Italian government took responsibility for rebuilding Afghanistan's judiciary. Neither government could muster the necessary resources, expertise, or will to adequately repair these institutions, both of which remained subject to infiltration and intimidation by CPNs. In time, the U.S. government established parallel efforts in both sectors, gradually increasing its training, mentoring, and advisory programs. By 2009, the newly elected Obama administration announced that reform and growth of the Afghan National Police would be a pillar of its new Afghanistan strategy. This led to a new influx of assistance and U.S.-led training programs. Much of this new American aid, however, went toward building the police as a paramilitary force trained to conduct combat operations, and new programs gave little attention to developing the police's law enforcement capabilities. European donors, in contrast, pushed for a less militarized community policing model, leading to a continued lack of coherence in training the Afghan police.[25]

Even as the resources devoted to capacity building grew steadily during 2005-10, failure to acknowledge the inherently political nature of institutional reform remained a fundamental obstacle. Because the international community failed to make disrupting CPNs a priority, criminal networks within the Afghan government thwarted many of the structural and administrative anticorruption reforms that international donors had advocated since the early years of the conflict, including merit-based hiring, pay and grade reform, and asset declaration policies for senior government officials.

The failure of pay and rank reform of the Afghan police in 2006 provides an example of the limits of institutional reform undertaken without attention to political influence of the criminal networks it sought to target. In 2006, in response to growing concerns about the Afghan police's criminal and factional behavior, international donors pressed the Ministry of Interior (MoI) to undertake a comprehensive program of rank and pay reform. The initiative aimed at rationalizing the Afghan police's rank structure, removing the glut of mujahideen-era senior officers, and introducing merit-based criteria for promotions, including exams and the vetting of officers with criminal records. The international community believed that these measures would cull corrupt officers from the force and reduce opportunities for corruption.[26]

The selection board created by the pay and rank reform process submitted its first

---

[25] Robert Perito, "Afghanistan's Police: The Weak Link in Security Sector Reform," U.S. Institute of Peace Special Report no. 227, Aug. 2009; Antonio Giustozzi and Mohammad Isaqzadeh, "Afghanistan's Paramilitary Policing in Context," Afghanistan Analysts Network, July 2011, www.afghanistan-analysts.org/wp-content/uploads/downloads/2012/10/AAAN-2011-Police_and_Paramilitarisation.pdf.

[26] Perito, "Afghanistan's Police"; author interviews with UN officials in Kabul, Jan. 20 and June 30, 2010.

list of 86 candidates for senior police appointments across Afghanistan in May 2006 and recommended removal of dozens of officers for corruption, criminal records, and poor performance. But days after the ministry made its recommendations, antigovernment riots, encouraged by Northern Alliance-affiliated politicians, erupted across Kabul. An insecure Karzai was unnerved by the riots, interpreting them as a threat from Fahim and other Jamiat-e Islami affiliates demanding a greater share of power in the Karzai government. To placate Jamiat and Fahim, Karzai unilaterally added 14 police officers, almost all of them Jamiat affiliates with reputations for extensive corruption, to the newly vetted list of senior police appointments. The move, driven by Karzai's political calculus, undermined the new pay and rank reforms by signaling that the palace and its mujahideen-era allies would continue to override selection boards at the Ministry of Interior, for political motives. While the international community eventually pressured Karzai to remove some of his 14 late additions from their posts, his disregard for the new selection board signaled that political elites would continue to intervene in appointing criminal allies to police command. MoI officials who tried to obstruct these appointments were often removed or intimidated into acquiescence. By relying on procedural fixes and not anticipating the interaction between national politics and institution building, the supporters of pay and rank reform in 2006 made limited progress in countering CPNs' influence within the MoI.

Because portions of Afghan ministries functioned as vertically integrated patronage networks, technical assistance and capacity building alone, without measures to counter CPNs' influence, could do little to prevent the growing dysfunction in state institutions. Moreover, CPNs have actively suppressed or sought to co-opt the junior, reform-minded officials trained by the United States and others in recent years, as well as the experienced—but politically vulnerable—technocrats operating within Afghan government bureaucracies.

Further limiting the capacity building approach was poor coordination between the countercorruption objectives established at major international donors' conferences, and the efforts of international agencies, embassies, and headquarters on the ground in Afghanistan. International donors established benchmarks for countercorruption reforms at the 2008 Paris Conference and the London and Kabul Conferences of 2010. In both cases, the broad benchmarks set by the international community focused largely on capacity building and institutional reform and, thus, had little impact on the underlying political causes of corruption. At the 2008 Paris Conference, the Afghan government announced the creation of a new High Office of Oversight to help implement donor nations' countercorruption benchmarks. But the new organization, designed to placate the international community, had an ill-defined portfolio that overlapped the authority of the Afghan Attorney General's Office and made no progress toward ending the impunity problem. International donors' conferences ultimately failed to provide leverage to hold the Afghan government accountable for disbursement of international funds, instead further complicating international countercorruption efforts in Kabul.

As the limits of capacity building became clear by 2007 and 2008, some U.S. officials focused on overcoming the impunity problem by encouraging the Afghan government to investigate, prosecute, and convict officials who committed serious crimes. The approach recognized that capacity building would fail to achieve lasting institutional reform unless Afghan officials faced real sanctions for serious crimes and for diverting state resources. In tackling the impunity problem, U.S officials depended on the cooperation of the Afghan Attorney General's Office, the state's prosecutorial arm. In 2007, the Bush administration made an early attempt to energize prosecutions by appointing Tom Schweich coordinator for counternarcotics and justice reform in Afghanistan, with ambassadorial standing. Schweich partnered with Afghanistan's attorney general, Abdul Jabar Sabet, to encourage prosecutions, but Sabet's investigations were heavily politicized, used as tools to pressure the palace's rivals. This provoked a backlash, particularly when Sabet targeted the country's fledgling press. Schweich ended his tenure disillusioned and wrote a *New York Times* op-ed characterizing Afghanistan as a narco-state where the Karzai government protected a class of criminal elites.[27]

In 2009, the Department of Justice (DoJ) in the newly elected Obama administration made a second, better-resourced attempt to secure prosecution of criminal Afghan officials. The most promising of the DoJ's initiatives was the creation of a vetted and FBI-mentored Major Crimes Task Force (MCTF) to investigate crimes by high-level government officials. With unprecedented protection from political interference, the MCTF achieved several successes in 2009 and 2010, including the successful prosecution of an Afghan police general and several police colonels for narcotics smuggling. It also opened investigations into several notoriously corrupt governors and members of parliament. Initial progress was undermined once again, however, by lack of attention to the realities of Afghan politics. The MCTF earned the ire of the presidential palace after it arrested Ahmad Zia Salehi, director of administrative affairs at the Afghan National Security Council, in July 2010. Salehi was apprehended for accepting a ten-thousand-dollar bribe to protect officials at the New Ansari banking network who were illegally moving money for government officials, CPNs, drug traffickers, insurgents, and terrorists.[28] Unbeknownst to the MCTF or the DoJ, allegedly Salehi was a key palace insider who moved money to facilitate Karzai's political agenda and was on the CIA payroll.[29] The presidential palace wrongly interpreted Salehi's arrest as an effort to undermine its influence and to target the president's key operatives. Seeing the MCTF as a threat to his political strategy, Karzai reined in the unit, reversing the momentum on prosecutions and stalling several high-profile cases.[30]

Refocused ISAF Countercorruption Efforts (2010-13)

By 2010, ISAF was increasingly aware that countering the strategic threat of corruption required lines of effort beyond capacity building, as well as greater international coordi-

---

[27] Thomas Schweich, "Is Afghanistan a Narco-State?" *New York Times*, July 27, 2008.

[28] Yaroslav Trofimov, "Karzai and U.S. Clash over Corruption," *Wall Street Journal*, Aug. 3, 2010; Joshua Partlow and Greg Miller, "Karzai Calls for Probe of U.S.-Backed Anti-corruption Task Force," *Washington Post*, Aug. 5, 2010; Dexter Filkins and Mark Mazzetti, "Key Karzai Aide in Corruption Inquiry Is Tied to C.I.A.," *New York Times*, Aug. 25, 2010; Dexter Filkins, "The Afghan Bank Heist," New Yorker, Feb. 14, 2011.

[29] Dexter Filkins and Mark Mazzetti, "Karzai Aide in Corruption Inquiry is Tied to C.I.A." New York Times, August 25, 2010.

[30] Yaroslav Trofimov and Matthew Rosenberg, "Karzai Targets Two U.S.-Backed Task Forces," *Wall Street Journal*, Aug. 5, 2010; Filkins, "The Afghan Bank Heist"; Maria Abi-Habib, "U.S. Blames Senior Afghan in Deaths," *Wall Street Journal*, Apr. 1, 2012.

nation. To regain the initiative on countercorruption, ISAF and U.S. Forces Afghanistan created the Combined Joint Interagency Task Force–Shafafiyat ("transparency" in Dari and Pashto) in coordination with the international community in the summer of 2010. Shafafiyat was designed to support the Afghan government, foster a common understanding of the corruption problem, plan and implement ISAF anticorruption efforts, and integrate the coalition's countercorruption activities with those of key interagency and international partners. From the outset, the task force engaged regularly with leaders from Afghan civil society and officials across the Afghan government to frame the problem of corruption from the perspective of those who had experienced it and to develop a shared understanding as a basis for joint action and reform. In partnership with senior Afghan leaders, Shafafiyat (which in late 2012 evolved into the Combined Joint Interagency Task Force–Afghanistan, or CJIATF-A) established a variety of structured forums in which ISAF, U.S. interagency partners, and international organizations could exchange information and work cooperatively with Afghan officials to develop and implement concrete anticorruption plans and measure progress. In its coordination with Afghan and international partners, ISAF sought consistently to illustrate the comparative long-term risks of *inaction* in order to persuade senior leaders that it was in their ultimate interest—and the interest of the Afghan state and people—to address the problem with a sense of urgency.

Starting in 2010, ISAF and the NATO Training Mission in Afghanistan (NTM-A), Regional Commands, and CJIATF-Shafafiyat, in partnership with the Afghan government and the international community, explicitly ranked the forms of corruption and organized crime presenting the greatest threat to the coalition's mission and to the viability of the Afghan state. ISAF also adjusted its approach to account for the challenges encountered by earlier countercorruption efforts. This section discusses the priorities that ISAF established after 2010, and their relative success in reducing corruption and countering the influence of CPNs.

Security Ministries and the Afghan National Security Forces

ISAF, through NTM-A, expanded its efforts to develop professional Afghan security forces managed by transparent, accountable security ministries. By 2010, it was clear that training, equipping, and professionalizing Afghan security forces required complementary efforts to promote internal accountability. So NTM-A, Shafafiyat, and other organizations across ISAF worked closely with senior officials from the Ministries of Defense and Interior, through multiple joint Afghan-led working groups and commissions, to develop detailed anticorruption recommendations and implementation plans. Joint measures included creation of insulated investigative, oversight, and adjudicative bodies within the security ministries so that the Afghan National Security Forces (ANSF) could enforce internal accountability while avoiding political interference and intimidation. ISAF regional commands and task forces, meanwhile, gave greater attention to the politics of Afghan security force appointments and to the patronage networks that protected corrupt army and police commanders. ISAF officers worked to build trust with their Afghan colleagues and expand intelligence sharing, in order to encourage leaders within the ANSF to act to remove corrupt officials.

ISAF efforts from 2010 onward made some senior ANSF leaders more willing to remove officers with known criminal records from their posts. In Eastern Afghanistan, for example, from mid-2011 to mid-2012, nearly 50 ANSF officers were removed for crimes, thanks to collaboration and information sharing between ISAF and ANSF leaders. Meanwhile, Bismillah Khan Mohammadi, during his 2010-12 term as minister of interior, removed a significant number of officials connected to CPNs from across the police (though he leaned toward removing officers from CPNs hostile to his own Jamiat Party). An increased willingness by some Afghan security leaders to remove CPN affiliates helped check CPNs' influence within the ANSF.[31]

But the removal of criminal officials often depended on a handful of bold Afghan leaders, and Afghan political elites continued to protect criminal affiliates, meaning that those officials removed from their positions were often recycled into new positions thanks to the influence of outside patrons. Progress in empowering internal accountability and oversight mechanisms—critical to the sustainability of countercorruption efforts—progressed more slowly.[32]

Rule of Law and the Judicial Sector

ISAF and its partners in the U.S. mission in Kabul worked to support development of Afghan law enforcement and judicial institutions responsive to the population's needs and to reliably enforce the rule of law. Efforts included creating a U.S. Rule of Law Field Force in September 2010.[33] ISAF strove to achieve sustained, rather than merely episodic, engagement with the Afghan judicial sector and to enable investigators, prosecutors, and judges to operate free from bribery, intimidation, and political interference. After the Salehi case in July 2010, ISAF officials were more attuned to the political dimension of key criminal prosecutions. They increased efforts to work directly with senior national-level Afghan leaders to lift protection from criminals, encourage prosecutions, and prevent officials removed for corruption from being reinstated elsewhere in the government.

Despite more sophisticated international efforts, the Afghan government never prosecuted enough criminal officials to reduce the CPNs' power. After the 2010 Salehi case, the Attorney General's Office constantly succumbed to intense pressure from political elites to suppress high-level prosecutions. President Karzai, Vice President Fahim, and other senior officials showed no change of heart and continued refusing to support investigations and prosecutions.[34] The palace allowed only a handful of relatively low-level prosecutions to proceed, in response to pressure from international donors. The

---

[31] Conclusions in this paragraph are based on authors' personal observations, assessments, and interviews at various times during 2010-12, while in Kabul.

[32] The Ministry of Interior, for example, established several Transparency and Accountability Commissions (TACs) at the regional and provincial levels in 2012, which brought police leadership together to collectively develop and commit to countercorruption initiatives and oversight mechanisms. It is unclear what impact the TACs had, however, after the transfer of Minister Bismillah Mohammadi from MoI to MoD in late 2012.

[33] NATO later created a Rule of Law Field Support Mission (NROLFSM) of its own in June 2011. For greater unity of effort, the US and NATO missions were eventually placed under a dual-hatted common commander.

[34] See, for example, Filkins, "Letter from Kabul"; Special Inspector General for Afghanistan Reconstruction, Quarterly Report to the United States Congress, Oct. 30, 2013, 137.

most notable of these was the prosecution of Kabul Bank owners Farnood and Fruzi, who were convicted and given mild sentences in March 2013 for their role in the Kabul Bank scandal. The verdicts followed—and were almost certainly the result of—credible threats from U.S. and European donors to suspend billions in aid payments to the Afghan government if Farnood and Fruzi should escape legal sanction for their role in Kabul Bank's collapse.[35]

Borders and Airports

The civil-military team in Kabul and senior Afghan officials worked together to expose and act against the criminal networks operating at Afghan borders, airports, and customs depots. These countercorruption efforts sought to enable the Afghan state to maintain credible sovereignty and achieve enduring security while collecting revenue sufficient to expand its licit economy and reduce its dependence on international assistance. Recognizing the integral link between fighting corruption at Afghanistan's borders and interdicting the movement of insurgents, weapons, bomb-making components, and narcotics, ISAF created the CJIATF-Nexus, a task force under operational control of CJIATF-Shafafiyat. Nexus coordinated efforts between Regional Commands, ISAF headquarters, and interagency law enforcement to target the intersection of CPNs, the narcotics trade, and the insurgency, with a particular focus on Afghanistan's borders, border crossing points, and airports.

Through effective integration, ISAF, the United States, and the UK supported Afghan law enforcement in steadily increased interdiction of insurgent materiel moving across the Afghan border. The international community and the Afghan government achieved far less progress, however, in dismantling CPNs within the police and customs offices at the borders. Because criminal activity along the borders was so lucrative, CPNs and their patrons were highly motivated and well enough financed to ward off attempts to investigate or remove criminal officials.

Countering Drug Trafficking and Transnational Crime

As the mutually reinforcing relationship between the narcotics trade, corruption, and the insurgency became increasingly apparent, U.S. and UK interagency and law enforcement organizations, along with ISAF, began targeting the intersection of these convergent threats. Likewise, as the United States, the UK, and ISAF got a closer look at the flows of money, narcotics, precursor chemicals, weapons, and other resources across Afghanistan's criminal networks, it became clear that corruption and organized crime in Afghanistan had a significant transnational dimension. As the 2011 U.S. Strategy to Combat Transnational Organized Crime explained, "Nowhere is the convergence of transnational threats more apparent than in Afghanistan and Southwest Asia."[36]

Thus, the application of international law enforcement actions and targeted financial

---

[35] Rosenberg, "Trail of Fraud and Vengeance."
[36] White House, "Strategy to Combat Transnational Organized Crime: Addressing Converging Threats to National Security," July 2011, www.whitehouse.gov/sites/default/files/Strategy_to_Combat_Transnational_Organized_Crime_July_2011.pdf.

sanctions became a critical way to degrade Afghanistan's criminal networks, creating a deterrent effect that the judicial system could not. From 2010 on, U.S. and UK law enforcement and ISAF also strove to better integrate law enforcement and military operations for synergy and mutually reinforcing effects. To facilitate this integration, the U.S. and UK governments established an Interagency Operations Coordination Center in Kabul. This expedited intelligence sharing and coordination between the counternarcotics efforts of ISAF, the U.S. Drug Enforcement Agency, and the UK Serious Organized Crimes Agency.

In 2010-12, joint international and Afghan counternarcotics efforts with military support achieved a high rate of interdictions and arrests, including the arrests of several high-level drug traffickers. Drawing on better intelligence support, coalition counternarcotics operations targeted drug traffickers who facilitated the insurgency. This had some effect in disrupting the movement of insurgent supplies, and a small (but not negligible) effect in squeezing insurgent finance.[37] Due to sustained UK mentorship, Afghanistan's Counter-Narcotics Justice Center functioned as the only judicial institution in the country that was insulated from bribery and political intimidation. This led to the successful conviction and sentencing of criminals and insurgents charged with narcotics trafficking.[38] But from 2011 to 2013, interdiction dropped sharply: interdictions of precursor chemicals fell 73 percent, and interdiction of hashish fell 79 percent. Further undermining counternarcotics progress, opium production in Afghanistan reached unprecedented levels in 2013, nearly doubling from 2012. The Department of Defense attributed the steep drop in interdiction to the drawdown of U.S. forces, and the resulting loss of operational enablers. The increase in opium production, meanwhile, occurred largely in areas with poor security and limited government presence. These trends demonstrate the tremendous importance of supporting and integrating counternarcotics efforts with a large military force, and the correlation between security and progress in counternarcotics.[39] But they also indicate that the Afghan government failed to develop sufficient capacity and will to carry on independent counternarcotics operations as U.S. force levels declined.

Contracting and Procurement

In 2010, recognizing the ongoing abuses of U.S. contracting by both Afghan and Western firms, the U.S. government, in tandem with CJIATF-Shafafiyat, created Task Force 2010, a component of U.S. Forces-Afghanistan. The new task force's mission was to coordinate, expand, and apply greater oversight and management to U.S. contracting, acquisition, and procurement processes, thereby denying criminal patronage networks and insurgents access to U.S. funds and materiel. Acknowledging that international spending has the potential to directly affect campaign objectives, U.S. forces increasingly in-

---

[37] Department of Defense, "Report on Progress toward Security and Stability in Afghanistan," Dec. 2012, 123-4, www.defense.gov/news/1230_Report_final.pdf.

[38] Joshua Hersh, "Afghanistan's U.S.-Funded Counter-Narcotics Tribunal Convicts Nearly All Defendants, Records Show," *Huffington Post*, June 10, 2012, www.huffingtonpost.com/2012/06/09/afghanistan-counter-narcotics-tribunal_n_1580855.html.

[39] Special Inspector General, Quarterly Report, 106.

tegrated procurement and contracting considerations into planning and operations at all levels. Task Force 2010 and others recognized that high-value construction contracts in insecure areas are the most difficult to oversee and administer, and the task force set its priorities accordingly. Additional contracting reforms included disaggregating large contracts to encourage more bidders and deter the emergence of monopolies; advertising contracts more widely to improve Afghan vendors' awareness of, and access to, the bidding process; and identifying intended subcontractors in the course of bidding. U.S. forces' efforts at contracting reform were coordinated with the Afghan government and civilian agencies operating in Afghanistan (such as USAID and the U.S. embassy). The embassy published its own COIN contracting guidance, similar to that issued by ISAF, in November 2010. By 2013, the special inspector general for Afghanistan reconstruction (SIGAR) had taken over primary responsibility for vetting and investigating contract and procurement fraud, while Task Force 2010 moved into a supporting role and relocated to Doha, Qatar.[40]

The U.S. government, by vetting and restructuring contracts and investigating contracting fraud, made significant progress in cutting the flow of its contracts and spending to CPNs. In debarring contractors and cutting funding, the U.S. government could take virtually unilateral action without the need to work through the Afghan government, enabling efficient progress in cutting funding to CPNs. By February 2012, Task Force 2010 had vetted 1,000 contractors and debarred or suspended over 125 Afghan, U.S., and international companies. Investigations by SIGAR, meanwhile, achieved 47 convictions, 61 suspensions, and 94 debarments of individuals and companies by October 2013.[41] Because the total magnitude of the misuse of U.S. funds is impossible to quantify, it is difficult to determine what proportion of total contract fraud was exposed and ended by Task Force 2010 and SIGAR. But the high number of debarments achieved from 2010 to 2013 had a clear effect of deterring the misuse of U.S. funds and signaling that anyone committing contract fraud faced a high risk of legal sanction. Moreover, the efforts blocked diversion of U.S. funds to some of Afghanistan's most noxious CPNs.[42]

From 2010 on, ISAF, the U.S. country team in Kabul, and the international community made some progress working with their Afghan counterparts to reduce CPNs' influence within the Afghan government. Success was greatest in the two areas where the international community had the greatest influence and access: international contracting and the ANSF. As a result of international efforts, since 2010 Afghan officials have increasingly acknowledged that the scale of corruption within their country's critical institutions is compromising the state's security, stability, economic health, and cohesion. Afghan leaders also express concerns about their country's international reputation and standing, acknowledging that corruption, organized crime, and the narcotics trade jeopardize the credibility of Afghanistan's sovereignty. Many leaders remain deterred, however, by what they perceive as the near-term political risks of acting against powerful criminal networks. Because of these fears, progress in countering corruption

---

[40] Special Inspector General, Quarterly Report.

[41] Ibid.; John Ryan, "Units Aim to Root Out Corruption in Afghanistan," *Army Times*, Feb. 16, 2012.

[42] For example, the United States successfully cut funding to the CPN built around commander Ruhullah, who had been funneling money to the Taliban and whose men had abused Afghan civilians. "Matthew Rosenberg, "U.S. Cuts Off Afghan Firm," *Wall Street Journal*, Dec. 8, 2010.

often stalled when it depended on leadership by Afghan political elites. The 2010 announcement by the United States and NATO that the ISAF mission would end in 2014 further undermined Afghan confidence in the long-term viability of countercorruption measures. This announcement led some Afghan leaders to intensify their factional and criminal behavior in an attempt to consolidate influence before a return to civil war or a reconciliation deal with the Taliban. These are deep, intractable challenges that the international approach to countercorruption, at times, failed to offset.

Recommendations and Conclusion

After the withdrawal of most international troops, Afghanistan's criminal patronage networks and organized crime remain poised to have disruptive effects on the country's future. As long as international resources have flowed steadily into the country, the major political players have retained an incentive to cooperate within the tenuous political order built by Karzai and his allies. Afghanistan has not had the strong state institutions necessary to enforce the rule of law and mediate internal conflicts (such as an independent and empowered judiciary and reliable police forces). This vacuum has left power brokers with few constraints, beyond the residual international-force presence, to prevent the same brand of violent competition that occurred in the 1990s. The new coalition government of President Ashraf Ghani, inaugurated in September 2014, has promised to lead Afghanistan toward a more inclusive, forward-looking political order. If Ghani is indeed committed to reform and accountability, the fragile coalition he has formed will face opposition from powerful CPNs inside and outside the Afghan state. The resilience of corruption and criminal networks will likewise challenge the new government's efforts to maintain security while sustaining the shaky power-sharing agreement arranged after the 2014 election, all in the face of significantly reduced U.S. military support. The patronage networks dominant within Afghan government ministries have consistently sought both profit and political advantage through graft and illicit activities as they consolidate power in anticipation of future instability. Networks with dual criminal and political objectives sponsor thugs to foment instability in rivals' home districts and provinces, entrench their own positions in the country's vast narcotics trade at the expense of competitors, intimidate or assassinate enemies, and, at times, collude with the insurgency and foreign intelligence services to do so. As access to foreign resources wanes in the coming years and the U.S. presence is further reduced, competition among the ethnic and tribal factions within the current government may manifest in more overt violence than before.

Recommendations

As the U.S. military and national security establishment looks back on the wars of the past decade to glean lessons in preparation for future conflicts, the Afghan anticorruption experience must be an essential area of focus. Few threats have cut as widely across the lines of effort undertaken by U.S. and international forces in Afghanistan as have criminal patronage networks. By hollowing out the critical institutions that the coalition and its partners have struggled to build, by undermining the legitimacy of the government that ISAF and the international community have sought to support, by preventing the population's mobilization against the insurgency, and by contributing to insurgent narratives, CPNs have jeopardized all that the United States and its international allies have set out to achieve in Afghanistan. The full transfer of security responsibilities to a newly elected Afghan government will test whether Afghanistan's factionalized elites can pull together and overcome the legacy of a decade of limited action against — and, at times, complicity in — corruption and organized crime. The efforts of the U.S. government and ISAF to engage constructively with Afghan leaders on the corruption issue — while at the same time assembling the necessary tools to address the challenge directly — have laid a foundation for progress against Afghanistan's CPNs in the coming years. Much work lies ahead, though. The future will increasingly depend on the will of senior Afghan officials, including the Ghani-Abdullah coalition government, to commit definitely to reforms. While daunting in the near-term, these reforms represent the country's best hope for a promising, peaceful future, which its people, after decades of conflict, surely deserve.

In anticipation of future missions of similar complexity, it will be essential to integrate the lessons emerging from the countercorruption experience in Afghanistan into U.S. forces' training, doctrine, and leadership development. Although future efforts will demand close civil-military coordination and unity of effort, U.S. and allied forces and other international partners must be prepared to exercise initiative in addressing the problems of a political economy based on corruption and organized crime in counterinsurgency and stabilization environments. The lessons and insights outlined below reflect the expectation that U.S. and allied forces and their various partners will again operate where illicit power structures are drivers of conflict and impediments to sustainable security, political progress, and economic growth. These lessons come with the understanding that the United States and the international community have at times inadvertently contributed to the problems faced today with Afghanistan's criminal patronage networks — missteps that we cannot afford to repeat in future conflicts.

Anticipate and respond swiftly to criminal patronage networks. In insecure states with underdeveloped institutions and weak rule of law, any massive infusion of international resources to build local capacity, if disbursed with inadequate oversight, is likely to be accompanied by a surge in corruption and organized crime.[43] International forces and their interagency counterparts conducting counterinsurgency or stability operations must anticipate this development and be prepared, in the earliest stages of their mission, to put in place mechanisms to mitigate and monitor the problem (e.g., tracking illicit financial flows and implementing vendor-vetting measures). At the same time, expectations for transparency and accountability should be articulated to officials in the supported government. In all these efforts, timing is critical. It is vital to launch

---

[43] See, for example, UNDP, "Fighting Corruption in Post-Conflict and Recovery Situations," June 9, 2010, www.undp.org/content/undp/en/home/librarypage/democratic-governance/anti-corruption/fighting-corruption-in-post-conflict---recovery-situations.html; Fredrik Galtung and Martin Tisné, "A New Approach to Postwar Reconstruction," *Journal of Democracy* 20, no. 4 (2009).

countercorruption initiatives before criminal networks and patterns of corruption become entrenched, before the population grows disillusioned with its government and international forces, and before the perception arises within the host government that impunity for politically connected criminals will be tolerated. The international community will also maximize its influence if it acts before its will to impose costs for corruption—whether through conditionality of aid or through international law enforcement actions—has been called into question.

Acknowledge the centrality of politics. International forces and their civilian partners must ground their efforts in a thorough understanding of the history and politics of the state where they are engaged. Interventions, whether counterinsurgency campaigns or stability operations, are fundamentally political endeavors. Illicit power structures are also fundamentally a political problem, closely linked to the balances of power among national elites. As a 2010 United Nations Development Programme (UNDP) study noted, "Effectively responding to corruption can be difficult because it nearly always requires taking political, economic, and social power away from those who benefit from the status quo."[44] With this in mind, international forces must understand the key leaders they engage with, in the context of their political, social, and cultural networks. Intelligence on business networks and financial flows between elites can likewise provide tremendous insights into the structure of power and the influence of CPNs in the host nation. The intelligence community has an important role to play here. Analysts may need additional training, however, to get a proper focus, not only on the composition of illicit political networks, but also on the historical affiliations, dynamic relationships, and the balances of power within them—as well as an understanding of their roles in any broader national political settlement. It is also important to note that host-nation officials' interests may not always align fully with those of international forces. Host-nation political actors may be motivated by narrow agendas driven by their historical, ethnic, and factional affiliations—as well as by a desire to maximize their political and financial positions before international forces' ultimate departure—rather than by a shared commitment to satisfy mutual goals. Therefore, countercorruption efforts stand to have the greatest effect when implemented in support of a carefully coordinated political strategy by international forces and their civilian counterparts, designed to marshal military, diplomatic, and economic tools and resources in pursuit of a clearly articulated set of political objectives.

Rank disrupting criminal capture of institutions above capacity building. Attention to the politics of the supported government and early implementation of joint anticorruption measures are critical to preventing the emergence of what has been called the "political-criminal nexus"—a mutually beneficial relationship of protection and profit between corrupt government officials and criminal networks.[45] Left unchecked, this dynamic can

---

[44] UNDP, "Fighting Corruption in Post-Conflict and Recovery Situations."

[45] Roy Godson, *Menace to Society: Political-Criminal Collaboration around the World* (New Brunswick, NJ: Transaction, 2003). See also Mark Shaw, "Drug Trafficking and the Development of Organized Crime in Post-Taliban Afghanistan," in *Afghanistan's Drug Industry: Structure, Functioning, Dynamics, and Implications for Counter-Narcotics Policy*, ed. Doris Buddenberg and William A. Byrd (Collingdale, PA: Diane, 2006);

lead to the criminal capture of critical state functions. Thus, the supported government's institutions become directed toward serving the interests of a narrow political elite and their criminal associates, rather than advancing and protecting broad national interests. In post-intervention states whose governments receive large sums of international assistance, there is an enormous incentive for criminal networks to infiltrate and co-opt fragile institutions newly flush with resources. The Afghan experience has demonstrated that international technical assistance and professionalization training are necessary, *but not sufficient,* for girding institutions against criminal infiltration and subversion. Rather than focus narrowly on capacity building, those providing international assistance must attune themselves to patterns of criminal activity. And they must work with key leaders in the supported government to develop coherent, broadly acceptable strategies to disrupt criminal networks and sever the relationships between political patrons and their criminal clients.

Understand the impact of international spending. The infusion of substantial international resources—whether development assistance or contracts—without sufficient oversight, into a contested or postpeace settlement state with an underdeveloped economy has the potential to empower some actors while disempowering others, thus generating unintended political, social, and security consequences. Development, procurement, and acquisition are military operational concerns that must be aligned with a comprehensive national or coalition political strategy. In future conflicts, rigorous vetting of vendors and sustained oversight for large logistics and development contracts will be crucial. And full integration across the civilian and military agencies involved is indispensable to achieving a common contracting operating picture. In Afghanistan, failure to judge local populations' development needs or accurately assess communities' capacity to absorb international aid generated extreme waste and opportunities for graft, corruption, and patronage while preventing the emergence of free-market entrepreneurs.[46] Instead, models of development focusing on host nations' nascent small enterprises and business sectors can serve as a check against large-scale corruption while setting the conditions for inclusive and responsive governance that deters systemic abuse of power.[47]

Promote transparency and accountability in security force development. With the expectation that future counterinsurgency and stability operations will be coupled with security force development missions, international forces can anticipate having significant access, agency, and leverage within the supported government's security sector. This access presents a critical opportunity to integrate countercorruption efforts within training and professionalization initiatives. The development of effective, professional, and ac-

---

Peter Andreas, "The Political Economy of War and Peace in Bosnia," *International Studies Quarterly* 48, no. 1 (2004).

[46] Greg Mills and Ewen Mclay, "The Path to Peace in Afghanistan," *Orbis* 55, no. 4 (2011); and the U.S. Senate Foreign Relations Committee, "Evaluating U.S. Foreign Assistance to Afghanistan," June 8, 2011, www.foreign.senate.gov/press/chair/release/foreign-relations-committee-releases-comprehensive-report-on-us-civilian-aid-in-afghanistan.

[47] See, for example, R. Glenn Hubbard and William Duggan, *The Aid Trap* (New York: Columbia Univ. Press, 2009); and David Kilcullen, Greg Mills, and Jonathan Oppenheimer, "Quiet Professionals: The Art of Post-Conflict Economic Recovery and Reconstruction," RUSI Journal 156, no. 4 (2011).

countable security forces is essential, of course, for the transfer of security responsibilities to the host nation. But in many developing countries emerging from conflict, control of the security ministries and their forces is a prize much sought after by elites and their networks as the political settlement develops.[48] As a result, security forces can become subject to factionalism, politicization, and corruption.

Therefore, international forces assigned to develop the supported government's security sector must be prepared to apply the same rigorous analysis to the political and factional affiliations of key leaders within the host nation's security forces as to those of other national figures. A security force development model focused strictly on capacity building and professionalization may not be sufficient for ensuring a politically neutral force or for adequately integrating former combatants into new national security structures. As the 2006 *Counterinsurgency Field Manual* makes clear, "the acceptance of values, such as ethnic equality or the rejection of corruption, may be a better measure of training effectiveness in some COIN situations" than simple "competence in military tasks."[49] In the long term, of course, there is no dichotomy between these objectives, since host-nation security forces rife with corruption will suffer greatly impaired operational effectiveness. The Afghanistan experience demonstrated the extent to which corruption consistently undermined a unit is leadership, morale, will to fight, readiness, and logistical sustainability. To the degree that the population sees host-nation security forces as professional and above ethnic, tribal, and political factionalism, they can add credibility to the supported government, serving as the locus of an emerging sense of national unity.

Integrate law enforcement, military, and information operations. As the war in Afghanistan has made clear, corruption, organized crime, and insurgency are interconnected problems that cannot be dealt with in isolation. In states in or emerging from conflict, an effective response to these converging threats requires integration of law enforcement, military, and information operations at the tactical, operational, and strategic levels in order to employ the full range of tools available to address these problems. Integration of these capabilities is also essential to countering the likely criminalization of the insurgency. This dynamic has been seen to varying degrees in Colombia, Iraq, and Afghanistan as insurgent groups that had engaged in illicit activities initially as a means of financing their operations became increasingly profit focused at the expense of their original ideological or political aims. International forces and their civilian partners can capitalize on this dynamic not only through information operations—calling attention to the insurgent group's venality and hypocrisy—but by mobilizing and empowering host-nation law enforcement assets through evidence-based operations against the insurgent group's criminal activities.[50] Because host-nation law enforcement and judicial

---

[48] See Andrew Rathmell, "Reframing Security Sector Reform for Counterinsurgency—Getting the Politics Right," in *Complex Operations: NATO at War and on the Margins of War*, ed. Christian Schnaubelt, NATO Defense College Forum Paper 14, July 2010, www.hks.harvard.edu/cchrp/maro/pdf/NATO_Defense_College_MARO_article_7_10.pdf.

[49] Department of the Army, *Counterinsurgency, Field Manual 3-24*, section 6-13, Dec. 2006, http://usacac.army.mil/cac2/Repository/Materials/COIN-FM3-24.pdf.

[50] See David Spencer et al., *Colombia's Road to Recovery: Security and Governance 1982-2010*, (Washington, DC: Center for Hemispheric Defense Studies, National Defense University, 2011); Gretchen Peters, "Crime

institutions often become targets for infiltration and subversion by criminal networks and their affiliates, international forces and their civilian partners must also help insulate and protect these institutions from intimidation and coercion.

Identify and operate against the transnational dimensions of the problem. As President Obama's "Strategy to Combat Transnational Organized Crime" makes clear, the problem can be particularly acute within weak and developing states where criminal networks "threaten stability and undermine free markets as they build alliances with political leaders, financial institutions, law enforcement, foreign intelligence, and security agencies."[51] Transnational criminal organizations exploit and destabilize weak institutions, internal divisions, and permissive security environments of states engaged in or emerging from conflict. If the Afghan experience and others are a reliable guide, key figures within indigenous criminal networks will also rely on links to the international financial system to launder their criminal proceeds and maintain licit business interests abroad. In these instances, the United States and allies have a range of tools and capabilities at their disposal to operate against the transnational dimension of corruption and organized crime while furthering counterinsurgency and stabilization objectives—by tracking illicit finance, initiating targeted coercive financial actions, pursuing sanction designations, and identifying opportunities for mutual legal assistance. In future conflicts, international forces and their civilian partners would benefit from the creation of a central, unified interagency strategic planning body with the capacity to manage and coordinate application of these tools and capabilities against transnational networks.

Develop a countercorruption narrative that demonstrates long-term commitment. Given the extent to which corruption undermines popular confidence in a supported government's effectiveness, legitimacy, and sovereignty, international forces and their civilian partners must find a means of presenting themselves as an honest broker between the population and the state, thereby avoiding perceived complicity in the host government's corruption—even as international forces continue to provide vital assistance to the supported state's leaders and institutions. In this, it is essential to consistently transmit a message of enduring international commitment at the strategic and tactical levels—commitment not only to end corruption but also to ensure durable security and advance the interests and aspirations of the population. Without a compelling narrative of commitment, a series of harmful hedging strategies can develop: criminal networks and their patrons will accelerate and expand their illicit activities, driven by a short-term maximization-of-gains mentality that anticipates the eventual departure of international forces and, with them, the easy access to international resources.

Promote civil society as a force for anticorruption advocacy and reform. As evidenced in Sicily, Colombia, Georgia, Mexico, and elsewhere, civil society groups can play a dra-

---

and Insurgency in the Tribal Areas of Afghanistan and Pakistan," report, Combating Terrorism Center at West Point, Oct. 2010, www.dtic.mil/dtic/tr/fulltext/u2/a536511.pdf; Phil Williams, *Criminals, Militias, and Insurgents: Organized Crime in Iraq*, monograph, Strategic Studies Institute, U.S. Army War College Press, June 2009, www.strategicstudiesinstitute.army.mil/pubs/display.cfm?pubID=930.

[51] Barak Obama, "Strategy to Combat Transnational Organized Crime: Addressing Converging Threats to National Security," report, White House, July 2011.

matic role in reversing the influence of organized criminal networks and the institutional corruption they enable.[52] When properly networked and empowered, social activists, educators, entrepreneurs, the media, religious leaders, and other moral authorities can together foster a critical mass of societal support for upholding the rule of law while stigmatizing corruption, thus generating positive social pressure for reform. International forces and their civilian counterparts can create the space for these groups to mobilize unimpeded, in part by realizing that civil society organizations, much like the host nation's judicial institutions, will become targets of intimidation and retribution from criminal networks and their political patrons. Supporting and engaging with civic and social organizations can directly advance fundamental counterinsurgency and stability objectives. A healthy, vibrant civil society is the foundation of a stable state whose institutions are responsive and whose leaders are accountable.

Employ incentives and disincentives. International forces and their civilian partners should be aware of the leverage they have to shape events within a host nation's political space. This leverage derives largely from the security assurances provided by international forces and from international aid, which may be the only reliable source of revenue for the government of a beleaguered state struggling to emerge from conflict. Although it must be carefully and strategically applied, this leverage can prove vital when pursuing countercorruption efforts, especially where host-nation officials have little appetite for reform. Leaders will thus need to be motivated to action. Incentives can include additional assistance to a given state institution (or military unit), linked to the execution of a desired reform or law enforcement action. Targeted coercive financial sanctions or international law enforcement measures are another means of leverage (although, again, they must be applied only after careful consideration of the political context). And finally, the international community can exert a more indirect form of influence by taking steps to integrate the host nation into international regimes and compacts related to corruption, transparency, and accountability. Encouraging compliance with international norms and standards appeals to host-nation leaders' concerns about the state's international reputation, standing, and sovereignty.

Conclusion

Since 2001, the United States and its NATO and ISAF partners have made tremendous investments in Afghanistan's reconstruction. From 2001 to 2014, the United States spent over $57 billion to rebuild and sustain the Afghan Nation Security Forces, along with $30 billion on development and governance projects.[53] This assistance left Afghanistan stronger in 2014 than at any point in its tragic post-1978 history. But Afghanistan's state institutions remain fragile and, in some cases, dysfunctional, so that it is not clear whether the new government under President Ashraf Ghani can secure the country against a

---

[52] See, for example, Leoluca Orlando, *Fighting the Mafia and Renewing Sicilian Culture* (San Francisco: Encounter, 2001); and Roy Godson, "Fostering a Culture of Lawfulness: Multi-Sector Success in Pereira, Colombia, 2008-2010," report, National Strategy Information Center, Oct. 2010, www.strategycenter.org/wp-content/uploads/2011/03/Fostering-a-Culture-of-Lawfulness.pdf.

[53] SIGAR, "Interactive Funding Tables," 2014, www.sigar.mil/quarterlyreports/fundingtables/.

resilient Taliban. The persistence of corruption and the power of CPNs has undermined or coopted many of the institution-building efforts launched by the United States and ISAF. Too often, the coalition pursued capacity building without attention to whether criminal networks were diverting foreign resources, hollowing out military units, and appropriating state power for factional and illicit ends.

Capacity building efforts, meanwhile, failed to emphasize the primacy of the rule of law, overlooking the importance of investigative, judicial, and internal accountability institutions. The U.S. experience in Afghanistan clarifies that countercorruption cannot be an afterthought in stabilization and counterinsurgency operations. Rather, it is the necessary foundation for economic and military assistance to achieve its full potential. And it is the best assurance that foreign assistance does not inadvertently strengthen organized crime. The case of Afghanistan also highlights the centrality of elite politics to stability and the rule of law. Afghanistan's political settlement protected and, at times, empowered the country's CPNs and rendered ineffective many of the coalition's governance and development efforts. Postconflict reconstruction is, at its core, a political endeavor, not a technical one.

As intractable as the problems of corruption and organized crime appeared during the United States and ISAF's 13-year engagement in Afghanistan, countering these threats is not impossible. The international community's failure to coordinate countercorruption and make it a priority in the first eight years after 2001 was costly, forfeiting critical opportunities and handicapping subsequent efforts. By 2010-11, the coalition had developed a focused countercorruption strategy, with tangible lines of effort. Despite the resistance of some of Afghanistan's senior leaders and their affiliated CPNs, a growing number of Afghan officials and civil society leaders embraced the cause of accountability and worked in close cooperation with their international counterparts to achieve progress where they could.

Criticism and public outrage against corruption, meanwhile, has become a mainstay of Afghan political discourse. For the coalition government of Ashraf Ghani and Abdullah Abdullah, tackling corruption and displacing CPNs from government institutions will be imperative in consolidating the gains their country has made since 2001. Should the new government possess the political will needed to achieve real progress, it can hope to build on the initiatives of the past four years. Success is possible, but only with the confluence of genuine resolve by Afghan leaders to confront the problem, and international willingness to support and sustain Afghan-led efforts against corruption and organized crime.

Map No. 3835 Rev. 6 UNITED NATIONS
July 2014

Department of Field Support
Cartographic Section

2. Jaish al-Mahdi in Iraq

Phil Williams and Dan Bisbee

The U.S. invasion of Iraq in March 2003 was an emphatic military success. The resulting occupation, however, proved far more problematic, with various groups and factions opposing the U.S. military presence. The invasion itself, combined with the ineptitude of the occupation authorities, created several distinct but powerful strands of resistance, which erupted as a complex insurgency. Moreover, in the chaos and anarchy that followed the toppling of the Baathist regime, the line between licit and illicit power was blurred—an ambiguity never fully appreciated by the United States. This set in motion a series of missteps reflecting a profound lack of understanding of Iraqi traditions and politics, a failure to realize that common sectarian identity was no guarantee of harmony, and a sense of bewilderment when U.S. forces were not universally treated as liberators rather than occupiers.[1]

The failure to prevent widespread looting encouraged a sense of lawlessness, and the disbanding of the Iraqi army created a power vacuum and a whole new set of enemies challenging the U.S.-led Coalition. Moreover, the inability to recognize, let alone deal with, "industrial-strength" criminal organizations involved in kidnapping and extortion fostered a climate of fear and insecurity. This seriously impeded reconstruction and development and permitted, encouraged, and even compelled competing groups and factions to take matters into their own hands.[2]

One of those factions was the Sadrist movement, which in turn spawned the militia known as Jaish al-Mahdi (JAM). This chapter looks first at the background that helps explain JAM's emergence as a major player in post-Saddam Iraq. It then identifies those characteristics of the Sadrist movement in general and JAM in particular that led to JAM's chaotic evolution. Its development could be described either as (a) a skillful and pragmatic approach that maintained multiple options, engaged in frequent course reversals, and adopted a balanced but precarious mix of cooperation and confrontation; or (b) a hapless flailing about in response to a mix of internal and pressures, and driven by immediate opportunities and needs rather than any coherent strategy.

We then examine the various phases of JAM's evolution. These phases were characterized by dramatic shifts between *pragmatic policies*, in which the Sadrists participated in the political process, and *confrontational policies*, emphasizing the use of violence. The shifts were driven largely by the efforts of Muqtada and the Sadrist leadership to maintain or regain control over their organization, which, from the outset, displayed unruly behavior and a tendency to fragment. From outside, Jaish al-Mahdi looked like a hierarchical and structured organization, but it was actually unruly, bedeviled by factionalism and defections, and terribly difficult to control, let alone direct. The diverse agendas within JAM at times threatened the movement's cohesion and challenged Muqtada's ability to achieve what appear, in retrospect, to be his primary aims: removal of the U.S. presence, and significant influence over Iraq's future via legitimate political pow-

---

[1] The authors wish to thank Andrew Terrill, Alex Crowther, and Michelle Hughes for several helpful discussions on the themes in this chapter.

[2] Toby Dodge, "A sovereign Iraq?" *Survival* 46, no. 3 (2004): 39-58.

er. Members and supporters of JAM engaged in predatory and criminal activities that threatened to undermine Muqtada's more legitimate aspirations.

Thus, the breadth of the Sadrist movement was at once JAM's greatest strength and its most serious liability. Jaish al-Mahdi fused multiple roles and contradictory tendencies: it was both a protector and a predator. It sometimes reached out to Sunnis across the sectarian divide but also engaged in vicious sectarian cleansing. It lost battles but always survived intact and able to fight another day. It was often seen as a pawn of Iran yet was also the most fiercely nationalist and independent of all the Shia groups and factions in Iraq. And it was devoutly Shia yet struggled against the Shia establishment and fought other Shia factions as intensely as it fought the Sunnis.

JAM violence contributed enormously to the anarchy in Iraq in the four years after the U.S. invasion. But shifts in the landscape were occurring. These included wiser yet more aggressive U.S. policies, the constantly evolving intra-Shia political feud, the al-Maliki government's increased effectiveness in bringing Iraqi security forces under state control, and internal developments within the Sadrist movement that brought about the truce in 2007 and the reorganization in 2008. All these developments dramatically altered the role of JAM within the overall context of Sadr's agenda to shape Iraq's future.

Background

Even at the height of its violent activities and fundraising through crime from 2003 through 2007, JAM was not simply an illicit power structure acting as a spoiler. Although many parts of JAM have been heavily engaged in illicit and violent activities, the underlying Sadrist impulse is based on an important, though sometimes muted, strand in the Shiite tradition that is politically active and socially responsible. Its leader, Muqtada al-Sadr, was heir to a family tradition that had courageously stood up to Saddam Hussein during the 1970s and 1990s and sought to provide welfare and support to the marginalized and socially excluded Shia populations in Baghdad's slums and in communities of southern Iraq. Understanding Muqtada al-Sadr, the Sadrist movement, and Jaish al-Mahdi is therefore impossible unless we understand the Sadr family's role during the Baathist era. As one study noted:

> Descendants of the Prophet, the Sadrs form one of those large, transnational and learned families that, from one generation to another, pass on the requisite attributes of power and legitimacy in the Shiite world: prestigious ancestry, knowledge and accumulated resources. More specifically, Muqtada largely owes his position to two crucial figures of Iraq's contemporary history, Ayatollah Muhammad Baqir al-Sadr, one of his father's distant cousins, and his father, Ayatollah Muhammad Sadiq al-Sadr.[3]

Ayatollah Muhammad Baqir al-Sadr, known as the first martyr, was a distinguished Islamic scholar but was also innovative in both thought and organization.[4] He not only advocated "implementation of Islamic law and establishment of the rule of God on earth," but also formed "a political organization with a strict hierarchy and governing

---

[3] International Crisis Group (ICG), "Iraq's Muqtada al-Sadr: Spoiler or Stabiliser?" Middle East Report no. 55, July 11, 2006, 1.

[4] Ibid., 2.

charter."⁵ In doing so, "Baqir and his supporters fundamentally challenged the centrality of the Hawza," the elite Shia religious establishment school of theological learning, which is largely nonpolitical.⁶ By also offering a voice to a Shiite community that had hitherto been marginalized, Baqir challenged the regime and was arrested several times during the 1970s. According to the International Crisis Group (ICG), "Baqir's intransigence toward the regime, his popularity among Shiites, together with the 1979 Islamic revolution in Iran magnified regime fears. In response, the regime killed hundreds of Baqir's followers and, in April 1980, Baqir himself."⁷ The Hawza failed to react, provoking a widespread sense of betrayal among Baqir's constituents.⁸

Around the same time, Muhammad Sadiq al-Sadr, a pupil and cousin of Baqir, was placed under house arrest, and he, too, received no support from the Hawza. In the aftermath of the Shia uprising of 1991, however, the regime decided that Sadiq could be useful—not least because he believed that Iran had too much influence over the Shia establishment in Iraq—and supported his claim to be grand ayatollah. The attempt to co-opt Sadiq succeeded in the short term, but then he challenged the regime itself. After using regime support to attack the quietists in the Hawza, Sadiq successfully mobilized the Shiite masses with his sermons about the Twelfth Imam, restoration of Friday prayers, and social and economic support for a population that had long been marginalized and excluded.⁹ Indeed, as Leslie Bayless reports, "[When the] 'sanctions-depleted government of Saddam Hussein cut back services to Shiites,' Mohammad Sadiq al-Sadr took it upon himself to dole out charity."¹⁰ Inevitably, his actions challenged the authority of the regime, which responded with efforts at intimidation. After continuing to speak out in the face of threats, in February 1999 Sadiq, along with two of his sons, was killed by persons linked to the regime. "Sadiq al-Sadr's murder sparked violent demonstrations, a testament to his success in building a popular base despite his initial affiliation with the regime and notwithstanding the climate of fear that prevailed at the time." ¹¹

Indeed, Sadiq combined "Shiite martyrdom and social revolution" in ways that resonated with much of Iraq's Shia population.¹² The legacy of Baqir and Sadiq, who became known respectively within much of the Shia community as the "first martyr" and the "second martyr," was a powerful grassroots religious, political, and social movement that inherently challenged the mainstream nonpolitical, or "quietist," thread in Shiite tradition. The quietists had been passive in their opposition to the regime, had failed to give the al-Sadr family much support, and had largely been content to operate in exile in Iran and Europe rather than suffer the consequences of standing up to Saddam Hussein.¹³

---

⁵ Ibid.
⁶ Ibid.
⁷ Ibid., 23.
⁸ Ibid., 3.
⁹ Ibid., 45.
¹⁰ Leslie Bayless, "Who Is Muqtada al-Sadr?" Studies in Conflict and Terrorism 35, no. 2 (2012): 139.
¹¹ ICG, "Iraq's Muqtada al-Sadr," 4.
¹² Ibid.
¹³ For a fuller analysis, see ICG, "Iraq's Muqtada al-Sadr."

Emergence of Muqtada al-Sadr and JAM

Muqtada al-Sadr, the fourth son of Sadiq, who was married to one of the daughters of Muhammad Baqir al-Sadr, unexpectedly became the heir to this tradition of political involvement. Although he lacked major religious credentials and was largely unknown before 2003, in the aftermath of the U.S. invasion Muqtada al-Sadr rapidly emerged as a powerful, charismatic figure. Yet he did so with an agenda fueled not only by resentment toward the Baathists but also by the sense that the Shia religious establishment in Iraq had betrayed his family. Unlike the al-Sadr family, most members of the Hazwa had sought safety outside Iraq and had displayed neither courage nor commitment to the Shia population. Not surprisingly, therefore, Muqtada disdained the mainstream Shiite clergy and the quietist tradition embodied by Grand Ayatollah Ali al-Sistani. For its part, the traditional Shiite establishment regarded Muqtada al-Sadr as an upstart and rabble-rouser and saw his relationship with the impoverished masses as a threat.[14]

Equally important was Muqtada al-Sadr's antipathy toward the United States. This had much to do with the consequences of the tough UN sanctions regime imposed on Iraq following its invasion of Kuwait in 1990 and maintained until after the U.S. invasion of Iraq in 2003. The sanctions, linked to Saddam's nuclear program, significantly degraded Iraq's technological and engineering capacity, rendering reconstruction after 2003 much more difficult than the United States had anticipated. Moreover, rather than hurting the Hussein regime, sanctions had hurt the poor and the marginalized—the same people the al-Sadr family had tended and who were now a key part of Muqtada al-Sadr's power base. Far from welcoming the invasion for toppling Saddam Hussein, Muqtada al-Sadr saw it as a continuation of a punitive policy that had hurt ordinary Iraqis, especially those in the Shia community.[15] Indeed, he felt enormous antipathy toward the U.S. military presence. This was compounded by U.S. missteps, especially the failure by Coalition Provisional Authority (CPA) Administrator Paul Bremer to understand the al-Sadr family history and treat al-Sadr as someone who could influence a crucial segment of Iraq's population. Instead of engaging with al-Sadr, the CPA focused on building bridges with al-Sistani and other Shiite elites. The CPA initially treated Muqtada al-Sadr as irrelevant and later viewed him as a serious threat—a shift that introduced elements of a self-fulfilling prophecy into the equation.

Muqtada al-Sadr moved rapidly from obscure cleric to major power broker in postinvasion Iraq. He could do so because he inherited enormous legitimacy from his father's combination of religious authority and social obligation.[16] Although the Sadrists imposed a strict code in areas under their control, making women wear the veil and prohibiting alcohol, their tradition of providing social welfare assumed critical importance during the massive economic upheaval that followed the invasion. In effect, Muqtada and the Sadrists provided alternative governance and collective goods for vulnerable populations underserved by a state in crisis. This was most evident in Saddam City—which was quickly renamed "Sadr City" in honor of Sadiq, the "second martyr"—and among many tribes in the south, especially those in and around Basra.

---

[14] Ibid., 16.
[15] ICG, "Iraq's Muqtada al-Sadr," 10.
[16] Ibid., 9.

Sadiq had also gained many followers among rural tribal leaders in the south, either directly or through missionary work among their clansmen in East Baghdad. After the regime fell, Muqtada al-Sadr and a Sadiq disciple, Muhammad Yaqubi, made a concerted and successful effort to spread the Sadrist movement in the Marsh Arab areas.[17] Those who had relocated to east Baghdad but were still linked to the countryside assisted.[18] After Saddam's fall, many Marsh Arabs who had embraced the Sadrist movement relocated to the city of Basra, bringing with them a tradition of social banditry.[19] This predilection, together with traditional respect for their weapons—an important part of their identity—would inevitably bring them into conflict with the Coalition and with other tribes and political parties competing for the control of rents associated with the diversion, theft, and smuggling of oil.[20]

Al-Sadr's appeal to the disenfranchised among Iraq's Shia community is closely linked to a second factor that helped bring him to prominence. The anarchic conditions after the U.S. invasion, combined with the vacuum of indigenous power and authority in the country, leveled the playing field for the Sadrists in a way that few other things could. Moreover, al-Sadr's opposition to the invasion and U.S. presence evoked an enthusiastic response among Shia who were concerned that the Baathist dictatorship was simply being replaced by foreign oppression. Thus, according to Bayless, "[Al-Sadr could] present himself as the sole protector of the Shi'a inhabitants of Iraq. This move was fairly easy; no one else had stepped forward to assume the position."[21] Thus, the July 2003 creation of JAM as a protective militia stemmed from the need for security amid conditions of chaos, confusion, and continuing uncertainty about the future.

Another factor that helped propel Muqtada al-Sadr to prominence was his and his advisers' excellent sense of what the U.S. military called *strategic communication.* Al-Sadr made use not only of Friday prayers but also of satellite television and other media outlets.[22] Sometimes, these things went hand in hand. In his first Friday prayer on April 11, 2003, for example, he asked Shiites to express their piety by undertaking a pilgrimage on foot to Karbala.[23] Coming on the heels of the regime's fall, the massive celebrations offered Shiites their first opportunity to sense their unprecedented importance in Iraq. According to ICG, "The Sadrist phenomenon benefited from the Shiite community's visibility and, given the media's particular interest in Muqtada, from unprecedented focus on its most destitute members."[24] One journalist suggested that "the emergence of the Sadrist current after the regime's fall essentially occurred through satellite televisions" and the satellite dishes that had hitherto been banned but were enthusiastically bought by many Iraqis.[25] Not only did the Sadrists publish a daily and a weekly newspaper,

---

[17] Juan Cole, "Marsh Arab Rebellion: Grievance, Mafias and Militias in Iraq," Jwaideh Memorial Lecture, Bloomington, IN: Department of Near Eastern Languages and Cultures, Indiana Univ., 2008, www-personal.umich.edu/~jrcole/iraq/iraqtribes4.pdf.

[18] Ibid.

[19] Ibid.

[20] Ibid.

[21] Bayless, "Who Is Muqtada al-Sadr?" 136.

[22] ICG, "Iraq's Muqtada al-Sadr," 9.

[23] Ibid.

[24] Ibid.

[25] Ibid.

but they also had a radio station, TV station, and website. Using all these outlets, the movement made excellent use of symbolism, ranging from Muqtada al-Sadr's black turban—worn to signify his descent from the prophet—to what one observer described as "a fusion of sectarian and nationalistic iconography and language."[26]

Perhaps most significant of all, however, was the way that the Sadrists differentiated themselves from the other Shia factions and groups by "the pervasiveness of violent resistance" in their images, symbolism, and discourse.[27] This emphasis on resistance certainly distinguished the Sadrists from most other Shia groups in Iraq. It also led observer to suggest, "Muqtada al-Sadr sought to model his organization on Lebanese Hezbollah, combining a political party with an armed militia and an organization providing social services. The Hezbollah model gave the Sadrist Movement levers of control over political life through representation in Parliament, control of ministerial offices, the ability to organize popular protests, and the possibility of taking up arms when necessary."[28]

Yet it was not clear at the outset that Muqtada al-Sadr had a comprehensive design or that he was consciously trying to emulate Hezbollah. As with most other groups in the chaos of postinvasion Iraq, improvisation seemed more important than long-term strategy. Moreover, although Muqtada and the Sadrist leadership were good at mobilizing the disenfranchised Shia, controlling the movement was another matter entirely. In spite of senior leadership's efforts, JAM was never a monolithic, hierarchical, disciplined organization, nor was the broader movement. This was evident early on. ICG emphasizes this:

> As spectacular as it was disorganized, the Sadrist phenomenon did not reflect the growth of an already-structured movement shedding its prior clandestine status so much as a series of often-uncoordinated initiatives. Young imams, invoking Sadiq al-Sadr's name, rushed to fill the vacuum created by the collapse of the state apparatus. Surrounded by armed volunteers, they seized control of mosques, welfare centers, universities and hospitals and, particularly in Sadiq al-Sadr's former strongholds, instituted forms of local governance. Sadr City was the logical power base, but the movement also showed strength in large swaths of southern Iraq.[29]

Identifying the exact role that Muqtada al-Sadr himself played within his movement is enormously challenging. He has often received credit where it was not due, and blame where it was not completely warranted. Clearly on display in his career are significant survival skills, both personal and political. On several occasions, he rescued political success from military defeat by shifting between different identities. At times, he was a bitterly militant proselytizer capable of enormous pragmatism, and a fervent nationalist and outspoken critic of the occupation. Nevertheless, when he found it expedient, he cooperated with the U.S. military to remove some of the unruliest elements of his

---

[26] Ibrahim Al-Marashi, "Sadrabiliyya: The Visual Narrative of Muqtada al-Sadr's Islamist Politics and Insurgency in Iraq," in *Visual Culture in the Modern Middle East: Rhetoric of the Image*, ed. Christiane Gruber and Sune Haugbolle (Bloomington IN: Indiana Univ. Press, 2013): 162.

[27] Ibid., 146.

[28] Marisa Cochrane, *The Fragmentation of the Sadrist Movement* (Washington, DC: Institute for the Study of War, 2009), 13.

[29] ICG, "Iraq's Muqtada al-Sadr," 7.

organization. Often described as a firebrand cleric lacking in sophistication, he pursued strategies that, at times, have been remarkably pragmatic and subtle and gave him an important, if intermittent, voice in Iraqi politics. Al-Sadr was simultaneously sectarian and nationalistic, ideological and pragmatic, a theological scholar (albeit at a relatively low level) and a skilled political operator. This resulted in decisions and policies that, at times, appeared inconsistent or even contradictory. Nevertheless, there was a consistency of objectives: removal of U.S. military forces from Iraq, provision of social welfare for marginalized Shiites in Baghdad and Basra, and reduction of the power of the formal Shia hierarchy, embodied most obviously by Ayatollah al-Sistani. This provided a coherence and continuity sometimes hidden by the tactical shifts, reversals, and compromises that were essential to keeping the Sadrist movement—and himself—alive.

This is not to imply that Muqtada al-Sadr avoided making serious mistakes. On the contrary, he made several major missteps, sometimes becoming involved in battles he could not win. He never gained complete control over a movement riven with divisions and subject to factionalism, and some of the factions continued to be manipulated by Iran. Cooperation and compromise between his movement and rivals were often the result of setbacks and weakness, not positions of strength. Yet he learned from his mistakes and ultimately managed to overcome many problems. Despite significant challenges, Muqtada al-Sadr navigated a path that took his movement from its initial formulation as simply an anti-occupation force, past a tipping point to recognition as a major legitimate political party in Iraqi politics.

JAM Phase 1: From Emergence to Confrontation, 2003-4

Muqtada al-Sadr's new movement was off to a good start when, on April 7, 2003, Ayatollah Kazem al-Haeri, a leading Iranian-based Iraqi, appointed Muqtada as his representative in Iraq and authorized him to collect the khums, or Islamic taxes.[30] This gave the Sadrists an important source of funding. According to Adnan Shahmani, a cleric in Sadr's Najaf office, the movement collected roughly $65,000 a month from these donations. Roughly half the funds were used to support the poor and religious students. The other half directly funded the Sadrist offices in Iraq—quickly reopened as the "Office of the Martyr Sadr."[31]

Despite the financial boost, it was not long before al-Sadr made a major error in May 2003, when he issued a fatwa that "looters could hold on to what they had appropriated so long as they made a donation (khums) of one-fifth of its value to their local Sadrist office."[32] Al-Sadr's fatwa provided post hoc legitimacy to the widespread looting that had destroyed the government infrastructure, had seriously hurt businesses, and was to complicate the task of reconstruction. To some extent, Muqtada's endorsement was understandable: Iraq has a long tradition of looting as a form of social expression and political protest, and he was simply tapping into this in a manner that was crassly expedient. Unfortunately, it was ultimately counterproductive to his movement. Although

---

[30] Cochrane, *The Fragmentation of the Sadrist Movement*, 11-12.
[31] Ibid., 12.
[32] Patrick Cockburn, *Muqtada: Muqtada al-Sadr, the Shia Revival, and the Struggle for Iraq* (New York: Scribner, 2008), 130.

it offered short-term benefits for an organization not endowed with resources, it further alienated mainstream Shiites, especially those with property, and encouraged additional criminal activities by many of his followers. Moreover, the fatwa legitimized a strand in the Sadrist movement that embraced criminality and "often drifted toward gratuitous violence."[33] The problem would become pronounced after al-Sadr created his own militia, the Mahdi Army, or Jaish al-Mahdi, in July 2003. Indeed, the absence of payment for militia members made crime particularly attractive as a source of revenue for them, and convenient, at least initially, for Muqtada al-Sadr.

In what was to be another recurring theme, defections plagued the Sadrist movement from the outset. Mohammed al-Yacoubi, a prominent Shia cleric, claimed that Sadiq al-Sadr had designated him the rightful heir to the movement because he had greater religious knowledge and authority than Muqtada.[34] When his claim was ignored, Yacoubi created the Fadhila (Islamic Virtue) Party in July 2003.[35] The Fadhila Party won 15 seats in the Iraqi parliament in the 2005 elections and became JAM's persistent rival, especially in what would become a major struggle to control rents from oil smuggling in Basra.

Other rivalries had more immediate outcomes. Another potential rival to al-Sadr in the Shia community was Sheikh Abdul Majid al-Khoei, who had recently returned from exile in London and was the son of the Shia cleric who had led the revolt against Saddam Hussein in 1991. In Najaf, a mob attacked and killed al-Khoei and "reportedly left him to die outside the house of Muqtada al-Sadr."[36] Sadrist supporters also besieged the home of Grand Ayatollah Ali al-Sistani in Najaf.[37] Regardless of whether Muqtada al-Sadr ordered these actions or simply encouraged them, he was clearly implicated. Because of this and his anti-U.S. stance, CPA did not invite him to become part of the Iraqi Governing Council, formed in July 2003.[38] The snub put the Sadrists and the CPA on an inexorable collision course.

After simmering through the latter half of 2003 and early 2004, the relationship between the Sadrists and the Coalition entered an even more hostile phase in March 2004. Near the end of the month, Coalition Forces shut down the main Sadrist newspaper, *al Hawza*.[39] Then, in early April, they arrested a prominent aide to al-Sadr, Mustafa al-Yacoubi.[40] JAM responded to what it viewed as provocations, with major uprisings in Karbala, Najaf, and Kufa.[41] JAM forces established positions near the holy shrines, complicating Coalition forces' ability to use heavy firepower and enhancing the militia's ability to control access to these shrines and the revenue they generated.[42] An uneasy cease-fire between JAM and Coalition forces was established in May 2004 but broke down in Najaf in August 2004.

As ICG pointed out, the breakdown was not entirely al-Sadr's doing:

---

[33] ICG, "Iraq's Muqtada al-Sadr," 8.
[34] Cochrane, *The Fragmentation of the Sadrist Movement*, 12.
[35] Ibid.
[36] Ibid., 13.
[37] Ibid.
[38] Ibid.
[39] Ibid., 14.
[40] Ibid.
[41] Ibid.
[42] Ibid.

Coalition forces, determined to resolve the crisis once and for all by detaining Muqtada and disbanding his militia, never offered him or his followers an honorable exit. His Shiite opponents backed and even reinforced this inflexible position: coalition forces could not have entered the holy cities without the implicit consent of Hawza leaders for whom the partial destruction of sacred sites was a price worth paying in order to reassert their authority. In short, the conflict was not simply a struggle against the occupation; it was set against the deep and deepening intra-Shiite confrontation.[43]

Consequently, Coalition forces interpreted al-Sistani's departure for London for medical treatment in early August 2004 as implicit permission to confront the Sadrists in the holy sites.[44]

The second battle of Najaf was disastrous for al-Sadr. Many of his followers were killed; disputes over strategy with Qais Khazali, one of his top lieutenants, led Khazali to break away, creating a major split in the organization; the fighting alienated the merchants, who lost a lot of money because of it. And, in a major humiliation for al-Sadr, Ayatollah al-Sistani arranged a cease-fire upon his return to Iraq.[45] Moreover, in the aftermath of the confrontation, Ayatollah al-Haeri, who had initially provided al-Sadr with great legitimacy and a source of funding, not only disavowed al-Sadr but also issued a fatwa instructing followers in Iraq to stop paying khums to al-Sadr.[46] This was a critical financial loss that almost certainly contributed to a growing emphasis on fundraising through criminal activities. As Cochrane observed, "The heavy losses of JAM fighters forced the movement to spend a great deal of money to help the families of dead, increasing their financial expenditures at a time when their funding sources were declining."[47] In effect, the series of reverses placed the Sadrists in a bind where they had little alternative to creating and expanding illicit revenue streams. But some gains came amid the losses. As one study noted, "The Mahdi Army, which originally only had a presence in Sadr City, was able to hold its own against the U.S. for months. The losses suffered were terrible, yet they contributed to the image of resistance. Muqtada al-Sadr no longer fears arrest, and any disarming of his militia remains very hypothetical."[48] These gains were probably little comfort for Muqtada at the time, however, and he kept a low profile for the rest of 2004.

JAM Phase 2: From Politics to Sectarianism and Racketeering, 2005-7

According to Marisa Cochrane, "Sadr emerged from his seclusion in early 2005, which was also Iraq's first democratic election year. He adopted a more conciliatory tone towards the Iraqi government and turned his attention to politics."[49] The Sadrists joined a coalition of Shia parties in November 2005 and won 30 seats in parliamentary elections held in January 2006. Moreover, the Sadrist political bloc gained control of the Ministries

---

[43] ICG, "Iraq's Muqtada al-Sadr," 11.
[44] Kirsten Aiken, "Departure of Ali Sistani Prompted by Coalition: Middle East Observer," *AM*, Aug. 7, 2004, www.abc.net.au/am/content/2004/s1171202.htm.
[45] Cochrane, *The Fragmentation of the Sadrist Movement*, 15.
[46] Ibid.
[47] Ibid.
[48] Michael Goya, quoted in ICG, "Iraq's Muqtada al-Sadr," 12.
[49] Cochrane, *The Fragmentation of the Sadrist Movement*, 16.

of Health, Transportation, and Agriculture. This enabled them "to provide jobs, services, and revenue for militiamen and loyalists."[50] Being part of government provided both a new source of income and an expanded patronage network.[51] At the same time, al-Sadr placed renewed emphasis on discipline and control within the organization. He also displayed conciliatory tendencies by reaching out to the Sunni community and advocating national unity.

Baghdad's municipal politics clearly illustrate the expanding political role of the Sadrists during this period. Elections (largely boycotted by Sunnis) in January 2005 gave Shia-affiliated parties a dominant voice over Baghdad politics. ISCI/Badr, a coalition of well-organized Iranian-sponsored parties, won 28 of 51 seats on the provincial council, and the majority clout to take over the key positions in Baghdad's municipal government.[52]

In the ensuing months, Baghdad's political landscape was transformed. Shia party loyalists from either the ISCI/Badr camp or the Sadrists replaced many nonpartisan and technocratic officials who had cooperated with early Coalition reconstruction efforts. A pattern emerged in which ISCI/Badr held the primary positions of provincial council chairman, governor, and amin (commonly referred to as the "mayor," or city manager of Baghdad) while the Sadrists held the deputy positions within each of these major institutions. This power-sharing arrangement reinforced a deep sense of sectarianism within Baghdad, with Shia parties dominant, Sunnis underrepresented (if at all), and independent or technocratic officials increasingly marginalized. It also illustrated the tidal shift from the Iranian-sponsored parties that had initially gained power in a top-down process through electoral victory, toward a Sadrist bottom-up approach that saw the movement's power on the street and in the neighborhoods increasingly converted into institutional authority at higher levels.

The Amanat Baghdad, or "city hall" of Baghdad, became a significant battlefield in the intra-Shia contest over power in Iraq. With the population of metropolitan Baghdad estimated at around seven million of Iraq's roughly 27 million (est. 2007) people, control over Baghdad's service delivery institutions and the associated budgeting resources, contracts, jobs, and patronage opportunities made the Amanat a crucial component of Iraq's governance apparatus. With an ISCI/Badr-appointed official serving as mayor, and the powerful deputy mayor position filled by a newly appointed Sadrist (who had risen rapidly from the local Sadr City district department and reportedly had ties to JAM and the Office of the Martyr Sadr), departments, and sometimes entire wings, of the Amanat were known within the institution as either "Badr" or "Sadr" territory. The Badr-affiliated governor of Baghdad (who had temporarily filled in as amin during a previous political crisis that accompanied the Shia takeover in early 2005) repeatedly told a U.S. official that millions of U.S. reconstruction dollars were going directly to the Sadrist offices because of the deputy mayor's authority over major Amanat contracts. The overall atmosphere of conflict between Sadrists and the other Shia parties, which frequently

---

[50] Ibid., 18.

[51] Ibid., 18.

[52] The Islamic Supreme Council in Iraq (then known as SCIRI, or the Supreme Council for the Revolution in Iraq) and the Badr Organization (previously known as the Badr Corps—an organization that maintained both political and militia wings tightly connected to supporters in Tehran).

erupted in violence on the streets of Baghdad, also intensified throughout this period.[53]

With greater representation within nominally legitimate positions of political authority, distinguishing between the licit and illicit strands of the Sadr movement became more problematic and illustrated the complexity of confronting illicit power structures amid conflict and political transition. Disparate strategies, ranging from espousing social welfare to pursuing extremist sectarian agendas, to profiting from criminality, were increasingly pursued under the same banner.

Some of al-Sadr's followers were unhappy about the new emphasis on political participation. Khazali, nominally still within JAM, was increasingly working with the Iranian Quds Force, forming what became known as the JAM "special groups" (even though they were, in fact, largely independent of JAM).[54] Another defector, who became known as Abu Dura, operated notorious death squads in Baghdad that were responsible for the kidnapping, torture, and murder of thousands of Sunni civilians during 2004-6.[55]

While both the Khazali network and the Dura forces were operating outside JAM, the JAM mainstream was also becoming involved in sectarian violence. After the attack on the Samarra shrine in February 2006, al-Sadr remobilized and rebuilt JAM.[56] Then, as Iraq appeared to be heading toward civil war, JAM underwent a major resurgence, first as a protector of the Shia in Sadr City and then as a relentless force for sectarian cleansing. According to Cochrane, the loosely knit nature of the Sadrist movement also facilitated the "emergence of a mafia-like system [that] undermined Muqtada al-Sadr's control over his commanders. As local commanders grew more powerful and financially independent, they became less likely to follow orders from Muqtada al-Sadr and the clerical leadership in Najaf."[57]

It is perhaps not surprising, therefore, that JAM benefited most obviously from its criminal activities in 2006-7, even though some of these activities had been providing money for the militia from the outset.[58] The first and perhaps best established of the revenue streams came from rents or taxes imposed on oil smuggling in Basra.[59] The theft and smuggling of crude oil from Basra became evident soon after the U.S. invasion. It was facilitated by the general anarchy and insecurity and also by the specific lack of gauges and measurements in the oil facilities, combined with the pervasive corruption and lack of oversight. Reports even suggested that oil bunkering—similar to the process in the Niger Delta whereby small boats were used to move stolen oil to the bigger tankers—had become common practice. Without any real state presence, let alone regulation and enforcement, and with the spoils shared among politicians, militias, and smuggling gangs, the potential for violence was enormous.[60]

The struggle over oil rents brought the Mahdi Army into conflict with two other Shiite militias. One belonged to the Sadrist breakaway faction that had become the Fadhila

---

[53] Author (Bisbee) interview with Baghdad governor, 2007. Bisbee served on the Baghdad Provincial Reconstruction Team in 2005-6 with the U.S. Army and in 2007-8 with the State Department.

[54] Cochrane, *The Fragmentation of the Sadrist Movement*, 6, 19.

[55] Ibid., 16.

[56] Ibid., 21.

[57] Ibid., 21.

[58] This section draws heavily on Phil Williams, *Criminals, Militias, and Insurgents: Organized Crime in Iraq* (Carlisle, PA: Strategic Studies Institute, 2009).

[59] Ibid., 235.

[60] Ibid., 75-76.

Party; the other to the Badr organization, which was linked to the main Shia political parties. The three militias became engaged in sporadic but often intense violence fundamentally related to the battle over payoffs from the smuggling of crude oil.[61] All three militias demanded a cut from the smuggling proceeds carried out by several tribes, while Fadhila at times also became more directly involved in the smuggling.[62] The conflict was further complicated by militia infiltration of the police and government agencies, so that militia violence in Basra occasionally involved different police units fighting against one another.

In a completely separate activity (though also linked to the oil industry), JAM had control over black-market sales in the forecourts of many gasoline stations. Ironically, the diversion of gasoline from the major Iraqi oil refinery at Baiji was largely under the control of the Sunni insurgents.[63] Somewhere along the commodity-smuggling chain, therefore, commercial transactions were likely, either directly between Sunni insurgents and JAM or via intermediaries, although how and where these exchanges took place is uncertain. Whatever the precise arrangements, this was far from the first case of tacit cooperation between enemies to fund themselves so they could wage war on each other more effectively. The same thing had happened in places as diverse as Sarajevo and Northern Ireland.[64]

Another important revenue stream came from JAM dominance over the Shia trade in propane gas canisters, which Iraqis use for cooking.[65] Sometimes, the militiamen sold the propane at a premium, and at other times, they sold at below-market rates to earn the goodwill of the poor.[66] Too much should not be made of this, however, since JAM had little hesitation in using its territorial control for extortion. Those who paid for "protection" avoided violence and kidnapping; those who did not pay became targets.

Indeed, according to one report, JAM "was involved at all levels of the local economy, taking money from gas stations, private minibus services, electric switching stations, food and clothing markets, ice factories, and even collecting rent from squatters in houses whose owners had been displaced. The four main gas stations in Sadr City were handing over a total of about $13,000 a day, according to a member of the local council."[67] One U.S. general even compared the situation to "the old Mafia criminal days in the United States."[68]

By 2007, reports from Baghdad also suggested that the Mahdi Army had obtained complete control over the Jamila market, the most important wholesale center in Baghdad, and the receiving point for millions of dollars of market-bound goods into the capital.[69] Until mid- to late 2007, Sunni truck drivers transporting goods from Jordan and

---

[61] Reidar Visser, quoted in Ben Lando, "Shia Parties Battle for Control of Oil-Rich Basra Region," UPI, Aug. 18, 2007.

[62] Williams, *Criminals, Militias, and Insurgents*, 81.

[63] Ibid., 90.

[64] See Peter Andreas, *Blue Helmets and Black Markets: The Business of Survival in the Siege of Sarajevo* (Ithaca, NY: Cornell Univ. Press, 2011).

[65] Jeffrey Bartholet, "How Al-Sadr May Control U.S. Fate in Iraq," *Newsweek*, Dec. 4, 2006.

[66] Ibid.

[67] Sabrina Tavernise, "A Shiite Militia in Baghdad Sees Its Power Wane," *New York Times*, July 27, 2008.

[68] Ibid.

[69] IraqSlogger, "Anbari Trucks, Mahdi Guns, & High-Volume Trade,"' Oct. 2, 2007, http://iraqslogger.powweb.com/index.php/post/4520/Anbari_Trucks_Mahdi_Guns__High-Volume_Trade?PHPSESSID=a9a8b1d85a59b3ab94543c9f6a1e83cb.

Syria to Baghdad adopted the practice of transferring their loads to haulers outside the city in order to avoid JAM.[70] Somewhere in the latter half of 2007, however, the truckers began to complete the trip to the Jamila market. This development led to rumors that an agreement had been struck between the wholesale merchants, truckers, and the Mahdi Army, providing another lucrative revenue stream. As *IraqSlogger* put it, "With the high volume of goods arriving at Jamila market on a daily basis, bound for Baghdad's millions of consumers, any arrangement allowing a militia to take a cut of the action in exchange for non-interference with shipping operations [would have paid] very well indeed."[71]

JAM elements in Basra were also implicated in smuggling cars into Iraq from Dubai, paying for them with the proceeds obtained from oil diversion.[72] Although this was a for-profit activity rather than using the cars as weapons, the profits were almost certainly used to support both JAM's military and social welfare activities.

None of these illicit fundraising activities was surprising. All had analogues in other conflicts and illustrate the complex but inescapable linkages between crime and conflict. The profit motive also became a key impetus for sectarian cleansing as the situation in Iraq in 2006 and 2007 degenerated into what was effectively a civil war. Although Muqtada al-Sadr had earlier called for unity between the Sunni and Shia communities, JAM became a key player in Baghdad's sectarian violence. Nowhere was this more obvious than in the Ministries of Health and Interior. Although control over the Ministry of Health was designed to give the al-Sadr movement greater control over—and greater credit for—service provision in Iraq, JAM also exhibited a vicious sectarianism that included attacking Sunni patients in hospitals, and the doctors who attended them. JAM control over health facilities also meant an attack on women's health services. In an effort to push traditionalist social practices of sexual segregation, male gynecologists and other women's health specialists were banned from practicing in several facilities in Sadr City, leaving female patients to the care of often undertrained female health workers.[73]

Involvement in sectarian killings became even more pronounced as JAM members infiltrated the police in 2006-7 and used the Ministry of Interior as a base for kidnapping and killing Sunnis.[74] JAM created a climate of fear and intimidation within the Ministry and leveraged its access to set up roadblocks, disguised as police checkpoints, at critical locations. When cars with Sunnis were stopped, the occupants were often kidnapped and tortured. After they were killed, their bodies were deposited at the initial location of the kidnapping, demonstrating the impunity that JAM enjoyed.

There was also a financial impetus for continuing and expanding such activities. As the Baghdad death squads became especially adept in kidnapping, torture, and murder, car theft became a form of funding for JAM. Death squads typically financed their operations by stealing anything of value from the victims, and often the most valuable things

---

[70] Ibid.

[71] Ibid.

[72] *Al-Sharq al-Awsat*, "Basra—Iraq's Second Largest City Controlled by Militias and Mafias," Nov. 11, 2007; Ghaith Abdul-Ahad, "When Night Falls, the Assassins Gather in Hayaniya Square," *Guardian*, Nov. 17, 2007.

[73] Author interview with medical professional and local council member, Baghdad, 2007.

[74] Williams, *Criminals, Militias, and Insurgents*, 209.

were their cars.[75] Reportedly, stolen cars sold on the black market for roughly half their fair-market price, although the typical price was in the $2,000-$2,500 range.[76] A similar impulse was at work when JAM cleared neighborhoods of Sunni residents. When Sunnis were evicted from Shiite-dominated areas, their houses were often taken over by Mahdi Army members, who then rented or sold them. In short, as the ICG put it, JAM found that "assassinating Sunnis also became highly lucrative."[77]

But Mahdi Army violence and the resulting profits created a self-perpetuating spiral that ultimately became counterproductive. As the sectarian violence peaked, many of those who had seen JAM as their best protection against Sunni insurgents came to regard it as a mixed blessing at best and highly pernicious at worst. As one October 2007 report noted, "In a number of Shiite neighborhoods across Baghdad, residents are beginning to turn away from the Mahdi Army, the Shiite militia they once saw as their only protector against Sunni militants. Now they resent it as a band of street thugs without ideology."[78] According to one Shiite, "We thought they were soldiers defending the Shiites . . . But now we see they are youngster-killers, no more than that. People want to get rid of them."[79] The difficulty, as conflict analyst William Reno has pointed out, is that protection and predation are two sides of the same coin, and it is easy to move from one to the other.[80] By late 2007, it appeared that JAM had flipped the coin.

This is not to deny that both service provision and protection of Baghdad's Shia population against the Sunnis added enormously to the legitimacy of al-Sadr and his movement. Some of the proceeds of criminal activity and ethnic cleansing also fed into service provision for the Sadr City poor and marginalized. ICG points out, "In a city virtually abandoned by the state, Sadrist offices in several neighborhoods became the last and only resort for Shiite residents in need of help."[81] Although JAM militants' violence pushed out the very state institutions whose absence the Sadrists condemned, the Sadrists capitalized on this. They provided shelter, food, and other staples to displaced and poor Iraqis.[82] They also housed families in vacant homes, many of which were obtained through sectarian cleansing, and provided heating and cooking fuel they had either stolen outright or obtained from government sources through corruption or intimidation.

But widespread criminality soon undermined this legitimacy. How much of the blame was Muqtada al-Sadr's remains uncertain, but most accounts agree that he had very limited control over his followers. As Cockburn notes, JAM "had always had a loose structure and its fighters were largely unpaid. Units often had their origin in locally raised vigilante groups that were never amenable to discipline from the center. And as the sectarian war got bloodier, local commanders became more independent and

---

[75] *IraqSlogger*, "Stolen Cars Finance Militia Operations. Fenced Autos Fetch $2,000 on the Black Market; Few Live to Complain," June 28, 2007.

[76] Ibid.

[77] ICG, "Iraq's Civil War, the Sadrists and the Surge," Middle East Report no. 72, Feb. 7, 2008, 6.

[78] Sabrina Tavernise, "Relations Sour between Shiites and Iraq Militia," *New York Times*, Oct. 12, 2007.

[79] Ibid.

[80] William Reno, "Protectors and Predators: Why Is There a Difference among West African militias?' in *Fragile States and Insecure People? Violence, Security, and Statehood in the Twenty-first Century*, ed. Louise Andersen, Bjørn Møller, and Finn Stepputat (New York: Palgrave Macmillan, 2007), 99-122.

[81] ICG, "Iraq's Civil War, the Sadrists and the Surge."

[82] Dean Yates, "Cleric Sadr Key Player in Helping Poor Iraqis—Report," Reuters, Apr. 15, 2008.

more powerful."[83] The Mahdi militia began to operate as a series of semiautonomous or even wholly independent units that answered not to the Sadrist leadership but to their own commanders. At times, JAM seemed little more than a brand name used by Shia militia factions heavily involved in criminal activities and political assassinations. According to the *Washington Post*, "Bands of young gunmen used the Mahdi army name as a cover for extortion, black marketeering, and other crimes."[84] Not surprisingly, therefore, as sectarian violence waned, so did Shia tolerance for JAM's abuses.

Along with an increasingly divisive and uncontrollable movement and disillusionment among segments of his support base, al-Sadr also faced a less hospitable political environment. In November 2006, after a split over the timetable for withdrawal of U.S. forces from Iraq, the Sadrists withdrew from the government. This left al-Maliki free to marginalize al-Sadr and allowed Coalition forces to go after JAM more directly even in Sadr City. Recognizing the dangers that he faced from a direct confrontation with U.S. forces that were finally proving increasingly adaptable and effective, al-Sadr adopted a more conciliatory approach.

JAM Phase 3: Factionalism, Cease-fire, and the Struggle for Control, 2007

Reports from late 2006 suggest that al-Sadr and his loyalists had already tried to clean house by giving rogue commanders' names to U.S. and Iraqi forces.[85] This led to death threats against al-Sadr.[86] In January 2007, increasingly beleaguered and facing a more aggressive U.S. strategy, al-Sadr ordered JAM forces to stand down rather than fight Coalition forces.[87] At this point, he seems to have realized that direct confrontation with Coalition forces would not only complicate his efforts to purge his organization but would also severely weaken his movement. His primary focus was to reestablish control over a movement and militia that were increasingly fractured and fractious.

As ICG reported, "Muqtada's objective was to improve his movement's reputation by imposing greater discipline. Seeking to distance himself from abuses, he blamed excessive violence on rogue elements and overzealous militants, claiming to be a moderate leader urging calm."[88] Although he went into seclusion, he and his close supporters sought to reestablish central control over JAM and punish or expel those who had committed gratuitous violence or unduly exploited the Shia population. During 2007, al-Sadr also sought to position himself as a representative of national unity and began to make overtures toward the Sunni tribal leaders, who were increasingly distancing themselves from al-Qaeda in Iraq (AQI).

Against this background, Marisa Cochrane has identified several distinct segments

---

[83] Cockburn, *Muqtada*, 184.

[84] Hamza Hendawi and Qassim Abdul-Zahra, "Al-Sadr Overhauling His Shiite Militia," Associated Press, Sept. 10, 2007, www.washingtonpost.com/wpdyn/content/article/2007/09/10/AR2007091000169_pf.html.

[85] Babak Rahimi, "Muqtada al-Sadr Steps into the Power Vacuum," *Terrorism Focus* 4, no. 19 (June 19, 2007).

[86] Ibid.

[87] See Stephen Biddle, Jeffrey A. Friedman, and Jacob N. Shapiro, "Testing the Surge: Why Did Violence Decline in Iraq in 2007?" *International Security* 37, no. 1 (2012): 740.

[88] ICG, "Iraq's Muqtada al-Sadr," 13.

or factions, with divergent though sometimes overlapping agendas, and very different reactions to Sadr's calls for a cease-fire: "The first group included mainstream JAM led by the Najaf-based clerical leadership and the Baghdad-based political leadership. This group believed the movement had been heavily infiltrated by criminal elements and Iranian influence."[89] For this reason, it had lost much of its legitimacy and status in Shia communities. While the JAM rank and file generally followed Sadr's orders for a cease-fire, other parts of the movement were unrulier. Consequently, the leadership formed an elite unit, known as the Golden Battalion, to restore discipline by punishing or eliminating rogue elements.[90] An al-Sadr loyalist acknowledged that members of this unit "conduct spot checks, and . . . deal harshly with troublemakers."[91] According to one report, the Golden Battalion, or Golden JAM, as it was sometimes termed, operated in northwest Baghdad as early as February 2007 but intensified its activities from April onward after some groups disregarded al-Sadr's orders.[92] The effort to restore control and restrain rogue elements of JAM became even more urgent after clashes on August 27, 2007, between JAM and the Badr Militia in Karbala led to pilgrims' deaths and further tarnished JAM's image. The next day, al-Sadr announced that all militia activities, including attacks on U.S. forces, were to be suspended for the next six months and "that his movement would purge all rogue factions and reorganize."[93]

One JAM commander, Abu Jaffar, admitted that soon after the freeze on violence, the Golden Battalion summoned him and blamed him for letting the hundred men in his unit commit crimes against civilians. As a result, he was fired. The names of others who had been demoted or expelled were read at Friday prayers and posted on walls and in flyers.[94] Reportedly, one such flyer was addressed to "All Mahdi Army Members" and reported the expulsion of one member because of his "immoral actions" and "use of the blessed name of the army to loot, kidnap and bargain."[95] Other militia members were dismissed for disobeying orders or going beyond the rules of engagement established by the Sadrist leadership.

The campaign was not simply one of name and shame; it also involved direct violence. In one instance, three men on motorcycles killed 25-year-old Saif Awad. Known as "the Assassin" and heavily involved in kidnapping and extortion in Baghdad's Hurriya neighborhood, Awad was killed while in one of his two new cars.[96] Such killings became increasingly frequent, reflecting the leadership's decision to do away with those "whose thuggish tactics have disgusted ordinary Iraqis."[97] The message was clear on what would happen to those going too far in their criminal activities and hurting JAM's reputation. In January 2008, another JAM commander, called Hamza, whose units had continued killing and kidnapping despite the freeze on violence, was given a death sen-

---

[89] Cochrane, *The Fragmentation of the Sadrist Movement*, 25.
[90] Ibid.
[91] Babak Dehghanpisheh, "The Great Muqtada Makeover," *Newsweek*, Jan. 28, 2008.
[92] Cochrane, *The Fragmentation of the Sadrist Movement*, 25-26.
[93] Ibid., 30.
[94] Amit R. Paley, "Sadr's Militia Enforces Cease-fire with a Deadly Purge," *Washington Post*, Feb. 21, 2008.
[95] Ibid.
[96] Dehghanpisheh, "The Great Muqtada Makeover."
[97] Ibid.

tence. A loyal Sadrist commander noted, "We were ordered to eliminate him and we did . . . This is how we have been cleaning the Mahdi Army."[98]

This process was not entirely smooth, however. The leadership frequently encountered complaints either that the cease-fire had made the Sadrists more vulnerable to rival Shia groups or that crimes such as robberies and car thefts had increased.[99] The purges also met with resistance and, sometimes, retaliation. In December, a dozen Mahdi Army fighters on motorcycles stormed into an ice factory in Baghdad's Tobji neighborhood and kidnapped its Sunni owner. When the family complained, the al-Sadr office told the kidnappers to return the victim or a complaint would go to the main Sadr office in Najaf. But the kidnappers made clear that they did "not take orders from anyone," and killed the owner.[100] In another instance, fighters who had exceeded their orders were dismissed. A Sadrist official said, "We sent people to talk to them, to inform them of Moqtada Sadr's instructions and abide by them, but they refused . . . We now consider them a splinter group. They do not belong in the Mahdi Army."[101] In response, a few days later, the fighters "attacked the Sadr office in Hurriyah with rocket-propelled grenades and machine guns."[102]

Whether because of this resistance or simply as a different tactical approach, a second but perhaps overlapping group of loyal Sadrists began working with U.S. forces to get rid of the unruliest elements in the movement. The group, Noble JAM, felt that the Mahdi Army could most effectively be purged of the death squads and other criminal elements by seeking the Coalition's help.[103] According to Cochrane, "Noble JAM operated in northwest Baghdad. In the Shi'a-dominated neighborhoods of Shula and Hurriyah, Noble JAM squads secretly cooperated with Coalition and Iraqi forces."[104] As a result, it was easier to target the criminals, the sectarian death squads, and militia members who were believed to be working closely with Iran. Noble Jam identified renegade Mahdi Army figures and provided sworn statements against them, thereby providing Coalition forces with a legal basis to take them off the streets.[105]

While some factions in JAM worked with Coalition forces, others continued to fight. Particularly troubling to the Coalition were the JAM Special Groups, also known as the Khazali network.[106] Khazali was arrested in March 2007 and was succeeded by Akram Kabi, who became head of a splinter militia known as Asaib Ahl al-Haq (AAH).[107] The Special Groups did not accept the freeze on hostilities, and with Iranian training and the use of explosively formed projectiles (EFPs) that were particularly lethal, they continued to pose a challenge for U.S. forces.[108] The Special Groups were never really under al-Sadr's control, and by 2007, they were completely outside his movement and militia

---

[98] Paley, "Sadr's Militia Enforces Cease-fire."
[99] Ibid.
[100] Ibid.
[101] Ibid.
[102] Ibid.
[103] Cochrane, *The Fragmentation of the Sadrist Movement*, 26.
[104] Ibid.
[105] Lauren Frayer, "U.S. Seeks Gains in Shiite Militia Rifts," *Washington Post*, May 15, 2017.
[106] Ibid., 26.
[107] Cochrane, *The Fragmentation of the Sadrist Movement*, 31.
[108] Ibid.

and probably much closer to Iran than he had ever been. In 2008, when al-Sadr announced the dismantling of JAM, Asaib Ahl al-Haq continued to operate as an independent militia with continuing attacks on U.S. forces.[109] Ali Mamouri reports, "Following the withdrawal of U.S. forces, the group turned to political participation and changed its name to the Ahl al-Haq movement." Nonetheless, as recently as 2013, it was involved in relatively minor clashes with al-Sadr loyalists.[110]

The criminal gangs within JAM also refused to stand down. As Cochrane noted, "These fighters operated local criminal rackets that generated large sums of money. They stood to lose personal power and wealth in the stand-down. Their extortion of the population also meant that they were prime targets of Coalition Forces and Noble JAM."[111] Nevertheless, many of their criminal enterprises and activities were deeply entrenched, and although the attacks significantly weakened them, they were certainly not destroyed.

By the end of 2007, the traditional Sadrist leadership had made serious efforts to reestablish control over JAM, with only partial success. Criminal activities and intermilitia rivalries continued, especially in Basra. The Special Groups had not been purged but had become overtly independent from JAM, whereas earlier they had remained nominally under the JAM umbrella even while pursuing their pro-Iran agenda.

The U.S. military, however, was now differentiating between "irreconcilable" rogue members of the Mahdi Army and "reconcilable" ones they could engage. As one officer could not resist putting it, there were "all sorts of different flavors of JAM." They included those who could be integrated into the political process, as well as irreconcilable elements "as bad as AQI," and "criminal elements that use JAM as their cover."[112] Indeed, in the last four months of 2007, there was evidence of an improved relationship with U.S. forces, and explicit though covert collusion in removing from the streets some of the more extreme and uncontrollable elements within JAM. In February 2008, Muqtada al-Sadr extended the cease-fire he had put into place in August 2007. A month later, however, JAM was the target of a major offensive by the Iraqi government and, subsequently, U.S. forces. It suffered a major military defeat that paradoxically allowed al-Sadr greater freedom of action politically.

JAM Phase 4: Military Defeat and Political Reconciliation 2008-14

In March 2008, faced with continuing militia unrest in Basra, Iraqi Prime Minister Nouri al-Maliki launched a military offensive to stabilize the city and restore order. The offensive, named the Charge of the Knights, began on March 25. At first, it appeared to be something of a debacle for government forces, with many deserting rather than fighting fellow Shiites. But with the introduction of additional forces and U.S. assistance in surveillance and air support, the tide gradually turned in the Iraqi government's favor. JAM resisted vigorously but had been significantly weakened by the time Iran brokered a truce. The outcome was a reduction in crime and violence, enhanced stability, and a significant dilution of JAM influence in Basra.

---

[109] Ibid., 39-40.

[110] Ali Mamouri, "The Rise of 'Cleric Militias' in Iraq," *Iraq Pulse*, July 23, 2013.

[111] Cochrane, *The Fragmentation of the Sadrist Movement*, 27.

[112] Department of Defense Bloggers Roundtable with Col. Martin N. Stanton, Chief of Reconciliation and Engagement, Multinational CorpsIraq, via teleconference, Federal News Service, Nov. 2, 2007.

In the aftermath of the war, there was much speculation about the Iraqi government's objectives in going into Basra. Critics saw it as a direct attempt to undermine Muqtada al-Sadr and his constituents before provincial elections. A more cynical view is that it was part of the continued struggle over who would control Basra and its resources. In the final analysis, though, it is hard to dispute the notion that the operation was a long-overdue effort by the Iraqi state to reassert centralized power and authority and diminish the power of the militias, especially JAM. By early 2007, Iraqi government authority was viewed as almost irrelevant, with the security institutions unable to control substate groups.[113] The Charge of the Knights was an attempt to rectify the situation in a city plagued by militia violence, criminality, and repression, and it succeeded.

The battle of Basra was soon followed with another offensive in Sadr City, this time provoked by persistent rocket fire, coming mainly from the Special Groups, against Bagdad's Green Zone.[114] Led by Coalition military forces using a mix of heavy armor, Apache helicopters, and unmanned Predators, the offensive penetrated deeply into the Sadrist stronghold. Despite strong resistance, about 700 JAM militiamen were killed, and on May 11, 2008, al-Sadr requested a cease-fire.[115] Negotiations were completed the following day, and a week later, Iraqi forces moved in to take control of Sadr City. The battle had reduced JAM's military strength, removed a safe haven from the militia, and severely undermined both JAM's social control and its dominance in criminal markets. And yet, the al-Maliki and Coalition offensive also created new opportunities for Muqtada al-Sadr. In effect, it allowed him to disengage from those parts of the movement where he lacked control, and reenergize and revitalize other elements, such as his social and religious programs. In retrospect, the setbacks in Basra and Sadr City made it possible for him to move directly into the political mainstream.

Al-Sadr's willingness to move in new directions was evident on June 13, 2008. After the Friday prayer in Kufa, a Sadrist cleric read a letter from Muqtada providing guidance for JAM. Under this guidance, the Mahdi Army would be largely transformed into a civilian movement dealing with religious, social, and cultural affairs.[116] Weapons were to be restricted to experienced fighters and pointed exclusively at occupying forces.[117] The letter contained words of defiance and promised continued resistance to the occupation, while also promising to disown "any JAM members who disobeyed."[118] Several months later, in November 2008, al-Sadr created the Promised Day Brigade (PDB) as his personal militia, which was also permitted to resist the occupation and fight against U.S. troops.

All this made it uncertain whether Muqtada al-Sadr had simply adopted a low profile to let JAM rebuild its strength or had chosen to emphasize the social role of the Sadrist movement and to operate primarily through the political process. This ambiguity would persist. The United States viewed the PDB as one of the Special Groups influenced by

---

[113] Adeed Dawisha, *Iraq: A Political History* (Princeton, NJ: Princeton Univ. Press, 2009), 636.

[114] For a fuller analysis, see David E. Johnson, M. Wade Markel, and Brian Shannon, "The 2008 Battle of Sadr City," RAND Occasional Paper, 2011.

[115] Ibid.

[116] Anthony H. Cordesman and Jose Ramos, "Sadr and the Mahdi Army: Evolution, Capabilities, and a New Direction," CSIS, Aug. 4, 2008, 13, http://csis.org/files/media/csis/pubs/080804_jam.pdf.

[117] Ibid., 23.

[118] Ibid.

Iran, but the new militia's import may have been more symbolic than substantive. Some observers noted, however, that the group became more active after May 2009, and in June 2011, the PDB claimed responsibility for 52 attacks against U.S. forces.[119] Significantly, the following month, "a senior Iraqi intelligence official" told U.S. policymakers that "in order to avoid antagonizing Washington, al-Sadr had ordered the Brigade to limit its attacks to 'hard targets'—installations and armored vehicles—to minimize the likelihood of U.S. casualties."[120] At the same time, al-Sadr posted on his website a statement that "the Brigade would have the 'mission' of 'resisting' U.S. forces if they were not all gone by the end of 2011" (the deadline set for withdrawal).[121]

Even though al-Sadr had become more dependent on Iran, he was never wholly compliant. The PDB fought against other Special Groups in Sadr City and, in October 2009, defeated Asaib Ahl al-Haq, or League of the Righteous—which was very close to Iran—in a struggle for dominance in the Shia stronghold.[122] In some ways, this was a continuation of the struggle that had been going on since the splits in the military command in 2004. It also suggests that although Iran may have been trying to shape the remnants or successors of JAM, its influence over al-Sadr remained limited. The main reason for this, as suggested above, is that al-Sadr was very much an Iraqi nationalist. Consequently, the decision to withdraw U.S. forces enabled him to achieve one of his long-term goals. Moreover, he did not want one outside influence simply replaced by another. Religious deference and the need for some Iranian support kept al-Sadr's opposition to Iranian influence in Iraq subtle and covert, unlike his direct, explicit opposition to the U.S. presence. Nevertheless, the opposition was there. At the same time, the prospective U.S. withdrawal also enabled al-Sadr to move more energetically into the political arena.

In the months before the March 2010 elections, al-Sadr became a positive force, urging Iraqis to participate at the polls in order to end the foreign occupation.[123] He also called on Iraqis "to opt for the candidates who would best serve the nation and work for Iraq's liberation."[124] Given the setbacks the Sadrists had suffered, they did remarkably well at the polls, winning 40 of the National Iraqi Alliance's total of 70 seats in the 325-member Parliament.[125]

According to West Point's *CTC Sentinel*, "The key to the Sadrists' electoral success was how they applied systematic polling methods such as databases with information on voters in all provinces and a cunning campaign strategy to win voters in the south. Along with anti-establishment and populist tactics . . . al-Sadr was able to present himself and his followers as the primary political force to defend the Shi`a population."[126] It

---

[119] See Stanford University, "Mapping Militant Organizations: Promised Day Brigades," 2015, http://web.stanford.edu/group/mappingmilitants/cgi-bin/groups/view/249.

[120] Ibid.

[121] Ibid.

[122] Ibid.

[123] Benjamin Isakhan, "Despots or Democrats? Sistani, Sadr and Shia Politics in post-Saddam Iraq," in *Islam, Islamist Movements and Democracy in the Middle East: Challenges, Opportunities and Responses*, ed. Rajeesh Kumar and Navaz Nizar (New Delhi: Vision, 2013), 171-92, http://dro.deakin.edu.au/eserv/DU:30057125/isakhan-despotsordemocrats-2013.pdf.

[124] Ibid.

[125] Ibid.

[126] Babak Rahimi, "The Return of Muqtada al-Sadr and the Revival of the Mahdi Army," *CTC Sentinel* 3, no. 6 (June 3, 2010).

is perhaps not surprising, therefore, that in April 2010, in response to continued Sunni attacks on Shia targets, al-Sadr announced the restoration of JAM[127] specifically "to support Iraqi security forces" and help protect Shia religious events.[128] This was not the JAM of old, and it appeared that al-Sadr had successfully engineered the transformation of his movement and militia into the political mainstream.

Babak Rahimi noted in 2010, "As long as al-Sadr remains a major political figure, operating within the Iraqi electoral process, it is unlikely that JAM will return to its combative roots and reactivate its military program. Al-Sadr's ultimate interest is to maintain his political prestige, with the possible ambition to one day become the country's first Shia cleric prime minister or perhaps a major Shia spiritual leader like his father."[129]

Indeed, from 2011, al-Sadr took notable steps to establish himself as a leader with broad appeal, speaking in favor of social justice and the rule of law, and against sectarianism.[130] During 2012 and 2013, as large groups of Sunnis protested the increasingly sectarian policies the al-Maliki regime, al-Sadr, too, repeatedly criticized al-Maliki and distanced himself from the block of Shia political parties giving al-Maliki his parliamentary majority. This stance against sectarianism was not easy, however. Tensions between Sadrist loyalists and breakaway militant factions highlighted the cost of Sadr's retreat from militancy, and the readiness of others to fill the gap.[131]

In February 2014, al-Sadr himself seemed to be vacillating between several strategies. He announced he was retiring from political life and shutting all his offices except his charities.[132] Within a few days, however, he delivered a fiery speech calling al-Maliki a dictator and tyrant, criticizing parliament, and condemning his own party for financial corruption.[133] Moreover, any intention to withdraw from political life was nullified by the rapid advances of Islamic State insurgents. The fall of Mosul in June 2014 threw Iraq into deep crisis, and al-Sadr joined the chorus of Iraqi political elites calling for an emergency government.[134] He dramatically reenergized the Mahdi Army by mobilizing Shia fighters in Sadr City, Najaf, and Basra following an appeal by Ayatollah al-Sistani for able-bodied Iraqis to join the fight against ISIS.[135] And yet, this revival of the Mahdi Army also reinforced the familiar rifts within the Shia community as Sadr's mobilization of an independent militia challenged al-Sistani's emphasis on supporting government forces.[136]

Even in the face of an existential threat to Iraq from ISIS, Shia forces remained divided.

Lessons

What lessons about illicit power structures can be drawn from this case? What should

---

[127] Ibid.
[128] Ibid.
[129] Ibid., 10.
[130] Mustafa al-Kadhimi, "The New Muqtada al-Sadr Seeks Moderate Image," *Al-Monitor*, Mar. 13, 2013.
[131] Ali Abdel Sadah, "Sadr Reconsiders Political Role, Mahdi Army," *Al-Monitor*, Aug. 30, 2013.
[132] Harith Hasan, "Is Muqtada al-Sadr Retiring or Repositioning?" *Iraq Pulse*, Feb. 21, 2014.
[133] "What next for the millions in the Sadrist movement?" *Middle East Online*, Feb. 21, 2014.
[134] Al Jazeera, "Al-Sadr Calls for Emergency Government," June 27, 2014.
[135] C. J. Chivers, "Answering a Cleric's Call, Iraqi Shiites Take Up Arms," *New York Times*, June 21, 2014.
[136] See Thomas Erdbrink, "Rifts among Shiites Further Threaten the Future of Iraq," *New York Times*, June 23, 2014.

and could have been done differently? The answers to these questions are not obvious, partly because of the nature of the Sadrist movement and the challenges arising when a legitimate social and religious movement overlapped a violent militia whose criminality helped finance the overall enterprise. When power structures are clearly licit or illicit, the course of action is usually clear. But in a complex political environment, this will seldom be the case, and in the twilight zone along the licit-illicit divide, policy choices are less obvious. Nevertheless, the experience with JAM offers some discernible lessons, about both what to avoid and what to do.

1. *Be aware of both national and local realities.*

The U.S. experience in Iraq highlights the critical importance, from the outset of any military intervention, of paying close attention to national and local realities. The poor planning for postwar Iraq both reflected and exacerbated intelligence gaps. Some of these gaps may have been inevitable, but there is little evidence that neoconservative decision makers wanted intelligence that would challenge the wishful thinking and naïveté with which they approached military intervention. Overly simplistic views about freedom and democracy could not be tarnished by serious consideration that U.S. military forces might not be universally welcomed as liberators.

At one level, the misunderstanding was simple: a failure to appreciate the power of Iraqi nationalism in the face of external intervention. At a more fundamental level, U.S. officials had little understanding of the Sadrists' key role in Saddam Hussein's Iraq, the sacrifices the family had made, or the popularity and legitimacy Muqtada al-Sadr enjoyed as the heir to this tradition. Nor did they understand the intra-Shia rivalries the United States had inadvertently become entangled in, or the consequences of siding with the exiles who had returned to Iraq after the regime's collapse, rather than with those who had stayed and incurred the wrath of Hussein. Bremer's targeting of al-Sadr transformed resentment into violence and helped enhance al-Sadr's reputation and status. Although al-Sadr suffered a major defeat by the Coalition in 2004, he also displayed a willingness to resist the occupation—a stance that increasingly appealed to many Iraqis as conditions of disorder and insecurity continued.

2. *Recognize and integrate all the stakeholders.*

Second, it is crucial to identify and understand the whole spectrum of stakeholders and deal with them equitably. Be cautious about preferential approaches to stakeholders. While some are obviously more important than others, snubbing and excluding particular groups and treating them as irrelevant to the country's future is a recipe for disaster. Even worse is the tendency for the United States to classify as a spoiler any group with a different vision for the country's future. Such an approach almost invariably creates a self-fulfilling prophecy: treating groups as irreconcilable merely pushes them into being so.

3. Develop more sophisticated policies for complex challenges.

Third, and an important concomitant to the preceding lessons, is the need for sophisticated and subtle policies to deal with ambiguous and complex challenges. The Sadrist movement emerged as, and remains, the advocate for a large, young, and hugely disadvantaged sector of the Iraqi population, whose grievances and concerns must be met for Iraq to have any hope of long-term stability. Although JAM was more problematic, for a while it protected a significant portion of the Shia population, acting with a high degree of legitimacy. But its actions intensified sectarian violence and undermined even the rudimentary efforts by the government and the United States to reestablish the rule of law. As the Iraqi government, with U.S. backing, gradually and incrementally restored a degree of order and a sense of security to the population, JAM became less crucial.

4. Reduce or minimize opportunities for illicit activities to flourish.

The Coalition's permissive and complacent attitude toward the looting that accompanied the fall of the regime was a profound mistake. It helped establish a culture of lawlessness that proved difficult to change. Similarly, in its response to kidnapping, the Coalition focused only on its own nationals and those Iraqis who provided critical support to the Coalition.

This permissive approach not only perpetuated a climate of insecurity but also clearly revealed the ways that political actors, whether militias such as JAM or insurgents such as AQI, could usefully appropriate organized-crime methodologies. Indeed, kidnapping and extortion, as well as the theft, diversion, and smuggling of oil, helped fund the violence in Iraq. Ironically, a team from the UN Office of Drugs and Crime had warned the United States about the growth of organized crime and its potential impact.[137] Unfortunately, as one member of the team noted, when the members briefed U.S. civilian and military leaders in Iraq, they met a mixture of hostility and indifference.[138]

5. Encourage the moderate factions in violent groups.

The preceding lessons, from the early period of the occupation, are about what to *avoid*. Several lessons from the later years, on the other hand, concern the kinds of approaches to *embrace*, to help reduce violence. One of these is that even in a group generally regarded as beyond the pale, there may still be differences between parts of the group that can be reconciled and brought into the political process, and those elements that remain irreconcilable. Encouraging and exploiting potential fissures can provide important opportunities. In the case of al-Sadr and JAM, this was made easier by the leadership's loss of control, and the obvious overreach with which zealous militant elements inflicted extraordinary predation on wide swaths of the Iraqi population. Moreover, some of JAM's activities were tacitly or explicitly repudiated by Muqtada al-Sadr and those close to him, who displayed a readiness to sacrifice the more extremist and criminal elements in

---

[137] UN Office of Drugs and Crime (UNODC), "Addressing Organized Crime and Drug Trafficking in Iraq: Report of the UNODC Fact Finding Mission," Aug. 5-18, 2003, Vienna.

[138] Author interview with member of UNODC team, June 2009, Vienna.

JAM. This was something that the U.S. military encouraged by employing a differentiated policy that distinguished between those elements of JAM that could be reconciled with developments in Iraq, and those that were irreconcilable and should be targeted. This policy was vastly more sophisticated and subtle than the ham-fisted initial Bremer approach.

6. Provide consistent governance at all levels.

Another set of lessons to be drawn from the experience in Iraq, while seemingly tangential to the problem of militant groups such as JAM, is, in fact, directly related. The plan to radically decentralize Iraqi governance was a flawed experiment with disastrous results. Locally created councils were supposed to influence the local delivery of services, but Iraq's service delivery institutions were merely branches of a network of centralized national ministries, each controlled by different factional elites with interests far removed from local concerns. The elites in charge of the national ministries had priorities wholly disconnected from those of the local leaders trying to voice local concerns. Mid-level ministry officials, who actually had the most ability to improve Iraq, kept their head down, trying not to run afoul of party elites, militants, and angry demagogues in the local councils. Thus, Iraq's new democracy was dysfunctional on three levels. Moreover, U.S. reconstruction efforts focused mainly on the elites and the street—the top and bottom levels—with reconciliation and grassroots democracy, and neglected the middle. This failure to put appropriate emphasis on the top-to-bottom functioning of Iraqi ministries produced an environment that enabled illicit power structures such as JAM to corrupt and subvert legitimate governance activities to their own ends.

An enormous mismatch existed between U.S. support at the elite levels of ministry leadership, and the efforts of local Coalition military units pursuing expedient measures, in the name of counterinsurgency, to assist local government institutions. U.S. efforts oscillated between the mutually exclusive aims of decentralization, in which institutional arrangements were radically changed, and stabilization, which aimed to prop up whatever seemed to be working at any given time.

7. Enhance, rather than undermine, security goals and policies.

The overall reconstruction effort often put long-term reconstruction and development goals at odds with military necessity. Even when these were not at odds, the focus on economic development sometimes had unintended adverse consequences. Iraqi contractors, for example, who were typically victims of extortion by militant organizations, built their extortion payments into their contract bids. Consequently, the United States indirectly helped fund the very groups it was fighting. Oversight of reconstruction and development assistance to ensure that funds are not extorted or diverted by hostile groups is essential.

8. Use military force to create incentives for armed groups to move into the political process.

Although military force can sometimes be used in ill-advised ways, it is an indispensable tool to counter illicit power structures. Military resources should be heavily focused on the irreconcilable elements of illicit groups, in ways that highlight the futility of those groups' efforts and compel hard-line factions to realize that nonviolent options might actually offer more benefits. In this connection, the reassertion of state power, albeit with U.S. support, in Basra and Sadr City in 2008 had a dramatic impact. Not only did the Charge of the Knights and the subsequent moves into Sadr City represent a pivotal point in limiting the power of militias in general and JAM in particular, these offensives also paved the way for restoration of stability, reduction of violence, and containment of criminal activities. Paradoxically, they also made it easier for al-Sadr to justify his transition from violent resistance against the occupation to more moderate forms of resistance, expressed through the political process. Perhaps the ultimate irony in Iraq is that in 2007-8, the U.S. military displayed a remarkable capacity for adaptation and strategic learning, which was matched only by a similar capacity in al-Sadr and those parts of his militia known as Golden JAM and Noble JAM. It is still uncertain whether this coincidence of flexible and subtle strategies on both sides was purely fortuitous or is something that can be replicated in other contingencies.

Conclusion

The demonstrable shift in JAM's status from irreconcilable spoiler to the U.S. occupation, to diminished armed wing of a Sadrist political movement aiming to move into the legitimate sphere of Iraqi governance, was, from the U.S. perspective, one of the success stories of the Iraq War. The operational shifts that accompanied the Surge—especially the targeting of JAM's most extreme irreconcilables—proved effective in reshaping the political calculations of one of the main factions responsible for the instability undermining Iraq. An illicit power structure, built on militant force and enriched by aggressive criminality, was eventually contained when it overreached in its predation on the population and miscalculated its military capability. Internal fissures within the Sadrist organization, coupled with the determined U.S. efforts to help the al-Maliki regime harness the state's security forces and stabilize Iraq, seemed to offer windows of opportunity for a new political bargain to emerge. Had these trends continued, this transition of the Sadrists from resistance to participation might have offered even more lessons about neutralizing a spoiler and constructing a lasting postconflict political order.

Unfortunately, many of these hard-won gains have dissipated following the U.S. departure from Iraq. Under al-Maliki, stability proved chimerical because his authoritarian tendencies undermined opportunities for a lasting political settlement between factions. The possibilities that the Anbar Awakening gave al-Maliki to reintegrate the Sunnis into more inclusive political arrangements and give them a real stake in the new Iraq were squandered. Instead, the exclusive and corrupt nature of the regime inspired little support or loyalty from the Sunnis. By the end of 2014, with the resurrection of AQI as Islamic State (or ISIS, ISIL, or Daesh), and the capture of Mosul, the de facto disintegration of Iraq, thought to have been prevented by the Surge, had become a reality. This breakdown of the state has seen the resurgence of Shia militia groups. Al-Sadr has reformed Jaish al-Mahdi, rebranded as "Peace Brigades," to join the fight alongside the forces of other Shia militias (many directed by Iranian Quds Force leadership), Iraqi security forces, and apprehensive Sunni fighters, to push back against Islamic State's advance into Iraq. But in its efforts to retake the major cities captured by Islamic State, this force seems as likely to engage in sectarian slaughter as to restore the authority of Baghdad. Muqtada al-Sadr has been one of the few Shia militia leaders sensitive to the dangers that another round of sectarian bloodletting presents to Iraq. Perhaps this is a measure of how far he has come since 2003. But if so, it is unfortunate that his concerns do not appear to be widely shared.

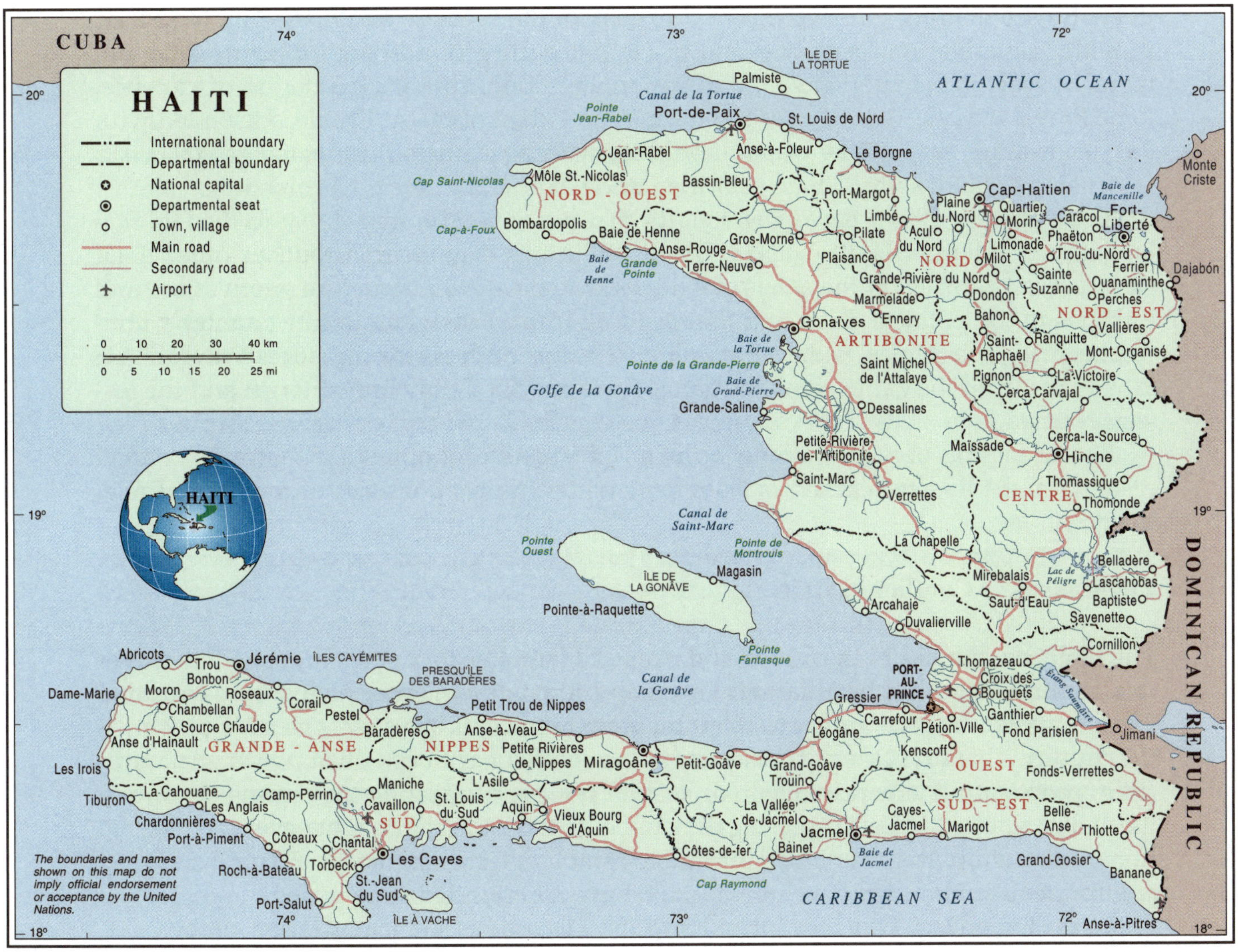

Map No. 3855 Rev. 4 UNITED NATIONS
June 2008

Department of Field Support
Cartographic Section

3. Haiti: The Gangs of Cité Soleil

D. C. (David) Beer

For Haiti, the new century began marked by instability, with Jean-Bertrand Aristide at the center of political controversy. In 2000, Aristide returned to power for his second term as president, but in 2004 a U.S.-led multinational force (MNF) intervened to quell a swelling tide of violence, and he was chased into exile. The United Nations returned to Haiti with another massive mission: to sustain the security established by the MNF, provide humanitarian assistance, and put in place the groundwork for future development. In 2006, in yet another round of national elections following a period of violence, René Préval was elected for a second term as president of Haiti. The decade ended with the devastating earthquake of January 2010, with a death toll estimated at 200,000, a refocusing of Haiti's immediate priorities, and yet another new government.

The period of 2004-7 saw gang activity that posed a profound threat to the UN Stabilization Mission in Haiti (MINUSTAH). While this examination touches other illicit structures, it zeros in on the gangs of Cité Soleil: their power, structure, motivation, and linkages to the political class; and the resulting impact they had on the country's politics, security, economics, and everyday life.[1] We then address the measures MINUSTAH took to confront the gangs, how the gangs were subdued only to reemerge, and the lessons learned.

Because issues of justice, crime, human rights, and economics are central to conflict resolution, the lessons learned in Haiti have wider implications for future international interventions.

The Causes of Conflict in Haiti

The violence that has both marred and shaped Haiti's history since independence over two centuries ago has been largely internal. Intrastate violence, fueled by corruption and motivated by race, religion, economics, or politics in basic struggles for power, is the country's history. Often at the center of the conflict have been illicit power structures (IPS), propagating violence to gain or sustain power, supporting the regime, co-opting the justice system, and endangering the public. In recent decades, whenever the international community has intervened in Haiti to establish security, furnish aid, and create an environment suitable for development, IPS have confounded progress.
François "Papa Doc" Duvalier (president 1957-71) and his son Jean-Claude "Baby Doc" (president 1971-86) exercised repressive dictatorial control through the formal structure of the military, and the informal structure of the Tonton Macoute. The latter was created specifically to counter any threat the military might pose to the dictator.[2] In the 1990s, as Haiti took its first steps toward democratic government, the Front for the Advancement and Progress of Haiti (FRAPH) emerged as a violent counterforce to Aristide's populist Lavalas Party, supporting the military regime that overthrew Presi-

---

[1] Cité Soleil is the most infamous "bidonville," or slum area, of Port-au-Prince.
[2] The Tonton Macoute later became the Milie de Volontaires de la Securité—as if the new name might confer some measure of legitimacy.

dent Aristide and terrorizing Aristide supporters during his exile.[3] In response, voodoo-inspired *zenglendos,* (a Creole term for bandits, robbers, rapists, and other violent criminals), emerged to counter the military and support the exiled Aristide. Apart from this IPS, however, organized violence was typically the domain of those supporting Haiti's dictators.

The end of the Cold War also ended Haiti's strategic importance as neighbor to Soviet-supported Cuba—a situation that had brought discreet U.S. support to both Duvalier governments. Jean-Bertrand Aristide emerged as a champion of the poor, challenging the country's political and economic status quo and playing an important role in ending the Duvalier regime. Aristide's popularity translated into political support for his Lavalas (Avalanche) Party, which swept to a landslide victory in 1990, in Haiti's first truly democratic exercise, following the U.S.-supported exile of Jean-Claude Duvalier. As president, though, Aristide quickly alienated the country's economic and political elites. In 1991, a military coup—Haiti's historically preferred method for transfer of power—halted Aristide's first term as president. He was deposed by General Raoul Cédras, commander of the Haitian Armed Forces, or Forces Armées d'Haïti (FAd'H), and spent most of his presidential term in exile. His overthrow initiated steps toward U.S. intervention.

Although the U.S. government no longer had a compelling strategic interest in Haiti, the invasion of Florida shores by refugee "boat people" escaping Haiti's poverty and violence at the end of the Duvalier era was catalyst enough for intervention. Preaching violence and anti-Americanism, Aristide thus began undermining the foundation of his foreign support, which had hoped for democratic success in Haiti, the western hemisphere's first black republic. After the Cédras coup, violence increased and the flood of refugees to the United States drew worldwide attention. Diplomatic intervention failed to remove Cédras from power, and the U.S. government considered stronger measures. Given Aristide's evident anti-Americanism, the U.S. security strategy amounted to keeping Haitians in Haiti.[4]

In October 1993, the USS *Harlan County*, a tank-landing warship, with military, police and governmental advisers aboard, was sent to Port-au-Prince, supporting the first step of a planned UN-sanctioned intervention. But the Clinton administration, facing the real possibility of an armed confrontation and after the recent loss of American lives in the "Black Hawk Down" incident in Somalia, ordered the ship back to sea the next day[5]. The incident, a diplomatic retreat, was politically embarrassing for the United Nations, the United States, and the international community generally. After months of UN-sanctioned embargoes and stalled negotiations, it seemed that no action short of armed force would see Cédras relinquish control and let the democratically elected president return to office.[6]

---

[3] FRAPH was a paramilitary organization with support inside the military and among political conservatives opposing Lavalas. Haitians understood FRAPH as merely the latest in the line of politically backed gangs that existed to gain and hold power by violence or the threat of it.

[4] The end of the Cold War changed U.S. strategic interest in Haiti. The impoverished "boat people" arriving in Florida certainly had an important economic and political impact, but Haiti was also ripe to become a staging area for South American cartels trafficking drugs to Miami, New York, and Montreal.

[5] Peter J. A. Riehm, "The USS Harlan County Affair," *Military Review* 77, no. 4 (1997), www.questia.com/library/journal/1P3-23556725/the-uss-harlan-county-affair.

[6] Walter E. Kretchik, "Planning for 'Intervasion': The Strategic and Operational Setting for Uphold

In July 1994, the UN Security Council passed Resolution 940, calling for the application of all necessary means to dislodge the Cédras regime. The resolution envisioned a multinational force composed primarily of U.S. military, followed by the UN Mission in Haiti (UNMIH). In September, after a diplomatic mission led by former U.S. President Carter negotiated the Cédras regime's departure, U.S. forces moved into Haiti, leading the MNF. With security restored, UNMIH relieved the MNF, intending to usher in a period of development, capacity building, economic recovery, and political stability.

Aristide spent most of his first term as president in exile, and though he was returned to office, the remainder of his term quickly expired. Constitutionally, he could not serve consecutive terms as president, and in the next election (1996), René Préval, prime minister under Aristide, was elected president. Aristide nonetheless managed to wield much influence and power. As he awaited his next political opportunity, Haiti's poor continued to recognize him as their real leader.

The Peace Process

The next few years seemed to promise peace. There were no external threats to security, and the United Nations had a significant military presence. America led bilateral donors that made significant contributions, and the international media were attentive and supportive. A new democracy seemed to be emerging. There was reason for optimism over Haiti's future.

But as the twentieth century drew to a close, optimism for sustainable progress waned. The Haitian government seemed frozen on important constitutional, justice, finance, and tax reform issues. Aristide, though out of office, maintained power and influence, as measured by the public adoration and media attention he attracted. As other crises arose globally and competed for aid and development money, the international donor community grew weary of the political inaction and wasted money and effort. A political paralysis set in, where all parties seemed to serve self-interest first, and all opposed the Aristide left. Aristide's aspirations for another term as president, leading the Fanmi Lavalas Party, promised ongoing confrontation and political blockage.

In 2000, Aristide was again elected president. The victory was protested as fraudulent, boycotted by opposition parties, and openly dismissed by Haiti's economic elites, presaging continued government paralysis and political confrontation.[7] As in his previous term, opposition to Aristide continued to grow, and by early 2004, political violence and armed uprisings were again threatening the country's security. Following another U.S.-led multinational intervention, Aristide resigned his position as president and was ushered from the country, this time to serve his political exile in South Africa. Aristide claimed that his resignation was only to quell the violence and avoid bloodshed, but

---

Democracy," in, *Invasion, Intervention "Intervasion": A Concise History of the US Army in Operation Uphold Democracy*, ed. Walter E. Kretchik, Robert F. Baumann, and John T. Fishel (Fort Leavenworth, KS: US Army Command and General Staff College Press, 1998), 35-40.

[7] Earlier in 2000, an overwhelming Lavalas Party victory in parliamentary elections brought charges of widespread fraud. This ignited the opposition against Aristide. When decisions were made restricting candidates qualified to proceed to the second electoral round, it all seemed contrived to limit challengers, and the political opposition intensified against Aristide.

later he would accuse the diplomatic community of orchestrating his "kidnapping."

With a backdrop of violence, corruption, and exploitation, Haiti was afflicted by clear security gaps. Aristide had disbanded the military in 1995; the police force, still in early development, was ineffective and fragile; the justice system simply did not function; and the coastline and border were largely open and unprotected. Haiti was and is the poorest country in the western hemisphere. Distribution of wealth overwhelmingly favors a tiny percentage of the population, racism (brown or "blancs" versus black) is prevalent, and economic opportunity remains absent. For the poor, these realities were underscored by the political isolation, once again, of their champion, Jean-Bertrand Aristide. All this was, in many ways, the outward manifestation of illicit power as wielded in Haiti.

Assessing Haiti's Illicit Power Structures

Although this chapter focuses on the gangs of Cité Soleil, an understanding of other groups that wielded illicit power is important to understanding just how they exerted power and how they affected the peace process and society generally.

The Former FAd'H

First among the other illicit structures is the former Haitian army, or FAd'H. The exile or decommissioning of its leadership in 1994 saw the FAd'H evolve as an illicit power structure, with particular support in the northern and central provinces. Government inaction in addressing issues of military pensions and retraining for unemployed soldiers caused grievances to fester. By 2004, the failed performance of elements of the Haitian National Police (HNP) became a cause for former soldiers, dressed in makeshift uniforms, to take up weapons, march publicly, and declare themselves responsible for local security.[8]

Ex-FAd'H activity was thus sometimes rationalized as filling security voids left by the fledgling HNP. But when peaceful protest and political lobbying failed, ex-FAd'H members turned to criminal activity. Their numbers actually grew as unemployed men, too young to have been part of the military that was disbanded years earlier, signed on in hope of opportunities with an organization that enjoyed some small measure of legitimacy. In time, they began to act illegally, committing thefts, robbery, and violence.

In Haiti's northern and central regions, the ex-FAd'H was prominent in the uprising of 2003-4, which led to Aristide's most recent exile and second unfinished presidential term. The visible presence of the ex-FAd'H as an armed force, even though of questionable capacity, challenged the Government of Haiti's (GoH's) legitimacy, particularly in outlying locations such as Cap Haïtien, Hinche, and Gonaïves.

The presence and activities of the ex-FAd'H, armed and threatening to future security, stimulated significant expenditure of time, effort, and funds toward strategies of disarmament, demobilization, and reintegration (DDR). The MNF and, later, UNMIH's

---

[8] The author, as UNMIH Police Commissioner with a mandate that included developing the Haitian security apparatus, was discreetly approached by Haitian political figures seeking support for a revitalized Haitian military, including a gendarmerie. Given Haiti's history of military violence and control, the idea was patently unacceptable to the international community.

security forces were often preoccupied with public demonstrations and criminal activity by the ex-FAd'H, and the Haitian government was routinely reminded of promises not kept, and embarrassed by its inability to control such an unruly group. Behind-the-scenes elements of the government politicked for the military's return. The weaknesses and corruption that gripped the HNP were held up as evidence that the experiment in civilian policing was staggering and that a revitalized military would be a quicker and more certain solution to national security. Though the ex-FAd'H represented a much less violent threat than urban gangs, this illicit power structure was an important political, economic, and security distraction.

Having failed to gain support for ex-Fad'H causes, its acknowledged public leader, a former sergeant known as Ravix, aligned with Cité Soleil gangs in attacks against MINUSTAH and HNP. When Ravix died in a 2005 gun battle with the HNP backed by MINUSTAH, public support for the ex-FAd'H faded, and for a time its significance waned.[9]

Drug Trafficking and Organized Crime

After the Duvalier regime, Haiti emerged as a preferred transshipment route by South American drug cartels moving narcotics to markets in Miami, New York, and Montreal. Haiti was ripe to be exploited. Only 12 hours by "fast boat" from Colombia to a vast unpatrolled coastline, it had unregulated remote air strips, undeveloped policing partnerships with regional neighbors, inadequate border control with the Dominican Republic, shipping to North America through Cap Haïtien and air transport connection from Port-au-Prince, a largely cooperative poverty-stricken population, and international security forces otherwise engaged and concentrated in urban areas.[10]

In the early post-Duvalier years, the North American transshipment link in Haiti was made by Florida- and Quebec-based organized-crime groups. In time, the North American groups would reduce their work and risk by engaging newly emerging Haitian criminal organizations. Some of these became well known by the broader Haitian community and the international community for their links to Haitian politicians. The emerging narco-economy was damaging to Haiti's reputation as the country struggled in the early stages of democratic development and hoped to encourage international aid and investment. But narcotic-funded corruption was soon established—a reality bound to influence politics, economics, and social development for some time to come.[11]

---

[9] A long history of human rights violations against its own people contributed to Fad'H's being decommissioned. For 20 years, any idea of reinstituting the Haitian military had no legitimate support. But today, after the 2010 earthquake, with the economy stagnant, UN-led development stalled, and HNP failing as a guarantor of public security, reestablishing the military has emerged as a subject of legitimate public debate. Haiti's younger population simply does not remember the Duvalier years or FAd'H's legacy of violence and fear.

[10] Because of its horseshoe shape, Haiti has a disproportionately long coastline, stretching 1,100 miles, with prominent peninsulas north and south. By comparison, Florida has a coastline of about 1,350 miles. See Country Studies Program, "Country Profiles: Haiti," 2013, www.mongabay.com/reference/new_profiles/908.html.

[11] In 2005, organized crime orchestrated a mass escape from the National Prison. Officials were bribed, cells were left open, and hundreds of prisoners simply walked away. The additional burden on police resources from HNP and MINUSTAH, already fully engaged with the gangs of Cité Soleil, was enormous.

Haitian National Police

HNP was crippled by Aristide's return to power in 2001. It was still struggling to develop but had made undeniable progress since its creation only six years before. Then suddenly, senior police leaders were fired or resigned as Aristide installed his own leadership. By 2004, HNP's fragile foundation had collapsed. Progress was erased, and the organization tumbled quickly as some units became little more than violent street gangs. Certain elements supported all manner of criminal activity, including drug trafficking, robbery, murder, and kidnapping. The police, even while receiving important international assistance, contributed to the insecure environment impeding development in every sector. Moreover, without capacity or credibility and with a deservedly negative reputation in the region, HNP could not call on its neighbors for assistance or collaboration.[12] In many respects, HNP itself was fast becoming an illicit power structure.[13]

Private Security

The business of private security was a constant reminder of the insecure environment and of who actually held the money and power. Virtually no business in the country and virtually no residence occupied by anyone of any means was without armed guards. Although there was no regulation and no recordkeeping, private security companies were surely the greatest importer of weapons, ammunition, and security equipment, and perhaps the country's most important industry. While the absence of regulations and controls alone does not support consideration of private security as an illicit power structure, it was common knowledge that some industrialists engaged criminal gangs as private security for their businesses and factories or simply compensated gang leaders for a "hands-off" approach. This was a particular strategy in Port-au-Prince and the neighborhoods and industrial areas around Cité Soleil.

Gangs of Port-au-Prince

<u>Gangs as instruments of power.</u> As a Catholic priest in the "Baby Doc" Duvalier era, Aristide was a visible and vocal antipoverty activist with a strongly committed following among Haiti's poor. In the 1980s, he became widely known internationally. In time, though, his international celebrity turned to notoriety as he became known for his public

---

[12] Policing and crime control, even on a national level, is a matter of "neighborhood watch." Haiti had a less-than-stellar reputation in its Caribbean neighborhood and did not have positive policing relationships with its neighbors; particularly the Dominican Republic but also the United States, Cuba, and the Bahamas. Nor did it have a good reputation in the Association of Caribbean Commissioners of Police. Without regional partnerships, Haiti was defenseless in matters of international criminal intelligence, narcotic traffic, money laundering, weapons traffic, and illegal immigration.

[13] When experienced senior police resigned or were fired after Aristide's return as president, their replacements were often appointees without experience or professional training and with suspect personal histories. When the United Nations returned in 2004 with a mandate to build capacity and aid the HNP operationally, vetting these suspect appointees was a top priority. Curiously, the HNP and GoH gave no cooperation. No one was removed from the HNP because they were inappropriately or illegally engaged, were not qualified, presented security risks, or were suspected of criminal history.

anti-Americanism, support for violent uprisings, and rumored involvement in the drug trade. He was defrocked by the Catholic Church, but his popular support among poor Haitians remained strong. Aristide facilitated organization of armed support among the poor, particularly in the slum areas of Port-au-Prince. This took the form of armed street gangs, led by young men who had grown up in the slums and under the influence of the priest Aristide. There has been little empirical research—and certainly no admissible evidence for any judicial proceeding—tying Aristide and Lavalas to the use of gangs as instruments of illicit power. Also, there is little admissible evidence tying prominent business leaders to the gangs. And yet, most who know much about Haiti regard both assumptions as truth—a reality that perhaps makes the gangs' origins a question of "chicken or egg."[14]

The gangs were characterized by unsophisticated leadership, opportunistic crime, and brutality. Promiscuous violence accompanying criminal activity was not uncommon and was intended to shock, threaten, and intimidate. While gang activity could be linked to public protests, general labor strikes, elections, or political events, the gangs' real power derived from violence or the threat of violence. They were motivated by turf protection and group loyalty but also displayed characteristics of political objectives and criminal entrepreneurialism.

According to Max Manwaring of Strategic Studies Institute, first-generation gangs are traditional street gangs with a turf orientation. When they engage in criminal enterprise, it is largely opportunistic and local in scope. Second-generation gangs are engaged in business. Entrepreneurial and drug centered, they tend to pursue implicit political objectives. Third-generation gangs are primarily mercenary, but many seek to advance explicit political and social agendas. Thus, third-generation gangs find themselves "at the three-way intersection among crime, war, and politics." While this categorization may oversimplify, it is useful as a broad picture of gang structure and development, emphasizing the complexity of the gang as illicit power in Haiti.[15]

Kolbe and Muggah aptly describe the relationship between Haiti's gangs and its chronic instability:

> What is most worrying, however, is that some of Haiti's gangs—particularly those affiliated with organized crime, paramilitary, and private security companies—are tied to the country's political elite. It is those with money and power who are most inclined to use gangs as a means of intimidating enemies and extending business interests . . . As everyone in Haiti knows, there are specific politicians, business leaders, and wealthy land owners who serve as the gangs' chief patrons. Essential to diminishing insecurity associated with gangs, then, is a better understanding of these relationships and exposing them for all to see.[16]

---

[14] For recent research on the gangs' motivation, the relationships holding them together, and public perceptions, see Athena Kolbe, "Revisiting Haiti's Gangs and Organized Violence," HASOW International Conference Discussion Paper, June 4, 2013, www.hasow.org/.

[15] Max G. Manwaring, "Street Gangs: The New Urban Insurgency," monograph, Strategic Studies Institute, 2005, 6, www.strategicstudiesinstitute.army.mil/pubs/display.cfm?pubID=597.

[16] Athena Kolbe and Robert Muggah, "Kolbe and Muggah: Haiti's Gangs Could Be a Force for Good," Ottawa Citizen, June 4, 2013.

In time, the gangs situated in and controlling slum neighborhoods of Cité Soleil came to be seen as the armed element of Aristide's populist Lavalas Party. If not created expressly to support Aristide and his Lavalas political movement, the gangs were certainly exploited for that purpose. Aristide had long known the five highest-profile gang leaders of 2004-5. The five, known as Amaral, Tu-Pac (2Pac), Bily, Labanye, and Dread (Dred) Wilme, were competitive and threatened one another but were known to fight together against the police, international forces, or rival gangs. A well-known and dominant gang leader in Gonaïves, Ti Will, was not known to venture openly into Port-au-Prince. Along with Ravix Ramissainte (the former FAd'H sergeant, with influence in Cap Haïtien and the interior provinces), these were the best-known gang leaders in the country.[17]

The gangs seemed to believe they responded to Aristide's direction as a counter to prevailing political and economic interests. Not that the gangs were actually controlled by outside forces, but they characterized themselves as antiestablishment champions of Haiti's poor. Gang activity included participation, or at least a presence, in political action and at public events. Wherever peaceful protests erupted in violence, or armed protesters suddenly appeared in a crowd, speculation arose that such acts were politically directed to provoke HNP into overreacting and initiating violence, thus discrediting the GoH and UNMIH.

Although the Cité Soleil gangs' origins were linked to the rise of Aristide and the politics of the Lavalas Party, and although the gangs were seen as an armed representation of Haiti's political left, they did not stand as one in support of any political or ideological position. Indeed, the gangs' motivation and what or who actually spurred them to act was not always clear. Much of what we think we know of the Port-au-Prince gangs' motivation comes from academic writings, media reporting, official reports of variously mandated international agencies, and statements and reports of the self-interested. An interesting source is to watch and listen to the gang leaders themselves using online and social media communication. Although the leaders do not share equal time and although some are more camera shy than others, the gangs' message and motivation becomes clearer.

At their foundation, the gangs exist as a social reality spawned by strong and influential personalities, abject poverty, and the lack of access or opportunity that is reality for most Haitians. There are few other options for young men of Port-au-Prince's bidonvilles. In the end, at least for now, the gangs of Port-au-Prince (and Cité Soleil) might best be described as politically *influenced* rather than politically *motivated*.

In an unusual cultural phenomenon, gang leaders gained a sort of "rock star" status in Port-au-Prince. Some local media catered to and fed this social status, creating a public profile and fostering reputations. The gangs' size, visibility, and, indeed, power gained from this public profile. Interviews, including live radio broadcasts, served no particular purpose but to boast of violent acts or threaten legitimate authority and frighten the public.

While gangs of some description were evident in all Haitian cities, it was the gangs of Port-au-Prince—and the gangs of Cité Soleil in particular, with a direct historical

---

[17] The source was an NGO employee with a history of work in Haiti, who knew the gang leaders as children and had occasional contact with them as young adults and gang leaders.

association to Aristide—that were responsible for the violence that affected the country's politics. Cité Soleil borders the Port-au-Prince industrial-commercial zone. The area, developed in the Duvalier era, became home to Haitians from around the country gravitating to Port-au-Prince in hope of finding work. Home to the city's most impoverished, it is a small peninsula recovered from the sea and hemmed in by the water, Route Nationale (the national highway), the city's industrial center, and the port. It is also very near the international airport. This geography makes the neighborhood of real tactical importance where internal security is concerned.

Aristide supporters saw the U.S.-led coalition that facilitated Aristide's departure in 2004, and the UN force, MINUSTAH, that replaced it as invading and occupying forces. Before 2004, gangs almost never molested the international community. UN Police, foreign military, and aid and development workers traveled the country and operated at minimal risk. This may be because at that time, the (U.S.-led) international community had returned Aristide to power, hundreds of millions in aid and development dollars were available, and the possibilities for a new democracy still shone. Meanwhile, the gangs openly targeted the newly created HNP. (In 1996-97, on average, one police officer was killed every week.) The international community orchestrated Aristide's 2004 exile; then came the MINUSTAH "occupation" of Haiti, previewing the international community as targets of the gangs.

Early in 2004, gang attacks against UN forces were tentative, probing hit-and-runs. By autumn, attacks became more frequent, much larger, and better coordinated and sustained, in the manner of guerrilla urban warfare. Small-unit assault techniques included ambushes, coordinated arcs of fire, and heavier small arms. When MINUSTAH did not respond with the same aggression as the MNF before it, gang attacks became much more aggressive, more frequent, and deadlier.

The gang structure. Street gangs were generally hierarchical structures with recognizable leaders, lieutenants, and soldiers. Leaders, born and raised in the neighborhoods where their gangs operated, were well known to the local population and were often identified by nicknames acquired in childhood. A unique characteristic of these gangs was the addition of young criminal deportees from the United States.[18] Having grown up in North America and often without economic means or French or Creole language skills, and sometimes without close family ties in the country, they were lost in Haiti and easily recruited into gangs. Often, their only assets were their criminal connections and experience in North America and their own brand of street violence.

The size of each gang seemed to rise or fall with the leader's notoriety. Jewelry, designer-labeled clothing, and cars separated gang leaders from the populations they controlled. Allegiances among the gangs were common as forces joined for attacks against police, international forces, or a rival gang. But alliances crumbled as fast as they formed, often ending in gang-on-gang violence. As often happens in the gang lifestyle, the greatest risk to gang members came not from the legitimate security forces but from ambitious subordinates and rival gangsters.

---

[18] Hundreds of Haitians were deported from America, and fewer from Canada. Deportations from both countries were low key and handled discreetly.

Illicit gang activities. Most gangs were associated with the geographic area they occupied or with the leader they followed. They thrived on local criminal activity. Using Cité Soleil as a safe harbor, gangs moved in and out of the neighboring commercial zones to commit crimes. Robbery and hijacking were typical activities, and kidnapping became a thriving cottage industry. Rape was commonplace. The sad irony is that the poor bore the real brunt of gang violence. Violence was often so intense that humanitarian agencies could not provide service in gang-controlled zones or would not put their people at risk by venturing into hot areas. Even outside the high-risk zones, humanitarian organizations risked theft or robbery since their humanitarian supplies were of value to the criminal element.

Opposition to the peace process, use of violence, irreconcilable interests. The gangs viewed MINUSTAH as an occupying force: an extension of the MNF, which had managed Aristide's departure. Moreover, they saw it as supporting the country's economic elite, the interim government, and HNP, and fundamentally anti-Aristide and anti-Lavalas. The United Nations was rationalized as the enemy, and its continued presence meant that Aristide was being denied an opportunity to return to power.

Not that the gangs had any definite political position, but since they were willing to fight UN security forces to control their own turf, it is fair to say the gangs opposed stabilization—or, at least, the UN version of the process. There was no unified position, manifesto, or constitutional plan for reform. Demands for UN withdrawal and Aristide's return were messages heard most frequently, but there was no consistency of message and certainly no rallying cry among the gangs, which were very much centered on their individual leaders. The gangs did occasionally join for common purposes, such as to attack UN forces, but they generally competed for local control, and intergang violence was common.

Gangs were mercenary, too, in the sense that their protection was for sale. Even the economic elites recognized the value of having a gang for security where there was no legitimate alternative. Said one gang member: "We have always gotten money and political support from [name of wealthy business owner]. So we are accountable to him." The businessman is not named, but any citizen of Port-au-Prince could fill in the blank.[19]

Gangs were routinely hired to protect business interests in particular neighborhoods.[20] That antiestablishment, Lavalas-inspired gangs sold their services to the hated *gros mangeurs* underscores the indistinct, opportunistic nature of their motivation.[21] They wielded power through violence, fear, and crime while professing to offer a measure of local "protection," authority, and control. With an array of small arms, Molotov cocktails, and, occasionally, grenades, the gangs waged a guerilla-style insurgency. Attacks against the HNP and UN Police and military were frequent, sustained, and deadly.[22]

---

[19] Kolbe, "Revisiting Haiti's Gangs."

[20] "Although ostensibly criminal in nature, the gangs of Port-au-Prince were an inherently political phenomenon. Powerful elites from across the political spectrum exploited gangs as instruments of political warfare, providing them with arms, funding, and protection from arrest." Michael Dziedzic and Robert Perito, "Haiti: Confronting the Gangs of Port-au-Prince," U.S. Institute of Peace, Special Report no. 208, Sept. 2008, 1.

[21] A Creole term meaning "big eaters," referring to the economic elite.

[22] Gangs were rumored to have mortars and rockets, but this was not verified. A 2004 grenade attack against the Provisional Electoral Council building was an isolated incident.

Analysis of the gangs' role in violence against journalists is also noteworthy. According to the Committee to Protect Journalists (CPJ), no suspicious deaths of journalists were reported when Aristide was out of office during 1994–99. At least eight journalists were killed during 2000-2007—the years he returned to official power, continuing into the period of his second forced exile (which began in 2004). These were the years of peak gang violence. All the journalists were killed violently (seven by gunshot and one by machete). It is suspected that at least five of those deaths were directly related to the victims' work as journalists. In the other three cases, motive was unconfirmed. While no clear evidence ties violence against journalists to their reporting on Lavalas, Aristide, or the gangs, the stories, as reported, are cause for reflection.[23]

For a period immediately following President Aristide's departure, the gangs' public image was elevated by the ongoing perception that they somehow represented the poor and Lavalas as the popular political movement. Over time, this view was discredited by the steady barrage of extraordinary and indiscriminate violence, with gang leaders murdering each other in turf wars or being killed or apprehended by local police and international forces. When the deaths of innocent victims increased right along with those of gang members, it seemed evident, especially to President Préval, that the gangs were little more than violent thugs whose interests were irreconcilable with the peace process—not political activists he wanted to negotiate with. After Préval's attempts to negotiate with gang leaders met with shocking episodes of violence in November 2006, he issued an ultimatum to "surrender or die." There were no rules, and the environment had devolved into anarchy.

Even though MINUSTAH was not operationally effective immediately, its deployment threatened the gangs' safe havens and the freedom of individual members while also challenging Aristide's power. The violence grew so bad that the United Nations could not deliver aid or implement quick-impact projects (projects intended to give a small measure of immediate humanitarian and economic relief to the poor while demonstrating the intent to develop longer-term projects). The gangs, in their violent opposition to the United Nations, directly hurt the poorest of the poor—those they claimed to protect and represent. Finally, the view that the gangs opposed, or were being used in opposition to, the peace process has added credibility when we consider that UNMIH's presence in Port-au-Prince had gone largely unopposed by the gangs. But when MINUSTAH deployed in 2004, the UN presence worked against, not for, Aristide's return to power—the opposite of UNMIH's impact.

The Cité Soleil Gangs' Relationship to the Peace Process

MINUSTAH was tasked with a complex mandate that included helping the GoH establish and maintain security, provide humanitarian assistance, and create an environment conducive to development.

---

[23] CPJ, analyzing events surrounding the deaths, cites the victims' role as journalists as the likely motive for their murders. CPJ, "5 Journalists Killed in Haiti since 1992/Motive Confirmed," www.cpj.org/killed/americas/haiti/.

Protection of Civilians

As with everything else in Haiti, the influence of gang activity varied depending on one's socioeconomic stratum. For the economic elite, impact was measured in terms of profit and productivity, freedom of movement, and the fear and risk of kidnapping. This small minority of the population reacted predictably. It was vocal in protest and active in its lobbying—in Haiti, of course, but also in the United States, Canada, and the United Nations. At home, groups such as the Chamber of Commerce discussed security concerns and protested UN ineffectiveness. When gangs threatened to fire on aircraft landing and taking off at the international airport, the threats had to be taken seriously.

But for the poorest people living in central Port-au-Prince, the impact was more direct. At the height of violence, local street markets did not open, and regional produce was not delivered. The few who could afford the meager cost of simple education kept their children home, and churches closed. Even those who stayed off the streets during times of extreme violence were not free of danger. Many were killed or injured when bullets ripped through the sheet-metal walls of slum dwellings.[24] Rape was a common tactic of power and control, and women were victimized and revictimized sexually, their only defense being to flee the zones of gang control. Many did flee, seeking assistance from international organizations outside their home neighborhoods, but where there were few opportunities to relocate to safer areas, victims were destined to return to their gang-controlled neighborhoods. Fearing retaliation and revictimization, they dare not identify their attackers.

Humanitarian and development assistance. In the poorest slum areas of Port-au-Prince, where the poverty-stricken population depended on aid and humanitarian organizations, any disruption of service owing to gang activity was bound to have a dire impact. For example, in 2004, gang violence meant that tons of emergency aid (135 containers, or 2,500 tons), much of it destined to hurricane-ravaged Gonaïves, could not be moved from the seaport without an armed escort.[25] Agencies with long-standing reputations for independence, such as MSF, tended to have less difficulty accessing gang-held territory, as did organizations with good local contacts. But general access was undeniably restricted, freedom of movement impaired, and distribution of critical goods and services limited. UN-associated humanitarian agencies could not venture into gang-held territory without an overwhelming police or military presence. Even then, security from sniper attack was never certain. Where security could not be assured, services could not be provided, food and water distributed, or medical assistance administered, and development programs could not proceed as planned. The result was that the people who needed assistance most—those forced to live in the worst possible conditions—suffered most from the gangs' power to limit access to their territory.

---

[24] Medecins Sans Frontieres (MSF) reports that admissions for gunshots wounds fell from 1,300 in 2006 to 500 in 2007. Numbers for stab wounds, rape, and beatings continued to rise. In 2007, 2,847 patients were admitted for violence-related trauma. Not reported by MSF was the constant tension surrounding HNP's entering and searching MSF facilities for gang members being treated. HNP largely ignored pleas from MINUSTAH about this. MSF, "US Annual Report 2007, (Project Support—Haiti)," 24-25, www.doctorswithoutborders.org/news/allcontent.cfm?id=31

[25] Kevin Sullivan, "Beheadings Mark Haiti's Latest Misery," *Washington Post*, Oct. 8, 2004, 26, www.washingtonpost.com/wp-dyn/articles/A16234-2004Oct7.html.

Governance. For the interim government, gang violence pushed security issues ahead of all political business. Exacerbating the situation, the HNP, the government's only security force, itself was a target of the violence, as was the United Nations, the main international player and partner of the GoH. Also, the government saw itself as restricted by its interim status, and largely powerless to make decisions that would reach beyond its limited tenure. The interim prime minister, Gérard Latortue, let the street violence overshadow his political responsibilities. He met with police and military leaders regularly and personally chaired security committee meetings. He was briefed routinely at the strategic level on plans, and at the tactical level on operations. He, rather than his ministers (of public security and justice), was the government's political face on security issues. The government and the United Nations were blamed for failing to take effective action, but in turn, the GoH blamed the United Nations for failing to be aggressive enough against the gangs and for failing to give HNP effective operational support.[26]

Rule of law. HNP's considerable early development (1994-2000) largely ended after the core of capable executives and senior managers was dismissed or resigned in the presence of the new Aristide government in 2001.[27] When untrained and unqualified leaders replaced them, a culture of corruption began to consume the organization. Operational elements of the police were turned by the lure of drug money, facilitating narcotics transshipment from South America and Jamaica to North American markets.[28] Others facilitated or actually participated in kidnappings, often collaborating with the gangs in the process.[29]

Even the legitimate HNP units cooperated with the UN partners only as it suited their purposes. The police, internally directed at the highest levels, often purposely evaded monitoring by MINUSTAH. HNP operations were often conducted without UN knowledge or sanction. In such situations, armed engagements with alleged "bandits" where a virtual certainty. Invariably, suspects were killed and civilians killed or wounded, but no known internal investigation or public accountability ever followed.[30]

On the other side, gangs hunted police and bragged publicly of their murders. HNP had been a target of the violence since its inception in 1994. In 1996-97, on average, one Haitian police officer was murdered every week. While not all this violence could be attributed to gang activity, much of it was, and HNP took extrajudicial vengeance on

---

[26] By design and agreement, HNP and the United Nations were to work together in security operations. This was meant to ensure coordination, of course, but it included the mentoring and oversight implied by working together.

[27] Pierre Denize (director general), Ernst Mompoint (chief of administration), and Jessie Coicou (head of forensic science) were among those forced to leave. Leon Charles and Mario Andresol left HNP as young officers but were, in succession, named HNP director general after Aristide left the country.

[28] The U.S. Drug Enforcement Administration (DEA), actively investigating drug transshipment through Haiti, was well aware of HNP complicity with drug traffickers. The DEA collaborated only with specially vetted HNP officers, where their presence was necessary to facilitate an investigative action.

[29] MINUSTAH, investigating kidnappings in Port-au-Prince, discovered cases where the HNP "kidnapping unit" had facilitated or actually committed kidnappings. In other situations, HNP elements inserted themselves into investigations or negotiations, attempting to inflate ransom demands and helping themselves to a cut of the action.

[30] Where the United Nations felt compelled to intervene, HNP gave no cooperation and certainly no access to records or witnesses. Simply put, HNP would not subordinate itself to external scrutiny.

reputed gang members. At the time, UN forces and the international community were almost never targets of gang violence.

In a particularly horrific event in 2004, three young HNP officers were murdered and their bodies later beheaded, in an event that gang members (and local media) hyped as signaling the transformation of Port-au-Prince into Baghdad. In 2003 postwar Iraq, insurgent activity included well-publicized high-profile kidnappings and executions, including beheadings. Haitian gangsters, hoping to intimidate the GoH and MINUSTAH, threatened to turn Port-au-Prince into another Baghdad. Photos, held out to be the victims, are still online. Evidence suggested that the three HNP officers' beheading was postmortem, though the gangs still achieved their objective of shocking the public and influencing local media. The intent, of course, was to strike fear in the public, the international community, and the Haitian political leadership.

International Strategy and Its Impact on the Gangs of Cité Soleil

No international agency or group was affected more by the gangs of Cité Soleil than MINUSTAH. Ill-prepared to deal with the level of violence, the mission immediately fell short of mandated objectives and public expectations of its ability to provide security. Security (in Port-au-Prince particularly) became the all-consuming priority, and effort and resources were diverted from other stated objectives of the multifaceted mandate that included aid, capacity building, and human rights issues. Coordination between the UN military and police was inconsistent at best, and the United Nations had no executive authority to act independently of the Haitian government on security matters. Thus, the United Nations struggled to quell gang violence, and its ability to protect the civilian population was repeatedly called into question.

This does not mean that the United Nations was unaware of the gangs as illicit power structures, their attachment to Aristide and Lavalas, and their mercenary willingness to do the bidding of whoever paid. But information about the gangs, though generally considered true on some level, was generally unsubstantiated and certainly had never been exposed officially through the courts or even with credible people making on-the-record declarations. Even accepting that the gangs could be externally influenced, Aristide had been removed from the scene, the country's economic elites supported the international intervention, and the history of recent missions indicated that while the gangs were a potentially violent and disruptive influence, they had never before been a direct threat to UN personnel.

In fact, the United Nations was unprepared for the level and nature of violence soon to emerge. The security situation was expected to be a matter of establishing public order, in political circumstances similar to the insecurity that had characterized Haiti during its first brushes with democracy 15 years earlier: public disorder, large demonstrations and protests, both planned and spontaneous, and the violence and disruption to daily life and economic activity that sprang from these. To meet this need, Formed Police Units (FPUs) were recruited from police-contributing countries. Responsibilities included operational support to the GoH and institutional development through training, mentoring, and advising. Outside Port-au-Prince, FPUs were deployed to the historical political hotspots, including Cap Haïtien, Gonaïves, and Hinche. The UN military component, four times the police strength and seen as the mission's security foundation, was concentrated in Port-au-Prince but deployed widely to represent MINUSTAH throughout the country.

Expectations were high that the United Nations would be immediately effective in maintaining security established by the MNF. But shortly after the mission's launch, gang violence erupted, and the United Nations was criticized, publicly by the media and privately by the GoH, for its inability to respond quickly and effectively. Meanwhile, the UN mission administration and rollout was slow, and months after the Security Council resolution had passed, only a small percentage of the security resources had arrived in country.[31] Not until six months after the mission began did the UN military determine that it had sufficient resources to assist in a major operation to establish a police presence in Cité Soleil. Though the military component outnumbered MINUSTAH police four to one, the military interpreted its role as limited to support and considered criminal gangs a police issue, outside primary military responsibility.

But the MINUSTAH police's capacity was also limited, and from their first patrols in central Port-au-Prince, they were under attack. Over a period of weeks, the intensity and sophistication of assaults increased. The FPUs were neither trained nor equipped for the urban guerrilla violence they faced. In the tight confines of urban Port-au-Prince, bulky armored personnel carriers were of limited value, and the FPUs had neither experience in "dismounted" armed patrol nor close-quarter battle skills required to fight gangs in densely populated neighborhoods.

The police component had been recruited as mentors and advisers and hailed from dozens of different countries. Thus, it was not a coherent operational force and had no equipment or training to deal with the insurgent-style violence. English and French language skills (the primary languages of the mission) among the police were limited, and, of course, MINUSTAH police operated without executive authority. At the same time, HNP often operated independently of MINUSTAH when it chose not to be constrained by an international presence. More generally, MINUSTAH was reluctant to engage gangs in pitched urban battles that were guaranteed to result in civilian casualties. International and local media criticism was relentless.

The violence was so extreme that police-contributing countries and the UN General Assembly representatives were concerned for the safety of their nationals, who were untrained and ill-equipped for the dangers of the mission. While some contributing countries considered withdrawing, and concern arose about whether renewed commitments were forthcoming, governments quietly put national restrictions on the roles for their contingents. These national caveats were rarely publicized but were not uncommon, even though restrictions generally contradicted the very mandate the contributing countries had pledged to support. Gangs quickly learned which contingents would not engage them, and they exploited this vulnerability with aggressive attacks.

The UN Security Council was soon preoccupied with the level of violence being encountered and the inability of the mission or the GoH to address it. Strong arguments were being made to change the MINUSTAH mandate and give it executive authority to

---

[31] The UN secretary-general reported 240 police and 2,755 soldiers in country four months after the mission began. Interim Report of the Secretary-General on the United Nations Stabilization Mission in Haiti, 30 Aug. 2004, www.un.org/ga/search/view_doc.asp?symbol=S/2004/698.

engage in law enforcement. In an unusual step, the Security Council assembled in Haiti to assess the situation firsthand. In an equally unusual move, the Department of Peacekeeping Operations (DPKO) ordered a review of the UN military and police response to the mandate, investigating reports of mission shortfalls, including failures to integrate operations effectively.

MINUSTAH now found itself in an untenable position. Tasked with a complex mandate that included humanitarian assistance, establishing rule of law, development, and security, it needed to reexamine its capacity to support the GoH in quelling gang violence, as a first step to fulfilling the mandate. The safety of personnel, the viability of the mission, and the United Nations' reputation were at stake. Change was needed. To confront this existential challenge to the mission, MINUSTAH needed to consider a strategy that emphasized mission integration, intelligence and information management, personnel with the requisite skills and experience, and an improved partnership with the GoH/HNP.

Except for developing rule of law, progress was made toward implementing changes to the mission structure and operations before the end of 2005.[32] The United Nations is a large, complex mechanism that does not change direction easily, however, and it would be over a year before most of the changes could be considered complete. Meanwhile, gang violence increased, and the United Nations' reputation became even more tarnished.

MINUSTAH and HNP undertook some halting operations in late 2006, but not until January 2007 was the MINUSTAH military contingent fully engaged in support of full-scale, coordinated offensive action against the gangs of Cité Soleil. The outcome of the new strategy to confront the gangs with overwhelming force aimed at crippling their ability to threaten the mission and the people of Cité Soleil is described in a Special Report published by the U.S. Institute of Peace:

> UN military and police units working with the Haitian National Police moved neighborhood by neighborhood throughout, arresting gang leaders or forcing them to flee. Once the United Nations established that it was prepared to use superior force, resistance from the gangs quickly diminished. Gang members deserted their leaders and sought to blend into the population . . . By March 2007, the United Nations had regained control of Cité Soleil. Once the gangs had been flushed from their sanctuaries, with support from police-led operations by UN Police and the HNP, some eight hundred gang members were eventually arrested, and all but one gang leader was either apprehended or killed.[33]

Finally, the operational processes and capabilities essential to confront the gangs were in place and successfully put into action. Each of the building blocks for the successful campaign against the gangs is described below.

---

32 Rule of law, an important component supporting justice development in parallel with the policing and corrections missions, was not advanced by MINUSTAH. It remains an important gap in the mission's accomplishments. In fairness, the GoH and Haitian judiciary have been entirely resistant to change—eager to accept development money but unwilling to venture toward public accountability and transparency.

33 Dziedzic and Perito, "Haiti: Confronting the Gangs of Port-au-Prince," 5.

Mission Integration

The integrated mission, a concept identified in the United Nations' Brahimi report and intended for use by MINUSTAH to maximize its collective capacity, had not materialized. A review panel was convened on March 7, 2000, before the Millennium Summit, and tasked to review UN peace and security activities and recommend change. The report, named the "Brahimi report" for the chairman of the commission that produced it, noted that there was neither a standing UN army nor a police force. (Both were envisioned in the UN Charter.) As a result, peace operations have been ad hoc coalitions of willing states. The report identified many dysfunctions of UN peace operations, including shortfalls in personnel, skills and training, and intelligence capacities (all shortfalls experienced at MINUSTAH's start-up). Following the report, the UN Security Council adopted several provisions related to peacekeeping.[34]

In 2004, MINUSTAH military and police were not working to a singular purpose and did not even interpret the gang threat similarly. The police did not have the capacity to deal with insurgent-like violence, and the military refused to consider the violence as anything but criminal behavior.

From the mission's beginning, efforts had been made to harmonize and coordinate security operations and provide support to the HNP. Committees were formed, meetings were held, and policy was written. Information exchanges, coordinated patrols, HNP and UN colocation strategies, and joint operations resulted, but with limited success. Mission integration was of such concern to DPKO that it dispatched General Maurice Baril, former head of the DPKO Military Division, to study the mission's security operations and integration.[35] Although the mission was on the path to integration, and policies and procedures were in place, the external review by Baril reinforced the need for a quicker pace and demonstrable commitment. In particular, the review proved to be the catalyst for the MINUSTAH military contingent to assume responsibility for a frontline role in confronting the gangs and for coordination, particularly with the UN police, to form an integrated mission response.

Inside the mission, better operational coordination was already in evidence with the creation and mixed staffing of the Joint Operations Centre (JOC) and the Joint Mission Analysis Centre (JMAC). These integrated units were tangible examples of integration emerging at the level of mission headquarters. But true integration was some distance off because national caveats continued to influence participation in offensive operations and because divergent opinions persisted about the nature of the gang threat. Simply stated, some contingents were willing to engage the gangs; others were not. And some military leaders continued to question the military's role in this nontraditional fight. And yet, in spite of the impediments, the mission was moving toward the desired integrated model.

---

[34] United Nations Rule of Law, Report of the Panel on United Nations Peace Operations (Brahimi Report), 2000, www.unrol.org/doc.aspx?n=brahimi+report+peacekeeping.pdf.

[35] Beyond the self-evident issues of mission integration, General Baril's study needed to consider some politically sensitive issues. Rumors of national caveats against armed engagement, and concerns over collateral risk and protection of civilians were widespread in the mission. Included was the fact of internal dissent where it was believed that certain contingents could not be counted on to protect mission colleagues during armed engagements. The results of Baril's inquiry were not widely published, but his visit and inquiries had the desired effect of changing attitudes among senior mission personnel.

In time, mission integration developed and matured. It would never be perfect, of course. (A mission of 50 or more contingent nations and diverse mother tongues, experience, and professional systems can never be perfectly integrated.) But as the months and years of the mission passed and new mandates were established, integration was no longer an evident impediment to progress against the gangs. Military and police resources (FPUs especially, with better equipment, training, and readiness to confront the violence of the mission) worked together in better harmony than during the inaugural mission mandate. With the arrival of a new force commander in early 2007, the potential for integrated military and police action against the gangs would be fully exploited.

Intelligence and Information Management

At inception, MINUSTAH had no intelligence and information management capacity. Typically, the United Nations has worked in countries where governments had used intelligence operations to torment civilian populations and prop up dictatorial regimes. Founded on the desire to maintain the reality and perception of neutrality, the United Nations had resisted engaging in intelligence collection. This impeded the capacity to conduct intelligence-led operations against threats from illicit power structures such as the gangs in Haiti.[36] In fact, though never sanctioned, intelligence collection has always been important to UN military operations and in supporting military-run joint operations centers. While the United Nations had officially resisted collecting intelligence, it was understood that military operations have always been supported by the collection of intelligence. Almost certainly, no mission executive ever refused to hear critical and perhaps lifesaving information merely because it may have been the fruit of military intelligence collection. UN Police collected operational-level intelligence unofficially, but with their limited resources and expertise, it was used discreetly and reported and exploited in very limited ways. During the 1990s, the UN Mission in Kosovo had successfully employed an intelligence capacity against criminals (and beyond military needs), but the success had not translated to doctrine, policy, or operational acceptance at the political level. Although mission intelligence was addressed in the Brahimi report of 2000, it may not have been until the 2003 attack against the United Nations in Baghdad that the United Nations awoke to the realities and need of protecting assets and conducting operations supported by real knowledge of the environment.

The same has generally not been the case for UN Police operations, however. This is due in part to the UN recruitment process, which, with the exception of FPUs, recruits individual police officers with generic skills, but not police units with collective capacities and organizational or operational systems.

The development of the JMAC in the MINUSTAH mission, beginning in late 2004, was a dramatic and progressive step that would have an almost immediate impact on

---

[36] "States have historically been opposed to granting the UN any intelligence-collection powers, fearing that such a role could lead to violations by the UN of national sovereignties, expressly protected by Article 2(7) of the Charter. UN officials adopted a similar stance during most of the UN's history, seeking to shield the UN from any activity that could be interpreted as espionage by any of its members." Melanie Ramjoué, "Improving UN Intelligence through Civil-Military Collaboration: Lessons from the Joint Mission Analysis Centers," *International Peacekeeping* 18, no.4 (Aug. 2011): 469.

the mission's ability to fulfill the mandate. The unit was one of the first instituted in a UN mission and was created in the integrated model, with military, police, security, political affairs, and other units contributing information, personnel, and expertise.

Little was made of the fact that development of the MINUSTAH model of JMAC was in the hands of a civilian, Michael Center, who was responsible for security information coordination. Interestingly, this may have been important to the quick acceptance and success of the project internally, since one criticism of the integrated model had been that integration tends to be dominated by military components. Where the military is usually the largest component and brings organizational and institutional capacities to the mission, it tends to be relied on to provide capacity where mission capacity might not otherwise exist. While this seems both logical and practical, integration can be difficult where civilians may not work well in a military-dominated environment or where the military may fail to consider important civilian aspects of projects. Center, after launching the JMAC at MINUSTAH, was named deputy chief, under Heiner Rosenthal, a senior civil affairs manager with extensive experience in Haiti. Together, they broke new ground in policy, procedure, and practice of intelligence-led UN operations, and perhaps even more importantly, they contributed to civilian-led integration of the mission's security effort. After a year of serving JMAC development, Rosenthal returned to his civil affairs role. Another senior civilian, Phil Menez, was named chief of JMAC. With Center as deputy, he oversaw development of intelligence target and evidence packages as the groundwork for successful mission operations against the gangs.

Though JMAC's work was initially strategic, there was confidence that it could produce an accurate and professionally analyzed intelligence product. At a time when the mission was under extraordinary pressure to make greater contributions to combating politically affiliated gangs, the ability to collect, analyze, and report intelligence was an important breakthrough.

Two years later, in an unprecedented step (but one made absolutely necessary by the level of violence and the gangs' widespread political impact), JMAC received the authority and budget to recruit, develop, manage, and pay human sources. This made for effective intelligence-led operations and tactical success against the gangs of Cité Soleil in 2007.

Dziedzic and Perito describe JMAC's crucial role: "JMAC provided sophisticated target packages detailing the obstacles that UN forces would encounter (e.g., tank traps and areas of fierce resistance), along with photos for the identification of gang leaders. To free kidnap victims or apprehend gang leaders, they needed actionable intelligence. Real-time tactical intelligence about the locations of gang leaders or concentrations of gang members allowed MINUSTAH to mount intelligence-led operations to arrest them."[37]

Rule of Law

Although "rule of law" refers to the full extent of law, regulation, systems, and administration of justice that serve a society, the focus here is strictly on the criminal justice system as it operated against illicit power in Haiti. In that sense and in dealing with the

---

[37] Dziedzic and Perito, "Haiti: Confronting the Gangs of Port-au-Prince," 8.

gangs specifically, the Haitian criminal justice system was entirely dysfunctional. The courts did not function routinely, and suspects languished in inhumane prison conditions, often without charge or even legal representation. The judiciary was widely considered corrupt. The codified law was archaic, lacking modern standards of procedure or evidence.

The HNP operated in a rule of law vacuum, without internal control, judicial coordination, or public oversight. The need to confront corruption and internal criminality and vet the force was ignored. HNP made searches and detained and incarcerated people without legal authorization. The GoH and HNP largely ignored calls to take corrective steps or even minimally demonstrate intent to be more accountable and transparent. Where the police were believed to have used illegal means or excessive force against suspects or to have caused the wounding or death of civilians, UN calls to investigate were ignored. In sum, HNP operated outside the bounds of legal control and without independent oversight. To the extent that MINUSTAH police and military were mandated to support the HNP operationally, and where calls for transparency and accountability by the GoH/HNP were simply ignored, MINUSTAH, too, operated outside the bounds of criminal justice.

In 2004-5, as MINUSTAH began its mission, there were no legally trained UN personnel to advise the mission leadership, support UN/GoH operations, or support development efforts under the heading "rule of law." Legal support to the mission, and especially to its police component, would have provided at least the appearance of legal accountability and the desire for transparency. Instead, the mission appeared tolerant of illegal or, at the very least, inappropriate GoH and HNP behavior and later even seemed supportive of the Préval "surrender or die" ultimatum. Although the presence of a legal team supporting the mission and, ultimately, the GoH would not likely have helped wrestle illicit structures to submission, it would surely have improved public perception, media reporting, and the UN organizational image.

A small team of judicial experts could have been deployed to provide legal advice in many situations. Selected prosecutions could have taken place in concert with Haitian legal experts, including judges, prosecutors, defense counsel, and court administrators. If the most heinous crimes could have been investigated and prosecuted, the strategy would have demonstrated the GoH's desire and ability to operate within a justice system, rather than a willingness to operate outside it. In a mission focused on security and rule of law as the foundations of future development, legal support was vital—if not exclusively to fight illicit power, then certainly for the mission's reputation and relationship with the GoH in that fight.

This need for a legal support team became even more relevant when, given the slow progress against the gangs, there were calls inside and outside the United Nations for the next UN mandate to include executive authority. In the absence of real operational capability and a functioning justice system, taking executive authority from a Haitian government that jealously guarded its sovereignty would have been entirely counterproductive to the mission's objectives and the United Nations' image. Only with a purposefully recruited and resourced rule of law program enjoying the full support of the Haitian Ministry of Justice could executive authority have been viable. In the end, the struggle against illicit power in Haiti was not supported with a rule of law capacity. The

UN Security Council resisted pressure to assume executive authority in Haiti, but it took no important steps to bridge the gaps where rule of law support was required. It became ever more difficult to demand accountability, transparency, and the primacy of human rights issues of the GoH and the HNP when the United Nations itself continued to work outside a justice system.[38]

UN Police Competence

Another issue of concern in dealing with the gangs and their criminal-political linkages was the UN Police's skills and competence. Police officers on mission are generally identified for their generalist experience and a set of basic skills (e.g., minimum number of years' policing experience, proficiency with their firearm where applicable, ability to drive a standard-transmission vehicle, proficiency in the mission language, and an unblemished human rights record). In MINUSTAH, there was no effort to recruit personnel qualified to lead special operations, gather criminal intelligence, or support complex investigations into sexual violence, corruption, kidnapping, homicide, or counternarcotics—all operational challenges in Haiti. Other unmet needs included operational support skills in strategic planning, communications, and forensics.

There were no immediate solutions to the shortfall in specialized policing expertise. Contributing countries would need to be canvassed, and individuals with specific skills identified, for future rotations of personnel.

A singular example illustrates the situation. During the summer of 2005, MINUSTAH struggled to help HNP investigate several dozen kidnappings simultaneously. Most were perpetrated by gangs or gang affiliates, but some of the crimes involved rogue HNP elements. Victims came from any family that could pay a ransom. Where victims/families had the means, cases might be discreetly negotiated by international intervention firms. MINUSTAH established a kidnap investigation and negotiation support unit that almost immediately had dozens of active reported cases (suggesting there were always unreported kidnappings, negotiated without official intervention). The mission's only trained investigator with any experience in extortion investigation was a junior officer from Canada with no negotiation experience and little experience in major case management.

Meanwhile, the UN Police initiated an interim solution. The idea of a standing availability of police experts—a group that could plan new missions and support ongoing missions—was being considered. While the project had no official status, a small group of experts representing a range of experience and skills was identified and deployed briefly to support the mission in areas badly needing specialized expertise. While the short duration deployment amounted to a limited contribution, it was a positive step where the UN Police Division recognized that every effort was important in gaining ground against illicit power structures.

---

[38] Where the HNP operated outside the justice system and HNP leadership and GoH officials refused to publicly investigate police wrongdoing (including allegations of murder), it was hard to dispel the notion that government officials were supportive, if not complicit. Where the United Nations did not insist on greater transparency and accountability (and take appropriate action, such as having a legal team in place), it, too, was suspect. The media speculated freely and criticized openly, and public confidence in the United Nations, GoH, and HNP suffered.

A specific example of a competency shortfall was the Formed Police Units. Trained and equipped to deal with public-order issues (historically Haiti's largest security problem) they were ineffective where urban combat, countersniper, and close-quarter battle were the tactical challenges. What the FPUs lacked were special weapons and tactics (SWAT) skills and equipment. The original MINUSTAH component of six FPUs proved insufficient to meet patrol commitments in the gang-held areas of Port-au-Prince and provide general security in Haiti's largest cities. But the FPUs were invaluable where UN military support to the police was inconsistent or absent because gang criminality was considered strictly a police matter.

By 2006, FPUs in mission increased to eight, but more importantly, the long-awaited capability of a 40-person SWAT unit was added. Finally, MINUSTAH had an experienced, trained tactical response to deal more effectively with gangs in the densely populated urban environment. Although protecting civilians remained a challenge in densely populated Port-au-Prince, this new expertise surely reduced the risk of collateral injury and death to innocent civilians. In the 2007 operation in Cité Soleil, FPUs performed a range of decisive roles that preserved the peace process. Dziedzic and Perito reported:

> In antigang operations involving MINUSTAH Police, FPUs performed a range of pivotal and often decisive roles, including crowd and riot control, hard entry, and high-risk arrest. The military contingent's initial foray into Cité Soleil on January 24 was placed in jeopardy when gang members organized a demonstration of unarmed civilians. An FPU with nonlethal riot control capabilities quickly dispersed the crowd . . . The forty-person SWAT team from Jordan that MINUSTAH incorporated within the FPU structure has been heavily employed in antigang operations. The vast majority of police-led operations involved the arrest of gang leaders or members as a prominent objective. FPUs were central to these highly successful operations, in particular the integrated use of MINUSTAH and PNH SWAT teams. Virtually all high-priority targets were brought to justice. For example, both the MINUSTAH and the HNP SWAT teams and an additional FPU platoon were assigned to Gonaives for two months to assist the FPU and other police assets assigned there in following up on the arrest of gang leader "Ti Will." All but one of the thirty most-wanted gang leaders from Gonaives were apprehended in that period.[39]

We can only speculate about MINUSTAH's success against the gangs had the SWAT capacity been available earlier in the mission. Clearly, having the correct tactical capacity—combined, of course, with actionable operational intelligence—helped reduce the gangs' impact as illicit power structures, at least in the short term.

The MINUSTAH Police / HNP Partnership

Where the mandate of a multilateral organization is to provide support to a host state, partnership is an important strategy for both program implementation and sustainability. Success is often measured by indicators of cooperation and collaboration, extent of buy-in, and evidence of leadership commitment and demonstrated will to pursue necessary reform. In confronting the illicit power of the gangs, the security partnership

---

[39] Dziedzic and Perito, "Haiti: Confronting the Gangs of Port-au-Prince," 11–12. "PNH" is the abbreviation for "Police Nationale d'Haiti" (Haitian National Police).

between MINUSTAH security forces and the GoH and HNP was perhaps the most important consideration. Framing this partnership was the reality of a broad mandate but narrow mission capacity and capabilities. Further weakening the partnership formula, MINUSTAH had no executive authority, HNP had very limited capacity, the GoH closely guarded its sovereignty, and both the GoH and HNP often acted without consulting their supposed partner.

Few insiders would dispute that MINUSTAH began without the firm ground of mutual trust and confidence that mark a flourishing partnership. Despite the usual steps of joint planning, coordinating committees, public displays of solidarity, and professional relationship building, evidence of cooperation and collaboration was absent. It was common for MINUSTAH-led joint police operations to be delayed or canceled when HNP failed to adhere to plans. There was animosity when HNP felt that MINUSTAH was not aggressive enough against the gangs, and conversely, HNP received widespread criticism for excessive force resulting in collateral civilian casualties. When HNP did not receive unquestioning UN fire support for its operations, it would accuse the United Nations of being nonsupportive. HNP showed clear intent to conduct independent operations, avoiding constraints that a UN presence might impose or imply. Continuous relationship building and maintenance were required to encourage mutual trust and confidence.

Continuity among senior managers on both sides of the arrangement was also an important consideration for partnership building and management. The large number of key positions and personalities leading the partnership and influencing security issues contributed to the complexity. On the UN side, these included the secretary-general's special representative, military force commander, and police commissioner. On the Haitian side, the key players were the interim prime minister, ministers of justice and public safety, and HNP director general. As a minimum expectation, the UN senior leaders were in place for over a year. Their Haitian executive counterparts, however, changed regularly, to the partnership's detriment. The HNP director general and the ministers of justice and public safety changed in the first year of the MINUSTAH-GoH partnership. The prime minister was the only Haiti executive in place for the entire first year of the mission, but as an interim appointment, he had limited mandate and authority.

No matter how professionally the officials involved may approach the task, the trust and confidence building required for partnership depend on a degree of familiarity unattainable if key individuals change regularly. Also, language issues and cultural and professional differences were barriers to overcome, underlining the reality that partnerships cannot simply be scripted in mandate letters. They are living relationships that must be planted and nurtured before they can bear fruit.

Further, in security matters, including crime control and the risk of penetration by criminal enterprise, an island's geography usually demands effective partnerships. The Dominican Republic (Haiti's immediate neighbor), the Association of Caribbean Commissioners of Police (the regional partner), and the International Association of Chiefs of Police (the transnational connection) were opportunities for partnership where introductions, engagements, and memberships could have led to earlier and more productive relationships. The GoH and HNP had neither the positive reputation nor the capacity to partner internationally. The United Nations could have done more in this respect by helping establish a measure of credibility, creating networking opportunities, and sourcing limited funding to bring HNP to the international table.[40]

Reintegration of Former Gang Members

The successful campaign against the gangs in early 2007 created the need to deal with gang members after their leadership had been either jailed or killed. Typically, in the wake of a peace agreement, the United Nations considers a demobilization, disarmament, and reintegration program for former combatants.

MINUSTAH planned for and attempted a DDR program for ex-FAd'H soldiers involved in the unrest that precipitated Aristide's departure, though the project got no support from the interim GoH and failed unceremoniously. At the same time, the GoH made amply clear that it had no taste for negotiating with criminal gangs or their leaders. While the international community held out some hope for DDR as a component of the solution for dealing with the gangs, the Haitian government showed little desire to see it implemented, let alone sustained or its promises fulfilled.

Having succeeded with the security initiatives of 2007 but lacking GoH support for DDR strategies, MINUSTAH was challenged to find alternative strategies to solidify the security gains. A useful innovation was reorientation of the DDR strategy to a plan for community violence reduction (CVR). The new strategy involved activities intended to provide a peace dividend in communities where MINUSTAH took assertive action to counter gang influence. It began with Security Council Resolution 1743, which urged MINUSTAH to reorient resources devoted to DDR, toward community violence reduction.[41] Targeting the environment where the gangs tend to thrive rather than just the gang members themselves, MINUSTAH developed a mechanism providing labor-intensive projects in violence-affected areas, addressing the gap between security operations and the arrival of humanitarian and development assistance and employment. Of the $3.47 million budgeted for CVR, $2.19 million was allocated to such labor-intensive stabilization projects.

The CVR strategy was a creative alternative to more established DDR ideas supporting the MINUSTAH mandate to create a stable environment conducive to future development. It provided tangible compensation to neighborhoods that had suffered terribly at the hands of the gangs, but it was also an interim solution intended to precede longer-term economic and job-creation strategies. Perhaps most importantly, CVR was a strategy that the GoH could support for its economic benefit but that (unlike DDR) did not directly reward members of criminal gangs by giving them access and opportunities unavailable to the average Haitian.[42]

---

[40] Naturally, where the issue of border security is concerned, the implications extend beyond narcotics and weapons trafficking, money laundering, and immigration to include control of the movement of goods and services, and collection of taxes and tariffs.

[41] UN Security Council Resolution 1743 (2007), Feb. 15, 1-4, www.refworld.org/docid/45fffb5f2.html.

[42] Dziedzic and Perito, "Haiti: Confronting the Gangs of Port-au-Prince," 11.

Summary, Recommendations, and Conclusions

Summary

The United Nations misread the threat posed by the gangs of Port-au-Prince and Cité Soleil, anticipating an environment of public disorder as the principle security threat. The gangs' potential to derail the peace process was unforeseen, and though gang violence against the government and HNP had been evident for some time, such intense violence against the international community and the United Nations in particular was a new reality in Haiti.

The gangs threatened the civilian population (with the most vulnerable suffering disproportionately), HNP, the GoH, and the international community. They attacked MINUSTAH openly, aggressively, and routinely. UN strategy and preparations failed to consider the range of illicit structures in the environment, their power to disrupt, and former president Aristide's political-criminal nexus with the gangs, which constituted an existential threat to the mission.

The gangs of Port-au-Prince were violently opposed to the UN presence and the peace process it represented, and their interests were irreconcilable. The persistent gang violence in Port-au-Prince constituted a real threat to the GoH, the UN mission, and the relationship between the two. The strategy ultimately adopted was to confront this threat with overwhelming force and cripple its ability to threaten the mission and the peace process, and to protect the civilian population. By 2007, most of the initiatives proposed two years before were largely in place, and MINUSTAH was increasing its operational effectiveness. Most significantly, the military component was finally prepared to engage the gangs operationally, not as uniquely military threats or criminal threats, but as security threats beyond the operational capacity of the police alone and pervasive enough to threaten the future of the mission itself.

With the military contingent coordinated with the police and, more importantly, prepared to act with the same muscularity displayed by the multilateral force that preceded MINUSTAH, gang leaders and members were arrested in impressive numbers. Gang activity was reduced, violent criminals were taken off the street, the local economy moved again, and the people of central Port-au-Prince were safer. Also, the UN mission got a boost, for it could claim success after being criticized so strongly. Demonstrable evidence of operational progress also bolstered HNP's commitment to the partnership with MINUSTAH.

After three murderous years on the streets of Port-au-Prince, the results against the gangs in early 2007 seemed impressive. Key gang leaders were arrested, and 800 gang members were reportedly apprehended and incarcerated.[43] This is impressive in terms of operational success against the gangs, but the number alone is cause for reflection. By any accounting, there were certainly not 800 gangsters attacking UN security forces and killing HNP officers in 2005. So how many of these 800 were merely unemployed, uneducated, poverty-stricken young men with no prospects, who were "going along to get along" with the gangs on the brutal streets of Cité Soleil? Moreover, how many of the gang members arrested in 2007 had in fact been deported from the United States

---

[43] These were not the same gang leaders in place in 2005, however. By spring 2005, those leaders had all been killed or apprehended and replaced by their lieutenants-in-waiting.

or Canada? How many of those arrested and jailed were ever actually charged with a crime? And how many were ever convicted?

Since criminal justice in the country was dysfunctional and the capacities of the police (HNP and UN) alone were insufficient to dismantle the linkages between political elites and the gangs, there seemed no other solution to the violence but to neutralize the gangs in a military sense: removing them as a threat to the security of the country and the mission.

But this strategy's success cannot be judged without considering that the GoH refused to consider DDR for gangs and that President Préval declared his "surrender or die" ultimatum in response to his failed political intervention. Was this capture-or-kill response the only strategy available? Even accepting that only a suitably aggressive response could stop the violence, would other strategies have been potentially useful to prevent a reemergence of the gangs, or renewed violence in the future. If we consider that perhaps only a small percentage of the gangs were hard-core and represented the real power through their linkages to the elites, perhaps, over time, the rank and file might have been coerced away with alternative life choices. With a view to the longer term, perhaps new opportunities—new lifestyle alternatives, in fact—might have been considered to dissuade the next generation of potential gangsters.

Two years after the gangs were put down, Haitians suffered extraordinarily in the earthquake of January 2010. An estimated 200,000 were killed and two million displaced, Port-au-Prince was largely destroyed, and the limited progress that had been made in improving the lives of Haitian citizens was gone.[44] In a related extraordinary situation, the prison population walked away from the National Penitentiary when guards abandoned their posts. But UN reports from the months immediately following the earthquake do not suggest increased gang activity resulting from these prisoners' return to the streets. Indeed, David Becker, then coordinator of the Haiti Stabilization Initiative, recounts that after the 2010 earthquake, the gang leader known as "Blade" (for his habit of torturing enemies with razor blade cuts) and a lieutenant returned to Cité Soleil after escaping from the federal prison, intent on reestablishing their control. As they confronted local citizens working in a community mobilization group, the two were attacked from behind by workers wielding shovels. Other onlookers soon joined in, and the two gangsters were beaten to death and their bodies dumped in front of the local police station.[45]

After the successes of early 2007, including the CVR economic program, we must consider that illicit power structures cannot be conquered merely by increasing operational capacity of security forces and getting gang members off the streets. Certainly, the operational focus on aid and assistance, and away from gangs and criminality, after the 2010 earthquake crated an environment favorable to gang reemergence. But reemergence of the gangs did not follow that event automatically.[46] By the fall of 2012, though

---

[44] This death toll was an early estimate, repeated often enough that it was soon reported as fact. A precise figure will never be known, though it would no doubt be staggering.

[45] David Becker, National Defense University (coordinator, Haiti Stabilization Initiative, a US DoD-funded project), author telephone interview, Mar. 12, 2015.

[46] Robert Perito drew a picture of uncertainty and renewed insecurity: donors avoiding assistance to the government because of the perception of general dysfunction, 147 civilians and 5 police killed in the

MINUSTAH reported the gangs of Port-au-Prince to be active, dangerous, and again the source of real insecurity in the country. They engaged in narcotics and small-arms trafficking, racketeering, and turf wars, employing murder, kidnapping, and robbery as their methods.[47] Women, children, and the poverty-stricken population at large continued to be at greatest risk.[48] Adding to this gang-generated insecurity was public unrest about politics and the failed economy, and a renewed debate over the future of a Haitian military. Apparently, the security situation had changed less than many had thought, largely because the strategy to confront the gangs of Cité Soleil with overwhelming force addressed only one side of Haiti's illicit-power-structure equation. Perhaps not enough attention went to the root causes of the illicit structures and the factors that favored their existence.

No effort was made to understand how the gangs were controlled for political purposes or exploited by self-interested political or economic elites.[49] Although the United Nations was initially unprepared to deal with the gangs as a violent threat to the mission and the civilian population it was mandated to protect, it has remained both unwilling and unprepared to deal with the external forces that influenced the gangs. It had no jurisdiction, no independent executive authority, and no legal capacity to pursue those who conspired to use the gangs. The Haitian justice system was clearly dysfunctional, but with an interim government protective of its sovereignty, unsure of its constitutional footing, and resistant to external scrutiny, a justice reform partnership may have been impossible. Judicial pursuit of Aristide at the early stages of the UN intervention would certainly have been trumpeted as a ploy designed only to keep him from power, and would likely have inflamed the political situation. On the other hand, investigating the economic elites that exploited the gangs was a low priority where those same elites' cooperation was essential to national economic recovery. Still, other priorities paled beside the need to get a firm grip on the security situation before next steps in any direction.

Six years after the operational successes against the gangs of Cité Soleil, the only real change is that the United Nations, better organized, equipped, and committed to confronting the gangs, is no longer their primary target. The operational success of 2007 must now be qualified because security—fundamental to a peaceful existence and a favorable development environment—is unsustainable unless Haiti's institutions develop the capacity to control gang activity and hold elites demonstrably accountable when they attempt to exploit gangs as instruments of political power. Deficiencies in the rule

---

first months of 2012, reports of rising crime in Port-au-Prince, demonstrations by Aristide supporters, and reports of armed paramilitary training in defiance of the government. Robert Perito, "On Haiti: Current Situation," U.S. Institute of Peace newsletter, Mar. 2012, www.usip.org/sites/default/files/USIP%20on%20 Haiti_March%202012.pdf.

[47] Report of the Secretary-General, 22 Feb. 2010, para. 3, 21-22, www.un.org/en/ga/search/view_doc.asp?symbol=S/2010/200. Many of those incarcerated were in prison for gang activities. When they walked away in the aftermath of the earthquake, escalating gang crime was anticipated, but it never really materialized. See also Report of the Secretary-General, 1 Sep. 2010, para 6, www.un.org/en/ga/search/view_doc.asp?symbol=S/2010/446. Nine months after the earthquake, the United Nations still reported crime rates as essentially unchanged from the months preceding the earthquake and prison escape.

[48] Ibid., paras. 24-26.

[49] Kolbe, "Revisiting Haiti's Gangs," 29. Kolbe makes the case for purposeful empirical research before we can really understand the relationships (historical and personal) that support and promote gangs, or know how to change the environment that feeds illicit power.

of law remain a critical vulnerability, systems of accountability need to be set in place, and systemic corruption must be tackled. Of course, free and fair elections will be a crucial indication that the country is on a better track.

Recommendations

In any conflict situation, establishing and sustaining a safe, secure environment is central to providing aid, humanitarian assistance, and a solid foundation for development. Where MINUSTAH's planning and preparations failed to recognize and then deal with the gangs as illicit power structures, the mission struggled.

The successful operations against the gangs of Port-au-Prince and Cité Soleil resulted from changes in intelligence collection and analysis, increased numbers and capacity of FPUs (particularly the SWAT team), more effective integration of operations (police/FPU and also JMAC and JOC), an improved partnership with HNP, and, especially, a committed military component. It is hard to imagine operational success if any one of these had been ineffective. While the Brahimi report addressed most of these features of mission structure, organization, and operations to some degree, they certainly bear mentioning again.

Intelligence-led mission planning. Threats in modern conflict continue to evolve in the dynamic environments where they emerge. These threats include illicit power structures, which appear in many forms and are motivated and operate differently. The first key to addressing such threats is to recognize their existence. One might say, "This is not your father's peacekeeping mission." And certainly in the case of Haiti, while we focused on the illicit power of the Cité Soleil gangs, we met with illicit structures in many corners. Mission planners must be alert to the dynamics of conflict environments, recognizing illicit power where it exists and understanding its impact. This is particularly sensitive where political or economic power brokers seek to exploit illicit power for their own purposes. Understanding the complex security environment and the sources and impact of illicit power demands current, accurate intelligence. Just as the 2007 success against the gangs required strategic intelligence to plan and tactical intelligence to achieve, mission planning requires current intelligence from all legitimate and available sources.

Skill sets appropriate to the tasks ahead. Once a mission's security environment is understood, recommendations can be made for recruiting, staffing, and operational planning. This suggests a new emphasis on finding the skill sets required by the job ahead. When we think about illicit power and the specific circumstances where it may manifest, as we found in Haiti, specific skill sets will be required if the mission hopes to respond effectively. Where missions are to include a patchwork of contributing countries, the selection of operational units, rather than unassociated individuals, from contributing countries must be a consideration. In dealing with the illicit power wielded by Haitian gangs, the utility of having a kidnapping, homicide, and sexual assault investigation capacity and a tactical SWAT capability from the start needs little elaboration. Where international police are not recruited as operational units or even for their special skills or expertise, it is unrealistic to suppose that the mission will ever have the complement of experienced personnel required to address critical security issues.[50]

Law enforcement accompanied by the full rule of law spectrum. Gang activity in Cité Soleil and Port-au-Prince caused considerable desperation for the GoH, the United Nations, and, of course, the victimized public. The result was tolerance for methods and actions generally unacceptable for security forces, including search without authority, arrest without warrant, and incarceration without showing cause. The security operations of 2007, in an environment of extraordinary insecurity, would not have been countenanced in an environment where the rule of law was in place. It was accepted that the law enforcement ends justified the means.[51] The United Nations and its HNP partner could not continue to operate without external scrutiny and oversight. Given its mandate to support and develop the HNP, the United Nations could not demand accountability and transparency from the GoH and HNP while itself continuing to operate outside the usual boundaries stipulated by the rule of law.

The United Nations should have deployed a justice system task force in support of operations, demonstrating that law enforcement is only one component of the broader and more essential objective: development of the rule of law. Effective investigation and public prosecution of high-profile cases would have served to demonstrate the supremacy of the rule of law. This action might even have included the judicial pursuit of politically motivated conspirators and individuals criminally exploiting the illicit power of gangs. At the same time, the GoH's passage of emergency-measures laws would have provided political and judicial authorization recognizing an extraordinary security situation and the need for special powers. Since the UN mission had no executive authority, the GoH would have been seen to lead, which would have emphasized the link between security and justice as important to peace, security, and future development.

In addition to the principles that must be respected if a UN intervention is to advance rule of law, there is also a practical rationale for taking a holistic approach. It has long been axiomatic that justice system development is not possible without parallel effort in all associated sectors (police, judiciary, corrections, and legal code). Operational success in establishing security, whether generally or against illicit power structures specifically, will not likely be sustainable when any one sector of the justice system is ineffective. At the same time, rule of law approaches must consider the legal questions that affect the population most directly. Surely, issues of land reform, tax reform, and systemic corruption must get early attention in the rule of law spectrum of priorities. The consequences

---

[50] As of this is writing, the Police Division of UN DPKO is establishing a Strategic Guidance Framework, leading police-contributing countries in the overhaul of UN mission policing under the headings of "command and control," "operations," "administration," and "capacity building." The purpose will be to raise the level of professional policing in missions. This initiative should begin to address shortfalls documented during MINUSTAH and other missions.

[51] In 2005, the author made a presentation on security issues and plans to the Port-au-Prince Chamber of Commerce. The economic elite of the country was well represented in a large audience that was clearly dissatisfied with "rule of law" approaches to combat the gangs. At the conclusion of the presentation, the audience gathered near the podium for follow-up questions and comments. A middle-aged woman, clearly of means, said in a calm but venomous tone, loudly enough for all to hear, "Why do not you just kill them all?" There was no shock, no surprise, and no reaction from anyone in the room but me.

of not taking a holistic approach to fostering the rule of law are well understood, and the international community's experience in Haiti since 1993 demonstrates this clearly.

The experience of justice system development (police, justice, and corrections sectors together as a system) in Haiti is a particular example. The international community's investments in police development are known to far outweigh the efforts and contributions elsewhere in the system. While there is still a long road to travel with police development, there has been virtually no progress in modernizing and reforming the justice sector and basically no investment at all in corrections. Where the Haitian police now operate in a justice vacuum, and where corrections operates a century behind the times, paralysis of justice system development may well be occurring, so that no sustainable progress in the broader legal system is possible until the sectors that lag behind can catch up.

Planning for long-term, sustainable success. Success against gangs and other illicit power structures in Haiti must be gauged with the measuring stick of sustainability. This will require a sturdy foundation: attention to fundamental rights and rule of law, a justice sector with transparent systems and accountability of process, fair and free elections, and an economy that provides life options and access to opportunities for the impoverished. The reemergence of gangs in Haiti, despite the tactical successes of 2007 and the UN operational changes that made success possible, signals that operational capacity alone—ability to "defeat" the gangs—is not enough to ensure sustainable success against illicit power.

MINUSTAH demonstrated a seemingly useful strategy in following up its operational success against the gangs with economic initiatives in the form of CVR projects. But since they had no sustainable impact (according to UN reports, gang violence has returned), do we now judge the CVR projects ineffective over time? Is it possible the CVR projects did not go far enough or, at least, were not tightly linked to mid-term and long-term community projects? It seems logical that any operational plan seeking to suppress illicit power over the long term must consider (a) the conditions that fostered the linkage between illicit structures and power (political and economic) and (b) how to alter those conditions going forward. Where participation in gang activity is related to security, economic access, and life opportunities, addressing those issues over the long term also seems vital to any strategy for reducing the power and impact of illicit structures.

The international community leading against illicit power structures. The purpose here has been to examine the Cité Soleil gangs' illicit power to threaten peace, and to identify the most appropriate strategy and the capacities required to combat them. The Haiti experience exposed weaknesses in the international response, including lack of capacity to conduct intelligence-led operations, shortfalls in skilled, experienced personnel, and inability to deliver integrated operations. At another level, the political players and economic elites who exploited the illicit power remained unchecked.

All this leads to a general lesson or, at least, a question for future consideration: what is the international community's responsibility in dealing with illicit power structures? Although MINUSTAH was initially unprepared for the existential challenge that gangs posed to the mission, over time it addressed the gaps that needed to be bridged to meet the threat. In some respects, MINUSTAH recognized those gaps as HNP development needs, but it did not recognize them as immediate security threats that the mission itself needed to confront. Intelligence capacity, border integrity, major crime investigation, gender violence investigation, and protection of civilians fell short. Where such shortfalls affect not only public safety but also the mission's effectiveness, is it not incumbent on the international community to take the lead, even if it is leading from behind, supporting weak and overwhelmed local institutions?

Conclusion

Where the international community proposes to support a state emerging from conflict, it must be prepared with expertise and capabilities to provide the necessary support. In Haiti, the UN mission was ill-prepared and caught unawares. Illicit power structures threatened peace and security so badly that the mission's pace, direction, and, indeed, credibility were affected. Although the mission addressed many operational and security gaps over time, early attention would have saved civilian lives and, moreover, created an environment where humanitarian assistance and sustainable development might be the actual focal points.

The potential for sustainable security was surely impaired by lack of progress in the wider context of rule of law. Also, following up the 2007 operational success, more could have been done to incorporate after-action strategies that considered economic conditions, access and opportunity, and sustainable development as factors influencing the very existence of gangs and other illicit power structures. Most critically lacking, however, was a strategy to fully understand and transform how hidden political and economic elites externally influence Haiti's gangs and exploit their power, contributing to insecurity and blocking development. Where such elites are held out to represent legitimate authority even as they are allowed to manipulate through a criminal underworld, the damage to public confidence, transparency, and the rule of law is devastating.[52]

In the end, faced with extraordinary violence and mayhem during 2004-6, the United Nations demonstrated that it could marshal the necessary resources and take the measures required to bring the rampant power of illicit structures under control and establish security where there was none.[53] What it has so far failed to do is to prevent those same illicit structures from reemerging as impediments to sustainable progress in security and development.

---

[52] Kolbe and Muggah, "Kolbe and Muggah: Haiti's Gangs Could Be a Force for Good."

[53] "The United Nations must be capable of mounting assertive operations to defend and enforce its mandates. The second lesson is that the United Nations has the means to succeed in these efforts, given the proper conditions, if it can summon the necessary will." Dziedzic and Perito, "Haiti: Confronting the Gangs of Port-au-Prince," 1.

ap No. 3775 Rev. 9 UNITED NATIONS
eptember 2014

Department of Field Support
Cartographic Section

4. Liberia: Durable Illicit Power Structures

William Reno

After a 15-year civil war that shocked the world with the scope and scale of its brutality, the UN peacekeeping and peacebuilding work in Liberia is a qualified success. The Accra Comprehensive Peace Agreement was signed on August 18, 2003.[1] National elections were held in 2005 and 2011. The winner of both presidential polls, Ellen Johnson-Sirleaf, also won the 2011 Nobel Peace Prize, awarded to her along with two other women in recognition of their roles in promoting peace, democracy, and gender equality. Peacebuilding comes at considerable expense, however, and Liberia retains drivers of instability behind a facade of stability.

The UN Mission in Liberia (UNMIL) has monitored a peace agreement since September 2003. UNMIL personnel peaked at about 15,000 in 2006, and foreign assistance to build government security and administrative institutions typically exceeded the country's annual domestic product throughout the first postconflict decade. The foreign reconstruction of Liberian state authority has required a complete overhaul of the army, police, and intelligence services. Despite this extraordinary effort, illicit networks associated with wartime rebel groups continue to influence the country's political and economic scenes. External intervention was essential to ending open conflict in Liberia. The impact of intervention is great, but it has not fundamentally altered the illicit social and economic relationships and political patronage networks that are the focus of this analysis.

The gradual withdrawal of foreign assistance raises concerns about the roles that illicit power may play in the future. One concern is that resurgent criminal networks will bring a return to a violent status quo ante. The more likely prospect is that illicit power will continue to adapt to the international stabilization process and reappear in new forms. Rather than a return to war, this interaction is likely to lead to a more authoritarian and corrupt Liberia, which diverges from the democratic model that stabilization was designed to create. International actors and Liberian reformers will find repeatedly that they must deal with officials and others whose real interests lie in these illicit power structures, some of which are indeed essential for providing local order. And in many instances, interveners will be forced to choose between this order and good governance.

The Conflict and Its Causes

The leaders of all the main armed groups in Liberia's 1989-2003 conflict were players in a violent prewar political system of personal rule that predated, but intensified under, Samuel K. Doe, the master sergeant who led a successful coup on April 12, 1980. Doe demanded the personal loyalty of his associates and used a system of material rewards and threats of violence to remain in power. That system dictated how Liberia was run before the war, and it continues to set the pattern for postwar politics. As Doe lost his

---

[1] UN Security Council (UNSC), *Peace Agreement between the Government of Liberia, the Liberians United for Reconciliation and Democracy, the Movement for Democracy in Liberia, and the Political Parties* (Accra, Ghana: United Nations, 2003), www.ucdp.uu.se/gpdatabase/peace/Lib%2020030818.pdf.

grip on power, his more enterprising subordinates organized their own supporters to seize power. In this sense, the conflict was a civil war, but not one that featured rival ideologies or political programs. The fighting was not tribal or sectarian at first, though it became so as the politicians-turned-rebel-leaders targeted their rivals' home communities for attack. This conflict was brutal, notable for the use of child soldiers, plunder of resources, and devastation of noncombatant communities. By the war's end in 2003, estimates of conflict-related deaths ranged from 150,000 to 250,000 out of a prewar population of about three million people. Ultimately, personal loyalties and patronage networks, which have always been important in the underlying logic of the country's politics, were the drivers of this conflict.

The Old Order

By the 1960s, President William Tubman (1944-71) reviewed any official expenditure of more than $250 to evaluate whether it reinforced the president's personal influence in the political system.[2] In the private sector, too, a tight-knit political establishment dominated commerce, including illicit trade and trafficking directly connected to a system of political domination that participants leveraged for personal profit. Historically, high-level officials used state security forces and armed supporters to attack critics and rivals and protect personal business operations. They used their positions in government to manipulate laws and regulations in ways that shielded their activities. Political supporters benefited from this selective targeting of investigations and prosecutions, while opponents faced bureaucratic obstacles and prosecution.

These networks of high-level corruption have included international outlaws. In one wartime example from the early 2000s, UN reports cited Russian citizen Viktor Bout's air transport companies, logging firms of Ukrainian-Israeli Leonid Minin, and Dutch businessman Gus van Kouwenhoven for transferring weapons and ammunition to President Charles Taylor in defiance of a UN arms embargo.[3] Arms suppliers needed to be paid for their wares, and to do this Taylor linked them into existing illicit commerce in diamonds, timber, and other resources that he personally controlled. This nexus of illicit commerce and wartime politics played a large part in the decisions of Security Council member states, particularly the United States, to support imposition of the arms embargo and, when Taylor continued to violate it, to support his prosecution for war crimes before the Special Court for Sierra Leone.

This situation illuminates a core element in Taylor's political logic. He needed the illicit source of weapons and commercial operations to survive in power and to organize fighters. For example, some especially trusted associates, including foreign business partners, were allowed to maintain militias within the country, for their personal use and to carry out tasks for their politician protectors. This illicit channel linking commerce and military supplies was essential for Taylor in a country that lacked a viable economy or any other means of exercising authority and financing his battles against rivals.

---

[2] Gus Liebenow, *Liberia: The Quest for Democracy* (Bloomington, IN: Indiana Univ. Press, 1987), 117.

[3] Paul Holtom, "Case Study: Liberia, 1992-2006," in *United Nations Arms Embargoes: Their Impact on Arms Flows and Target Behaviour* (Stockholm: Stockholm International Peace Research Institute, 2007), 12-13.

These criminal syndicates, both before and during the war, were integral to the exercise of political authority in Liberia. Access to any substantial source of income, licit or otherwise, depended on accommodations with members of the president's clique and important government officials. This system of control, with the threat of violence by state security forces and regime favorites never far away, gave Liberia's political leaders the capacity to co-opt critics and challengers, hound opponents, and discipline supporters. Even black-market operators and others who seemed beyond the grasp of tax collectors and regulators had to seek protection from the same politicians who were the real middlemen in a wide range of both legal and illicit activities. For example, even the lowest-level black-market currency trader or vender of smuggled goods needed a protector and, therefore, had to come to an accommodation with this power structure. For the top levels of this trade were dominated by political insiders whom Taylor (and Doe before him) allowed to exploit these business opportunities in return for political and financial support. That is, the country's political leaders and their business partners were also the top black-market operators and the biggest smugglers.

By the 1980s, it was hard to distinguish state officials from criminal syndicates. In an illustration of this blurred distinction, a task force set up in 1985 to recover arrears of $150 million owed to government corporations found that most of the debtors were government officials, including two heads of then-president Samuel Doe's security services.[4] This investigation occurred because Liberia's foreign creditors insisted on it. The reality was that the entire system of governance rested on this dense system of misappropriated funds, insider scams, and illicit commercial activity under the protection of the country's political leaders, up to and including the president.[5] A Liberian activist summed up this prewar system of governance as "years of rape and plunder by armed marauders whose ideology is to search for cash and whose ambition is to retain power to accumulate and pocket wealth."[6]

The Rise of Charles Taylor

Charles Taylor, the future head of the National Patriotic Front of Liberia (NPFL), was one of the officials involved in illicit activities in the 1980s. Samuel Doe gave him the job of running the General Services Agency, the government's procurement office. He saw Taylor as a rising political actor, in part because of Taylor's activities in the 1970s as a student organizer among the Liberian diaspora in the United States. Also, Taylor had a close personal relationship with Thomas Quiwonkpa, one of Doe's confederates in the coup that brought Doe to power.

In 1983, Taylor fled to the United States to avoid prosecution in Liberia. His crime was not that he had engaged in corruption, but that he was an ally of the former head of the army, who was plotting to overthrow President Doe. Doe then requested that U.S. officials extradite Taylor, claiming that Taylor had embezzled $900,000 from the

---

[4] *Africa Confidential*, "Liberia: Towards Collapse," April 10, 1985.

[5] USAID, "Final Report of the Liberia Economic Stabilization Support Project," USAID, Dec. 1989, http://pdf.usaid.gov/pdf_docs/PDABP324.pdf.

[6] Jimmy Kandeh, "What Does the 'Militariat' Do when It Rules? Military Regimes, the Gambia, Sierra Leone, and Liberia," *Review of African Political Economy* 23, no. 69 (Sept. 1996): 396.

procurement office.[7] In any event, in 1985 Taylor escaped from a Massachusetts federal prison, where he was being held pending a decision on the Liberian government's request for his extradition to Liberia. About a year later, after moving around West Africa in search of supporters to help him overthrow Doe, Taylor traveled to Libya with help from Burkina Faso's ambassador to Ghana, Memuna Qattara, and the ambassador's cousin, Captain Blaise Compaoré.[8]

These contacts were especially useful after Compaoré's October 15, 1987, coup, which put Taylor in a better position with his Libyan patrons and positioned Burkina Faso to play an important role as intermediary in future arms transfers to Taylor's NPFL.[9] Taylor's mastery of this regional network of ambitious political operators who were organizing to overthrow their governments enabled him to prevail among the numerous factions of Liberian dissidents who were plotting against Doe's government. And it established him as the head of the NPFL. Then Taylor and about 160 men—Liberians and others he had met either in Libya or during his stays in West Africa—entered Liberia from Côte d'Ivoire on Christmas Eve, 1989.

Ivorian President Félix Houphouët-Boigny's stake in allowing Taylor to organize in his country ultimately involved a family matter. The 1980 coup leaders murdered, among others, Liberian President A. B. Tolbert's son, who was married to Houphouët-Boigny's stepdaughter, Daisy Delafosse. She then married Captain Compaoré—another reason why it was so easy for Taylor to find shelter in Côte d'Ivoire and to launch his invasion from that country. Thus began his bid to overthrow President Doe and install himself as the next president—an event that sparked Liberia's 13-year civil war.

A key point in understanding the root causes of Liberia's war is that the "rebels," whether in Taylor's NPFL or in the half-dozen other groups that opposed the NPFL at various points in the war, were not ideologically driven. They merely reflected the efforts of enterprising members of Liberia's political establishment to control this intersection of political and commercial networks. This competitive ambition—not rebellious youth, masses of unemployed people, or marginalized communities—drove the conflict. This underlying continuity from the "peace" of a violent and corrupt government to the tumult of civil war pointed up the major roles of smuggling rackets and partnerships with shady foreign "businessmen" in the political strategies of prewar and wartime politicians.

Taylor associate Guus Kouwenhoven managed timber companies that helped President Doe finance arms imports, and then used this enterprise to do the same for Taylor.[10] Many Liberians who were business partners with officials under Doe's government also found places in Taylor's network. This transferable nature of intermingled criminal trafficking organizations and political authority continues to define Liberian politics.

---

[7] "In the Matter of the Extradition, Charles M. Taylor" (D. Mass., Magistrates Docket, 1986) [court docs in author's possession].

[8] Alberta Davies, *Raw Edge of Purgatory: I Survived the Liberian Pogrom* (Bloomington, IN: Xlibris, 2011), 26.

[9] Mark Huband, The Liberian Civil War (London: Frank Cass, 1998), 51-55.

[10] UNSC, *Report of the Panel of Experts Appointed Pursuant to United Nations Security Council Resolution 1306 (2000), paragraph 19, in Relation to Sierra Leone*, Dec. 2000.

Other Actors in the War

Other ambitious political actors appeared on the wartime scene from 1990 onward, reinforcing the dominance of the prewar political model in defining the wartime environment. These included Prince Yormie Johnson, a U.S.-trained soldier who served as aide-de-camp to General Thomas Quiwonkpa, the commander of Liberia's army until both fled into exile in 1983. Johnson started out as Taylor's partner but split in early 1990 to form his own Independent National Patriotic Front of Liberia (INPFL). He was notable early in the war for his hand in the torture and murder of Doe (which happened when Doe was in Johnson's custody)—an event that was captured on videotape and can easily be located on the Web. Johnson is a durable figure, serving as a senator in postwar Liberia and taking third place in the 2011 presidential election, with 11.6 percent of the vote.[11]

President Doe's information minister, Alhaji G. V. Kromah, emerged in September 1991 as leader of the United Liberation Movement for Democracy, which brought together former army officers and government officials. He received 2.8 percent of the vote in his 2005 run for Liberia's presidency and lost the 2014 Lofa County Senate race. Kromah's attempt to continue in postwar electoral politics ultimately failed, but his transformation from Doe's time to Taylor's and then into postwar politics reflects how prominent figures in Liberia refashion old connections and resources to remain relevant in the country's political scene.

George Boley, President Doe's former presidential secretary and assistant secretary of education, formed the Liberia Peace Council—another armed faction despite its misleading name. He was deported from the United States in 2012 under the provisions of the Child Soldiers Accountability Act of 2008, for his role in recruiting and using child soldiers in Liberia's war.[12] He then ran for the Grand Gedeh County Senate seat in 2014, receiving 12.5 percent of the vote. Although he lost the election, Boley's performance shows how even the widespread knowledge of his past misdeeds was not sufficient to deter the one-eighth of the voters in Grand Gedeh who supported him in the election. Given the underlying logic of Liberian politics, it is reasonable to suppose that Boley's greater capacity to mobilize wartime connections to people and resources accounts for his better electoral showing than Kromah's.

Political Patronage Networks and Conflict: The Integral Connection

Wherever an armed group controlled territory and people, its method of governance reproduced the tight integration of political authority and illicit power rooted in the local and regional economy. Taylor created a unit within the NPFL to invite timber firms based in neighboring Côte d'Ivoire, many of which had done business in Liberia under Doe's rule, to operate in NPFL-controlled territory.[13] These firms were required to

---

[11] National Elections Commission, "2011 Presidential and Legislative Elections," NEC, Nov. 8, 2011, www.necliberia.org/results2011/.

[12] US Immigration and Customs Enforcement, "Liberian Human Rights Violator Removed from US," Mar. 29, 2012, www.ice.gov/news/releases/liberian-human-rights-violator-removed-us.

[13] Letter from member, NPFL's Special Economic Committee, to the NPFL commander of River Cess,

make ad hoc "contributions" and provide logistical support to NPFL members for their personal use, in a practice that mimicked prewar relations between politicians and loggers.[14] Initially, violent, predatory criminal operators responded to the invitations to do business in this risky environment. Taylor made deals with them to come to the port and load logs that other loggers had left there when they fled or that Taylor's local business partners had cut, but then some of these foreigners would simply disappear without paying, or abscond after threatening local NPFL agents.

"Everybody he sold a cargo of logs to, they'd take the logs and that was that; they'd never come back," said an American businessman referring to some of the first foreign operators who responded to Taylor's invitations to do business. The high risks of doing business with armed rebels tended to attract criminals who were accustomed to risky operations. "And the reason for that was, why should they come back? I mean, they'd make twenty, thirty million or whatever they were making off a cargo of mahogany logs."[15] This violent commercial environment troubled some of the legitimate operators from Doe's time, who returned to Liberia only to find that they now had to fear for their personal safety amid the mix of unpredictable rebel fighters and rogue shippers accustomed to using violence whenever it gave them an edge in business transactions.[16]

This development is important because it showed that even well-armed experts who were well connected to regional illicit power structures were vulnerable to getting ripped off by international seafaring criminals; thus, they had to establish some sort of system of stable relations with their commercial partners in these illicit operations. NPFL officials dealt with this problem by implementing procedures to identify ships that arrived at the port and compelling those allowed to dock to prepay for port services and rebel-imposed fees and taxes.[17]

This close integration of armed groups, state administration, elite politics, and the illicit economy continued throughout the war. This type of politics was familiar to Liberians while continually presenting a challenge to international actors seeking to end the conflict. Taylor spoke directly about this integral relationship, warning Liberians not to interfere with his "pepperbush"—a colloquial term for a personal possession that is dear to the owner. At least once after becoming president, Taylor used this expression to refer to the Oriental Timber Company, a firm that UN investigators cited as violating international sanctions against trade in "conflict timber."[18] Such cozy relationships between politicians and criminal elements in the economy remains a staple of the Liberian postwar political scene.

Grand Kru, Sinoe, and Maryland Counties, written in San Pedro (Côte d'Ivoire), Nov. 6, 1990 [in author's possession]. (Many of the smaller logging operators took refuge in San Pedro, Côte d'Ivoire, close to the Liberian border, to avoid the violence.)

[14] For example, Forestry Development Authority, "Request for Payment," Buchanan, Liberia, July 18, 1991 [in author's possession].

[15] Interview with U.S. logger, Nov. 2001, Chicago.

[16] Author interview with member of the Liberian Timber Association, Aug. 1994, San Pedro, Côte d'Ivoire.

[17] "Payment for Services at the Ports," Associated Development Company, Buchanan, Liberia, Nov. 23, 1991 [in the author's possession].

[18] Global Witness, "Taylor-made: The Pivotal Role of Liberia's Forests and Flag of Convenience in Regional Conflict," report, Global Witness and ITF, Sept. 2001, 25, www.globalwitness.org/sites/default/files/pdfs/taylormade2.pdf; Republic of Liberia, *Truth and Reconciliation Commission, Volume 3, Title III: Economic Crimes and the Conflict, Exploitation and Abuse* (Monrovia: Truth Commission, 2008), 24-27.

The Peace Settlement

The 2003 Accra Comprehensive Peace Agreement created a National Transitional Government of Liberia (NTGL), which included elements of the major armed factions, leaders of civilian political parties, and leaders of major civic groups.[19] Charles Taylor was notably absent from this arrangement because the two main rebel groups opposed to him, Liberians United for Reconstruction and Democracy (LURD) and the Movement for Democracy in Liberia (MODEL), declared that an agreement would be impossible if he continued in office. The parties included in the agreement received cabinet positions in the NTGL. This was an interim arrangement since the agreement called for elections in 2005 to establish a post-settlement democratic government.

Two months after the agreement was signed, Gyude Bryant, a Monrovia businessman with no significant ties to any of the warring parties, took office as chair of the transition government. Removing Taylor from office also addressed U.S. government concerns that Taylor's connections to criminal networks implicated him in support for international terrorist groups. A 2001 *Washington Post* article reported that men named by the FBI as al-Qaeda operatives purchased diamonds from Sierra Leone's Revolutionary United Front (RUF), an armed group that by then was widely thought to have close ties to Charles Taylor.[20] Even though Taylor and people central to his regionally disruptive relationship with rebels in neighboring countries were excluded from the postwar settlement, the incorporation of the political networks of wartime leaders into the interim government ensured that the illicit power structures would continue to play significant roles in postwar Liberian governance.

The Accra Agreement cleared the way to establish the UNMIL peacekeeping force, with a peak strength of 15,000 uniformed personnel, to oversee disarming the warring factions and to guarantee the new government's security. The agreement reserved no place for rebel units or elements of the old Liberian Army (AFL) to join a restructured army, specifying instead that recruits be "screened with respect to educational, professional, medical and fitness qualifications as well as prior history with regard to human rights abuses." The agreement also stipulated a leading role for the United States in restructuring the army, while the police and security forces previously under Taylor's personal control were to be restructured under UN auspices.[21] The 2003 agreement had no provisions for dealing with the serious problems of corruption in Liberia's political establishment, beyond a recommendation to convene a reform commission. That advice was ignored, although a truth commission, also provided for in the peace agreement, sat from 2006 to 2009 to investigate human rights violations and economic crimes.

Although its recommendations were ignored, the commission was significant for its reflection of public sentiments. Initially planning to interview 34,000 people—about one percent of Liberia's population—it met with about half that number. Testimonies of

---

[19] UNSC, Peace Agreement.

[20] Douglas Farah, "Al Qaeda Cash Tied to Diamond Trade: Sale of Gems from Sierra Leone Rebels Raised Millions, Sources Say," *Washington Post*, Nov. 2, 2001. This analysis is reflected in UNSC, *Report of the Panel of Experts Pursuant to Security Council Resolution 1343 (2001), paragraph 19, Concerning Liberia*, Oct. 26, 2001.

[21] Accra Agreement, Article VII, Article VIII.

alleged perpetrators and companies associated with these crimes led the commission to recommend that many incumbent politicians, including President Ellen Johnson-Sirleaf, be barred from public office for 30 years.[22] These recommendations captured the popular sense that most of Liberia's politicians were corrupt, and the inclusion of the category "economic crimes" accurately reflected the tight connections between the country's political establishment and the illicit economy. This also showed how real reform in Liberia would be tantamount to a social revolution, completely disrupting the existing order—something that neither the incumbent local establishment nor international actors desired.

The peace settlement seemed to signal a radical break with the politics of the civil war. It differed from the dozen or so preceding agreements in not specifying offices for leaders of armed groups to occupy in a new government, instead leaving it to them to win elections if they wished to remain in politics. Also, international actors—such as U.S. diplomats involved in mediating the transition—had the power to pressure notable politicians and businessmen, threatening to use their previous human rights violations and violations of UN sanctions to exclude them from politics if they should act as spoilers to the settlement. Liberians had to take these threats seriously, given the advent of prosecutions for war crimes and other abuses after conflicts in the former Yugoslavia, and the promise to do the same in West Africa. Taylor was excluded entirely, leading to his exile from Liberia. Defying and, in many cases, frustrating this agenda of a radical break from the past was the reality that many wartime leaders really did have popular bases of support that they could mobilize to win elections. For these people, politics still included engagement with criminal enterprises as sources of income to build patronage networks supporting their business and political ambitions. For example, Charles Bennie, a former LURD spokesman, contested with Senator Roland Kaine, a former NPFL member who defeated him in a 2005 senate race, to control valuable land. Bennie hired armed men at $25 a day to clear Kaine supporters from the land.[23] About 3,700 LURD ex-combatants settled on Guthrie rubber plantation under the protection of Sumo Dennis, a former LURD commander.[24] The peace agreement grossly underestimated the durability of the underlying Liberian political system and of many of the principal actors from the civil war.

Rebel Networks

Illicit power was integral to the rebel organizations in much the same way that it had been essential to the operation of Liberia's formal government. Even before the NPFL's 1989 push into Liberia, the group's organization rested on regional illicit power. At that time, Libya's leader, Muammar Qaddafi, one of Taylor's hosts before the 1989 NPFL invasion, provided a fertile environment for multiple illicit power structures.

---

[22] Lansana Gberie, "Truth and Justice on Trial in Liberia," *African Affairs* 107, no. 428 (2008), 455-65.

[23] Othello Garblah, "The Land Wars: Rising Dispute over Scarce Land," *New Democrat*, June 11, 2008, 2-3.

[24] Christine Cheng, "The Emergence of Extralegal Groups after Civil War and Liberia's Rubber Industry," Univ. of Oxford, Apr. 15, 2008, 12-14, www.yale.edu/macmillan/ocvprogram/conf_papers/cheng.pdf.

A Regional Rogues' Gallery

Qaddafi's support for rebels and opposition groups from several African countries linked Taylor with opposition figures who went on to fight in Sierra Leone's civil war or to participate in a coup in Gambia.[25] At first, what would later become the NPFL was little more than a network of opponents to President Doe. But Taylor leveraged his contact with the Libyan leader to convince exiled Liberians and politicians, political dissidents and aspirants, and businessmen from throughout West Africa that Taylor had the best chance of overthrowing Doe and installing a new political network in the capital, with himself at the top.

These preinvasion contacts played important roles in the construction of Taylor's security apparatus once the NPFL was inside Liberia. For example, Taylor's chief of operations, Mohamed Adams, and the deputy commander of his paramilitary Anti-Terrorist Unit, Abu Suleimana, were Ghanaians. The NPFL's vice president in 1990, Kukoi Samba Sanyang, was a Gambian, and the NPFL was assisted by hundreds of troops from Burkina Faso.[26] These networks simply replaced similar networks that Doe had constructed to tap resources and assert his political authority. In this sense, the rebel war in Liberia represented continuity in political practice (albeit in a more violent form), rather than an abrupt break from a previous order.

Foreign Business Connections

As a key part of his political strategy, Taylor set out to make deals with as many foreign businesses as he could to generate the income he needed to build his own armed patronage network. Some that made deals included companies with existing assets in Liberia, such as U.S.-based Firestone. After the NPFL had seized Firestone's rubber plantation headquarters, it negotiated with the company, encouraging it to return and pay taxes to the rebels.[27] The company then imported rice to Liberia, which it "donated" to the NPFL as one of the costs of doing business.[28] The foreign firm's infrastructure provided Taylor with resources and a base from which the NPFL was able to stage its October 1992 Operation Octopus invasion of Monrovia.

This practice replicated the prewar arrangements between President Doe and foreign timber companies, in which the president acted simultaneously as a private business partner and head of state security forces while the foreign investor provided resources and logistics that the president needed to train and equip armed forces that answered to him personally.[29]

---

[25] Special Court for Sierra Leone, *Prosecutor of the Special Court v. Charles Taylor*, Transcript of 7 Feb. 2008, 3418-22.

[26] Stephen Ellis, "Charles Taylor and the War in Sierra Leone," Dec 5, 2006, 5 [Exhibit P-478, Document prepared for the Special Court for Sierra Leone].

[27] "Memorandum of Understanding" and "Firestone Restart Timetable," Gbarnga, Liberia [NPFL headquarters], Jan 16, 1992 [mimeograph copies in author's possession].

[28] U.S. Embassy, "Liberia Situation Report," Monrovia, Oct. 4, 1991, www.propublica.org/documents/item/1362170-dos-1991-october04-monrov-07468.html.

[29] *Africa Confidential*, "Liberia: What to Do with Doe?" Oct. 21, 1988, 24. See also PBS episode "Firestone and the Warlord," *Frontline*, Nov. 18, 2014, www.pbs.org/wgbh/pages/frontline/firestone-and-the-warlord/.

To bolster his authority, Taylor outdid even his predecessors in letting others conduct illicit trade in timber and diamonds. A U.S. diplomat estimated that during 1990-94, Liberia exported diamonds—many of them smuggled from Sierra Leone—valued at US$300 million, timber valued at over $50 million, and rubber valued at $27 million, mostly from areas controlled by Taylor's NPFL.[30]

These operations built on the long practice, among Liberian politicians, of seeking out foreign investors to organize the exploitation of natural resources and then using at least some of the proceeds to train paramilitary forces and import weapons. The typical ground-level relationship involved bringing a local politician into the firm as a partner and giving him a portion of the profits in exchange for his protection. Taylor's desperate need for resources early in the war pushed him and his associates to solicit a wide array of foreign logging operators to generate revenue and provide logistical services to NPFL fighters, such as letting fighters use their premises as bases and lending vehicles and other equipment to support NPFL operations.[31]

The Warlord's Trappings of Respectability

The intersection of political patronage networks and international mediation of the conflict were central to Taylor's ascension to Liberia's presidency in 1997. The 1995 Abuja Accord, the thirteenth peace agreement since 1990, set up a six-member Council of State, which included the heads of the main rebel groups and allocated ministries and public corporations to them as a prelude to a general election in 1996 (later postponed to 1997).[32] This agreement essentially codified and sought to regulate the competition between rebel leaders in grabbing resources with which to build rival political networks. By 1996, as head of the largest and most successful rebel group, Taylor reportedly exercised personal control over an annual income of $75 million, with which he paid commanders and fighters who understood that he paid them based on their loyalty and usefulness to him.[33] These resources gave him an advantage over his rivals, many of whom he killed once he was Liberia's president.

Once elected, Taylor set to work using the prerogatives of his office to sustain and expand his personal networks. Unlike during his days as a rebel leader, he was now in a position to give official Liberian government logging concessions to local and foreign firms. These included one of up to 1.44 million hectares, to Oriental Timber Company, to export timber in defiance of a UN embargo.[34] The company allegedly maintained a

---

[30] William Twaddell, "Statement of William Twaddell, Hearing on Liberia, before the House International Relations Committee," U.S. State Dept., June 26, 1996, 12, http://dosfan.lib.uic.edu/ERC/bureaus/afr/960626Twaddell.html.

[31] NPFL, "Guidelines for Reactivating Logging Companies," Office of the Chairman, Special Economic Committee, Gbarnga, Liberia, Nov. 5, 1990 [memo in author's possession].

[32] "Abuja Agreement to Supplement the Cotonou and Akosombo Agreements as subsequently clarified by the Accra Agreement," USIP Peace Agreements Digital Collection, www.usip.org/sites/default/files/file/resources/collections/peace_agreements/liberia_08191995.pdf.

[33] U.S. State Dept., "Testimony by William Twaddell, Acting Assistant Secretary of State for African Affairs," Washington, DC, June 26, 1996, http://dosfan.lib.uic.edu/ERC/bureaus/afr/960626Twaddell.html.

[34] *The Perspective*, "Investigative Report on Oriental Timber Corporation," Mar. 20, 2000, www.theperspective.org/otc.html.

private militia of 2,500 armed men, which operated in support of Taylor's government and paid various "public relations" contributions and "advance taxes" to Taylor.[35] The operation, under the direction of Guus van Kouwenhoven, was reported to include involvement in arms imports in violation of a UN embargo, involving transactions with many businesses across several continents.[36]

Taylor Stirs the Sierra Leone Pot

Along with the timber export revenues available for Taylor's personal use, his resources increasingly came to include revenues from diamonds smuggled from parts of Sierra Leone under RUF control. The importance of this transnational trafficking network in Taylor's domestic political strategy, and his support for the RUF, which controlled Sierra Leone's diamond mining areas, pushed the UN Security Council (UNSC) to form a panel of experts to investigate whether there was a link between the illicit diamond trade and the supply of weapons to the RUF. The expert investigators estimated that an illicit trade in diamonds to Liberia, on the order of $25 million to $50 million annually (exceeding official Liberian government revenues) linked Taylor to the RUF and that "a Liberian is said to be President Taylor's representative in Kono [the center of Sierra Leone's diamond mining industry], with a mandate to supervise diamond operations."[37] This network included foreign air transport companies and South African mercenaries.[38]

The collapse of a peace agreement in Sierra Leone, and RUF attacks on peacekeepers there focused international attention on the links between Taylor's government and the conflict in Sierra Leone. "Taylor's role emerged more clearly than ever as pivotal," wrote the U.S. ambassador to Sierra Leone, who concluded that "as long as Liberia sought access to Sierra Leone's diamonds and offered refuge to RUF fighters, the war would continue."[39] European investigators concluded that Taylor harbored two al-Qaeda operatives. He received $1 million, allegedly to hide the men in Camp Gbatala, near his private farm, which also served as the base for Liberia's elite "antiterrorism" unit and the mercenaries accused of training the unit.[40]

The Juggling Act

Such networks of domestic commercial operations, support for a neighboring country's rebel group, and foreign suppliers and logistics was central to the Liberian president's exercise of authority. This system built on old linkages that politicians had built in cross-border trades in diamonds and other illicit commerce. Though more violent in wartime,

---

[35] Andrew Feinstein, *The Shadow World: Inside the Global Arms Trade* (New York: Farrar, Straus and Giroux, 2007), 123.

[36] Marlise Simons, "The Dutch Try One of Their Own over Links to Liberia," *New York Times*, May 3, 2006; UNSC, *Report of the Panel of Experts on Liberia Submitted Pursuant to Paragraph 5 (c) of Security Council Resolution 1792 (2007) Concerning Liberia* (New York: United Nations, June 12, 2008), 27-28.

[37] UNSC, *Report of the Panel of Experts Appointed Pursuant to Security Council Resolution 1306 (2000)*, 18.

[38] *Africa Confidential*, "Godfather to the Rebels," June 23, 2000, 1-2.

[39] John Hirsch, *Sierra Leone: Diamonds and the Struggle for Democracy* (Boulder, CO: Lynne Rienner, 2001), 89.

[40] Douglas Farah, "Report Says Africans Harbored Al Qaeda," *Washington Post*, Dec. 29, 2002, A1.

this network was well entrenched by the middle of the twentieth century and played important roles in the politics of Sierra Leone and Liberia.[41]

In addition to providing political leaders with patronage resources, personal control over these economic opportunities enabled them to sustain their own armed militias on the basis of personal loyalties. Taylor took advantage of this opportunity and, after his election in 1997, maintained about a dozen militias, which he used to protect commercial operations, hound rivals, and provide personal security. They were staffed and supported in ways that forced them to compete with one another for their leader's favor, reducing the risk that their commanders might collaborate to overthrow Taylor. The command structures of the security forces underscored the intertwined nature of these illicit networks and the state, since many of those involved in these networks simultaneously occupied official offices, were businessmen, and had personal followings as important wartime commanders.[42]

The Fall of Taylor

Taylor's personal control over these networks began to unravel in 2003 as two rebel groups, LURD and MODEL, gained ground in their fight to seize power. He also became the target of prosecutors for the Special Court for Sierra Leone, which had been established in 2002 to try those deemed most responsible for crimes against humanity connected to Sierra Leone's war. Given Taylor's relationship with the RUF, he was a major focus of the prosecution, which issued an indictment against him in 2003. Taylor resigned on August 11, 2003, and left Liberia for exile in Nigeria. His Nigerian government hosts allowed him to settle there, provided that he not interfere in Liberian politics. Illicit power structures still provided him with levers to influence politics, however, and UN Secretary-General Kofi Annan reported to the Security Council that Taylor's "former military commanders and business associates, as well as members of his political party, maintain regular contact with him and are planning to undermine the peace process."[43] Taylor was arrested in 2006, while attempting to flee from his exile in Nigeria, and was delivered to the court for trial.

At the time of the peace settlement in 2003, significant elements of this illicit power structure were violently opposed to the peace process. And in the long run, the underlying logic of patronage-based governance and the violent pursuit of commercial resources remained intact.[44] These linkages retained their importance in Liberian politics after Taylor's exile in 2003 and during the international intervention to promote peacebuilding in Liberia. This persistent influence of illicit power up to 2003 created a conundrum for international actors in how to respond to these networks and how to conduct large-scale interventions to halt conflicts more generally.

---

[41] H. L. van der Laan, *The Sierra Leone Diamonds: An Economic Study Covering the Years 1952-1961*, (New York: Oxford Univ. Press, 1965).

[42] Global Witness, *The Usual Suspects: Liberia's Weapons and Mercenaries in Côte d'Ivoire and Sierra Leone*, (London: Global Witness, 2003), 14-15.

[43] UNSC, *Sixth Report of the Secretary-General on the United Nations Mission in Liberia* (New York: United Nations, 2005), 18.

[44] François Prkic, "The Phoenix State: War, Economy and State Formation in Liberia," in Klaus Schlichte, The Dynamic of States (Burlington, VT: Ashgate, 2005) 115-36; Douglas Farah and Shaoli Sarkar, *Following Taylor's Money: A Path of War and Destruction* (Washington, DC: Coalition for International Justice, 2005).

In short, the risk was that the more the international actors did to build the state, the less their local partners would contribute to this effort. Local networks could easily adapt to exploit the opportunities associated with the huge volume of resources needed for this task. And this they did, siphoning off aid money without helping the foreign contributors build an effective state. Yet effective state building is the only long-term answer to the problem of illicit power in Liberia. Unfortunately, state building in such an environment is a slow, violent process. The international response to these networks from 2003 on, the subject of the next section, provided a good test.

International Strategies and Their Impacts on Illicit Power

Liberia in 2003 looked like an ideal place to test the international community's political will and capacity to target illicit power in a coordinated attempt to implement extensive postwar reforms. This was a revolutionary state-building project, since it envisioned changing the core logic of how the country was run, permanently eliminating illicit networks from governance and the economy and fundamentally changing the relationship between the government and the citizenry. If it could be done anywhere, Liberia was a good candidate for success, for these reasons:

- its small population (about four million people),
- its long historical ties to the United States (which has a sizable skilled Liberian diaspora community) and to Europe,
- concern among officials in West Africa and beyond that an unresolved conflict in Liberia would destabilize other countries in the region,
- and indications that Liberia's population would tolerate significant foreign intrusion in domestic affairs (distinguishing this from most international involvement, such as in Somalia in the early 1990s).

This urgency to transform Liberia's political economy occurred against the backdrop of wider international responses to the al-Qaeda attacks on New York and Washington on September 11, 2001. As noted above, Taylor was indicted before the Special Court for Sierra Leone in March 2003, charged with crimes against humanity linked to his wartime association with Sierra Leone's RUF. He was convicted of these crimes and sentenced in May 2012 to 50 years, which he is now serving in a prison in Manchester, UK.

Taylor was the first former head of state convicted by an international tribunal, and the first head of state to be prosecuted for the use of rape and child soldiers in war. By bringing him to justice for crimes against humanity and war crimes, the Special Court for Sierra Leone removed a major player from Liberia's political scene. But it is not evident that even this deep intervention actually transformed the underlying political economy of Liberia, including the integral role of illicit power structures.

Seeking a New Order

Tentative international efforts under the Accra Agreement to create an alternative to corrupt governance began in October 2003, with the appointment of the Transitional National Government (TNG). The TNG was intended to break with conflict-era politics while also recognizing the reality that anti-Taylor armed groups and the various security agencies and paramilitaries loyal to Taylor and his associates could act as spoilers if left out of the distribution of state offices. Therefore, the TNG included some members of wartime rebel groups, but its mandate was limited to preparing the way for national elections in 2005.

The October 2005 elections promised more opportunity to break with the past and, ultimately, appeared free from the influence of rebel leaders. The two main presidential candidates were a football star and a former international banker.[45] Ellen Johnson-Sirleaf, the international banker, became Africa's first elected female head of state. Her winning the 2011 Nobel Peace Prize underscored what appeared to be a significant break with the politics of the past. At the same time, she had her own connections to this past in her efforts at raising money early in the war to support the NPFL's campaign to oust President Doe. While she admitted to this in her memoir and in testimony before the Truth Commission, many Liberians suspect that her early ties to the NPFL went deeper. Moreover, her admitted record was at odds with the Nobel Prize committee's justification for the award as including nonviolent struggle.

Timber Extraction: Business as Usual?

Other reform efforts focused on severing the connection between state power and the exploitation of Liberia's resources to reward political clients and perpetuate the power of a corrupt political class. After a UN panel of experts called for a review of forest management practices, the Liberian government reviewed links between the illicit export of timber and the maintenance of personal militias. The government's 2005 review found that most forestry concessions did not comply with basic regulations and involved numerous people who were targets of UN travel bans for their suspected roles in Liberia's civil war.[46]

Numerous networks of wartime vintage continued to play a role in the timber industry despite UNSC sanctions from 2003 to 2006 on timber exports. For example, a large concession granted to foreign investors included Eddington Varmah as their Liberian partner. When Varmah was justice minister in Taylor's government, "he signed an agreement with Exotic Tropical Timber Enterprise, owned by Taylor confidant and business partner Leonid Minin, whereby the government paid ETTE $2 million for undisclosed services."[47] Using his Odessa criminal contacts, Minin played a central role in connecting Ukrainian small arms suppliers to Taylor as well as to Côte d'Ivoire's dictator, Robert Guei, before Guei was overthrown in 2000.[48]

---

[45] David Harris, "Liberia 2005: An Unusual African Post-conflict Election," *Journal of Modern African Studies* 44, no. 3 (2006): 375-95.

[46] Government of Liberia, *Forestry Concession Review, Phase III* (Monrovia: Forestry Concession Committee, 2005), www.fao.org/forestry/lfi/29659/en/.

[47] Africa Confidential, "Undue Diligence in the Timber Sector," July 2009, 6.

[48] Ian Traynor, "The International Dealers in Death, *Guardian*, July 9, 2001, www.theguardian.com/world/2001/jul/09/armstrade.iantraynor.

Although the UN peacekeeping force and security sector reform (detailed below) played critical roles in drastically reducing violence, the underlying link between timber exploitation and political corruption remained intact through new means well after the war's end. Adaptations such as "private use permits" enable officials of the Forestry Development Authority to collude with logging companies by signing exploitation agreements outside the regulatory framework that reviews concession agreements. This loophole, originally justified to allow private landholders to exploit their own assets and as a response to foreign pressure to open up the forestry industry to greater investment, gives officials and their commercial partners a cover to appropriate communally held land for private gain. NGOs report that land deeds are often suspicious or irregular, and long-drawn-out government efforts to address these problems suggest a lack of incentive or political will to deal with the problem.[49] The durability of this illicit extraction scheme points to the participants' ability to seemingly comply with international norms while finding new ways to shield themselves from the intended effects of reforms.

Drug Trafficking in the New Liberia

UN investigators started to find evidence of a growing problem of cocaine and heroin trafficking through Liberia. In 2014, they reported that "a considerable number of those individuals involved in this trafficking as couriers were former combatants and currently serving personnel of the military and police forces."[50] Reports that South American traffickers have used Liberia as a transit point and tried to bribe Liberian officials suggest that traffickers view corrupt politicians in Liberia as potential partners.[51]

The scale of resources involved in drug trafficking risks turning a small, poor country into a narco-state and will require a significant international role to compensate for Liberia's lack of enforcement capacity. In 2010, for example, U.S. Drug Enforcement Agency (DEA) officials, based in U.S. embassies in Monrovia and several other countries, managed a joint undercover operation to prevent Colombian and Venezuelan drug traffickers from using Liberia as a transit point to move four metric tons of cocaine under the protection of the Colombian rebel group FARC (Fuerzas Armadas Revolucionarias de Colombia), valued at over $100 million, to Europe and the United States. The traffickers attempted to bribe officials in Liberia, promising cash and a portion of the cocaine to traffic in their own operations.[52]

Liberian government cooperation with DEA officials was real, with the president's stepson and head of Liberia's National Security Agency, Fombah Teh Sirleaf, playing a

---

[49] Global Witness, *Logging in the Shadows: How Vested Interests Abuse Shadow Permits to Evade Forest Sector Reforms* (London: Global Witness, 2013), 16-18.

[50] UNSC, *Midterm Report of the Panel of Experts on Liberia Submitted Pursuant to Paragraph 5 (b) of Security Council Resolution 2128 (2013)*, Apr. 25, 2014, 20.

[51] UN Office on Drugs and Crime (UNODC), "Transnational Organized Crime in West Africa: A Threat Assessment," press release, Feb. 25, 2013, 32, www.unodc.org/unodc/en/press/releases/2013/February/transnational-organized-crime-continues-to-affect-vulnerable-west-african-countries-says-new-unodc-report.html; Yudhijit Bhattacharjee, "The Sting: An American Drugs Bust in West Africa," *Guardian*, Mar. 17, 2015, www.theguardian.com/world/2015/mar/17/the-sting-american-drugs-bust-liberia.

[52] U.S. Attorney, SDNY, "Manhattan U.S. Attorney Announces Unsealing of Charges Arising from Historic Joint Undercover Operation in the Republic of Liberia," June 1, 2010, www.justice.gov/usao/nys/pressreleases/June10/operationrelentlessliberiapr.pdf.

key role in the sting operation. The closely coordinated work of U.S. government agencies played a decisive role in the operation.

Drug traffickers fit well with Liberia's culture of illicit power, and resources from drug trafficking can be huge. UN officials estimate that traffickers ship 18 tons of cocaine through West Africa each year. Just one ton has a value exceeding the military budgets of most countries in the region.[53] This puts some Liberian officials in ideal positions to incorporate drug trafficking in their own patronage networks. Thus, while high-level officials cooperated with the U.S. antinarcotics effort noted above, UN investigators found that "senior officials of the Government of Liberia have prevented the arrests of heroin couriers on at least two occasions in 2013." They concluded that even though the deputy director of Liberia's Drug Enforcement Agency was fired for violating policies, "networks of higher-level Liberian government officials continue to be influenced by criminal networks smuggling narcotics."[54] One danger for international actors in this environment is that counternarcotics efforts that appear to be based on cooperation may, in fact, consist of foreigners and a few Liberians who actually do the work of maintaining the facade of state capacity. The foreign partners publicly give credit to the Liberian agencies relevant to the effort, but the underlying structure of governance goes merrily on with business as usual.

The Persistence of Insider Networks

International efforts to dismantle illicit power structures after the 2006 transition to an elected government required considerable intrusion into domestic governance, in many ways. Given its external dependence, Liberia's elected government had little option but to agree to the Government and Economic Management Assistance Program's (GEMAP's) overseeing official fiscal and financial accountability. GEMAP put foreign financial experts into nearly every Liberian government agency that spent donors' resources. Expenditures required countersignatures of the foreign experts, and the program was overseen by a panel headed by President Johnson-Sirleaf and the U.S. ambassador.

Even though GEMAP concluded in 2010, public perceptions of insider networks remained. For example, one of the president's sons, Robert Sirleaf, served as chairman of the board of the National Oil Company. A former National Oil Company head claimed in an open letter to the president that Robert Sirleaf used his position to advance his political and economic interests. The president denied this allegation, but since the president's son contested the 2014 Montserrado senate election, it appeared to many that nepotistic insider networks were at work, regardless of the professional qualifications of the candidate.[55] While it is not surprising that in such a small country, two well-connected men people should hold the same position in succession in the oil company and then face each other in an election, this intertwining of elite networks feeds the popular perception that personal relationships trump merit in determining one's position or of-

---

[53] UNODC, "Transnational Organized Crime in West Africa: A Threat Assessment," Sept. 2013, www.unodc.org/documents/data-and-analysis/Studies/TOC_East_Africa_2013.pdf.

[54] UNSC, *Final Report of the Panel of Experts on Liberia Submitted Pursuant to Paragraph 5(f) of Security Council Resolution 2079 (2012)*, Nov. 21, 2013, 9-10.

[55] UNSC, *Twenty-eighth Progress Report of the Secretary-General on the United Nations Mission in Liberia*, Aug. 15, 2014, 2.

fice, and that hidden agendas drive such people's behavior. It is worth noting that three other sons of the president served in government posts, including deputy internal affairs minister, deputy governor of the Central Bank, and, as noted above, head of the National Security Agency.[56]

Wartime political networks found places in Liberia's political system after the peace settlement. The speaker of the National Assembly, George Dweh, a former leader of LURD and cousin of President Doe, continued to exercise considerable personal authority. Rather than resign after being accused of embezzlement, he appeared in the legislature with armed guards and held other legislators hostage until UNMIL soldiers intervened.[57] Dweh did not return to the legislature despite his protests that he had nothing to apologize for.

The 2005 elections saw the migration of wartime patronage networks into post-settlement governance. Senate election winners included Jewel Howard-Taylor, Charles Taylor's former "first lady," who won reelection in 2011, and Prince Johnson, the former INPFL leader. Johnson also ran for president in the 2011 election, placing third with 11.6 percent of the vote, and had to be satisfied instead with his 2014 reelection to the senate. Another senator, Adolphus Dolo, known as "General Peanut Butter," faced a UN travel ban aimed at wartime commercial and political actors. After MODEL commander Kia Farley, whose nom de guerre was "White Flower" (also the name of Taylor's Monrovia mansion), was elected to the House of Representatives, fighters under his command occupied a rubber plantation and were in contact with fighters in neighboring Côte d'Ivoire.[58] Speaker of the House of Representatives Edwin Snowe appeared on the UN travel ban because of his ongoing ties to the exiled Taylor. He resigned in 2007 but was reelected to the House of Representatives in 2011.[59] Another representative had previously served as deputy chief of police in Taylor's government.[60]

It is not remarkable for prominent wartime personalities to continue to play important roles in politics after peace settlements and political transitions. The issue in Liberia, however, is whether wartime figures continue to exercise political authority through personal control over proceeds from illicit enterprises, in collusion with state officials responsible for making and enforcing laws. Ample evidence shows that this mode of politics persists in Liberia at the expense of impartial bureaucratic procedures and policies that are the objective of foreign engagement. Illicit power plays a major role in shaping how citizens relate to state power. Thus, many Liberians continue to regard politics, the (often illicit) accumulation of personal wealth, and violence as closely related endeavors.

---

[56] *Africa Confidential*, "Liberia: Feted and Berated," Nov. 1, 2013, 10; *Africa Confidential*, "Family Business under Pressure," May 2, 2014, 8-9.

[57] UNMIL, "Liberia: UNMIL Humanitarian Situation Report no. 3," Reliefweb, Mar. 22, 2005, 1, http://reliefweb.int/report/liberia/liberia-unmil-humanitarian-situation-report-no-3; International Crisis Group (ICG), "Liberia's Elections: Necessary but Not Sufficient," ICG Africa Report no. 98, Sept. 7, 2005, 12, www.crisisgroup.org/en/regions/africa/west-africa/liberia/098-liberias-elections-necessary-but-not-sufficient.aspx.

[58] Ibid., 41.

[59] UNSC, *List of Individuals Subject to the Measures Imposed by Paragraph 4 of Security Council Resolution 1521 (2003) concerning* Liberia, Dec. 15, 2006.

[60] Backgrounds of legislators elected in 2005 appear in Konrad Adenauer *Stiftung, A Profile of Members of the 52nd Legislature of Liberia* (Monrovia: Konrad Adenauer Foundation, 2006), www.kas-benin.de/liberia/Profiles_2005_52nd_Legislature-Liberia.pdf.

The International Footprint

Resources involved in the international effort to reform Liberia's politics have been immense when measured in relative terms. In 2003, the GDP of Liberia stood at about US$500 million, with government revenues at about $80 million, while three-quarters of Liberians subsisted on less than a dollar a day.[61] In such a small economy, the international material presence was enormous. UNMIL cost $6.2 billion from 2003 to 2013.[62] Official overseas development assistance over the first six years of the intervention amounted to almost $3.2 billion.[63] NGO assistance contributed millions more.

GEMAP's insertion of foreign experts to oversee the inner workings of Liberia's government administration was intended to transform how the government related to its citizens. These resources and policy reforms did promote high economic growth rates—above eight percent annually from 2011 to 2013—though these figures also represent substantial growth of investments in extractive enterprises, such as iron ore mining and timber exploitation, that continued to be either insulated in enclaves or implicated in illicit power. Even though poor infrastructure and limited access to credit constrain local job creation, Liberians benefit from a vastly improved security situation compared to the years before 2003.[64] Yet in this sector, too, international intervention leaves significant elements of illicit power intact, as detailed in the next section.

Strategies for Security Sector Reform

In 2003, Disarmament, Demobilization, and Reintegration (DDR) and Security Sector Reform (SSR) were core elements of the international strategy to weaken the dominance of illicit power and assert the state's monopoly over the use of force. One major objective was to dismantle the militias and create a cohesive army and police, under government control, that would protect the citizenry.

Impediments to SSR

After the 2003 settlement, SSR faced a contradiction. The transitional government included the still-powerful leaders of the armed groups that opposed Taylor's government, as well as top commanders of Taylor's NPFL. These leaders still claimed considerable popular support from their home communities. This situation differed from neighboring Sierra Leone's more thorough SSR, happening at the same time, largely because insurgents in Sierra Leone eventually met a decisive defeat and lacked any meaningful

---

[61] International Monetary Fund (IMF), "Liberia: Report on Post-Conflict Economic Conditions and Economic Program for 2004/05," IMF Country Report no. 04/408, Dec. 2004, www.imf.org/external/pubs/ft/scr/2004/cr04408.pdf.

[62] Calculated from quarterly UNSC Reports of UNMIL operations. See United Nations, "United Nations Documents on UNMIL: Reports of the Secretary-General," 2003-15, www.un.org/en/peacekeeping/missions/unmil/reports.shtml.

[63] Global Humanitarian Assistance, "Liberia: Key Figures 2012," 2012, www.globalhumanitarianassistance.org/countryprofile/liberia.

[64] IMF, "Liberia: First Review under the Extended Credit Facility Arrangement," IMF Country Report no. 13/216, July 2013, 6, www.imf.org/external/pubs/ft/scr/2013/cr13216.pdf.

post-intervention political structure. (Sierra Leone's military, while hardly a standard-bearer of civilian protection during that war, lacked the legacy of deep complicity in the pervasive preconflict violence that characterized Liberia in the 1980s.)[65] Thus, it was difficult to begin serious SSR before the inauguration of the democratically elected government in late 2005.

The 2003 Accra Peace Agreement confronted a situation in which two groups, Taylor's NPFL and the coalition of LURD and MODEL forces, were still well armed and mutually hostile. Consequently, the agreement called for DDR and SSR programs to neutralize this threat. It gave several LURD leaders positions in the transitional government, as well as management of the Liberian Petroleum Refining Corporation. By 2004-5, that company proved well suited to illicit business ventures built around manipulating government imports of fuel, and fuel prices, to maximize personal profits, despite GEMAP oversight.

These illicit operations sometimes involved cooperation that crossed old wartime divides. A high-profile corruption case involved Edwin Snowe, the Speaker of the House of Representatives, noted above. MODEL occupied key timber-exporting coastal towns and was given control of the Forestry Development Authority. These decisions, deeply flawed from the perspective of eliminating illicit power structures and reforming Liberia's security sector, reflected the reality that elements of established rebel armies continued to be influential. Co-opting them was essential to gaining their acquiescence to deployment of the international peacekeeping force.

Prominent leaders of anti-Taylor armed groups from the 1990s found post-settlement positions in state security as well. Brownie Samukai, the head of the Black Berets, who fought alongside ECOMOG in the 1990s, became defense minister in 2006. Samukai proved able to run the ministry in a professional fashion. Nevertheless, his wartime service to the state as a leader of a paramilitary force rather than as a regular army commander highlighted the extent to which institutions of the Liberian state had collapsed during the war. While individuals such as Samukai had specific military skills, they also had experience running political networks. In Liberia in the 1990s, this meant that because a network's members were not paid, it had to combine its exercise of violence with illicit revenue-generating activities. During the 1990s, the AFL had essentially been disbanded and replaced with armed groups held together by personal loyalty to their commanders. These groups also included the Anti-Terrorist Unit, Jungle Lions, Marines, Special Strike Force, Special Security Service, Special Operations Division, and numerous smaller units, most of which were involved in illicit businesses as a means to support themselves and, for some of them, to set up other businesses.[66] This meant that the implementers of security sector reform, even if they found competent local professionals to work with, had to operate in a context where even the state security forces assumed the fragmented personalist structure that is a key characteristic of conflict in failed states.

The political trajectories of some wartime leaders deemed threats to security illustrate the considerable popular support some of these men enjoyed years after the civil war. For example, Benjamin Yeaten, the head of Taylor's Special Security Service, lost

---

[65] For an analysis of Sierra Leone's SSR, see Peter Albrecht and Paul Jackson, eds., *Security Sector Reform in Sierra Leone, 1997-2007: Views from the Front Line* (Hamburg: LIT Verlag, 2010).

[66] ICG, Liberia: Security Challenges," Africa Report no. 71, Nov. 3, 2003, 8, www.crisisgroup.org/en/regions/africa/west-africa/liberia/071-liberia-security-challenges.aspx.

that job when the militia was disbanded. UN investigators noted, however, that in 2011 he was also involved in recruiting mercenaries to fight in neighboring Côte d'Ivoire's conflict.[67] Although a UN monitor's report paints a picture of minimal mercenary activity and effective government surveillance, Yeaten could not be apprehended, because he could call on supporters to conceal him.[68] The head of the Special Strike Force was elected to the Senate. He mobilized his fighters to take up business in his hometown, combining his patronage for these ex-fighters with new licit business opportunities to build up his popular support base. By 2007, four years after the start of the SSR program, a RAND report still found that "rebel group structures and command chains have not been eradicated and remain a concern."[69]

Evidence of armed militias continued well into what was supposed to be the drawdown of the peacekeeping force. In 2014, for example, a team of UN experts investigating a series of attacks on UN peacekeepers in Côte d'Ivoire—which included Liberians among the attackers—reported "a much broader and more sophisticated combatant network than previously known." The investigators linked these combatants, including the Liberians among them, to illicit regional networks that brought violence to neighboring countries, much as Taylor's network had done during Liberia's war. The investigators concluded that the attacks "revealed evidence that they had been, at least in part, organized, planned and financed by the political and economic elite linked to the former President of Côte d'Ivoire," who had played a significant role in wartime Liberian networks.[70] The situation demonstrates the need for significant resources, particularly for increasing the capacity of the border control force and national police to sustain effective security sector reform. More problematic is the persistence of interests among local actors to limit state capacity in the security sector.

For more than a decade after the 2003 settlement, Liberia has seen periodic incidences of violence as former fighters and their political patrons occupy rubber plantations and engage in illicit logging. These events showed how wartime connections between commanders and fighters continued in other ways, too. Clashes over land ownership illuminated some of these ties. In 2008, a long-standing land dispute between a former NPFL leader who had become a senator, and the former spokesman for LURD (who were distantly related!) resulted in clashes that killed at least 16 people. This shows how wartime networks melded with postwar commerce and long-standing local tensions—in this case, a generational dispute over who owned a productive piece of land—to influence postwar politics.[71] This influence extended to the capital since the former LURD official was a patron of Monrovia street hawkers, providing some with employment in return for their political support. A critical concern is whether these networks will reas-

---

[67] UNSC, *Final Report of the Panel of Experts on Liberia Submitted Pursuant to Paragraph 6(f) of Security Council Resolution 1961 (2010)*, Dec. 7, 2011, 32, 50.

[68] UNSC, *Final Report of the Panel of Experts on Liberia Submitted Pursuant to Paragraph 5 (f) of Security Council Resolution 2079 (2012)*, Oct. 22, 2013, 67, 24.

[69] David Gompert, Olga Oliker, Brooke Stearns, Keith Crane, and K. Jack Riley, *Making Liberia Safe: Transformation of the National Security Sector* (Santa Monica, CA: RAND, 2007), 11.

[70] UNSC, *Final Report of the Panel of Experts on Liberia Submitted Pursuant to Paragraph 5 (b) of Security Council Resolution 2128 (2013)*, Nov. 24, 2014, 11, 16.

[71] UNSC, *Seventeenth Progress Report of the Secretary-General on the United Nations Mission in Liberia*, Aug. 15, 2008, 3.

sert their influence once the international peacekeeping force completely withdraws—an event scheduled, at the time of this writing, to occur in 2016.

Implementing SSR

Given the density and persistence of illicit power, SSR in Liberia entailed building a new army and police force from scratch and revising a justice system that had largely ceased to function by the war's end.

Reforming the army. Initially, the view of the UN special representative of the secretary-general, Jacques Klein, was that Liberia did not need an army, because an army would just "sit around and play cards and plot coups."[72] The start of SSR also had to wait for the U.S. government, the partner in the international coalition specified in the 2003 agreement as leader on SSR matters, to engage DynCorp and Pacific Architects and Engineers, two private firms that work in coordination with U.S. military personnel, through lengthy State Department contracting procedures, to implement SSR measures.[73] Thus, the training of new noncommissioned army officers did not begin until April 2007. Even then, DynCorp had to refurbish much infrastructure and remove former AFL soldiers squatting in AFL facilities, including the Ministry of Defense building. When these people departed, they stripped the buildings of all valuable materials, including roofing.[74]

Reforming the police. Liberian police training under UNMIL auspices proceeded at a faster pace than the military training program, and the National Police Academy opened in July 2004. Within five years, the training program had produced almost 3,500 police officers—close to the mandated level.

Although police reform looked good on paper at that point, the police chief of Lofa County reported in 2009 that he had one vehicle to patrol an area with a quarter-million inhabitants. This encouraged local people to rely on vigilante units, some of them created out of old wartime command structures.[75] Five years later, a researcher reported that police headquarters in Bong County, a major interior center, lacked secure radio communication and had no budget for fuel, photocopying case documents, or feeding detainees. Lacking vehicles, police had to use motorbikes or taxis to transport suspects to court or prison. These logistical constraints prevented police from making regular patrols, leaving large areas of the region essentially inaccessible to the police, and the

---

[72] IRIN, "Liberia: Klein Urges New Government to Abolish Army," Nov. 5, 2003, www.irinnews.org/report/47067/liberia-klein-urges-new-government-to-abolish-army. See also Ezekiel Pajibo, "Why Demilitarization Makes Sense," *Perspective*, Nov. 19, 2003, www.theperspective.org/demilitarization.html.

[73] See Mark Malan, *Security Sector Reform in Liberia: Mixed Results from Humble Beginnings* (Carlisle, PA: Strategic Studies Institute, U.S. Army War College, 2007).

[74] See ICG, "Liberia: Uneven Progress in Security Sector Reform," Africa Report no. 148, Jan. 13, 2009, www.crisisgroup.org/en/regions/africa/west-africa/liberia/148-liberia-uneven-progress-in-security-sector-reform.aspx.

[75] Ana Kantor and Mariam Persson, *Understanding Vigilantism: Informal Sector Security Providers and Security Sector Reform in Liberia* (Stockholm: Folke Bernadotte Academy, 2010).

people reliant on their own devices.[76] In 2015, a year before UNMIL's scheduled departure, a UN report concluded that police capacity was "emerging" amid continuing problems of criminal behavior and misconduct among senior leaders.[77]

Reforming the judicial sector. Police are still accused of harassment and extortion, slow response times, and a weak presence that encourages citizens to take the administration of justice into their own hands with vigilante violence.[78] Judicial sector reform was supposed to address these problems, since a reliable judicial sector that was responsive to public demands would deter police from engaging in criminal behavior. But judicial reform has been problematic. The bulk of disputes in Liberia are adjudicated through customary courts that are state sanctioned but lack mechanisms for judicial review of judgments and abuses of power and are subject to executive authority.[79] An Independent National Commission on Human Rights was established in 2010. Under the terms of the 2003 agreement, this body was to oversee implementation of the truth commission's recommendations on wartime crimes and abuses of power. But the continuing influence of many who were identified as perpetrators of crimes during the war has prevented implementation of these recommendations.

Integrating SSR. International actors continue to adjust strategies to address problems. One response was the 2013 creation, with UN Peacebuilding Support Office funding, of the first justice and security hub in Gbarnga, the wartime headquarters of Taylor's NPFL. This, the first of five hubs, combines justice and security operations in a single location to coordinate government responses to citizens' complaints of abuses of power and to provide an alternative to seeking protection through vigilantism. The hubs concept, extended to the rest of the country in 2015, is designed to gain support for government reform efforts and shift popular reliance away from political patrons with roots in illicit networks as citizens turn to the state for security and services. The concept also shows the seriousness of the continuing impediments facing security sector reform, given that it still imitates classic counterinsurgency strategies. More than a decade after the formal end of the civil war, the hubs strategy aims to extend state services and security to the citizens in an effort to win them over from power structures outside the formal institutions of the state. This task is difficult when target populations believe that the government lacks the political will to make these changes and that many officials benefit from the insecurity and corruption that this consolidated institutional initiative is meant to address. This was tested in 2014-15 during the Ebola virus epidemic, "which revealed the depth of public distrust and weakness in national institutions"—a problem that faces the security sector as much as it does the rest of the government.[80]

---

[76] Trine Nikolaisen, "Decentralising Liberia's Security Sector: The Role of Non-governmental Actors in Justice and Security Delivery," *Conflict Trends*, 2014, 52.

[77] UNSC, *Twenty-ninth Report of the Secretary-General on the United Nations Mission in Liberia*, Apr. 23, 2015, 11.

[78] U.S. State Department, "2012 Human Rights Reports: Liberia," Apr. 19, 2013, 12, www.state.gov/j/drl/rls/hrrpt/2012/af/204136.htm.

[79] No Peace Without Justice, "Making JUSTICE Count: Assessing the Impact and Legacy of the Special Court for Sierra Leone in Sierra Leone and Liberia," Sept. 2012, 18, www.npwj.org/content/Making-Justice-Count-Assessing-impact-and-legacy-Special-Court-Sierra-Leone-Sierra-Leone-and.

[80] UNSC, *Twenty-ninth Report*, 9.

Recommendations and Conclusion

Illicit power remains integral to Liberia's politics and economy. It serves as an attractive tool for individuals and groups to defend or improve their positions in a hierarchy of power. This happens, for example, when politicians use their office and commercial partnerships to gain access to "vacant" land through their control over the exercise of eminent domain. This power makes them attractive as political patrons who can offer access to economic opportunities such as timber. These political strategies are manifest in bureaucratic incapacity, which is partly intentional, and in the parallel political-commercial networks. Thus, the Liberia Extractive Industries Transparency Initiative indicated that 60 of the country's 68 concessions in the mining and timber industries were not in compliance with national laws.[81] Illicit power also provides options to people who are otherwise marginalized or disadvantaged. In 2013, for example, Liberian government officials estimated that 1,500 children left school to work in illicit mining operations.[82] These vertical patron-client links play an important role in stabilizing political order, at least in the short term, since the disadvantaged discover that their most feasible survival strategy is to integrate as best they can and find a capable protector. In sum, patron and client alike may recognize the long-term benefits of deep reform at the same time that they oppose it for the threat it poses to their immediate survival strategies.

The central role of illicit power in Liberia's political economy translates into a significant moral dilemma for international actors that promote state building. This happens when those actors' resources and efforts end up facilitating Liberians' involvement in the very illicit power structures they were intended to suppress. This is not uniformly true, but many Liberians discover ways to manipulate and adapt to pressures, and to channel new resources into illicit networks. The outcome is a hybrid arrangement: a more functional postwar state with persistent illicit power structures. To some international actors, it can seem as though they are trapped in an open-ended position of having to manage critical tasks that the state is supposed to perform. This is particularly true of tasks associated with keeping regional and domestic order, such as suppressing transnational crime and keeping the peace inside Liberia, which are also primary interests of the international community. It is true that Liberian officials cooperate in counternarcotics in high-profile ways. The government participates in institutions such as the West Africa Cooperative Security Initiative and facilitates joint efforts, but at the same time, its own Transnational Crime Unit is inactive.[83] Similarly, the multilateral Peacebuilding Fund supports regional security hubs outside the capital, which are central to international actors' strategies, at the same time that Liberian government allocations to them decline.[84]

The international community cannot do for Liberians what Liberians are unwilling to do for themselves. International capabilities are great enough to keep a lid on the problem as long as the political will to do so remains. But in the relationship between

---

[81] Liberia Extractive Industries Transparency Initiative, *4ᵗʰ EITI Report for Liberia* (Monrovia: Ernst and Young, 2013), www.scribd.com/doc/152662365/LEITI-4th-Reconciliation-Report.

[82] Jennifer Lazuta, "Liberian Children Quitting School to Mine Diamonds," Voice of America, Feb. 11, 2013, www.voanews.com/content/Liberian-children-quitting-school-to-mine-diamond/1601253.html.

[83] UNSC, *Twenty-ninth Report*, 11.

[84] Ibid., 13.

personal interest and political power, the Liberian people and their leaders have to decide how they are going to change the culture. Some Liberians have made this decision, but the evidence insists that many have not. The main point is that foreigners can help the Liberians and partner with them every step of the way, but they cannot do it all for them.

There are indications that some problems associated with illicit power structures may be self-fixing outside the policy framework of international intervention. Economic changes across the region may encourage some patrons to shed costly clients as they find that it is easier to get ahead by focusing on enterprise efficiency rather than by working through political patronage networks. They do not end their relationships with supporters (including former fighters) or with state officials, but instead commodify their relationhips with followers, turning them into employees rather than clients.[85] This changed "business model" has taken the form of commanders redeploying ex-combatants as employees in private security firms, poultry factory farms, and fisheries; and, in one instance, giving scholarships to former soldiers.[86] Ultimately, promoting some of these changes means turning the illicit networks into something else based on the self-interest of key figures, rather than directly confronting every aspect of illicit power as a security issue.

Recommendations

The recommendations below fall in a practical order, starting with those that are most urgent, have greatest potential for positive impact, are most feasible, and are least likely to produce unintended negative consequences.

1. Maintain an implicit U.S. guarantee of the security of democratically elected governments. It is imperative that democratically elected Liberian officials who are otherwise committed to reform not believe that they must distribute state resources to threatening groups to lower the risk of their violent removal from power. The decision of other West African governments to condemn coups in the region is a positive development. A coup leader in Liberia would encounter difficulties governing in the face of sanctions and political isolation. Clear signals of U.S. commitment would increase the willingness of key Liberian politicians and policymakers to remain committed to difficult and often unpopular reforms.

2. Privilege security sector reforms that create an environment where Liberians can sort out their own problems. Police are the common point of contact between citizens and coercive agencies of the state. A 2013 Transparency International survey of Liberians found that 94 percent of those surveyed considered police to be "corrupt" or "extremely corrupt."[87]

---

[85] See Mariam Persson, "Demobilized or Remobilized? Lingering Rebel Structures in Post-war Liberia," *African Conflicts and Informal Power: Big Men and Networks*, ed. Mats Utas (London: Zed, 2012), 101-18.

[86] Mats Utas, Anders Themnér, Ilmari Käihkö, "The Informal Realities of Peacebuilding—Military Networks and Mid-Level Commanders in Post-War Liberia," Nordic Afrika Institute, 2014, www.nai.uu.se/research/finalized_projects/the-informal-realities-of/.

[87] Transparency International, "Global Corruption Barometer 2013: Liberia," 2015, www.transparency.org/gcb2013/country/?country=liberia.

Continued reform of police institutions ought to address low salaries—police earn, on average, about $100 per month—to deter collusion with criminals and illicit power structures for additional income. Ways need to be found to enable police officers to keep the income that they earn, without being pressured to pay commanders for their positions and support themselves with bribes collected from citizens.

3. *Continue reform of the army while maintaining its small size.* Many Liberians share Jacques Klein's skepticism that the army is good only for playing cards and plotting coups. Reform of the army has been slow and expensive, and some Liberians engaged in this effort resented that the United States sent corporations rather than "the real army" to assist in the effort. Since 2010, U.S. military engagement has focused on using Army Special Forces to train and advise the Armed Forces of Liberia. This has been effective in depoliticizing the force and refocusing it on useful tasks such as participation, since 2013, in the multinational stabilization mission in Mali.

4. *Focus on land tenure issues.* The trend toward freehold land tenure in rural Liberia is part of a continental trend and accords with economic policies intended to promote private enterprise. But where "private enterprise" is closely linked to illicit power networks, land tenure reform can have the unintended effect of increasing these networks' influence and generating wider insecurity. Reform needs to take local conditions into greater account and be exposed to wider public debate as this transition proceeds.

5. *Accept that some tasks, such as transnational crime prevention, are likely to require open-ended international oversight.* Suppression of drug trafficking and related transnational threats is critical to the success of other reforms in Liberia. Indeed, suppression is of sufficient U.S. and West African interest to justify external provision of this collective good to Liberia, regardless of the local resources committed to the effort. Local buy-in should be encouraged, perhaps through selective Liberian participation in regional organizations and operations.

Conclusion

International actors cannot fix everything in Liberia all at once, and they will not be effective without substantial Liberian buy-in. A more measured response to the problems of illicit power has its downsides, however. First, foreign officials who hold back will face criticism from the threat-inflation chorus. Second, some illicit power structures could pose a threat to local and regional stability. In any event, the core tensions in international strategy concern the persistence of deeply rooted corrupt practices in Liberian politics and society, and the difficulties in disrupting these. Also at play is the insidious dynamic that the more that foreigners do, the less their Liberian partners do. Effective state building is the only long-term solution to these problems, and state building will be effective only if it is driven by locals.

AFGHANISTAN

5. Traffickers and Truckers: Illicit Afghan and Pakistani Power Structures with a Shadowy but Influential Role

Gretchen Peters

Ever since the Soviet invasion of Afghanistan, transnational smuggling groups and Pashtun trucking organizations based in Pakistan have played a prominent though often unrecognized role in shaping and sustaining three and a half decades of conflict. These networks later became closely integrated with the Taliban leadership and also with contemporary Afghan and Pakistani government officials at the national and subnational levels. They moved both licit and illicit goods. For smugglers, continued instability in Afghanistan, a weak central government and judiciary, and malleable police, customs, and border officials were good for business. War in Afghanistan resulted in billions of dollars in transport contracts with the NATO Coalition, which had to import everything needed for the war effort, from jet fuel to toilet paper. For the trucking networks and smuggling groups—many of them defined by familial or tribal ties that cross the Durrand Line—corrupt state institutions on both sides of the Pakistan-Afghanistan frontier facilitated a multibillion-dollar transnational gray transport economy.[1] Such illicit power structures fostered violence as they enriched themselves through transactions in gray and black markets, corrupted state institutions, and fostered a culture of impunity. Worse, they impeded the emergence of a functioning, responsive state in Afghanistan.[2]

Leaders of these smuggling groups and Pashtun trucking organizations do not constitute a recognized constituency with any sort of official role in the reconciliation process. And yet, they have exerted surreptitious influence and could play a spoiler role if they perceived that an end to the Afghan conflict would hurt their bottom line. This chapter will examine how, over three decades of war, the "Pashtun trucking mafia," as it is locally known, and transnational trafficking groups linked to the Taliban emerged as illicit power structures (IPS) and how they have interacted with both state and insurgent power brokers in Afghanistan. Distinct illicit power structures connected to trucking and trafficking also operate in the north and in western Afghanistan, but they are not the focus of this chapter.

It is important to recognize that the impact of smuggling groups and Pashtun trucking organizations was a double-edged sword. Smuggling and the drug trade have no doubt harmed Afghanistan, but they also have provided a livelihood for millions of people there. Illicit markets sustained and prolonged the war and also fostered the emergence of a wealthy, well-connected new elite. The drug transport networks' penetration of the licit and illicit economies across a transnational area of operations has directly impeded the emergence of a stable, responsive state with the capacity to collect revenue. Thus, these industries remained a key obstacle to durable regional peace. In light of these informal networks' powerful destabilizing role, any sustainable peace in

---

[1] Arum Rostum, "How the U.S. Funds the Taliban," *Nation*, Nov. 11, 2009, www.thenation.com/article/how-us-funds-taliban.

[2] Kaysie Studdard, "War Economies in a Regional Context: Overcoming the Challenges of Transformation," IPA Policy Report, March 2004, www.peacekeepingbestpractices.unlb.org/pbps/Library/War%20Economies%20in%20a%20%20Regional%20Context.pdf.

Afghanistan will require not just a political settlement and an end to fighting. It will also take a regionwide economic transition that fosters the self-sufficiency and accountability of regional governments and cultivates viable alternative livelihoods to narcotics and other smuggling. Put another way, a peace deal will remain unlikely unless the political economy of conflict in Afghanistan, both licit and illicit, is taken into account.

Political-Economic Causes of the Afghan Conflict

Before the Soviet Invasion

Afghanistan has been at war since an April 1978 military coup, known as the Saur Revolution, in which the Afghan Communist Party seized power from President Mohammed Daoud, sparking a rural resistance. The outbreak of civil war disrupted historic patterns of economic activity across Afghanistan, already a poor country dependent on farm output for most of its GDP. Agriculture accounted for 60 percent of production in the 1970s, and the weak central government depended heavily on foreign aid to pay its security forces and bureaucrats.[3] Afghanistan has produced opium for centuries, but Soviet forces invaded in 1979, there was only small-scale, traditional production, primarily for local or regional consumption. During the 1970s, the poppy crop had begun to increase somewhat, thanks to a decline in poppy output in Southeast Asia's Golden Triangle, as well as a U.S.-led effort to stamp out opium in Pakistan's border areas, which pushed poppy production across the Durand Line.[4] By the end of the decade, more than half of Afghanistan's 28 provinces cultivated poppy, producing 250 metric tons of opium annually for export.[5]

Another significant prewar development was the growing economic might of the Pashtun trucking sector, which had capitalized on a 1965 transit trade agreement enabling landlocked Afghanistan to import selected commodities duty free.[6] Truckers soon began exploiting the transport pact to evade high Pakistani taxes. In what became known colloquially as the "U-Turn Scheme," Pashtun-owned trucking firms based in the port city Karachi carried electronics, auto parts, and other highly taxed cargo across Pakistan and into Afghanistan in sealed containers, then smuggled the same goods back into Pakistan for sale on the black market.[7] By the 1990s, according to World Bank estimates, the illegal trade in licit goods that flowed through the region was worth $2.5 billion a year.[8]

The Soviet Invasion

In December 1979, the Soviet Union invaded Afghanistan to replace the faltering leftist leadership and prop up the successor regime that the Soviets had installed. This prompt-

---

[3] Barnett Rubin, *The Fragmentation of Afghanistan* (Lahore: Vanguard, 1996), 19.
[4] Jonathan Goodhand, "Frontiers and Wars: The Opium Economy in Afghanistan," *Journal of Agrarian Change 5*, no. 2 (2005): 191-216; Gretchen Peters, *Seeds of Terror* (New York: St Martin's Press, 2009), 42.
[5] Anthony Hyman, *Afghanistan under Soviet Domination, 1964–1991* (London: Macmillan, 1992), 36.
[6] Goodhand, "Frontiers and Wars."
[7] Loretta Napoleoni, (*Terror Incorporated* (London: Penguin, 2003), 119-20.
[8] Z. F. Naqvi, *Afghanistan-Pakistan Trade Relations* (Islamabad: World Bank, 1999).

ed a Cold War escalation in the resistance. A multiparty, multinational insurgent force, known as the mujahideen, received military training, logistical support, and safe haven in neighboring Pakistan, as well as billions of dollars in covert funding from the United States, Saudi Arabia, and other countries. The resistance lasted nine years, during which fighting destroyed much of the country's limited agricultural infrastructure and roadways, prompting growing numbers of poor farmers to plant poppy, a sturdy crop that requires little irrigation and does not rot. Since the rebel mujahideen depended on a foreign-funded pipeline of military supplies, Pakistan-based Pashtun trucking networks could capitalize on the emerging drug trade as well as the war effort, bringing weapons and other war supplies into Afghanistan and hauling narcotics back out. Haji Ayub Afridi, one of the leading smugglers, typified the new transport elite that emerged during this era. He collaborated with the Pakistani government to truck weapons to the Afghan rebels, while also running a multimillion-dollar heroin empire from his enormous fortress in the Khyber Pass.[9]

Inside Afghanistan, mujahideen groups augmented the foreign assistance they received by extorting commodities, including narcotics, that traversed their areas of operation.[10] Certain commanders sought to establish financial independence by deepening their involvement in the heroin trade. For example, Nasim Akhundzada, whose control zone included the fertile Helmand Valley, enforced opium production quotas for poppy farmers and trucked drug shipments into neighboring Iran, where he could fetch higher wholesale prices.[11] Gulbuddin Hekmatyar and Yunis Khalis, two fundamentalist commanders in eastern Afghanistan, operated heroin labs in the border areas and ran transport and bus companies as a cover for moving narcotics and cash.[12] Rebel commanders routinely clashed among themselves over control of the drug business.

Emergence of the Taliban

Following the 1989 Soviet withdrawal from Afghanistan, foreign aid to the mujahideen decreased rapidly, and local commanders deepened their involvement in illicit activities in a realm of highly fragmented power and no effective central state.[13] As fighting between rival mujahideen factions spread across the Afghan countryside, "predation by commanders, opium cultivation by peasants, and smuggling to Pakistan and elsewhere constituted adaptions to this high-risk environment."[14] The violence and predation imposed high costs on commerce, interrupting the transit trade. This prompted a coalition of trucking firms, drug traffickers, ultraconservative religious leaders, and Pakistani intelligence officials to foster the emergence of the Taliban—a movement inextricably linked, from its inception, to the region's illicit economy.[15]

---

[9] See "Heroin in Pakistan: Sowing the Wind," a CIA report leaked to Pakistan's Friday Times newspaper and published in full on Sept. 3, 1983. Afridi was later convicted and jailed in the United States on narcotics trafficking charges.

[10] Peters, *Seeds of Terror*, 35-42.

[11] Rubin, *The Fragmentation of Afghanistan*, 263.

[12] Peters, *Seeds of Terror*, 34-35; author interviews with U.S. officials, 2007, Washington, DC.

[13] Barnett Rubin, "The Political Economy of War and Peace in Afghanistan," *World Development* 28, no. 10 (2000): 1789-1803.

[14] Ibid.

[15] Ibid. For more detail, see Peters, *Seeds of Terror*, 82.

With operational support from Pakistan's military spy agency, a steady supply of young fighters plucked from Pakistani madrassas, and seed funding from powerful traders and traffickers, the Taliban consolidated power over nearly all of Afghanistan's roads, airports, population centers, and major border crossings by 1998. The Taliban regime was thus positioned at the center of a predominant regional illicit economy.[16] Opium output in Afghanistan grew from 2,248 metric tons in 1996 to 4,581 metric tons just three years later.[17] The Taliban government collected taxes from poppy farmers, from laboratories processing opium into morphine base or heroin, and on narcotics shipments and other transit goods destined for export. It was a weak rentier state that depended on narcotics, the transit trade, and foreign donors for revenue.[18] Major traffickers, including Haji Bashar Noorzai and Haji Baz Muhammad (both of whom were later incarcerated in the United States on narcotics charges), sat on the Taliban's ruling council.[19]

Another major influence on the Taliban, and a key source of its financial support, was the truck transport mafia based in Pakistan's western Balochistan province. The trucking firms were drawn from the same tribes as the Taliban leadership and were also linked through business interests and intermarriage.[20] Within months of the Taliban's taking control of the southern province of Kandahar, leaders of the trucking mafia urged Taliban leaders to capture the western city of Herat in order to control key arteries into Iran and Turkmenistan.[21] In March 1995, the Taliban raised $450,000 from trucking networks in western Pakistan in just two days, to fund a failed assault on Herat. It was the Taliban's first major military defeat.[22] Despite this setback, relations remained strong between the Taliban leadership and the transport mafia, with key Taliban leaders also owning their own trucking firms.[23] "The cross-border smuggling trade has a long history in Afghanistan," writes Ahmed Rashid, "but never has it played such an important strategic role as under the Taliban."[24]

Post-9/11, the U.S. Intervention

Illicit trafficking continued to nourish the Afghan conflict in the post-9/11 phase—a period characterized by a high degree of state capture and an insurgency that diversified its criminal business portfolio and expanded into Pakistan. The ensuing decade was highly lucrative, both for traffickers and for transporters. According to the United

---

[16] Rubin, "The Political Economy of War and Peace in Afghanistan"; Michael Griffin, *Reaping the Whirlwind* (London: Pluto Press, 2001), 142.

[17] Peters, *Seeds of Terror*, 34.

[18] Peters, *Seeds of Terror*, 83-85; Rubin, "The Political Economy of War and Peace in Afghanistan."

[19] U.S. State Department, Embassy Islamabad, "Finally a Talkative Talib: Origins and Membership of the Religious Students' Movement," Cable, Feb. 20, 1995, http://nsarchive.gwu.edu/NSAEBB/NSAEBB97/tal8.pdf.

[20] Ahmed Rashid, "Pakistan and the Taliban," Nation, Apr. 11, 1998, www.rawa.org/arashid.htm.

[21] Ibid.

[22] Peters, *Seeds of Terror*, 81.

[23] Rashid, "Pakistan and the Taliban.

[24] Ahmed Rashid, *Taliban: Militant Islam, Oil and Fundamentalism in Central Asia*, 2nd ed. (New York: I. B. Tauris, 2010).

Nations Office on Drugs and Crime (UNODC), Afghanistan produces some 90 percent of the world's illicit opiates, and the increase has been precipitous. As of 2011, the opium trade had an estimated export value of $2.4 billion, and opium profits accounted for 15 percent of Afghanistan's GDP. By 2014, production had increased dramatically—up by 36 percent from 2012-13, which itself was a record year. Profitability was another matter. By 2014, the estimated export value was $.85 billion, accounting for four percent of Afghanistan's GDP—down 17 percent from the previous year. The reduction was not a good-news indicator. Instead, the Afghans had overproduced, and prices fell by 23 percent nationwide as a result.[25] More Afghans were involved, thus siphoning valuable agricultural resources from the licit economy. More troubling was the fact that the link between insecurity and opium cultivation, observed in the country since 2007, continued to be a factor in 2014. The bulk of opium poppy cultivation—89 percent—was concentrated in nine provinces in the southern and western regions, which include the most insecure provinces in the country.[26] In parts of Afghanistan where drugs are less dominant, smugglers tied to the Haqqani network and other illicit groups also export natural resources, including timber, marble, and rare earth metals, earning millions more. U.S. military and intelligence officials also report that some Taliban commanders in the south diversified their criminal earnings portfolio after counternarcotics operations led to losses in earning. New activities included illegal mining and gray-market smuggling.

The dozen years of foreign occupation were also highly profitable for the trucking industry, in particular those firms that transported goods for the U.S.-led coalition and the development community. U.S. military contracts with local trucking firms and security firms totaled nearly $2.2 billion annually, while international aid groups imported tens of millions of dollars' worth of supplies and building materials. Added to this was the country's enormous transit trade.[27] Even during the height of the Coalition "Surge" in 2010-11, transport, security, and trafficking made up as much as one-third of Afghanistan's annual GDP.[28] While not all transport and trade was unregulated and untaxed, black and gray markets thrived in an environment of insecurity, no rule of law, and a weak, corrupt state. Powerful figures in these industries interlinked closely with the insurgency, warlords and their militias, and senior officials in the Afghan government.

The Afghan insurgency has three separately commanded factions: the Quetta Shura Taliban, named for the city in Pakistan where the leadership often takes refuge, the Haqqani network, which operates in the Loya Paktia region straddling the Afghan-Pakistan border, and Hezb-e-Islami Gulbuddin (HIG), a dwindling faction that operates in Afghanistan's east. All three factions have deepened their integration in the black and gray markets since 2001. Taliban commanders in the south and southwest levy taxes

---

[25] See UNODC, "Afghanistan Opium Survey," report, 2014, http://www.unodc.org/documents/crop-monitoring/Afghanistan/Afghan-opium-survey-2014.pdf.

[26] UNODC, "Afghan Opium Crop Cultivation Rises Seven percent in 2014; while Opium Production Could Climb by as Much as 17 percent," Nov. 12, 2014, www.unodc.org/unodc/en/frontpage/2014/November/afghan-opium-crop-cultivation-rises-seven-per-cent-in-2014-while-opium-production-could-climb-by-as-much-as-17-per-cent.html.

[27] Rostum, "How the U.S. Funds the Taliban."

[28] The International Monetary Fund (IMF) estimated Afghanistan's nominal GDP during the 2010 and 2011 (the "Surge" years) at $15.9 billion and $18.4 billion respectively. IMF, "IMF Country Report no. 11/330," Nov. 2011, www.imf.org/external/pubs/ft/scr/2011/cr11330.pdf.

on poppy farmers, the convoys carrying raw opium, and the labs that process it into higher-grade morphine base or crystal heroin. Following the arrests of leading regional drug traffickers based in Pakistan and Iran, senior Quetta Shura commanders increasingly run their own drug-processing centers, and some have also gotten into the export business.[29]

An oft-cited example of this trend is Mullah Naim Barich, the Taliban's shadow governor of Helmand, the southern province that produces most of Afghanistan's opium. According to the U.S. Treasury Department, which designated him a narcotics kingpin in November 2012, Barich was involved in the heroin trade across many levels.[30] He held routine meetings with tribal leaders and large farm owners to set poppy production quotas in the province and issued a written decree detailing procedures to be adopted by subordinate Taliban commanders to combat planned government-led eradication operations in Helmand.[31] The decree stated that all measures, including planting improvised explosive devices, engaging in combat with coalition forces, and bribing state officials, were acceptable actions to protect the poppy harvest, which he called a critical source of funding for the Taliban. Barich also met with regional traffickers and heroin lab owners in Pakistan and coordinated the transport of narcotics consignments that he controlled. One consignment of processed white heroin, owned by Barich, was sent from a narcotics trafficker's compound in Girdi Jangal, Pakistan, to Salawan, Iran, and then on to the Turkish border for further distribution.[32]

Barich may have been the most powerful Taliban commander in the south to establish for himself a steady, independent income from the transport, processing, and export of opiates, but he was hardly the only one. The fact that more Taliban commanders began processing and exporting heroin in the past decade, rather than simply collecting taxes from poppy farmers, marks a critical juncture in the insurgency, indicating that the Taliban network was becoming as much a profit-driven drug trafficking franchise as it was a political organization.[33] At one point in the counterinsurgency campaign, U.S. law enforcement and military officials were tracking more than three dozen separate smuggling operations in Afghanistan, more than half of which answered directly to the Quetta Shura.[34] The immense scale of Taliban drug operations became apparent in 2009, when NATO and Afghan troops launched an offensive to clear militants out of an opium market in the town of Marjah, in Helmand province.[35] After three days of

---

[29] Gretchen Peters, "Crime and Insurgency in the Tribal Areas of Afghanistan and Pakistan," report, Combating Terrorism Center, U.S. Military Academy, Oct. 15, 2010, 24, www.ctc.usma.edu/posts/title-1; Matthew Rosenberg, "Taliban Run into Trouble on Battlefield, but Money Flows Just the Same," *New York Times*, June 13, 2014, www.nytimes.com/2014/06/14/world/asia/for-the-taliban-modest-success-in-battle-but-opium-trade-and-illicit-businesses-boom.html?emc=edit_tnt_20140614&nlid=65470243&tntemail0=y&_r=0.

[30] See U.S. Treasury Dept., "Treasury Targets Taliban Shadow Governor of Helmand Afghanistan as Narcotics Trafficker," press release, Nov. 15, 2012, www.treasury.gov/press-center/press-releases/Pages/tg1768.aspx.

[31] Ibid.

[32] Ibid.

[33] Peters, "Crime and Insurgency," 24.

[34] Author interviews with U.S. government officials, June 2009, Washington, DC.

[35] "Afghanistan's Narco-War: A Report to the Committee on Foreign Relations, U.S. Senate," Aug. 10, 2009, 19, http://fas.org/irp/congress/2009_rpt/afghan.pdf.

fighting, 60 Taliban were dead, and the Coalition had seized a staggering 92 metric tons of heroin, opium, hashish, and poppy seeds, as well as hundreds of gallons of precursor chemicals, making it the second-largest drug haul in world history. Indicating how closely opium merchants and insurgents were collaborating, the market housed a Taliban command center complete with elaborate communications systems, suicide vests, and a large weapons cache.[36]

Other insurgent networks have capitalized on the lack of effective regulation of Afghanistan's natural resources to tap into gray market smuggling opportunities for extractable resources, such as timber and marble. Hezb-e-Islami Gulbuddin (HIG), for example, protected a bustling timber trade in the eastern Korengal Valley—an enterprise that received a boost after the government of then-President Hamid Karzai banned timber exports.[37] The Haqqani network, meanwhile, protected the smuggling of chromite, which was mined illegally in Loghar and Khost provinces and trucked into Pakistan, then shipped to China.[38]

In addition, all three Taliban factions engaged in kidnap for ransom, collaborating with the Pakistani Taliban to hold high-value victims in Pakistan's Federally Administered Tribal Areas. Another major source of income for insurgents, particularly in areas of little or no poppy production, was the extortion of the transport industry, development and construction projects, and telecom firms. Directors and employees of trucking firms have reported that to protect their goods against Taliban attacks, they must pay hundreds—and often thousands—of dollars per container brought into Afghanistan.[39] A special U.S. task force estimated that $360 million in U.S. contracting funds were lost to the Taliban, criminals, and power brokers with ties to both.[40] U.S. military officials said that at least 10 percent of the Pentagon's logistics contracts consisted of payments to insurgents.[41] Some have suggested that the percentages went higher. One Kabul-based businessman in the construction and service industries, for example, reported paying 16 percent of his gross revenue in "facilitation fees," mostly to protect shipments of valuable equipment coming from the border.[42] The U.S. government and military, therefore, indirectly fueled the conflict and financed the very enemies they sought to defeat. This raises the possibility that greed-driven insurgent actors, not to mention state-allied power brokers who also earned "security fees" to protect the roads, could perceive an economic benefit to prolonging the conflict rather than ending it. Thus, any successful peace initiative will likely have to include a "peace dividend" for such actors.

---

[36] Heidi Vogt, "Troops Make Large Drug Seizure in Afghanistan," Associated Press, May 23, 2009.

[37] Peters, "Crime and Insurgency," 39.

[38] Gretchen Peters, "The Haqqani Network: The Evolution of an Industry," Combatting Terrorism Center, U.S. Military Academy, 2012, www.ctc.usma.edu/.

[39] Author interviews with executives and owners of three trucking companies, and several truckers, Kabul, Dec. 2010. See also Rostum, "How the U.S. Funds the Taliban."

[40] Richard Lardner, "Official: US Dollars Ending Up in Taliban Hands," Associated Press, Aug. 15, 2011, www.google.com/hostednews/ap/article/ALeqM5g_56qcWPsv5yff21MkGeD-BVWt0Q?docId=7911bdf124a649f392428fa74cf4ec4c.

[41] Rostum, "How the U.S. Funds the Taliban."

[42] Aryn Baker, "How Crime Pays for the Taliban," *TIME*, Sept. 7, 2009.

Assessing Afghanistan's Illicit Power Structures (IPS)

Power structures. A key factor making the Afghan conflict intractable was the high number of conflict players and the often murky linkages between them. Officially, Afghanistan is an Islamic republic, headed by a president, with a bicameral National Assembly consisting of an upper and a lower house, and a separate judicial branch. In actuality, informal backroom deals between powerful tribal leaders and warlords with private armies have dominated political and economic life since 2001. For example, the man considered by many observers to be the most powerful figure in Afghanistan's south—until his July 2011 assassination—was not the Kandahar governor, but Ahmed Wali Karzai, the president's half brother, who was widely believed to have close ties to the opium trade. Other powerful strongmen, such as Ismail Khan in Afghanistan's west and Rashid Dostum in the north, held military and economic sway over key fiefdoms, controlling important transport and smuggling routes.

The Taliban insurgency was also Balkanized around territorial domains, with Mullah Mohammad Omar recognized as the official leader of the umbrella movement. He commanded the Quetta Shura Taliban, operating primarily in the country's south and west and closely connected to powerful Pakistan-based drug trafficking networks. Senior members of the Shura, including the Taliban's financial commissioner, Gul Agha Ishakzai, and Barich, the shadow governor of Helmand, commanded major smuggling operations in their own right.[43]

The Haqqani network, a semiautonomous arm of the Taliban, operates in the country's southeast and has been implicated in a number of high-profile attacks on the capital, including a September 2011 assault on the U.S. embassy. Another semiautonomous faction, the smaller HIG, operates in Afghanistan's north and east, where it profits from smuggling narcotics, gemstones, and timber. None of these factions operates like a typical Western military, with a clear command-and-control structure. Rather, the insurgent networks are diffuse, with varying degrees of strategic, operational, and tactical coordination between factions.[44] Taliban factions have occasionally clashed—usually over the right to pocket criminal spoils—and there have been indications of a trust deficit between senior leaders.[45] But on balance, the three networks have shown a remarkable capacity to collaborate on military, political, and economic matters since the Taliban fell from power in 2001. Foreign terrorist groups, mainly al-Qaeda, have also collaborated with the Taliban, particularly with the Haqqani faction, which operates as a force multiplier and a training organization.

Emergence of a new criminalized governing elite. Criminal smuggling groups have played a key facilitating role, financing illicit activity in Afghanistan, providing cover to import and export needed commodities, and also bridging the gap to transact deals

---

[43] Author interviews with U.S. officials, June 2011, Washington, DC. Ishakzai was officially designated a financier of terrorism. See also UN Security Council, "Security Council Committee established pursuant to resolution 1988," 2011, www.un.org/sc/committees/1988/NSTI14710E.shtml.

[44] See Seth Jones, "Counterinsurgency in Afghanistan," RAND Security Studies Program Seminar, Nov. 7, 2007, http://web.mit.edu/ssp/seminars/wed_archives07fall/jones.htm.

[45] Peters, "Crime and Insurgency," 43; Peters, "The Haqqani Network."

that require indirect cooperation between insurgents and state officials (who remain officially at war). The complex web of relationships built by Haji Juma Khan, formerly Afghanistan's premier heroin trafficker, illustrates how political-criminal linkages cut across battle lines, proving that drug profits trump ethnicity, tribe, and politics in the Afghan war. Khan, who was arrested in 2007 and brought to the United States to stand trial, was until then the dominant smuggler in Afghanistan's southwest, where he subcontracted Taliban fighters to protect his poppy fields, drug convoys, and heroin processing plants.[46] To ensure that Afghan security forces did not hold up his shipments, Khan reportedly forged a deal with Ahmed Wali Karzai to coordinate heroin trafficking in the south.[47] He also paid off intelligence and border agents in Pakistan and Iran and was even on the payroll of the CIA.[48] According to U.S. officials who tracked the network, Khan's nephew Hafiz Akhtar took over and has continued to export multiton shipments of narcotics from Afghanistan, using Taliban gunmen to protect his consignments. Akhtar was briefly arrested in 2009 by U.S. forces, who handed him over to the Karzai government in order to comply with an agreement that NATO forces not hold detainees indefinitely. In a move suggesting persistent collaboration and collusion between drug traffickers and Afghan officials, the Karzai administration had Akhtar released within 72 hours. He appeared so unconcerned about a subsequent arrest that he returned to using the same satellite phone he had used before, even though apparently aware that authorities were tracking it.[49]

Major trafficking and transport operations in Afghanistan also integrated closely with powerful figures in the federal and provincial governments and the insurgency. Official investigations and media reports indicated that opium traffickers bought off hundreds of police chiefs, judges, and other officials, including senior members of the Kabul administration, in deals that cut across enemy lines.[50] President Karzai's own late half brother, as mentioned, reportedly conspired with Haji Juma Khan, a drug trafficker closely aligned with the Taliban, to take control of opium operations that opened up in the south after the U.S. government arrested Haji Bashar Noorzai, another leading smuggler.[51]

Connections between the transport industry, powerful warlords, and the Kabul regime also caused the emergence of a new and powerful Afghan elite with a significant financial stake in prolonging the status quo and, with it, the conflict. The powerful Watan Group, for example, run by two of Hamid Karzai's cousins, had a multimillion-dollar contract with the U.S. military until a yearlong investigation uncovered indications that the firm was operating an illicit protection racket funneling cash to the insurgency.[52] An-

---

[46] Peters, *Seeds of Terror*, ch. 5.

[47] James Risen, "Propping Up a Drug Lord, Then Arresting Him," *New York Times*, Dec. 11, 2010, www.nytimes.com/2010/12/12/world/asia/12drugs.html.

[48] Ibid.

[49] Author interview with senior U.S. military intelligence official, Jan. 2010, Arlington, VA.

[50] Thomas Schweich, "Is Afghanistan a Narco-State?" *New York Times*, July 27, 2008, www.nytimes.com/2008/07/27/magazine/27AFGHAN-t.html?pagewanted=1&ref=drugtrafficking; Graeme Smith, "Afghan Officials in Drug Trade Cut Deals across Enemy Lines," Globe and Mail, Mar. 21, 2009, http://aol.theglobeandmail.com/servlet/ArticleNews/aolstory/TGAM/20090321/AFGHANDRUGS21.

[51] Risen, "Propping Up a Drug Lord."

[52] Lardner, "Official: US Dollars Ending Up in Taliban Hands"; Rostum, "How the U.S. Funds the Taliban." See also Watan Group, corporate website, www.watan-group.com/.

other firm that attracted scrutiny was NLC Holdings, owned and operated by Hamed Wardak, the U.S.-educated son of the former defense minister, General Rahim Wardak. As part of the military's "money as a weapons system" doctrine, NLC and five other local trucking companies saw their contracts surge sixfold in 2008, to $360 million a year, even though NLC itself never owned a single truck.[53] Instead, Wardak perched atop a murky pyramid of subcontractors who provided the trucks and safeguarded their passage.[54] U.S. military officials said the system worked and convoys got through, but conceded that they knew little about how payments were disbursed.[55] How the U.S. and NATO drawdown will affect this market—and the illicit Afghan economy generally—remains unclear, but the large sums that key power brokers have earned raises the possibility that some might feel motivated to keep U.S. forces in the region.

Provincial and district-level power brokers have also appeared to benefit directly along Afghanistan's dangerous highways. Matiullah Khan, an illiterate former highway patrolman in the dangerous southern province of Oruzgan, for example, established a private army that earned millions of dollars annually protecting NATO convoys and that also conducted operations alongside U.S. Special Forces.[56] Foreign military officials said that people such as Matiullah helped fill a vacuum in areas where state security services were lacking—though at the price of undermining the very institutions that NATO was trying to build.[57] One day a week, Matiullah and his men lined a particularly perilous stretch of highway linking Kandahar with Oruzgan and declared it open, charging $1,200 for safe passage of each NATO cargo truck, or $800 for smaller ones. Matiullah Khan's income, according to one of his own aides, came to $2.5 million a month; U.S. officials suggested he was also protecting opium shipments in his control zone.[58]

Warlords such as Matiullah may have kept the roads safe for military trucks, but NATO officials said that data they collected suggest these networks often schemed to extend their contracts. One U.S. military study found that attacks on roads protected by paid gunmen rose, on average, more than threefold in the final three months of their NATO contract.[59] Applying the collective-action logic, strongmen such as Matiullah Khan no doubt perceived a benefit to prolonging the chaos along Afghanistan's highways and may even have colluded in or coordinated attacks on their own operations to sustain the status quo.[60] Illicit revenue from narcotics and other smuggling, and the troublesome issue of protection fees along Afghanistan's highways, far outweighed the potential income that war profiteers such as Matiullah Khan could likely earn from a peacetime economy. The challenge in Afghanistan has remained that the country's il-

---

[53] Rostum, "How the U.S. Funds the Taliban"; Karen DeYoung, "Afghan Corruption: How to Follow the Money?" *Washington Post*, Mar. 29, 2011, www.washingtonpost.com/wp-dyn/content/article/2010/03/28/AR2010032802971.html.

[54] DeYoung, "Afghan Corruption."

[55] Ibid.

[56] Dexter Filkins, "With U.S. Aid, Warlord Builds Afghan Empire," *New York Times*, June 5, 2010, www.nytimes.com/2010/06/06/world/asia/06warlords.html?pagewanted=1.

[57] Ibid.

[58] Ibid.

[59] Author interviews with U.S. and NATO military officials, Kabul, Dec. 2010.

[60] Mancur Olson, *The Logic of Collective Action: Public Goods and the Theory of Groups* (Cambridge, MA: Harvard Univ. Press, 1965).

licit economy is critical to the existing balance of power. Thomas Schweich, the former U.S. drug czar in Afghanistan, wrote, "The trouble is that the fighting is unlikely to end as long as the Taliban can finance themselves through drugs and as long as the Kabul government is dependent on opium to sustain its own hold on power."[61]

The Taliban's Path to Power. Control over, and protection of, illicit resources and industries has provided insurgents and associated illicit power structures with a degree of legitimacy that allows them to mobilize fighting forces and other public support.[62] We see evidence of this phenomenon in Afghanistan, where civilians and small businesses have turned to the Taliban as contract enforcers and where the insurgents have provided key "public" services, such as madrassa education and dispute resolution. Insurgents have bolstered their status in some rural communities where they continue to be perceived as more efficient and less corrupt than the state. But predation and violence by the insurgents has prevented the Taliban from capitalizing fully on whatever political capital the insurgency gained from protecting illicit activities. In surveys and field interviews, the Taliban remain deeply unpopular.[63]

Taliban leaders have at times appeared to recognize the strategic risk of alienating the populace. In 2009, for example, the Quetta Shura issued a new code of conduct in an apparent attempt both to exert control over violent and unruly Taliban field commanders and to improve relations with ordinary Afghans, by establishing a civilian shadow government at the local level. Under the new structure, the Taliban regulated tax rates that commanders could charge the public, and also created provincial-level commissions where Afghans could present their requests or complaints to a local council of religious scholars, who then had to answer to the executive council in Quetta.[64]

Strategic and financial motives appeared to be behind the overhaul. "The reason they changed their tactics is that they want to prepare for a long-term fight, and for that they need support from the people; they need local sources of income," said Wahid Muzhdah, a former Taliban official who later tracked the insurgency.[65] Apparently, the code was spottily enforced, although Afghan civilians have often praised the Taliban shadow justice system for being swift and fair and have reported that Taliban fighters extorted them less than corrupt state officials did.[66] But some civilians also said they cooperated with the insurgency simply out of fear. Taliban "night letters," distributed in rural communities—and backed by actual assassination campaigns—routinely threatened dismemberment or a grisly death for those who worked with the government or foreign forces.[67]

The Haqqani network operated as a shadow state in its control zones, running police forces and madrassas and managing a system of dispute management that was well

---

[61] Schweich, "Is Afghanistan a Narco-State?"

[62] Vanda Felbab-Brown, *Shooting Up: Counterinsurgency and the War on Drugs* (Washington, DC: Brookings Institution Press, 2010).

[63] Peters, "Crime and Insurgency."

[64] Ibid, 15.

[65] Alissa Rubin, "Taliban Overhaul Their Image in a Bid to Win Allies," *New York Times*, Jan. 21, 2010, www.nytimes.com/2010/01/21/world/asia/21taliban.html?pagewanted=all.

[66] Peters, "Crime and Insurgency," 19.

[67] Ann Scott Tyson, "In Helmand, Caught between U.S., Taliban," *Washington Post*, Aug. 15, 2009, www.washingtonpost.com/wp-dyn/content/article/2009/08/14/AR2009081403568.html.

regarded—and feared—at the local level. Haqqani leaders have intervened on key occasions in recent years to resolve disputes among other militant groups in Afghanistan and Pakistan, most notably helping ensure an orderly transition of power in the Pakistani Taliban after the group's former emir, Baitullah Mehsud, was killed in an August 2009 drone strike.[68]

The local business community and rival tribes have also engaged the Haqqani shadow justice system. Local sources reported that the Taliban collected fees for resolving disputes and would also hold "surety bonds" while disputes were being negotiated. The Taliban would occasionally confiscate this bond money if one party did not live up to its end of the bargain.[69]

The Haqqanis also have operated more than 80 madrassas and training bases. These serve as guesthouses and way stations for foreign fighters and terrorist groups and, therefore, could be seen as a breeding ground for cooperation among jihadi organizations.[70] The madrassa system typifies the interlocking web of political and economic bargaining that has supported interactions between the Haqqanis, other militant actors, the illicit business community, and the populace. The madrassas have served as key conduits for cash couriers moving illicit proceeds out of Afghanistan, and for network leaders sending monthly salary payments to fighters inside the country.[71] Also, the madrassas offer free room, board, and education for poor young men, thus providing a vital service to poor families who cannot otherwise feed and school their children. This has helped supply a seemingly endless stream of young militants to the network, which has been able to sustain monthly losses of about 150 men and still regenerate its forces.[72] Local sources have suggested that the Haqqanis incurred high daily costs feeding and housing militants—costs that were believed to be covered in part by their partners in the local smuggling and transport industries.

Other power brokers across Afghanistan have also provided the services of a shadow state, building roads, schools, and even hospitals for the local communities.[73] For example, the smuggler Haji Juma Khan financed road and power projects in the southwest, as well as desert reclamation projects to help poppy-farming communities expand into parched territory.[74] Another Pakistani-based smuggler, Sakhi Dost Jan Notezai, has operated a four-story hospital where Taliban fighters wounded on the battlefield could receive free medical care. The fact that public services run by illicit networks have supported and interacted with insurgent groups should not suggest that these groups were one and the same, or that their political goals were always consistent. Rather, deepening levels of cooperation suggest a symbiotic dependency. A key element that both depended on was weak governance.

There are growing indications, however, that rising levels of organized crime, coupled with high levels of terrorist violence, have undermined public support, particularly

---

[68] Peters, "Crime and Insurgency," 83.

[69] Peters, "The Haqqani Network."

[70] Shuhrat Nangyal, "Afghanistan's Jihad and the Virtuous Victory," Manba al-Jihad (in Pashto), no. 8 (Jan. 1991). Nangyal also wrote several articles for the group during the 1980s and 1990s.

[71] Peters, "The Haqqani Network."

[72] Joshua Partlow, "Haqqani Insurgent Group Proves Resilient Foe in Afghan War," *Washington Post*, May 29, 2011.

[73] Peters, *Seeds of Terror*, chs. 45.

[74] Ibid., ch. 5.

since local communities have been the primary victims of both. Militants may have protected some illicit economies, thereby winning a degree of public support from community members seeking to protect their income source. But militants' interaction with civilians was also predatory by nature, both in the taxes and protection fees the militants charged and because the instability they created hampered development of licit alternatives. Members of the local community cooperated with the militants at times as a coping strategy, but this does not mean they embraced the militants as a popular force.

It is difficult to characterize the Afghanistan insurgency generally, made up as it is of various shifting parts. But broadly speaking, we can describe the insurgency as an illicit power structure that has employed violence to oppose the peace process. Still, at least some, if not most, factions have demonstrated negotiable interests at times and, at certain points, may even have been amenable to a grand bargain. Indeed, various factions appeared to be more open to negotiation immediately after the 2001 invasion, meaning that circumstances have, in fact, deteriorated since then. Indeed, the most significant change between 2001 and 2015 is that there are now overlapping Taliban and state-linked IPSs, which collaborate to profit from the ongoing conflict—a factor that NATO strategy has completely failed to address. This has produced an environment where power brokers on both sides of the battlefield have significant financial incentives to spoil any peace process. The drawdown of international forces will inevitably have an effect on their economic opportunities, although it may simply cause an increase in illicit activity, particularly in drug trafficking.

The Insurgency's Relationship to the Peace Process

Early informal reconciliation efforts by insurgency and traffickers are squandered. In the months after the U.S.-led intervention in Afghanistan began, there were tantalizing indications that leading figures in the Taliban were prepared to reconcile with the U.S.-backed Karzai government and that major narcotics traffickers were supportive, even willing, to help facilitate the peace process. Senior members of Mullah Omar's inner circle reportedly sent a letter to President Karzai, claiming to have permission from the Taliban leader to surrender.[75] But under pressure from Washington and senior officials in the anti-Taliban Northern Alliance, Karzai reportedly ignored the overture, and the Taliban officials remained in Pakistan.[76] Separately, Ibrahim Omari, a brother of Jalaluddin Haqqani, surrendered to the Kabul government soon after the Taliban government fell, whereupon the Ministry of Defense, which was dominated by the anti-Taliban Northern Alliance, swiftly detained him.[77] Whether or not Omari's time in Kabul involved any talk of reconciliation for his brother's fighting forces, nothing concrete developed, and he was later quietly allowed to return to Pakistan.[78]

---

[75] These included the movement's number two leader after the collapse of the Taliban regime, Mullah Baradar, the Taliban's defense minister, Mullah Obaidullah, and the interior minister, Mullah Abdul Razzaq.

[76] Thomas Ruttig, "Negotiations with the Taliban: History and Prospects for the Future," *New America Foundation*, May 2011, 67, www.newamerica.net/sites/newamerica.net/files/policydocs/Ruttig_Negotiations_With_The_Taliban_1.pdf.

[77] Thomas Ruttig, "Talking Haqqani," *Afghan Analysts Network*, Jan. 7, 2010, http://aan-afghanistan.com/index.asp?id=873.

[78] Ibid.

A few months later, U.S. warplanes bombed a convoy of tribal elders allegedly traveling from southeastern Afghanistan to attend President Karzai's inauguration.[79] Dozens died in the air strike, with news reports suggesting that some were former Taliban, among them prominent members of Haqqani's Zadran tribe.[80] Around the same time, another group of former Taliban tried to present itself as a moderate alternative. The group was allowed to return to Kabul and was put up in a government guesthouse, but the U.S. doctrine of not talking to terrorists prevented any concrete role for the group in the new regime.[81]

Other efforts to reconcile with the insurgents amounted to little. A foreign-funded program aimed at enticing Taliban fighters to switch sides by offering financial incentives ended in failure and accusations of corruption. Assessments found that few if any known Taliban, let alone any senior leaders of the three main factions, were among the 4,634 people who had joined the program by October 2007.[82] The Saudi and Emirati governments and the United Nations facilitated a separate series of exploratory diplomatic initiatives with people close to the Taliban. Nothing concrete evolved from these early encounters, either.[83]

There were also opportunities to make deals with leading drug traffickers. Haji Juma Khan was briefly detained by U.S. forces in late 2001, whereupon he reportedly promised to help them track down terrorists.[84] He would allegedly remain on the CIA's payroll until his arrest in 2008 on narcotics trafficking charges.[85] Haji Bashar Noorzai, another prominent Afghan trafficker and Taliban financier, made contact with U.S. troops soon after they arrived in the southern province of Kandahar. He turned in truckloads of Taliban weapons and offered to negotiate the surrender of leading figures in the network.[86] Both Juma Khan and Noorzai wanted a role in the new government and economy, and amnesty for their ties to narcotics and the Taliban.[87] In other words, they had negotiable interests.

Negotiations with the smugglers continued, off and on, over the next decade, with Juma Khan even making a 2006 trip to the United States as a guest of the government.[88] Eventually, U.S. law enforcement arrested both men: Noorzai in 2005, and Juma Khan in 2008.[89] Noorzai was detained in New York City after flying there thinking he had an amnesty deal with U.S. authorities. Whether these two smugglers ever had any intention of quitting the narcotics trade remains an open question. What is pertinent is that both men appeared prepared to reconcile, so long as they could escape jail and maintain their stature in the community.

---

[79] Reuters, "Afghan Elder Warns Karzai over Convoy Bombing," Dec. 23, 2001, www.chron.com/news/article/Afghan-elder-warns-Karzai-over-convoy-bombing-2073821.php.

[80] Ibid.

[81] Ruttig, "Negotiations with the Taliban," 7.

[82] Ibid., 7.

[83] Ibid., 78.

[84] Peters, *Seeds of Terror*, 158.

[85] Risen, "Propping Up a Drug Lord"; author interviews with U.S. officials, 2009, Washington, DC.

[86] Peters, *Seeds of Terror*, 202.

[87] Ibid, 158, 202; author's interviews with U.S. officials, 2007, Washington, DC.

[88] Risen, "Propping Up a Drug Lord."

[89] Peters, *Seeds of Terror*, chs. 5 and 7.

Taliban returns to the fight; relationships with traffickers deepen. Another open question is whether the Taliban leadership would have embraced—or even entered into—a reconciliation process, had there been one after they fell from power in 2001. But when no prospects for doing so emerged, Taliban commanders and leading traffickers regrouped in the southern countryside and in Pakistan, some raising money by selling opium stockpiles.[90] The Taliban began launching low-intensity attacks in 2003, ambushing foreign aid workers and setting off bombs. Still, they did not cause major disruptions to the 2004 presidential election, and newly elected President Karzai responded by declaring that any Taliban who wanted to come back and live as normal citizens would be welcome.[91]

In the rural south, opium output continued to grow, soaring to a high of 193,000 hectares of poppy cultivated in 2007, according to UNODC figures.[92] Taliban actors expanded and diversified their ways of profiting from the drug trade. At the village level, Taliban subcommanders earned money from taxing poppy farmers and transporters. Over time, commanders able to consolidate power in poppy-producing regions in the south expanded into the business of running heroin and morphine processing labs and even exporting narcotics consignments. Until about 2009, trafficking networks, based mainly in Pakistan and Iran, dominated the business of exporting drug consignments from Afghanistan. But this began to change, too, in part because foreign counternarcotics forces (mainly British and American) arrested key traffickers, opening a space in the market for new entrants. Powerful Taliban commanders, including Mullah Naim Barich in Helmand, stepped into that gap. Barich represented a new generation of Taliban commanders, who were not simply collecting taxes on poppy crops and drug consignments but actively meeting with district-level officials and tribal leaders to set opium production quotas, organizing military campaigns to attack government-led eradication programs, and coordinating exports of drug consignments.[93] Barich and others like him gained increasing wealth and autonomy, making the insurgency more fragmented but also better funded and better armed.

The changing, more fragmented shape of the insurgency and the widening drug market brought extra layers of complexity for would-be peacemakers. Most importantly, the narcotics trade had a deeply corrosive effect on the Afghan government, corrupting senior officials, hollowing out already weak state institutions, and creating financial disincentives for officials to work toward stabilizing the country. Meanwhile, with widening insecurity in the south and southwest, where fighting and poor transport conditions limited many farmers' access to markets for licit crops, poppy cultivation became

---

[90] Ibid., 114.

[91] *Economist*, "Going Straight," Dec. 2, 2004, www.economist.com/node/3447071?story_id=3447071&subject=Afghanistan.

[92] In 2000, the Taliban implemented a nationwide ban on growing poppy. This one-year ban on poppy farming is often cited as evidence that the Taliban sought to suppress the narcotics trade. Some have argued that the Taliban banned farmers from growing poppy, because large stockpiles of opium had built up and they sought to bolster sagging farm prices for the drug. Foreign counternarcotics officials noted that the Taliban did not implement a simultaneous ban on morphine and heroin labs and continued to collect taxes on drug exports during this period.

[93] *RS News*, "Taliban Leader Designated as Drug kingpin," Nov. 15, 2012, www.isaf.nato.int/article/isaf-releases/taliban-leader-designated-as-drug-kingpin-nov-15.html.

"a low-risk crop in a high-risk environment."[94] Corruption, narcotics, and poor security hampered the emergence of a healthy licit economy while a U.S.-funded eradication campaign enraged poor farmers and drove some to support the insurgency in order to protect their livelihoods. Diplomatic and political initiatives revived the peace process, then faltered.

After coming to power in 2009, the Obama administration tried to pursue peace talks with the Taliban leadership in hopes of engineering a grand peace bargain with the insurgents. Obama appointed Richard Holbrooke, an architect of the 1995 Dayton Peace Accords in Bosnia, to oversee the effort, which centered on fostering talks between Kabul and the Taliban. The plan also called for increased foreign aid aimed at luring Pashtun communities with the promise of jobs and development.[95]

Many Afghans were dubious of the strategy, and discord arose locally and internationally over the shape and speed of the process. Some non-Pashtuns and former leaders of the Northern Alliance, which had fought the Taliban, complained that funneling millions of dollars into Taliban-dominated regions in the Afghan south would unfairly benefit ethnic Pashtuns and reward those who had fought the government.[96] Many Afghans, observers, and analysts doubted that the Taliban would agree to renounce violence, sever ties with al-Qaeda, and respect the Afghan constitution—the three core demands of the Karzai government. Meanwhile, women's groups and human rights organizations both inside and outside Afghanistan feared that accommodating the Taliban leadership would lead to a reversal of women's rights gains made since 2001 and would dissolve hopes of holding Taliban leaders accountable for human rights abuses. "I think it is just legalizing impunity," said Sima Samar, chair of the Afghanistan Independent Human Rights Commission. "Nobody is accountable, not for the past crimes and not for future ones. Anybody can come and join the government and they will be protected."[97] Even Taliban officials expressed skepticism, saying no real progress could be made while U.S. forces remained on the ground across Afghanistan.

Within the U.S. government, sharp differences arose over the framework and purpose of talks with the Taliban, and different government agencies often seemed to be working at cross purposes. U.S. military leaders suggested that a surge of U.S. troops could pummel the Taliban into a weaker bargaining position and prompt more low- and mid-ranking insurgents to desert.[98] U.S. military intelligence assessed the Taliban leadership as weak, divided, and no longer able to exert command and control over its

---

[94] David Mansfield and Adam Pain, "Evidence from the Field: Understanding the Changing Levels of Opium Cultivation in Afghanistan," briefing paper, Afghanistan Research and Evaluation Unit, Nov. 2007.

[95] Joshua Partlow, "Karzai's Taliban Reconciliation Strategy Raises Ethnic, Rights Concerns at Home," *Washington Post*, Feb. 4, 2010, www.washingtonpost.com/wp-dyn/content/article/2010/02/03/AR2010020303737.html.

[96] Sanjeev Miglani and Hamid Shalizi, "In Afghanistan's Panjshir, Disquiet over Taliban Reconciliation," Reuters, Sept. 8, 2011, www.reuters.com/article/2011/09/08/us-afghanistan-massoud-idUSTRE7872A920110908.

[97] Karen DeYoung and Joshua Partlow, "In Afghanistan, Karzai's Invitation to Taliban Creates Discord and Confusion," *Washington Post*, Mar. 3, 2010, www.washingtonpost.com/wp-dyn/content/article/2010/03/02/AR2010030204101.html.

[98] Author interviews with U.S. military and State Department officials, 2010, Washington, DC.

forces inside Afghanistan.[99] Leading diplomats, meanwhile, believed they could draw in senior Taliban commanders and privately doubted that the Pentagon's much-vaunted counterinsurgency campaign was making much headway.[100] A series of high-profile leaks and media reports exacerbated the U.S. debate over how to pursue reconciliation, fueling distrust between the White House and the Pentagon and eventually leading to the dismissal of General Stanley McChrystal, the commander President Obama had personally selected to lead the mission there.[101] U.S. law enforcement added another layer of complexity to the process by arresting and designating as targets key traffickers and Taliban officials. Some U.S. diplomats grumbled that the designations would limit options for negotiations since U.S. law prohibits negotiating with individuals designated as major narcotics traffickers.[102]

Other states in the NATO coalition did not always see eye to eye on the substance or pace of reconciliation efforts. While U.S. military leaders wanted more time to pound the Taliban into submission, British and European allies, facing strong domestic disapproval over the long-running war, appeared more eager to see negotiations begin regardless of the conditions.[103] These internal disputes, which played out in the global media, prevented unity of effort at both national and international levels. Moreover, there were virtually no substantive discussions on underlying economic issues that fed the violence—specifically, the narcotics trade—let alone offers from the international community to provide economic support and farm subsidies that might help wean Afghanistan's economy off heroin.

Despite all these differences, the reconciliation effort briefly appeared to be moving forward in 2010. Karzai affirmed that talks with the Quetta Shura Taliban were ongoing, and appointed 70 tribal and political leaders to his High Peace Council, which was intended to lead future negotiations. NATO leaders confirmed that they had facilitated the talks by providing security and technical support.[104] Around the same time, the Kabul government also reached out to HIG, and two senior HIG officials met Karzai in Kabul for talks, although those conversations ended without significant breakthroughs.[105] Even the Haqqani network's leader signaled that he was prepared to reconcile. In a rare media interview in 2011, Sirajuddin Haqqani, son of Jalaluddin and de facto leader of the network, said his group would take part in peace talks with the Kabul government and the United States, as long as the Quetta Shura approved the process.[106]

---

[99] See Jeffrey Dressler, "Reconciliation with the Taliban: Fracturing the Insurgency," Institute for the Study of War, June 13, 2012, www.understandingwar.org/backgrounder/reconciliation-taliban-fracturing-insurgency.

[100] Author interviews with U.S. State Department officials, 2010, Washington, DC.

[101] See Michael Hastings, "The Runaway General," *Rolling Stone*, June 22, 2010, www.rollingstone.com/politics/news/the-runaway-general-20100622; Max Fisher, "Leaked Cables Raise Questions on Kabul-Washington Tension," *Atlantic*, Nov. 12, 2009, www.theatlanticwire.com/global/2009/11/leaked-cables-raise-questions-on-kabul-washington-tension/26442/.

[102] Author interviews with U.S. State Department officials, 2009, Washington, DC.

[103] DeYoung and Partlow, "In Afghanistan."

[104] Ruttig, "Negotiations," 1.

[105] Ibid., 3.

[106] Reuters, "'No Sanctuaries in Pakistan': Haqqani Network Shifts Base to Afghanistan,"*Express Tribune*, Sept. 18, 2011, http://tribune.com.pk/story/254368/no-haqqani-network-sanctuaries-in-pakistan-sirajuddin/.

Significantly, the first reported high-level meetings between the Kabul regime and the Quetta Shura brought together Karzai's half brother Ahmed Wali and Mullah Baradar, the Taliban's second in command and a key figure in coordinating opium production and heroin exports, just a year before Wali was killed.[107] Britain's *Telegraph* newspaper reported that the two met in the border town of Spin Boldak, a home base for major Pashtun trucking networks, and that their meeting focused on the country's lucrative heroin trade. "Baradar controls a big stake in the drugs trade in Kandahar," said an unnamed source quoted in the *Telegraph*. "It is a peace deal in a sense, reducing the conflict between the government and the Taliban, according to private interests."[108] It is significant that economic matters and trade—not political or ethnic differences, development funding, women's rights, or a resolution to the fighting—ruled the agenda in the first high-level meeting between the Karzai regime and the Taliban. These power brokers wanted to make sure that underlying economic factors—the drug trade, specifically—were discussed before other political grievances and core issues came up.

The process was quickly interrupted by a regional spoiler. The Pakistan government responded to news of the Spin Boldak meeting by arresting Baradar and a half-dozen other senior Taliban officials and commanders.[109] Islamabad thus showed its hand and illustrated that it had the capacity to round up Taliban officials sheltered in its territory whenever it wished. As Thomas Ruttig, a leading analyst on Afghanistan and the reconciliation process, wrote, "With these [arrests], the Pakistani military de facto claimed a veto on all negotiations with the Taliban and therefore on Afghanistan's political future."[110] Perhaps more than anyone else in the region, Pakistan has had the ability to spoil any reconciliation process, especially since the same networks that transported both licit and illicit goods to and from southern Afghanistan were based in its territory.

The peace effort continued to face major challenges, indicating that spoilers (it was never clear precisely who) intended to disrupt the reconciliation process even before it got under way. In late 2010, it emerged that a man involved in secret meetings with Afghan officials, who had received as much as $500,000 in payouts from NATO, was, in fact, an imposter. He was later identified as a shopkeeper from Quetta, Pakistan, and not a representative of the Quetta Shura, as negotiators had first believed.[111] This deception deeply discredited the effort and brought into sharp focus the complexity of dealing with such secretive, fragmented insurgent networks. Then, in September 2011, an assassin detonated the explosives hidden in his turban as he embraced Burhanuddin Rabbani, the chair of Karzai's High Peace Council, in a classic and devastating spoiler attack.

The suicide bomber, who posed as a Taliban emissary carrying conciliatory messages, dealt a ruinous blow to Kabul's most productive effort to date on reconciliation with the Taliban. The assassination of Rabbani, an ethnic Tajik and a former president of Afghanistan, not only threatened to splinter fragile alliances between the country's eth-

---

[107] Dean Nelson and Ben Farmer, "Hamid Karzai Held Secret Talks with Mullah Baradar in Afghanistan," *Telegraph*, Mar. 16, 2010, www.telegraph.co.uk/news/worldnews/asia/afghanistan/7457861/Hamid-Karzai-held-secret-talks-with-Mullah-Baradar-in-Afghanistan.html.

[108] Ibid.

[109] Ruttig, "Negotiations," 1.

[110] Ibid., 10.

[111] Ibid., 3; Author interviews with U.S. officials, June 2011, Washington, DC.

nic groups, it also illustrated clearly the capacity and the will of anonymous spoilers to disrupt the peace effort. According to Matt Waldman, a fellow at Harvard's Belfer Center who had contacts with the insurgent leadership, Rabbani's killing was not approved by the Taliban leadership council. But that does not rule out the possibility that people inside the Taliban, perhaps with Pakistani backing, were involved. The Peace Council never recovered the momentum it had before Rabbani's killing, although proposals to make the effort relevant again included revamping the council to include former Taliban officials, negotiators experienced in conflict resolution, and members of Afghan civil society.[112] It is clear that throughout the process, Taliban interests remained negotiable and at least some insurgent leaders were motivated to preserve their economic interests while also gaining amnesty and a role in the future power structure.

International Strategy and Its Impact

Ultimately, it was always difficult to ascertain the center of gravity for the U.S. effort in Afghanistan. The Obama administration appeared to be conducting a two-pronged strategy: simultaneously reaching out to Taliban officials with one hand while trying to decimate their networks militarily with the other.[113] It can be argued that the reconciliation effort suffered particularly as a result. Even as recently as 2014, officials, academics, and journalists close to the Afghan reconciliation effort say that little of concrete value has been achieved. Key parties remain divided over what course of action to pursue, nothing yet can even be called a reconciliation process, and Pakistan continues to assert its role as arbiter and spoiler.

The history of reconciliation efforts during 2011-12 is illustrative. In 2011, according to diplomats and others close to the process, a series of productive secret meetings did take place between U.S. officials and the Taliban, but very little of the conversation touched on any substantive issues or the potential framework for a peace process. "From what we know, there were probing, exploratory sessions to exchange opinions and viewpoints and to develop ideas for moving forward," said Waldman. Discussants did not confront underlying factors that fed the violence, such as endemic poverty, regional militancy, the Indo-Pak rivalry, or the narcotics trade and other smuggling.

In early 2012, the Taliban appeared to be moving toward formal talks with the U.S.-led coalition in Afghanistan. The Taliban announced that it had struck a deal to open a political office in Qatar that could conduct direct negotiations over the endgame in the Afghan war.[114] But it was less clear whether the Taliban wanted to work toward a comprehensive peace settlement or simply wanted NATO to end its operations in Afghanistan in 2014, as scheduled—thereby removing a major obstacle to the Taliban's return to power in all or part of the country. The Taliban wanted the return of high-ranking prisoners held in Guantanamo Bay, Cuba, and reportedly held meetings to discuss this with U.S. Special Envoy Marc Grossman, who had replaced Ambassador Holbrooke

---

[112] Ibid., 15.

[113] Ibid., 3.

[114] Matthew Rosenberg, "Taliban Opening Qatar Office and Maybe Door to Talks," *New York Times*, Jan. 3, 2012, www.nytimes.com/2012/01/04/world/asia/taliban-to-open-qatar-office-in-step-toward-peace-talks.html?pagewanted=all.

upon Holbrooke's death in 2010.[115] By March 2012, however, the Taliban announced that it was withdrawing plans to open the Qatar facility, and ending further "pointless" talks with the Americans. In a statement posted on a Taliban website, the group blamed the breakdown on the "alternating and ever-changing [U.S.] position."[116]

The general impression was that the U.S. administration perceived reconciliation efforts as an extension of the military campaign, aimed at dividing and weakening the insurgency rather than working toward a genuine peace deal. Under President Obama, the number of reported drone strikes into Pakistan's tribal areas increased dramatically, as did targeted killings of Taliban by U.S. Special Forces. Some analysts blamed this two-pronged targeting approach for pushing the Taliban back from the bargaining table. Perhaps more significantly, it also cost public support in Afghanistan by causing a perceived increase in civilian casualties.[117] The U.S. strategy no doubt produced tactical results: from mid-May to mid-August in 2011 alone, 350 mid-level Taliban commanders were reportedly killed or captured.[118] What is unclear is whether those results actually impaired the Taliban militarily or politically. Some diplomats and intelligence analysts perceived the Taliban to be weakened, demoralized, and increasingly divided as a force.[119] But the strategy did not appear to slow the use of improvised explosive devices to kill and maim NATO troops, nor did it affect Taliban recruitment efforts and shadow-government operations.[120] Meanwhile, the Taliban responded to the stepped-up U.S. war effort with its own assassination campaign against public officials, aid workers, and civilians. Insurgents also launched high-profile attacks in Kabul, which claimed mainly Afghan civilian victims.

Divisions and paranoia within the Afghan government further stymied progress. Senior members of the former Northern Alliance, which fought the Taliban in the 1990s, remained hostile to any talk of concessions to the Taliban. And the Shia Hazara community, who had suffered ethnic massacres at the hands of the Taliban in the 1990s, expressed particular security concerns about any deal permitting the Taliban's return. Meanwhile, because of distrust between the Karzai government and the Americans, Washington officials insisted they would meet with the Taliban only if the Kabul government attended. President Karzai seemed at times to rely on his Peace Council to handle issues concerning reconciliation. At other points, he seemed to lean on members of his inner circle. Said one Western diplomat at the time, "The Karzai government at once wants to be at the center of things but has not demonstrated the capacity to conduct anything approaching a reconciliation process."[121]

There appeared to be no further efforts to engage major narcotics trafficking organizations or the trucking community to support the peace process. "These would be among the issues that you would want to discuss in the course of the process, but it never got that far," said Waldman.

---

[115] *Telegraph*, "Taliban Break Off 'Pointless' Qatar Talks with US and Karzai," Mar. 15, 2012, www.telegraph.co.uk/news/uknews/defence/9145709/Taliban-break-off-pointless-Qatar-talks-with-US-and-Karzai.html.

[116] Ibid.

[117] Ibid., 4.

[118] Ibid., 10.

[119] Author interviews with U.S. government officials, 2012, Tampa, Washington, DC.

[120] Ruttig, "Negotiations," 10.

[121] Author interview with Western diplomat, 2012, New York.

In early 2013, just as the Obama administration announced plans to withdraw 34,000 U.S. troops by the end of the year, there seemed to be renewed efforts toward reconciliation. After a stall in late 2012 as the domestic presidential election consumed U.S. attention, diplomats and political leaders from eight countries began mounting a concerted campaign to bring the Afghan government and its Taliban foes together to negotiate a peace deal.[122] In early February, UK Prime Minister David Cameron met with President Karzai and President Asif Ali Zardari of Pakistan in calling for fast-track peace talks. Weeks earlier, in Washington, Karzai himself committed publicly to have his representatives meet a Taliban delegation in Qatar to start the process. Officials cited a growing consensus that regional stability demanded some sort of broad settlement with the Taliban, given that most foreign forces would be pulling out within the coming 12 to 18 months. Still, few officials thought they would achieve even the limited goal of bringing the Afghan government and Taliban leadership to the table before the bulk of U.S. fighting forces left Afghanistan in 2014.[123] For the insurgents, the imminent departure of foreign troops was surely an incentive just to wait it out. There was also a growing concern that the departure of foreign forces would cause a sharp decline in economic activity, leaving tens of thousands of truck drivers, guards, and other local staff who served the military effort out of work. A 2012 World Bank study projected that GDP growth would drop from nine percent in 2010/11 to closer to five percent on average until 2018 and may decline further in the long term.[124] This prompted fears that the drawdown would invigorate the drug economy and other illicit smuggling, undoing whatever fragile gains had been made in reducing poppy cultivation in recent years.

Conclusions and Recommendations

Conclusions

Negotiating peace is especially challenging wherever key conflict actors have a major financial stake in sustaining disorder. It would require a regionwide approach to confront or co-opt conflict actors, such as drug trafficking networks, who have a regional capacity to finance disorder and corrupt political processes. Unfortunately, there has been no such regional strategy in Afghanistan. In fact, all signs point to an international community that wishes to be less, not more, engaged in Afghanistan. The regional illicit economy, not to mention growing rates of opium and heroin addiction, will be self-reinforcing, whereas strategies for disentangling these problems will be slow, expensive, and complex. It is hard to imagine the international community even being able to agree on a comprehensive strategy to reduce narcotics supply and demand in Central and South Asia. Therefore, any efforts to reduce or resolve the conflict will have to take the drug economy into account.

---

[122] Alissa Rubin and Decland Walsh, "Renewed Push for Afghans to Make Peace with Taliban," *New York Times*, Feb. 16, 2013, www.nytimes.com/2013/02/17/world/asia/pressure-for-peace-with-taliban-is-renewed.html?emc=tnt&tntemail0=y.

[123] Ibid.

[124] World Bank, "Afghanistan in Transition: Looking Beyond 2014, Volume 2," report, May 2012, http://siteresources.worldbank.org/AFGHANISTANEXTN/Images/305983-1334954629964/AFTransition2014Vol2.pdf.

Opportunities were lost in the years after the Taliban fell from power. The international community and the new Afghan government could have approached illicit power structures and potential spoilers in a variety of ways. One track would have been to co-opt and reintegrate them into the licit economy. Or they could have been formally designated as criminal groups and brought to justice.[125] It would have been useful for NATO commanders and the international community to identify which power brokers they considered beyond reconciliation, which ones could be considered supportive and uncorrupt, and which ones fell into a gray area in between. Then a strategy could have been developed to target the irreconcilables or degrade their capacity to do harm, and to bolster those who brought stability and progress. Such actions might have influenced those actors in the "gray zone" to clean up their behavior and become responsible, responsive brokers in the new regime. Meanwhile, regional economic development programs to address transnational shadow networks could have improved regulatory effectiveness. And regional trade regulatory programs designed to limit the flow of narcotics and other conflict commodities would have supported such a strategy.[126] In Afghanistan, that would have required greater commitment than the international community appeared willing to make, and the current drawdown suggests that traffickers and truckers will have considerable influence over that region's economy and politics for the foreseeable future.

Recommendations

As NATO nations continue to withdraw the bulk of their troops in 2015, it is hard to imagine the final chapter of the Afghan conflict being written anytime soon. Although it is far too early for a postmortem on the reconciliation effort, we can make four broad conclusions about lessons learned.

Recognize and address the illicit political economy's impact on the conflict. The peace process would be further along if the intervening powers had made the political economy of conflict in Afghanistan a primary focus. This is not to suggest that reconciliation was even best option—law enforcement strategies and counterinsurgency tactics, if properly applied from the outset, might have significantly weakened the insurgency. But the peace process was bound to falter since it did not embrace political-economic issues as central. Afghanistan is a place of widespread scarcity, but there is no doubt that the period from late 2001 through 2014 was immensely profitable for the conflict elites, both within the state and within the insurgency. It is rational for those who reap great profit from a war economy to seek to prolong the conflict rather than end it.[127] At the very least, war profiteers will tend to seek a conflict resolution that maintains their grip on

---

[125] Studdard, "War Economies."

[126] Ibid.

[127] David Keen, "Incentives and Disincentives for Violence," in *Greed and Grievance: Economic Agendas in Civil Wars*, ed. Mats Berdal and David Malone (Boulder, CO: Lynne Rienner, 2000); Kristina Höglund, "Violence in War-to-Democracy Transitions," *From War to Democracy: Dilemmas of Peacebuilding*, ed. Anna K. Jarstad and Timothy D. Sisk (Cambridge, UK: Cambridge University Press, 2008).

power and resources.[128] Efforts since 2001 to start a reconciliation effort have made little headway, and this is tied partly to the fact that initial talks mainly focused on tactical issues—exchanging prisoners and stopping attacks—rather than on financial and economic matters. It is worth pointing out again that the highest-level reported meeting between the Taliban and the Karzai regime focused not political issues but on the heroin trade. For Afghans, divvying up spoils of the drug trade appears to have been a primary concern.

Adopt a flexible approach to reconciliation with illicit power structures. A second lesson is that, distasteful as it may seem, it is sometimes better to co-opt or reconcile with illicit power structures than to seek their destruction. In the emotional months after the 9/11 attacks, when U.S. troops first invaded Afghanistan, U.S. officials had little appetite for bargaining with the Taliban, the Haqqanis, and the various drug traffickers who turned themselves in to Coalition forces or tried to probe options for surrender. At the local level, too, there were calls for justice that likely would have prevented reconciliation. However, 13 years later, the Afghan state and the NATO coalition have made little headway fighting the insurgency, and some insurgent factions have grown far richer and more powerful through their ties to transnational organized crime. They also harbor 13 years' worth of mistrust of U.S. actions and intentions. As Waldman points out, "Unwillingness on the part of the U.S. to engage the Taliban until 2011 meant there were deep misgivings within the Taliban about coming to an understanding with the Coalition."[129]

It is possible that major traffickers and transporters could have been engaged as part of a coalition to support the peace process in return for a promised peacetime dividend if they gave up illicit activities and began paying taxes into state coffers. Of course, we can never know the truth of their intentions, but it is worth recalling that two major traffickers, Haji Bashar Noorzai and Haji Juma Khan, offered to support the reconciliation process but were instead arrested by the U.S. government. Since they were not recognized as having negotiable interests, they were not embraced as potential agents for reconciliation. The irony is that arresting them made no difference in reducing the size of Afghanistan's opium economy; rather, opium output continued to grow, and widening insecurity meant that transport costs skyrocketed as well. As a result, trucking and trafficking networks enjoyed a highly profitable dozen or more years and may fear that an end to the conflict will bring about a decline in their wealth and power. Things could have been vastly different if negotiators had actively sought to separate the Taliban factions with negotiable interests from those without. Then they might have provided alternatives to violence and criminal activities to advance those negotiable interests.

Never dismiss organized crime and corruption as "secondary issues." A third, related lesson learned from the past 13 years in Afghanistan is that efforts to stamp out terrorism and militancy should not be given priority over efforts to fight corruption and organized crime. Afghanistan remains broadly unstable, not because its government is engaged in militancy, but because the state remains highly corrupt and tied to the narcotics trade, which in turn has fueled continued militancy. Meanwhile, the insur-

---

[128] Keen, "Incentives and Disincentives for Violence."
[129] Author interview with Matt Waldman, 2013, Washington, DC.

gency was also financed by narcotics and crime. In stability operations, U.S. officials, whether military, law enforcement, or diplomatic, needed to be intolerant of corruption and criminal behavior by their local counterparts. They needed to support the emergence of healthy state institutions, rather than looking the other way when those they considered allies were involved in illicit activities. NATO would have benefited from the existence of an interdepartmental task force with the responsibility to investigate the nexus of terrorism, insurgent organizations, and organized crime in Afghanistan, Pakistan, and Central Asia; the will to expose connections between the criminal underworld and business/government elites; and the authority, using international law, to oppose their criminal activities. Afghanistan illustrated how a foreign intervention ensures its own defeat when it turns a blind eye to local corruption and drug trafficking. Accommodating corruption costs the international community more in the long run because fragile states remain aid dependent and fail to evolve into stable, self-sustaining nations that can become durable partners.

Work to prevent transnational criminal networks from playing a spoiler role. Finally, the Afghanistan experience highlights why a peace process must identify all the relevant constituencies that could play the role of spoiler, and engage them in the pre-reconciliation process. This would have meant engaging Russia, India, Iran, the United States, and especially Pakistan—a tall order indeed. Not engaging some of these key regional actors meant that they could take on a spoiler role or, at least, influence local constituencies to spoil the process. At the local level, there were efforts to bring Afghanistan's myriad tribes, ethnicities, and political groups on board. But scant attention was paid to the economic constituencies, specifically the trafficking and transport sectors, that wielded tremendous influence over other political and ethnic factions in the country. It would never have been easy to achieve balance between these myriad actors and constituencies. But not engaging them helped guarantee that the process would fail.

This chapter focuses on a shadow constituency that has no name and cannot be formally identified as a single group, which nonetheless has played a critical role in shaping contemporary Afghanistan. This constituency is transnational and profit driven, and its members may fear that peace in Afghanistan will not be good for business. Because this constituency has proven capacity to distort and thwart peace efforts, any sustainable process will require not just a political settlement and an end to fighting. It will also require a regional economic transition that fosters good governance and cultivates durable economic alternatives to narcotics and gray-market smuggling. It is hard to imagine a solution to the Afghan war unless the political economy of the conflict is taken into account.

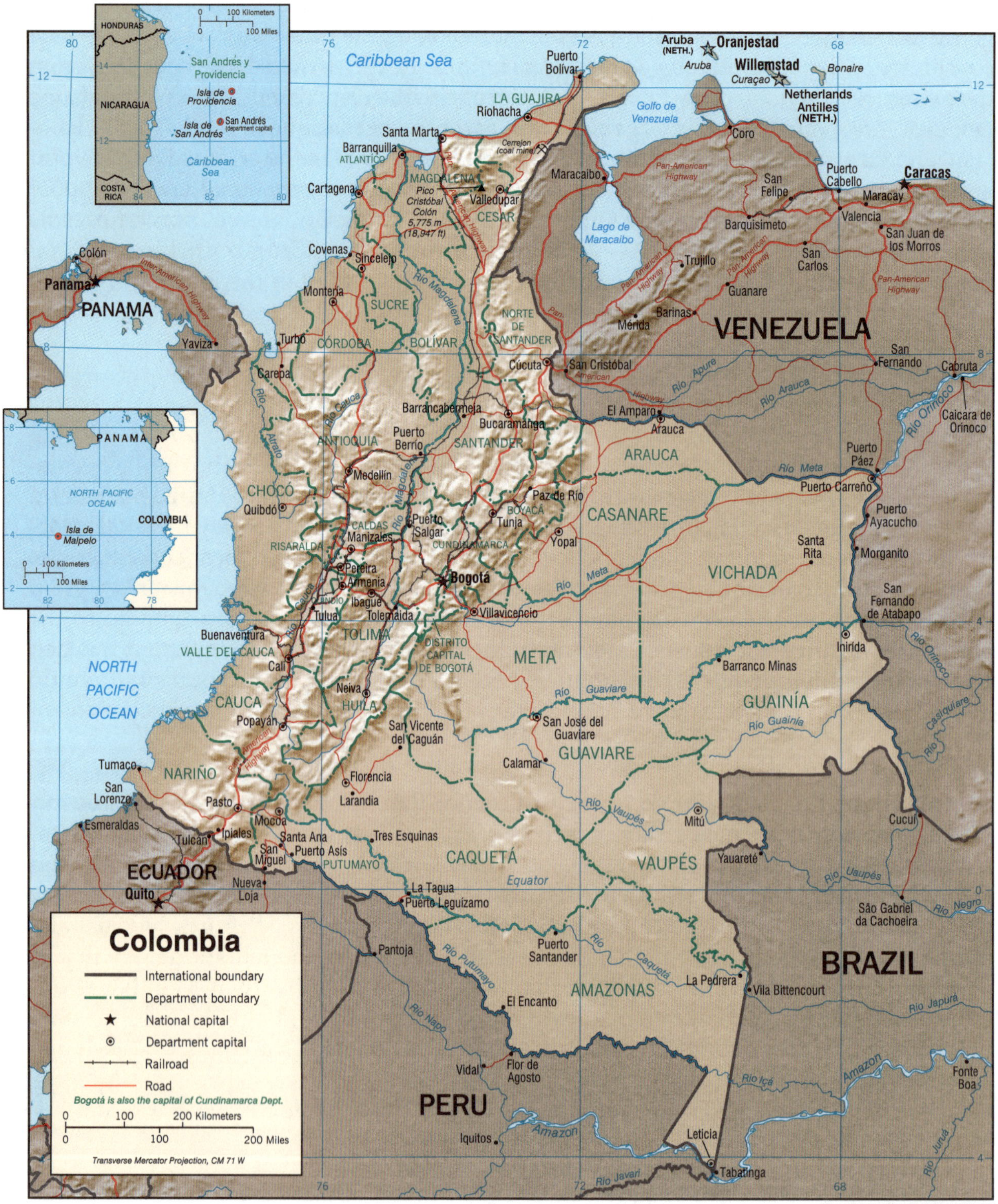

Base 803338AI (R00820) 1-08

6. Colombia and the FARC: From Military Victory to Ambivalent Political Reintegration?

Carlos Ospina

In the immediate aftermath of World War II, Colombia entered a decade of civil war from which it has never fully emerged. A key contributor to the violence throughout has been las Fuerzas Armadas Revolucionarias de Colombia (Revolutionary Armed Forces of Colombia), or FARC, an organization that reached its peak during the presidency of Andrés Pastrana (1998-2002) but declined precipitously thereafter when decimated by a military-led national resurgence. This occurred, in the first instance, during the initial term of President Álvaro Uribe (2002-6), with consolidation continuing in the second Uribe term (2006-10).

Then the first term of President Juan Manual Santos Calderón (2010-14) saw a startling reversal of the Santos electoral pledge to continue Uribe's policies and, in its place, a commitment to an open-ended peace process, which is ongoing at the time of this writing. Santos was narrowly reelected to a second term (2014-18) but finds himself faced with a host of economic, social, and political challenges aside from FARC's stubborn refusal to commit definitively to ending the conflict. This has placed Santos and his administration in an awkward position of needing a deal at almost any cost to validate their abandoning what was seen in 2010 as a model counterinsurgency, and embracing instead "conflict resolution" as advocated by an array of internal and external actors.

The Roots of Conflict

The origins of the current conflict in Colombia date back nearly to independence in the early nineteenth century, yet fighting continues between the government and insurgent groups even today. This long and bloody struggle can be traced to several root causes, or drivers, described briefly below.

Political fissures opened shortly after the Colombian War of Independence in 1819, when Simón Bolívar and the Liberator Army defeated the peninsular troops of King Fernando VII. As early as 1839, civil war erupted, dividing those forces previously united in the independence struggle. Seven more civil wars were fought between 1839 and 1899.[1] All were politically driven by the emerging and competing Liberal and Conservative political parties, each struggling for power and defending its own political and economic interests.[2] The last of these civil wars, the Thousand Days' War of 1899-1902, was especially costly, with an estimated 200,000 deaths and widespread destruction of infrastructure. The result was enduring poverty for the survivors. This conflict had two

---

[1] Manuel Santos Pico, *Historia militar del ejército de Colombia* (Bogotá: Sección Publicaciones Ejército de Colombia, 2007), 149.

[2] Colombia's long history since the original Spanish intrusion into the Amerindian world on the fringes of the Inca empire has resulted in a population that is largely ethnically, culturally, and linguistically homogeneous. Unlike neighboring Peru, Colombia has no substantial indigenous population (officially one percent versus Peru's more than half, which gave the Sendero Luminoso insurgency a large ethnic recruiting base). Colombia's four percent black population is clustered in Choco department; thus, any political movement there draws from it.

well-defined phases. The first, in which regular armies fought each other, ended with the Conservative Party's victory at the battle of Palonegro. The second, in which Liberal insurgents who survived Palonegro fought on with guerrilla actions for two more years, ultimately ended in their defeat and the conclusion of a peace treaty, bringing the conflict to an end. And yet, this guerrilla legacy lived on.

Panama, at the time still a Colombian department, having tired of the seemingly perpetual violence and destruction—and with the armed support of a United States interested in building the Panama Canal—seceded and became independent. Meanwhile, the increasingly visceral hatred between Liberals and Conservatives festered. This acrimonious relationship lasted more than 50 years, and even though actual civil war did not reoccur, societal developments "failed to resolve the tensions between these political elites who supported a strong central government and those who supported strong regional governments."[3] This enduring tension became a wellspring for future violence and the emergence of FARC in the mid-twentieth century.

Moreover, a tradition of rebellion and guerrilla warfare was born in certain regions of the country, such as el Cauca in the south and Santander near the Venezuelan border. In 1948, the historical accumulation of societal friction ignited an explosion when Liberal leader Jorge Eliécer Gaitán was assassinated in Bogotá. Liberals blamed the Conservative government and rebelled, even attacking the presidential palace with support from defecting police. The army was called in, and after three days the situation locally was brought under control—at a cost of 3,000 casualties and a capital core in ruins.[4] Worse was to come, for the urban violence in Bogotá and other cities spilled into the countryside. There, peasants affiliated with the two parties continued to battle, directed by their leaders in the cities. It was civil war, with the traditional working tool, the machete, the offensive weapon of choice. It is estimated that 300,000 Colombians perished during the terrible days remembered as la Violencia (the Violence). The memory of the sheer horror became an enduring part of the Colombian psyche, to be exploited in particular by those who claimed to speak for the victims or as the authentic voice of history.

To protect their families, armed self-defense groups (*autodefensas*), with Liberal or Conservative affiliations, soon were organized. And, of course, they, too, fought each other with escalating intensity and violence. The Conservative groups were loosely related to the police, at that time typically a local rather than a national force. Liberal groups followed their own urban leadership, while those on the radical margins turned to the Communist Party of Colombia (PCC), founded in 1930. It urged self-defense forces not only to "resist" but to declare their autonomy from the state—in the Marxist lexicon, to liberate their social as well as their political formation.

The PCC interjected itself into the crisis to implement the strategy inherent in the model provided by the Communist Party of the Soviet Union. Its goal as a self-proclaimed vanguard of the revolution was to exploit the situation in order to mobilize people across a wide range of sectors. In fact, the party had already organized cells in the countryside as an extension of its urban networks. When violence exploded in the countryside, these groups became quite active against the Conservatives, in time joining

---

[3] Robert D. Ramsey III, *From el Billar to Operations Fenix and Jaque: The Colombian Security Force Experience, 1988-2008*, Combat Studies Institute Press, Occasional Paper no. 34, 2009, 5.

[4] Santos Pico, *Historia militar del ejército de Colombia*, 268.

forces with the Liberals and attempting to co-opt them. In 1953, however, the political leadership of the country turned to a military man, General Gustavo Rojas Pinilla, and handed him the reins of power from June 1953 to May 1957 — the only instance of military rule in Colombia during the entire twentieth century. He implemented a compromise, pardoning all who had been caught up in the political violence, and moved the country to a political pact, the National Front (*Frente Nacional*), from 1958 to 1974, whereby the two major parties, Liberals and Conservatives, would share power by alternating in four-year terms of office (keyed to the presidency). The PCC, however, was not included and, hence, resumed its struggle.

Ironically, the PCC was (and remains) a legal political party and used the guerrillas as only one of its many tools, though it had no operational control over them, providing only political guidance. This resulted in the increasing estrangement of the PCC guerrilla leader, Manuel Marulanda, who was joined by the PCC's own field representative, Jacobo Arenas. This led to the formation of FARC. (Most sources use 1963 as the actual founding date, and 1964 as the date the founding was formally announced.) Eventually, the estrangement led to a de facto split with the PCC in 1982, at FARC's seventh party conference. It was at this conference that FARC first broached the idea (realized in 2000) of forming an alternative to the PCC: a "Clandestine PCC" (PCCC), with FARC as that party's "popular army." Although the PCCC would have to wait, FARC thereafter styled itself el Ejército Popular (the People's Army, or FARC-EP). Its autonomy became complete with its embrace of fundraising through taxing the drug trade (at least, initially — later, it integrated itself into the complete cycle of production).

The relationship with drugs, agreed upon at the same 1982 conference, gave FARC the financial means to seek implementation of its military project, which called for seizing state power. From the moment it entered into this relationship, despite recognizing the risk, FARC increasingly became a captive of its own means. The military and political cause served as the movement's motivating force, to which social and economic issues, though rhetorically significant, were subordinated. FARC embraced the struggle against rural poverty, discrimination and exclusion, and social class structure in the abstract and made them components of its narrative, but it took few tangible steps to address these challenges, even in isolated areas where it held sway.

Leading figures in local areas responded to FARC's depredations — and the government's lack of counteraction — by falling back on the historically tested response: forming *autodefensas*. Although popularly manned and supported in the beginning, these eventually followed a trajectory similar to that of FARC itself: recruiting manpower from outside threatened localities and raising money through involvement in the drug trade. Just as importantly, the same atrocities normally associated with FARC — and by Ejército de Liberación Nacional (National Liberation Army), a much smaller but still active Marxist-Leninist insurgency after the Cuban model — became staples of the *autodefensas*. Since self-defense groups were legal before the mid-1990s, a subsequent designation of illegality did not dissuade at least some individuals in the military and the police from continuing their support for such groups. In time, these grew and became yet another factor contributing to the endemic violence, with the peculiar characteristic that many of their actions were directed against the local populace (who supported FARC) and included massacres, assassinations, rapes, and other forms of violence.

Rise and Fall of the FARC Insurgency: Adrift in a Sea of Coca

FARC's evolution as an insurgent movement, and its transformation into a trafficking and terrorist organization, was a direct consequence of adopting involvement in the narcotics trade to finance its would-be revolution. During FARC's early period of growth, its financial means came from kidnapping for ransom and extortion, which resulted in a weak organization, poorly equipped and trained, with no real strategy or significant military capacity. But with the inflow of drug trafficking revenue, this situation changed dramatically, and FARC turned into a powerful force with enough capacity to sustain a protracted war and to attack military units of battalion and even brigade size.

Illicit Finance

FARC's financial support structure has evolved in line with the organization's history. In its initial incarnation in the 1960s, it was a genuine peasant organization waging a real insurgency; thus, as a movement it was poor. Support came from the people of those regions where FARC was present, because the combatants were seen as *los muchachos* (the boys), who were fighting for a better future for all Colombians. But the financial resources collected in this way were insufficient to meet the organization's aspirations. The PCC, while not an advocate of protracted war or "useless guerrilla fighting," saw these FARC groups as a tool for provoking the government in the rural countryside and also as a means to advance its conspiratorial strategy of arraying all possible means under the guidance of a Leninist vanguard party (as emerged from the PCC Tenth Party Congress in 1964). To the PCC, the guerrillas (soon to establish themselves as FARC) would be part of the strategy of insurrection—not the main part, but merely a component acting in the countryside.

Support from the PCC was therefore very limited, and FARC throughout the 1960s and 1970s remained a small organization with limited assets and limited local, rural objectives. During this early phase, then, the PCC mainly influenced and stimulated FARC through ideological support but had no operative or tactical control and, in fact, was not involved in FARC operations.[5] This could therefore be described as a politically "naive" phase in FARC's evolution; its illicit means and networks were not nearly as developed as they would be years later.

Eventually, FARC leaders realized that under this arrangement, they would never accomplish their objective of seizing power through prolonged war, and, moreover, that such a goal was not what its "mentor," the PCC, wanted. They came to the realization that FARC was merely a sideshow—merely a component part of the PCC's strategy of struggle. Thus, they decided that FARC must fight its own war by its own means, independently of any external support. This signaled the end of the purely "insurgent phase," when FARC depended completely on the rural people in its areas of operation for sustenance and security. It did remain in touch with the PCC, but the party was no longer a factor in FARC's decision making.

---

[5] Carlos Ospina, *A la cima sobre los hombros del diablo* (Madrid: Editorial Académica Española, 2012), 910.

With this distancing from the PCC, and loss of that source of financial support (meager though it surely was), sustainability became a pressing issue. FARC adjusted to its new financial circumstances by adopting innovative new methods of extraction, including kidnapping and extortion. Thousands of Colombian citizens were abducted, taken into captivity, and, in some cases, killed by FARC kidnapping squads. Soon, though, a new and even more lucrative means of financial sustenance would emerge and become dominant: drugs.

At the end of the 1970s, drug trafficking had arrived in Colombia.[6] The new commerce was rapidly adopted and became widely used by a range of illegal organizations. First, it was marijuana, but soon it became clear that this product required substantial physical inputs—not only space for cultivation but also vehicles for transportation. To generate profits, hundreds of shipments of product had to be moved—ultimately an inefficient business model because making a profit required transport of great quantities. Moreover, competition in the marijuana business was already quite brutal, with places such as California producing significant quantities for the U.S. market. In contrast, cocaine was superior on all counts. It was through exploiting the enormous potential inherent in the cocaine trade that a figure such as Pablo Escobar built, in Medellín, the cartel that was to emerge as the leading organization in the drug trade. Indeed, in 1989-91, it grew powerful enough to challenge the state directly. In Cali, the Rodríguez brothers established a less brutal (though possibly more profitable) rival cartel.[7]

The FARC leadership was not interested in "dirty money." Until that time, strict regulations prohibited the use of drugs in any way inside the organization. But a subordinate column chief, Argemiro, began extorting drug traffickers within his area of operations.[8] He charged by the number of kilos produced in the labs, the number of hectares under cultivation, number of workers, and in many other ways. Soon, his column became wealthy, and he became very rich. National-level FARC leaders—the Secretariat—were horrified. Arrested by Secretariat forces and tried in a revolutionary court, Argemiro was expelled from the organization and, some months later, assassinated, likely by his old companions.

But the idea of using drug money to finance operations began to make sense at the Secretariat level. The logic went that if FARC could tax all aspects of the cocaine cycle, with zero tolerance for individual use or enrichment within the movement itself, it could secure a sustainable source of financial support while improving its fighting capabilities. This would allow the conflict to move to a more intense phase, from guerrilla action to directly challenging the state. At the next FARC conference (the seventh, in 1982), the Secretariat approved taxing drug traffickers in FARC areas in the same manner that Argemiro had, but with the difference that revenues should be forwarded to the Sec-

---

[6] See James D. Henderson, *Colombia's Narcotics Nightmare: How the Drug Trade Destroyed Peace* (Jefferson, NC: McFarland, 2015).

[7] Felia Allum and Renate Siebert, eds, *Organised Crime and the Challenge to Democracy* (New York: Routledge, 2003), 89.

[8] FARC, *Seventh Conference Final Report*, June 1982; copy examined in mimeo (unavailable online). For a sanitized (i.e., no mention of drugs or terrorism) version of FARC's Marxism-Leninism, see Emilio Salgari, *Marulanda and the FARC for Beginners*, FARC-EP, 2011, https://farc-epeace.org/pdf/Marulanda-and-the-FARC-EP.pdf.

retariat, in this way preventing the accumulation of wealth in any individual guerrilla leader's hands.

Subsequently, various FARC regional commands were tasked with controlling coca cultivation in their areas, charging the authorized taxes, and sending the money to the Secretariat. The Secretariat in turn distributed funds according to the different regional requirements for weaponry and war materiel as well as sustainment of the operational tempo dictated by FARC strategy. In particular, the Sixteenth Front (one of 67 such rural geographic "fronts of war" that also functioned as unit headquarters with integral combatant formations), near the Venezuelan border; the Fifty-Seventh Front, near the Ecuadorian border; and the Fifty-eighth Front, near the Panamanian border, were assigned this responsibility. This was the Secretariat's theory and intent. And indeed, by adopting this new operating concept, FARC grew in strength. At the same time, however, individual guerrilla leaders were also lining their own pockets, and soon personal enrichment became a common practice for individuals with access to illicit funds. Some deserted, absconding with substantial personal fortunes.

Adoption of cocaine trafficking marked a significant turning point. With such a sustainable and highly lucrative revenue stream, many of FARC's strategic projects became feasible, and the idea of a full-scale war against the state took shape. With the organization of larger columns and "regularization" of formations (i.e., turning purely guerrilla units into light infantry), FARC could jump to the next stage in its "people's war" strategy of insurgency. This war of movement, in which FARC-EP "military" units attacked similar units of the army, was the high tide of the organization's revolutionary project. Initial assaults against military forces in small bases began as early as 1994, but the August 30, 1996, overrunning of a 120-man company at Las Delicias, eastern Putumayo, on the Río Caquetá began a series of devastating FARC attacks that often featured multiple-battalion strength and even improvised armor. More than 400 members of the military and the national police were killed, and some sources warned that FARC was winning the war.[9]

With the huge drug profits flowing in, FARC could also purchase new weapons. A shipment of 10,000 AK-47 rifles with the associated accessories and parts, for instance, was purchased in 1999 using falsified documents provided by corrupt senior Peruvian officials, including Vladimiro Montesinos, then the national security adviser to Peruvian president Alberto Fujimori. Simultaneously, the Secretariat decided to invest in legal businesses to build a sustainable support infrastructure. This was done using contacts in other countries, particularly in Colombia's self-proclaimed "Bolivarian" (neo-Marxist) neighbors. It was estimated by some observers that through shell companies, FARC ran a network of transportation companies, small industries, and similar businesses in Venezuela and Ecuador. The Secretariat also used coca receipts in neighboring countries to hire services, such as medical personnel (doctors and nurses), and hospital facilities to treat their wounded and sick personnel.

Despite the Secretariat's efforts to prevent leakage of illicit revenue flows within the organization, newly rich cells began diversifying into new businesses in their regions. As a consequence, their chiefs became wealthy. They began to distinguish themselves

---

[9] Douglas Farah, "Colombian Rebels Seen Winning War; U.S. Study Finds Army Inept, Ill-Equipped," *Washington Post*, Apr. 10, 1998, http://colombiasupport.net/archive/199804/wp41098.html.

and show off their affluence, wearing upscale wardrobes and eating better food. They had their own personal money—something previously forbidden within the organization. They created their own local networks of coca traffickers and learned the nuances of the trade, even imposing price controls and regulating the market in ways that caused resentment among the peasants and non-FARC local drug lords.

As a consequence of these developments, FARC has not needed direct external funding support. Nor has it received material support from abroad. On the contrary, with the huge revenues from its drug trafficking business, it has the resources to provide financial assistance to foreign friends and followers. From the FARC files seized in 2008 by Colombian forces during an operation (Fénix) in Ecuador, it seems clear that FARC provided financial support to the political campaigns of sympathetic Bolivarian supporters Hugo Chávez, in Venezuela, and Rafael Correa in Ecuador.[10] Both of them were ultimately elected—Correa is the incumbent president of Ecuador, and Chávez died of cancer in March 2013, while still in office.[11] The seized computers contained the files of Raúl Reyes, a member of the FARC Secretariat, who was killed during the raid. There are also reports from the government of Paraguay, of FARC financial support to the Paraguayan People's Army (EPP), a recently organized guerrilla force responsible for the kidnapping and execution of Cecilia Cubas, daughter of former Paraguayan President Raúl Cubas.[12]

Thus, over time, FARC's involvement in drug trafficking allowed it to become a very different organization, one that continued to use terrorism and guerrilla tactics but could also field light infantry columns structured and armed much like the Colombian forces themselves. Indeed, given the identical uniforms, equipment, and even weaponry, only tactical modifications (e.g., differently colored arm bands) distinguished one from the other. This was true also for the National Liberation Army (Ejército de Liberación Nacional, or ELN) forces (the "other" guerrilla group) and the AUC (the self-defense groups, which, at their peak, had roughly as many combatants as FARC). FARC and the AUC had become integrally connected with the drug trade, while ELN relied primarily on extortion for its funding. Not discussed here but also requiring state attention and application of resources were the numerous "petit cartels"—successors to the crushed major drug cartels. But FARC retained the focus of government concern because only FARC could field regular forces that threatened the very existence of the state. In this, it was aided by sanctuary and assistance from Venezuela and Ecuador. Gradually, too, a division of labor saw coca handed off to the Mexican cartels for distribution to the U.S. market, which brought these astonishingly violent actors into the larger field of government concern. A similar development in expansion of distribution routes occurred in West Africa, with Europe the ultimate destination (and profits from that market segment at one point nearly equaling those of the United States). In this theater of operations, though,

---

[10] See IISS, *The FARC Files: Venezuela, Ecuador and the Secret Archive of "Raúl Reyes,"* (London: IISS, 2011).

[11] Both presidents denied the allegations, which are based on overwhelming evidence, and have said that the captured documents were part of an elaborate conspiracy by the Colombian government (at the time, headed by Álvaro Uribe).

[12] The victim was found dead on February 15, 2005, five months after she was kidnapped. For recent discussion of the group and its connections with FARC, see David E. Spencer, "The Paraguayan People's Army 2010-2011: Strike-Counterstrike," *Security and Defense Studies Review 13* (2012): 51-62, http://chds.dodlive.mil/files/2013/12/pub-SDSR-v13.pdf.

Colombians continued to remain much more active, with FARC itself engaged in distribution (e.g., in Spain).

Growth of the FARC Threat to the State

As it became a very different movement in capacity and capability, FARC carefully organized itself to implement its strategic plan for seizing power. In warfighting, its strategic task was to penetrate the western 40 percent of the country, where 96 percent of the population lived (areas of mountains and coastal plains). Meanwhile, it used as its base areas the eastern 60 percent of the country, where only four percent of the population lived (but where most of the drugs were produced). It projected its power into local space by establishing "fronts of war" (*frentes de guerra,* or *cuadrillas*) — rather like beachheads in an assault landing, except that they were clandestine. The *frentes* could expect support from FARC's base areas in the eastern llanos (tropical plains), and they combined command and control with geographic domination and efforts to mobilize the population. In reality, mobilization of any consequence never occurred. Thus, the image of a beachhead being captured is apt, for popular allegiance was not *won*; normally, the people were captured. There was no set size (either in manpower or geography), but as an insurgency, FARC was always trying to become as powerful as possible.

During most of the conflict, there were 67 fronts in the rural areas and just four in urban areas. To orchestrate their actions, the fronts were grouped under seven *bloques* (blocks), though in reality two were "candidate *bloques.*" As with the *frentes, bloques* had no established number of assigned units. Just as the military had regular, territorially assigned units with the strike power in special counterguerrilla battalions and mobile brigades, so FARC had its equivalent units. Fronts contained numerous "columns" (actual combatant units of varying strength and armament), most of them assigned in an area, but others that were mobile. This was true even at the highest levels of command and control, with several "mobile *bloques*" that operated alongside the seven territorial *bloques.*

For much of its history, command and control of the seven *bloques* was exercised by the seven members of the Secretariat (with one member assigned to each *bloque*). National Guerrilla Conferences, of which there have been nine (the last was in 2007), comprise representatives from the organizational elements of FARC — the *bloques,* fronts, columns, and various other entities — and theoretically set the strategic direction for the movement, with the Secretariat then implementing the instructions. In reality, as in all Marxist-Leninist organizations, the mechanics of power and institutional patronage dictate that the Guerrilla Conferences are subordinate; they not only meet infrequently but, when they do meet, essentially serve only as a rubber stamp for decisions already prepared and staffed. Still, democratic centralism (behind-the-scenes debate but strict party discipline once a decision has been reached) dictates that diverse opinions may have their say — within the limits set by an armed organization with a track record of eliminating those who run afoul of party discipline.

Secretariat members themselves are ostensibly elected by the National Conference. As might be expected, in reality, the original FARC leadership has replenished its ranks by selecting replacements (as required by death or party discipline) from what, in standard Marxist terminology, would be termed a Central Committee, with the conferences merely giving "assent." The Central Committee is known in FARC as the "general staff" (*estado mayor*) and has some 32 members. These constitute the politico-military brain trust of FARC. They are effectively responsible for implementing the insurgency campaign against the Colombian government. Incapacitated staff members are immediately replaced by other guerrillas, so that the directive capacity of the Secretariat and staff is not interrupted. There is no fixed term for any of these individuals. Members of the Secretariat, in fact, have generally served until death. The de facto chairman of the Secretariat is FARC's leader. Historically, this person has had near-absolute authority and control over the entire organization and makes the major decisions. Thus far, FARC has had only three such chairmen, or maximum leaders.[13]

As noted above, it was the seventh conference, held in 1982, that adopted exploitation of the drug trade (or so it was intended to be) as the primary revenue earner for the organization. This occurred as an initiative of the Secretariat, authorized by the chairman. The consequences, as indicated previously, have been severe for the movement in terms of its intended close relationship with the people.

Ironically, FARC had specifically rejected the "guerrillas lead" approach to mass mobilization (associated with Che Guevara), whereby the guerrillas themselves are to be a focal point (*foco*) for sparking action through launching inspirational attacks. The intent, though, is not the attacks themselves but their role in mobilizing the people. (Hence the term for the strategy: *focismo*.) In contrast, the "guerrillas support the political organizers" approach, if it may be crudely put, is that associated with the "people's war" of Mao Tse-tung or the Vietnamese theorists, notably Truong Chinh.[14] It was this model that FARC adopted. And yet, its conduct negated the approach. In particular, recourse to drug trafficking, kidnapping, and extortion for funding meant there was little need to win the allegiance of the people when one could simply coerce them. Further, manpower came not from the purported social base but from marginalized youth who, in effect, were joining a gang called FARC. Image was everything, substance little (although, as with any left-wing organization, FARC could always attract support from "useful idiots" — to borrow from Lenin's purported terminology — particularly in European radical circles).

The result was that FARC's fronts, which should have contained a variety of units and types of manpower, were essentially *focos* in search of a mass base that never really appeared. Their main tasks were to fight government forces, recruit new members, and obtain local finance and support. What developed was that small units carried out the

---

[13] Manuel Marulanda was the organizer and first FARC leader. He died of a heart attack in 2008, after forty years of leadership. His successor, Alfonso Cano, was killed by the army during a military operation in 2010. The current leader, Rodrigo Londoño Echeverry (aka Timochenko), initiated the current peace process with the Colombian government.

[14] For historical discussion, see Thomas A. Marks, *Maoist People's War in Post-Vietnam Asia* (Bangkok: White Lotus, 2007), 21-55; for analytical discussion, see Thomas A. Marks, "Insurgency in a Time of Terrorism," *Desafíos, Bogotá* 12 (1st semester 2005): 10-34, http://revistas.urosario.edu.co/index.php/desafios/article/viewFile/671/601.

recruiting and financing, while the bulk of the front assets sought combat. And even this last term should be used with some caution, because FARC's actions were overwhelmingly terrorist in form and essence. They consisted primarily of attacking the people and their livelihood. Typically, FARC units were deployed in remote areas, where government was mostly absent and living conditions for peasants were primitive. In such a setting, the national challenge was always how to incorporate the local populations into the state. Thus, FARC became the only governing authority, neither improving nor particularly diminishing the local inhabitants' livelihood (though always using violence as a normal mechanism for dispensing justice).

Still, some three-quarters of Colombians nationwide lived in urban areas and towns. This meant that the small population segments that had been captured did not provide an adequate basis for building either immediate combat power or the "new world" lauded in deceptive FARC propaganda. The local populations were an inadequate foundation if ever FARC was to expand and make a bid for state power. This had the consequence of forcing FARC still deeper into the drug trade just to hold on to its minimal gains. It functioned, in other words, like any other criminal organization that must amass power to dominate its "turf."

The absence of any viable capacity for governance, then, has been glaring. Even during the Pastrana presidency (1998-2002) peace talks, when FARC was granted 42,000 square kilometers as a demilitarized zone, it did not provide more than the most basic services to the roughly 100,000 people who remained in the region (at least part of which coincided with FARC's historic "liberated zone" declared decades earlier, during la Violencia). The major service provided during this period was the improvement of perhaps a thousand kilometers of secondary dirt roads. But even this service had the ulterior motive of facilitating the movement of the columns implementing the struggle against the government—that is, for use as mobility corridors. The local population was authorized to use these roads only under certain circumstances, and yet, they were forced to maintain them during the winter rainy season. Some FARC units provided sporadic minimal medical assistance to the local population, though less than might be expected given that FARC had its own jungle hospitals for treating its wounded combatants.

FARC Slips, Stumbles, and Falls

Even as FARC followed "people's war" doctrine—an approach that was supposed to be built on interaction with the population—the organization's structure for projecting power increasingly operated in the clandestine manner demanded by involvement in the drug trade and in widespread use of terrorism. A pernicious inversion occurred whereby FARC's estrangement from the population increased in proportion to its increasingly symbiotic linkage with criminality. Rather than embracing the people, it brutalized them. By mid-1998, FARC, far from waging people's war, had become a predator.

It was this reality that allowed the Colombian state to regain the initiative. During the Pastrana administration (1998-2002), FARC's predatory nature and nearly total reliance on combat power financed through criminality exposed it to a carefully designed response executed by a dramatically enhanced military. FARC's structures—that is, its *bloques* and *frentes* with their constituent forces in columns—were attacked systematically even as the support network of the drug trade was disrupted. If this approach is reminiscent of Vietnam, it is because the armed forces were keenly aware that FARC, while claiming in essence to be like the Viet Cong, living amongst the people, in reality had become like the North Vietnamese Army in the border areas with Laos and Cambodia, where the combatants occupied bases far from the population.

Colombia's close relationship with the United States after World War II, which included contributing troops to the UN effort in Korea, meant that the military had a viable reservoir of knowledge about the Vietnam War. This was enhanced by specific strategic input, such as commissioned memoranda written by U.S. personnel present in Colombia.[15] Further, by 1998, FARC had thoroughly absorbed the Vietnamese variant of people's war, as transmitted through the FMLN insurgents of El Salvador. As with the Vietnam case, Colombian forces had access to ample U.S. input, enabling them to assess FARC's misuse of FMLN's warfighting approach—which, like Hanoi's, dictated that "all forms of struggle" form a seamless, symbiotic whole. Instead, FARC privileged combat to the near exclusion of all else. This made it vulnerable, particularly because its institutional hubris prevented it from recognizing the capabilities possible in a modern military that embraced adaptation.

Simultaneously—and certainly getting most of the publicity both domestically and abroad—the United States made an important contribution through Plan Colombia. This was an expansive "whole of government" approach to strengthening the Colombian state and its defensive capabilities. Only the counternarcotics aspect was realistically funded, however, and this in turn, for reasons of U.S. policy and law, was kept separate from operations against the counterinsurgency. Regardless, the U.S. role was important in reversing misguided earlier efforts by Washington to motivate Bogotá through "sticks" rather than "carrots" for its supposed lack of will in addressing its numerous challenges—notably, the growth of the narcotics trade.[16]

In the intense fighting during the Pastrana administration, FARC's effort to shift to mobile warfare (also known as *maneuver warfare* or *main-force warfare,* in the terminology drawn from the Soviet military) was defeated. To borrow another term apparently first used in a warfighting sense by the Soviets, the *correlation of forces* had shifted decisively against FARC. Astonishingly, though, the Colombian military strategy took place divorced from national strategy. For nearly all President Pastrana's four-year term, the government negotiated with FARC while ordering the military to pursue its own strategy. Only the military emerged strengthened from the process.

A remarkable feature of the first generation of FARC leaders was its permanent isolation from the quotidian realities of everyday Colombian life. These guerrilla leaders, hiding in the deep jungle, had little interest in establishing processes of strategic communication. For them, the key to the struggle was the armed confrontation with the army. All their efforts were directed toward achieving military advantage on the

---

[15] The most important of these memoranda made clear the deficient nature of the campaign architecture with which FARC sought to implement its people's-war strategy.

[16] See Thomas A. Marks, "A Model Counterinsurgency: Uribe's Colombia (2002-2006) vs FARC," *Military Review 87*, no. 2 (Mar.-Apr. 2007). For a negative but fascinating discussion (which claims extensive Colombian military involvement with the self-defense groups), see Winifred Tate, *Drugs, Thugs, and Diplomats: U.S. Policymaking in Colombia* (Redwood City, CA: Stanford Univ. Press, 2015).

ground. When former president Pastrana invited them to hold peace talks in 1998, they saw it as an opportunity to advance their military project. In their own terms, the peace process was a subterfuge for the development of their strategic plan. Their strategy was to debate endlessly while strengthening their military capabilities and expanding drug trafficking activity, recruitment, and arms purchases. Meanwhile, talks continued. Pastrana offered a demilitarized zone to which the state had no right of entry. FARC took advantage of this naive gesture and continued to strengthen itself while simultaneously unleashing attacks against the armed forces and the national police.

FARC participation in the peace process, then, was transparently a line of effort within the overall insurgent strategy. Its actions were disingenuous and calculating, counting on government hopes for peace and on Bogotá's gullibility. Insurgent violence was constant, and FARC launched seven major offensives during the three years of the peace effort (1999-2002). Of about 20,000 active combatants, some 35 percent were involved in these attacks. Initially, the government interpreted this as part of the negotiation strategy, assuming that FARC was attempting to show a strengthened military capacity that would fortify its bargaining position. But military intelligence managed to pierce the deception and discern FARC's intentions. According to captured FARC documents and prisoners, the plan was to prolong the talks as much as possible while at the same time preparing for future military offensives.

FARC's general approach was simple. It called for open-ended talks between the government, FARC's representatives, and representatives of groups that it could expect to be sympathetic to its positions and cant. What the government tabled as a limited agenda, FARC turned into an expansive agenda with blurred milestones and diffused responsibilities. There was no time limit, so meetings lasted for days as FARC used dilatory negotiating techniques to divert talks in many different directions. Colombian media gave wide coverage to the meetings, and guerrilla leaders had every opportunity to appear and address a national audience. Months went by fruitlessly, with no hope of achieving any agreement, and initial optimism gave way to increasing pessimism. Moreover, from the first moment, FARC declared that it would never surrender its weapons or disband its militias. After all, it felt that it was winning the war, for the government had come to them to urge peace talks after the astonishing series of battlefield reverses that began in 1996 and continued through mid-1998.

The general agenda of the talks included 12 points: (1) political solution to the armed struggle; (2) human rights; (3) integrated agrarian policy; (4) exploitation and preservation of natural resources; (5) social and economic structure; (6) reforms in the judicial system, and the fight against drug trafficking and corruption; (7) political reforms to improve the democratic system; (8) changes in the model of the state; (9) international humanitarian law; (10) military reform; (11) international relations; and (12) formalization of the agreements.

During the more than three years of discussion, not even one of these points was resolved. The Colombian government asked for an international verification commission to document and verify the way the talks were implemented, but FARC refused with the excuse that such a commission would have no authority. Trying to reinvigorate the peace talks, the Colombian government organized travel to several European countries to discuss the peace process and examine social-democratic modes of governance. This effort, too, came to naught as the talks continued to go nowhere. Ongoing attacks by FARC forces, political assassinations, and the hijacking of a domestic commercial airliner drove Pastrana to lose faith in the process and eventually, on February 23, 2002, to order the armed forces to recover the demilitarized zone and capture FARC's main leaders. (The first occurred, but not the second.) After talks that had occupied most of his term in office, President Pastrana realized that he had been deceived. It was a dispiriting experience for Colombians and was to play a key role in the unprecedented first-round electoral victory, in 2002, of a third-party candidate, Álvaro Uribe.

By the start of Uribe's two terms as president (2002-10), the security forces had regained the strategic initiative and presented him with options that simply had not been there for President Pastrana. Uribe took advantage of this opportunity to implement his spectacularly successful Democratic Security and Defense Policy. It enhanced still further the shield provided to democratic society by the security forces (military and police) while galvanizing social and economic mobilization through political initiatives. FARC was hammered to perhaps a third of its peak strength and completely lost initiative at the strategic, operational, and tactical levels. Even its efforts at a transition to terrorism were largely ineffective.

Whereas Democratic Security was explicitly the strategic approach during Uribe's first term (2002-6), Consolidating Democratic Security was the approach during the second term. The new approach differed in that it reversed the kinetic-versus-nonkinetic emphasis, with joint task forces focusing on FARC remnants, and special operations being used to target leadership figures. Successes were as impressive as those of the earlier Uribe term, but strategic focus suffered. Joint task forces were ultimately just an operational equivalent of the tactical "whack-a-mole" approach that had proved so ineffective before the Pastrana presidency. The search-and-destroy focus, ostensibly to eliminate FARC remnants (and, to a lesser extent, the still very small ELN), was an American-inspired approach that led to undue emphasis on body counts. It also laid the groundwork for scandal (the "false positives" episode, wherein noncombatants were killed and reported as guerrillas).

Imperfections aside, by 2010 the security forces' demonstrable operational progress left FARC with few options. Showing a sophistication surprisingly lacking for the previous several decades, FARC leadership recognized the need to emphasize nonkinetic approaches, notably seeking to alter the frame and narrative through information warfare, and recruiting Lenin's "useful idiots" in promising Colombian sectors. Externally, the movement managed to establish reasonably secure bases in Venezuela and Ecuador so that FARC could survive no matter what blows it suffered on its own soil. Now FARC could turn its internal focus to mobilizing coca growers, or *cocaleros* (as had been conspicuously successful in Bolivia, where coca is legal), marginalized members of organized labor, and alienated left-wing elements, such as radical professors and students. Colombia has a vibrant legal left wing, so both fellow travelers and recruits were there to cultivate.

FARC and the Peace Process

Thus, FARC mutated from an insurgent movement into a hybrid organization where political and criminal actions and personnel intermixed. This intermixing led to questions about the organization's true nature, goals, and motivations. While political and criminal activities may be linked in the field, the FARC leadership, through the Secretariat, is experienced and deft in managing this dichotomy. The objective ("ends") to which all FARC activity ("ways") is directed, whether political or criminal tactically, remains ideological and political: seizing state power. For many years, FARC leaders thought this goal could be reached only through force and a protracted guerrilla war funded through criminality, particularly the drug trade—a connection, it is worth noting, that FARC continues to deny.[17] As already noted, FARC's "means" (fundraising) swallowed its "ways," turning it into a hybrid narcoterrorist organization. Now it appears willing to consider other approaches.

Just how this shift has come about remains a matter of speculation. On the one hand, it is known that FARC's leadership has grown increasingly concerned at the damage that its involvement in the drug trade has inflicted on the "ideological project." On the other hand, whatever its bravado, FARC has, in fact, been decimated and presently numbers an estimated 7,500 combatants. Its leaders have increasingly been eliminated by targeted killings, both by precision-guided munitions launched from aircraft and by more traditional special operations. Still another factor is that the previously conducive regional context provided by the Bolivarian states, led by Venezuela and Ecuador, has altered dramatically. This shift began visibly after Colombia's March 2008 killing, inside Ecuador, of FARC second-in-command Raúl Reyes, by precision-guided munitions, in the previously mentioned Operación Fénix (Operation Phoenix), and the subsequent capture of what was essentially a duplicate of FARC's master computer files, by the special-operations unit that swept the ground after the strike. FARC vigorously denied the contents of the files. But the find nonetheless provoked considerable diplomatic controversy and pressure because it demonstrated conclusively the involvement of the Bolivarian states with terrorism and drug trafficking through their support of FARC (and other similar groups). Later, the March 5, 2013, death of the Bolivarian linchpin, Venezuelan President Hugo Chávez, combined with a rapidly deteriorating economic position, only accelerated the Bolivarians' desire to be rid of an increasingly problematic adventure.

Ten years after the unsuccessful talks of the Pastrana presidency, with a very different strategic reality and tactical situation on the ground thanks to the Uribe presidency, then-newly elected President Juan Manuel Santos (2010-present—he is now in his second term) extended secret peace feelers to FARC. By this time, after years of decline, the FARC leaders appeared to realize that their armed struggle had no prospects of success. As a component of their revised emphasis on political aspects of the struggle, they conditionally accepted new peace talks but remained determined to obtain as much advantage as possible by exploiting the government's eagerness to obtain peace through negotiation and compromise. The Secretariat again had some specific points:

---

[17] During the peace talks in Havana in 2013, Iván Márquez, the leading FARC negotiator, has denied such links and denounced the allegations as false.

- FARC would turn into a political party and become an active part of the legislature, with representatives in both houses.
- No FARC member would be imprisoned even a day or be tried by the Colombian justice system.
- The Colombian justice system would have no moral authority or jurisdiction to judge FARC's deeds.
- FARC cadres would not surrender their weapons to the government.
- Colombia's rural regions would enjoy political and economic autonomy. (This point was not included in the negotiating agenda.)[18]

These proposals make evident FARC's concept and goals in the peace negotiations. FARC has two main objectives: First, it seeks to obtain legitimacy before the Colombian people and the international community by showing goodwill in its pursuit of peace and a political settlement to the conflict. (This will be useful later on during the call for elections.) Second, if FARC's conditions can be realized, it will have control over important rural regions, especially in the southern part of the country, where it has long been active. This will allow the Secretariat to have its own political and geographical capital, as well as access to resources in those regions.

With these objectives in mind, FARC leaders think they will have better chances to gain political power through elections. Retaining their weapons, they will act as the main authority in those areas. Moreover, in time, FARC leaders hope that they will be able to change the nature of the state and turn Colombia into a socialist polity resembling the Bolivarian Republic of Venezuela with its neo-Marxist ideology. FARC leadership has not abandoned its Marxist-Leninist goals, only cloaked its ideology with language appropriate to the twenty-first century, when ignorance of communism is ubiquitous.[19]

The current round of peace negotiations between the Colombian government and FARC is taking place in Havana and is being facilitated by Cuba and Norway. While there has been some progress, there are no substantive agreements yet. The talks are organized around the main grievances cited by FARC over the years to justify its long insurgency. The talks were preceded by a secret agreement between the Colombian government and FARC, establishing the framework for an agenda. According to this framework, the Havana peace talks should be limited to those items identified in the secret agreement. To further shape the agenda, the government and insurgent leaders held a series of secret meetings for a year and a half. Once the agenda was ready, yet another agreement was signed, committing each party to limit the talks to those items identified in the framework. The purpose of this commitment was to avoid a repetition of the 1999-2002 experience, when FARC introduced new issues and allegations that were disruptive and counterproductive to actual negotiations. Such combative methods prolonged the process, and ultimately, the negotiating effort proved fruitless. Indeed, as already noted, FARC took advantage of those years to reorganize, expand, and train

---

[18] Iván Márquez, FARC's chief negotiator, issued a statement in Havana during a press conference. His statement included these points.

[19] Nothing in such an assessment of FARC need be treated as mere opinion. Anyone may access the group's extensive online presence and that of its fellow travelers. Indeed, FARC essentially boasts of its Cold War approach. See, for example, the official FARC website, http://farc-ep.co/.

its cadres, as well as to expand its involvement in the drug cycle. In retrospect, it is clear that FARC's objective was never to reach an agreement, but to buy time for a general offensive, which it attempted in the seven major assaults that were defeated by the revitalized security forces.

In this second effort, the government and FARC appeared committed to reaching an actual agreement, but this was to occur within one year of the August 2012 official agreement on an agenda, which was announced in Oslo, Norway.[20] Having learned from experience, the government stated that it would make no truce with the guerrillas. The agenda would consist of five basic points:

- rural development focused on mitigating the unequal distribution of land in rural areas, and the economic disadvantages of poor farmers;
- FARC's participation in Colombian political life after its demobilization as a guerrilla force and its organization as a legal political party;
- an end to the fighting, and the laying down of arms by the guerrillas;
- an end to drug trafficking by FARC;
- recognition of the rights of the conflict's victims.

Initially, and in breach of the agreement that there would be no cease-fire during the talks, FARC ordered a two-month cease-fire by all its members and invited the government to do the same. The government rejected the request. Fighting continued, though at lower levels of intensity, and FARC units continued recruiting. In certain regions, such recruitment included children under 17 years of age. Essentially, FARC continued to operate in the same way that it had for decades, and nothing on the horizon justified the belief that a change was taking place in the Secretariat's mind-set as a result of the peace process. The number of combatants, though diminished in its ability to fight the military directly, was sufficient to keep up operations throughout the country, using terrorist attacks to disrupt infrastructure. In this way, FARC could show the state the price for refusing to accommodate FARC positions at the negotiating table. That this was precisely the approach used so successfully by Hanoi in the Vietnam War—and taught directly to El Salvador's FMLN—seemed to escape Bogotá's negotiating team.

At the time of this writing (August 2015), no agreement had been reached on the most important issues: actual demobilization of FARC as an armed organization, and its integration into the nonviolent political process. Despite the government's best intentions, FARC has gradually turned Bogotá from its effort to use a relatively restricted, structured negotiating protocol. Now, because of the peace process's long duration and the excessive hopes raised, the government finds itself in the position of being gradually made to give way. FARC's position throughout has been to claim that the inequities and brutality of the state compelled it to wage its insurgency. It claims to speak for a broad social base and simply denies the extent to which it has, for decades, privileged assault on the innocent as its principle methodology for waging war. There is no crime that it has not committed: from torture and murder to laying extensive (and, normally, unmarked) minefields throughout the country, to kidnapping and rape, to drug trafficking. All these crimes it simply denies, insisting instead that the facts of history be

---

[20] For timeline and details, as understood by FARC, see FARC-EP, "Diálogos de paz" (in English), Mar. 1, 2011, http://farc-epeace.org/index.php/timeline.html. Needless to say, FARC would disagree with much of this analysis.

decided by various truth commissions and international panels. The vocal group of left-wing fellow travelers within Colombia who will lend themselves to such projects all but guarantees the turning inside out of events as detailed in this chapter. The state becomes the enemy of the people.

Apparent cases of indiscipline that have surfaced (such as the "false positives" scandal), have not helped the government's case. These have been overstated, however, and have obscured a much more important point: disorientation and uncertainty within the military. Ironically, the conscious effort to avoid the errors of the previous negotiations has ended up replicating the same structural disconnection between national and military strategies that was the Achilles' heel of that period. The talks have not only been conducted largely in secret but have been accompanied by the unrelenting FARC attacks on the security forces, to which the government has been loath to respond lest it disturb any progress being made. Rather than setting the agenda, as planned, the government has repeatedly been forced onto the defensive, all but acquiescing in the more extreme claims of FARC and cause-oriented groups who, throughout the conflict, have sought to portray the military as little more than a state-sanctioned death squad. The toll on military morale has been heavy.

This has been accompanied by a government obsession with maintaining a consistent public front, thus going to extreme lengths to ensure that no information contrary to government positions emerges from the security forces—who are the most reliable source of such information. Since evidence concerning FARC actions and motives frequently contradicts the government's publicly stated positions, the recourse of the Santos administration has been to ensure (through regular replacement of individuals) a compliant chain of command. The result has been a stifling of professional opinions and judgment.

The public is aware of this situation. Both the peace process and President Santos personally stand low in the opinion polls (Santos at consistently below 30 percent at the time of this writing). But Congress (and the major outlets of mainstream media) remains in the majority hands of the party of President Santos, National Unity, which results in control of the legal pathway for threatened and actual prosecution. The result has been an unhealthy imbalance in civil-military relations. This has been particularly dangerous because FARC has demonstrated a growing awareness of the security vacuum created by the failure of state armed capacity to empower democratic mobilization. Thus, FARC has thus dramatically increased its efforts to mobilize *cocaleros*, marginalized indigenous elements, and the extreme left wing of labor and of the political spectrum (e.g., students). These efforts, accompanied by a robust information warfare campaign, have allowed FARC to interject itself into national politics in the same manner as Hezbollah or any other political party that also fields its own armed forces.

FARC, then, with the assistance of antisystemic elements within the polity, is using the long-drawn-out peace talks to cultivate the position that Colombian democracy is not "authentic"—indeed, not even legitimate. Faced with this relentless effort, the government has simply not offered a response, let alone turned to the security services. The government cites a tactical count of incidents as proof that the security situation is healthy. But this misses the point that FARC has shifted its emphasis to a different line of effort—away from "violence leads" to forming front organizations in order to wage political warfare, while solidifying its position abroad. Its intent is to emerge as a strong political party leading a rewriting of the constitution, which was the methodology for bringing the communist M19 group into the political mainstream in the late 1980s. (The new constitution was ratified in 1991.) FARC's enhanced political effort has occurred even as global economic challenges have buffeted the Colombian economy. The situation has caused FARC leadership again to feel that it can seize the moment, that "history" favors it. FARC's orientation is one not of reintegration into democratic politics, but of yet another plan to destroy democracy in favor of its Marxist-Leninist (Bolivarian) alternative.

In demanding that the government support it in seizing the moment, FARC continues to claim that its astonishing record of terrorism never occurred and that its ongoing connections with the drug trade are pure fiction. It has signed minor agreements committing the organization to work with the government in clearing minefields throughout the country and in ending the drug trade but has portrayed both as a consequence—if not the creation— of the democratic system. At each moment of pressure in the talks, terrorism has increased dramatically to press the government for concessions. And these the government has given, in the latest instance committing to a bilateral cease-fire (effective at the time of this writing). Always, there is the persistent spinning of a narrative that portrays the violence of the country as "civil war." This is roughly the equivalent of framing the societal turmoil caused by the Red Brigades of Italy in the 1970s and early 1980s as evidence of a country divided against itself—which was not at all the case there, and much less so in present-day Colombia.[21]

Conclusion

The peace process is ongoing. At some point, hard negotiations will be needed, balancing the competing requirements of justice—for victims from both sides of the conflict—and peace. But because the discussion has spent years barely progressing beyond what ultimately are peripheral concerns, no trade-offs have yet been made between stabilization and justice. Worryingly, President Santos has announced that certain accommodations will likely be required, giving measured immunity to former guerrillas. The general prosecutor registered his agreement with the president and announced that justice must benefit the peace agreements and serve reconciliation. FARC has unequivocally stated that should a peace agreement be reached, no FARC members will be turned over to state justice, and that any measures applied to FARC must also be applied to the military. Colombians have little confidence in the government—a consequence of its obsessive secrecy concerning the peace negotiations, which are a matter of central national concern. This low level of trust has raised concerns within Colombia that the government's position might end up in de facto—or even de jure—impunity for terrorists, regardless of the atrocities they have committed. FARC is determined to achieve this through its campaign of equating isolated instances of security forces' indiscipline with its own blatant flouting of international law and adoption of terrorism, including torture of prisoners, execution of kidnap victims, and assaults on innocent civilian populations.

Indeed, most of the core issues driving the long conflict have yet to be addressed. They are part of the agenda of the peace negotiations, but they have not yet been moved

---

[21] On the Red Brigades, see Donatella della Porta, "Left-Wing Terrorism in Italy," in *Terrorism in Context*, ed. Martha Crenshaw (University Park, PA: Pennsylvania State Univ. Press, 1995), 105-59.

to a position of resolution. Both sides seem, in a sense, to be looking to the United States, which was asked to become involved late in the process, to produce a "package" that will somehow cut the Gordian knot. That the United States was "successful" in the earlier El Salvador case is often held up as a model. But that case, as is evident in even the most cursory examination of media reporting, has proved a disaster precisely *because of* the compromises made between security and justice. (There, too, cause-oriented groups played a leading role in demanding gutting of the security services, to the extent that the country was all but helpless before the onslaught of the drug cartels and other criminal gangs.)

One critical point that will prove extremely contentious will be the profits from drug trafficking that have poured into FARC's coffers in the past decades and have been the fuel powering this illicit power structure and the struggle in Colombia. Crimes such as kidnapping and extortion also wait to be addressed in the process, for FARC has stated emphatically that they were simply acts of war, for which it claims immunity from punishment.

Among the many countries worldwide grappling with illicit power structures, Colombia is distinct, perhaps even unique. For not only is an "ideological project" — FARC — being dismantled, but other projects, political (e.g., ELN) and criminal (the drug cartels), remain on the field of battle. Through strategic and institutional adaptation, Colombia has moved to an exceptional point in its history, when it is more integrated and more robust in its socioeconomic-political structure than ever before.[22] FARC has not been a part of this transformation and, indeed, has been passed by, continuing to live physically and psychologically in a Colombia of another era. It points to the imperfections of society, such as the high levels of rural poverty, but offers statist solutions that, time and again throughout history, have proved, at best, impractical, and at worst, devastating to human life. In this, FARC is supported by cause-oriented groups (the loudest voices being foreign) that see Colombia as akin to El Salvador or Nicaragua of the Cold War and claim that "anything" is better than war. Solutions for building are not a part of their agenda — a shortcoming that has ever been the gap in FARC's soul. That FARC's illustrations of "just" societies are the human and physical disasters seen in Cuba and Venezuela highlights the point.

Out of hubris, Colombia abandoned an approach that had brought it to the very brink of victory. Clichés (especially that "all insurgencies have been settled through negotiation") were substituted for strategic planning, and hope for method. The decisions made cannot be undone. But the methodology of tactical surrender in the expectation of strategic victory is destructive of the polity's democratic essence. To equate systematically applied terrorism with the imperfections and missteps (no matter how criminal in individual cases) of the state is to negate all that the Democratic Security and Defense Policy tapped: the willingness of an empowered citizenry to move beyond its parochial concerns and fight to establish a more promising, just order. It is that order, with its tenacity and deep democratic roots, that will lead Colombia forward.

---

[22] See Carlos Ospina and Thomas A. Marks, "Colombia: Changing Strategy amidst the Struggle," *Small Wars and Insurgencies* 25, no. 2 (2014): 35471, www.tandfonline.com/doi/pdf/10.1080/09592318.2014.903 641; David E. Spencer, "Importancia estratégica de una nueva cultura de accountability," in *Fortalezas de Colombia*, ed. Fernando Cepeda Ulloa (Bogotá: Banco Interamericano de Desarrollo, 2004), 391-415.

PHILIPPINES
Physical Map
120°E
125°E
20°N
15°N
10°N
5°N
BASHI CHANNEL
Itbayat I.
Batanes Islands
Batan I.
BALINTANG CHANNEL
LUZON STRAIT
Babuyan I.
Calayan I.
Babuyan Islands
Dalupiri I.
Fuga I.
Camiguin I.
BABUYAN CHANNEL
LAOAG
VIGAN
CORDILLERA CENTRAL
Cagayan
TUGUEGARAO
VALLEY COVE
PHILIPPINE SEA
SOUTH CHINA SEA
LINGAYEN GULF
MT. PULOG 2842
BAGUIO
SIERRA MADRE
Luzon
ZAMBALES MTS.
Agno
BALER BAY
Pampanga
Polillo Islands
Polillo I.
MT. PINATUBO 1660
Patnanongan I.
Jomalig I.
SUBIC BAY
MANILA
BATAAN PENINSULA
LAMON BAY
MANILA BAY
SAN MIGUEL BAY
LAGUNA DE BAY
TAAL VOL. 100
Alabat I.
MT. BANAHAW 2177
Lubang I.
BICOL PENINSULA
RAGAY GULF
Catanduanes
MAYON VOL. 2421
LAGONOY GULF
MT. HALCON 2585
Marinduque
LEGAZPI
SAN BERNARDINO STRAIT
MINDORO STRAIT
Mindoro
SIBUYAN SEA
Burias I.
Sibuyan I.
Ticao I.
Busuanga I.
Tablas I.
Calamian Group
Masbate
Semirara Islands
Samar
Culion I.
Coron I.
VISAYAN SEA
Biliran I.
Linapacan I.
CUYO WEST PASS
CUYO EAST PASS
Panay
Bantayan I.
TACLOBAN
Leyte
Cuyo Islands
Spratly Islands (claimed by the Philippines)
IMURUAN BAY
ILOILO
BACOLOD
Camotes Islands
LEYTE GULF
Homonhon I.
Guimaras I.
Cebu
Dumaran I.
PANAY GULF
CANLAON VOL. 2450
CEBU
Dinagat I.
Panaon I.
PUERTO PRINCESA
HONDA BAY
Bohol
Siargao I.
Negros
Bucas Grande I.
Palawan
Cagayan I.
DUMAGUETE
BOHOL SEA
Camiguin I.
Siquijor I.
SULU SEA
ILIGAN BAY
CAGAYAN DE ORO
LIANGA BAY
Agusan
SINDANGAN BAY
ILIGAN
Bugsuk I.
ZAMBOANGA PENINSULA
MT. RAGANG 2815
CATEEL BAY
Balabac I.
BALABAC STRAIT
SIBUGUEY BAY
ILLANA BAY
Mindanao
Cagayan de Tawi-Tawi
Pilas Group
DAVAO
Samal I.
MT. APO 2954
MAYO BAY
ZAMBOANGA
MORO GULF
Pulangi
Pangutaran Group
Basilan I.
DAVAO GULF
Turtle Islands
Sulu Archipelago
Samales Group
GEN. SANTOS
Jolo I.
SARANGANI BAY
MALAYSIA
Tapul Group
Sarangani Islands
Sibutu Group
Tawi-Tawi I.
CELEBES SEA
0
100
200
300
500
1000
2000
3000
0
50
100
150
200
250 mi
0
100
200
300
400 km

7. The Philippines: The Moro Islamic Liberation Front - A Pragmatic Power Structure?

Joseph Franco

Conflict in the Southern Philippines has traditionally been seen as a historical struggle between contending religious and ethnic identities: the majority Christian population and the Muslim minority. This discourse was tacitly accepted by both belligerents and victims of the conflict. Unfortunately, such a dualistic perspective obscures the complex roots of insecurity in Mindanao and overemphasizes the role of *ideational* rather than *material* factors in sustaining armed violence. Filipino Muslim secession, stripped of its overlaid ethnic and religious signifiers, can be seen more holistically as a manifestation of postcolonial tensions that resulted in a dysfunctional state-building process. These inequitable beginnings gave rise to an illicit power structure, the Moro Islamic Liberation Front (MILF).

Thus, looking at MILF as an illicit power structure first, and an ideological group second, enables a more useful perspective on the decades-long conflict. Filipino Muslims in Mindanao have a long history of contesting homogenizing influences from Manila—during its nearly four centuries as the center of the Spanish colonial administration, afterward, as the seat of the post-1898 U.S. occupation, and today, as capital of the Republic of the Philippines.[1] While various non-Muslim indigenous ethnic communities have resisted central authorities with varying degrees of success, these initiatives were mostly through nonviolent means.[2] The Filipino Muslims, however, have a long history of armed resistance, starting with almost incessant warfare against Spanish colonial rule and continuing through much of the post-1898 U.S. occupation.[3]

The Bangsamoro Conflict and Its Causes

From 1903 to 1913, Moro resistance against the Americans was centered in the Sultanate of Sulu on Jolo Island, off the coast of western Mindanao.[4] Jolo is known for the warrior tradition of its Tausug ethnic population. Tausug poetry abounds with vivid imagery of warfare and a well-developed discourse on waging a just war.[5] Accomplished seafarers, the Tausug of Sulu carved out a prosperous sphere of influence—the Sulu Zone—with

---

[1] "Mindanao" may also refer to the entire region spanning Mindanao Island and the provinces of Basilan, Sulu, and Tawi-Tawi. This chapter adopts this more expansive definition.

[2] For example, the Philippine government created the Cordillera Administrative Region on July 15, 1987, to improve administration over the indigenous highland communities of northern Luzon. It was the direct result of the Mount Data Peace Accord between the communist Cordillera People's Liberation Army and Manila. See Office of the Presidential Adviser on the Peace Process (OPAPP), "CPLA Peace Table: Background," www.opapp.gov.ph/cpla/background.

[3] W. K. Che Man, *Muslim Separatism: The Moros of Southern Philippines and the Malays of Southern Thailand* (Singapore: Oxford Univ. Press, 1990), 46.

[4] Jolo Island is one of the traditional strongholds of the Abu Sayyaf Group (ASG). The al-Qaeda-linked ASG was the focus of the US Joint Special Operations Task Force-Philippines (JSOTF-P) during the Balikatan military exercises with the Philippine military, starting in 2002.

[5] Gerard Rixhon, ed., *Voices from Sulu: A Collection of Tausug Oral Traditions* (Quezon City: Ateneo de Manila Univ. Press, 2010).

Jolo as one of the great centers of maritime trade in Southeast Asia.[6] Slave raiding and trading were also a significant source of the Tausugs' wealth, along with pirate attacks against Spanish-established settlements in the Central Philippines, north of Mindanao. It was only after the pivotal battle of Bud Bagsak (1913) that organized Moro resistance against the Americans ended and Muslim Mindanao was pacified.[7]

This history suggests that the conflict in the region owes more to economics than to interethnic differences. Indeed, the overarching driver of conflict in Mindanao all throughout its recorded history was the maintenance of specific economic rights. Discourses of religious and ethnic strife were the *effect* rather than the cause of conflict. The pejorative term "Moro," affixed to Muslims in Mindanao, alluded to the Moors the Spaniards had fought on the Iberian Peninsula. It was a means for the Spanish colonial government to cast Mindanao's Muslims as the subordinate "other." "Bangsamoro" (Moro Nation) was appropriated by Filipino Muslims as a badge of identity to underscore the existence of their communities long before the establishment of the modern Philippine state.

The establishment of the Third Philippine Republic, after World War II, did not bode well for the Bangsamoro.[8] The outbreak of a Soviet Marxist-inspired insurgency on the plains of Central Luzon by the *Hukbong Mapagpalaya ng Bayan* (People's Liberation Army), or the Huks, occupied the attention of the fledgling Philippine Republic. Political representation for Filipino Muslims was limited to the appointment of Manila-based governors. Along with this marginalization came large-scale divestment of land and agrarian rights—a legacy of the discriminatory policies enacted during the Commonwealth period. Commonwealth Act no. 141 of November 1936 declared all Moro landholdings public lands, with Moros limited to owning four hectares (compared to 24 hectares allowable for non-Moros. The defeat of the Huks also ushered in a more systematic resettlement of Christians in Mindanao as surrendered Huks were granted parcels of Mindanao land under the 1950 Homestead Act. A clash ensued between the formalized, legalistic notions of land ownership among Christian settlers and the informal, communal practice of usufruct by Filipino Muslims. The combination of state-sponsored homestead and resettlement policies led to an irreversible demographic shift that made Muslims a minority even in Mindanao.[9] Land-grabbing became the fixture of intersectarian relations, along with incidents involving plain banditry and cattle rustling.

It must be stressed that manifestations of crime at the *barangay* (village) level were not limited to Mindanao. What differed from the banditry occurring in Luzon was how the discourse of Muslim identity overlay the criminal acts. The escalation of violence led

---

[6] James Francis Warren, *The Sulu Zone, 1768-1898* (Singapore: National Univ. of Singapore, 2007).

[7] Charles Byler, "Pacifying the Moros: American Military Government in the Southern Philippines, 1899-1913," *Military Review*, May-June 2005, www.au.af.mil/au/awc/awcgate/milreview/byler.pdf.

[8] The Philippines declared independence from Spain in 1898, although the U.S. occupiers did not recognize this. After purchasing the Philippine Islands from Spain under the 1898 Treaty of Paris, the United States fought with the nascent republic. The end of the Filipino-American War (1899-1902) led to U.S. colonial rule. The Philippine Commonwealth was established on November 15, 1935, in response to Filipino nationalists' demands for independence. The Third Philippine Republic followed the formal grant of full independence from the United States in 1946. See "The Commonwealth of the Philippines," www.gov.ph/the-commonwealth-of-the-philippines/.

[9] Joseph Franco, "Violence and Peace Spoilers in the Southern Philippines," Middle East-Asia Project Essay Series on Sectarianism in Muslim Majority/Minority Countries in the Middle East and Asia, July 15, 2015, www.mei.edu/content/map/violence-and-peace-spoilers-southern-philippines.

to the establishment of both Christian and Muslim militias. Some, such as the Christian *Ilaga,* were known for their brutality.[10] Latent Muslim dissent in the Philippines was catalyzed after the March 18, 1968, Jabidah Massacre, where paramilitaries of Filipino Muslim descent were massacred by their Philippine Armed Forces (AFP) training cadre.[11] The massacre occurred after a mutiny of the Muslim trainees, reportedly to protest poor living conditions in camp. The Jabidah trainees were preparing for a subversion campaign against Malaysian-occupied North Borneo, or Sabah.[12] Had the Jabidah commando unit been successfully formed, it would have been the spearhead of an unprecedented expansionist military operation by Manila against another sovereign state.

Filipino Muslim youths angered by the massacre found their champion in Nur Misuari, a Tausug political science professor at the University of Philippines. With Malaysian training and funding, Misuari founded the Moro National Liberation Front (MNLF), the first organized Muslim secessionist group in the Philippines.[13] Misuari proclaimed his vision of an independent Bangsamoro and launched devastating attacks against the military in Jolo, Sulu, and Cotabato in central Mindanao. The well-organized MNLF inflicted significant losses on the Philippine military. Jolo was razed to the ground to prevent its capture by MNLF units. In Cotabato, the situation became so dire that at one point, government forces were left holding on to just the perimeter of the Awang military airport. Had Awang and the adjacent Central Mindanao Command headquarters fallen, Mindanao would have been an independent state.[14] Widespread fighting ceased only with the signing of the Organization of Islamic Conference (OIC)-brokered 1976 Tripoli Agreement, which promised the "establishment of autonomy in the Southern Philippines" within the bounds of the Republic.[15]

Misuari's lofty goal of using the discourse of the Bangsamoro to unite everyone who identified themselves as Filipino Muslim was not wholly successful. The sudden empowerment of the Bangsamoro through the MNLF and the Tripoli Agreement reawakened long-dormant notions of traditional leadership. Muslim traditional elites argued over who best represented the Muslim masses. The Muslim clans of central Mindanao, who trace their genealogy to the pre-Spanish sultanates, reemerged to challenge the dominance of Misuari, whom they viewed as a secularist young upstart. One such leader was Hashim Salamat, a member of a central Mindanao-based MNLF faction, the Kutawato Revolutionary Committee.[16] Salamat, a member of the Maguindanao ethnic group, decried how Misuari's MNLF appeared to mimic the secular strategy of the communist

---

[10] "*Ilaga*" is Visayan for "rat." Settlers from the central Philippine region of the Visayas made up most of Mindanao's settler population.

[11] Marites Vitug and Glenda Gloria, *Under the Crescent Moon: Rebellion in Mindanao* (Quezon City: Ateneo de Manila Univ. Press, 2000).

[12] Lela Gardner Noble, *Philippine Policy toward Sabah: A Claim to Independence* (Tucson: Univ. of Arizona Press, 1977); T. J. S. George, Revolt in Mindanao: The Rise of Islam in Philippine Politics (Kuala Lumpur: Oxford Univ. Press, 1980).

[13] For a bibliography on Malaysian sponsorship of the MNLF, see Joseph Franco, "Malaysia: Unsung Hero of the Philippine Peace Process," *Asian Security 9*, no. 3 (2013).

[14] See Fortunato Abat, *The Day We Nearly Lost Mindanao: The CEMCOM Story* (Manila: FCA, 1999).

[15] The 1996 Final Peace Agreement between the MNLF and the Philippines would be signed only after a protracted twenty-year negotiation process punctuated by low-intensity conflict.

[16] Vitug and Gloria, *Under the Crescent Moon.*

insurgency waged by the New People's Army, abandoning the Islamic traditions of the Moros.[17] By 1984, Salamat renamed his MNLF faction the Moro Islamic Liberation Front. It was composed of MNLF defectors from central Mindanao. MILF numbers would be augmented by further defections once the MNLF signed the 1996 Final Peace Agreement (FPA) with the Philippine government.

The Peace Negotiations

In 2014, 30 years after MILF's formation, peace in Mindanao appeared imminent with the March 27 signing of the Comprehensive Agreement on the Bangsamoro (CAB). But the agreement came only after a protracted negotiation process, punctuated by periodic eruptions of violence. To understand the major role that Mindanao's political economy played in the conflict requires a nuanced, nonideological approach. Cultural inequity and injustice are the effects of a flawed political economy, and not the other way around, as would be the case if the Bangsamoro conflict were ideologically based. The last part of the CAB preamble is a telling acknowledgment of the conflict's material and nonideological roots: "The Parties acknowledge their responsibilities to uphold the principles of justice. They commit to protect and enhance the right of the Bangsamoro people and other inhabitants in the Bangsamoro to human dignity; reduce social, economic and political inequities; correct historical injustices committed against the Bangsamoro; and *remove cultural inequities through agreed modalities aimed at equitably diffusing wealth and political power* for the common good [Emphasis added]."[18] Salamat's intent to highlight the difference between the "religious" MILF and the "secular" MNLF appeared to be reversed by his successor, Murad Ibrahim, who signed the CAB.

It was clear that good intentions could not sustain the peace. For example, the 1997 Agreement on the General Cessation of Hostilities (AGCH), signed during the term of President Fidel Ramos, was a broad statement of principles rather than an actionable cease-fire. A general cease-fire could work only through making peace at the community level, in the various barangays throughout Mindanao. The unintended consequence of the AGCH was a spike in armed violence between Christian and Muslim militias. Cease-fire mechanisms may have restrained the Philippine military and the MILF, but they did not resolve community-level tensions in Mindanao. In effect, violence devolved from the large AFP-MILF skirmishes to brush fires between contending networks of militias and private armed groups. Communal violence became the default method of resolving nonideological disputes over land, family differences, and even financial debts. Government intelligence estimates from the period point to the occurrence of around 400 skirmishes during 1997-99—an increase from earlier levels of violence involving MILF.[19] The tipping point in the conflict was a communal dispute over the use of a small (30-hectare) coconut plantation in Lanao del Norte province, in which MILF fighters

---

[17] Alvaro S. Andaya, *Philippine Mujahideen Mandirigma: Struggles against Imperialism and Colonialism* (Manila: AFP, 1994).

[18] See GPH, Comprehensive Agreement on the Bangsamoro, *Official Gazette*, March 27, 2014, www.gov.ph/2014/03/27/document-cab/.

[19] See Cesar Pobre, ed., *In Assertion of Sovereignty: The 2000 Campaign against the MILF* (Quezon City: AFP, 2009).

mobilized to support their clansmen against the opposing Christian militia.[20] Skirmishes escalated into open warfare when MILF forces across Mindanao attacked power transmission lines, plunging 16 provinces into darkness, and raided Kauswagan municipality, taking 294 civilians hostage.[21] In response, on March 21, 2000, the AFP, under orders from President Estrada, launched a major military campaign that became known as the 2000 All-Out War. The nearly seven-month campaign led to the capture of all 46 MILF camps, including the headquarters in Camp Abubakar. By the time the Philippine flag was hoisted on July 10, 2000, at the center of Abubakar, there were severe doubts over the feasibility of continuing peace negotiations.

The stalemate after the capture of Camp Abubakar would not be broken until the next presidential administration. In 2001, President Gloria Macapagal-Arroyo decided to internationalize the negotiations, bringing in Malaysia as a third-party facilitator. The gradualist and informal nature of the talks nonetheless yielded consensus on strategies to prevent the resurgence of conflict, so that by 2005, only the question of ancestral domain remained unresolved. But the question of resources and wealth remained a recurring hurdle to resolving the Mindanao conflict. On July 27, 2008, the draft Memorandum of Agreement on Ancestral Domain (MOA-AD) was completed in Kuala Lumpur after successive rounds of closed-door talks between the Philippines and MILF. The MOA-AD sought to establish a Bangsamoro Juridical Entity, with the "ultimate objective of entrenching the Bangsamoro homeland . . . [with] a system of governance suitable and acceptable . . . [to Filipino Muslims] as a distinct dominant people."[22] What followed was a cascade of opposition, initially from Mindanao Christian politicians, who sued for a temporary restraining order from the Supreme Court.[23] On the MILF side, rogue commanders Ameril Umbra Kato, Abdullah "Commander Bravo" Macapaar, and Solaiman Pangalian launched raids against civilian targets, scuttling the talks.[24]

As with prior episodes of the conflict, it would take another change in executive leadership for a breakthrough to occur after the MOA-AD debacle. President Benigno Aquino III secretly met with MILF chairman Murad Ibrahim[25] on August 4, 2011, in Tokyo to secretly discuss a new substate proposal for Mindanao. This was met by a renewed MILF demand for Malaysia again to take on the third-party facilitating role. Taking the lessons from the aborted MOA-AD talks, both the Philippine government and MILF engaged local stakeholders from the beginning of negotiations. The result was a draft Framework Agreement on the Bangsamoro (FAB), which recognized "the status quo as unacceptable and that the Bangsamoro shall be established to replace the

---

[20] Francisco Lara Jr. and Phil Champain, "Inclusive Peace in Mindanao: Revisiting the Dynamics of Conflict and Exclusion," International Alert, June 2009, endnote 30, www.international-alert.org/sites/default/files/publications/Inclusive_Peace_in_Muslim_Mindanao_Revisiting_the_dynamics_of_conflict_and_exclusion.pdf.

[21] Pobre, *In Assertion of Sovereignty*, 15.

[22] GPH and MILF, "Memorandum of Agreement on Ancestral Domain," "Concerns and Principles" section, Aug. 5, 2008, http://pcdspo.gov.ph/downloads/2012/10/MOA-%E2%80%93-Ancestral-Domain-August-5-2008.pdf.

[23] Supreme Court [Philippines] G.R. no. 183591 En banc decision, Oct. 14, 2008.

[24] Pobre, *In Assertion of Sovereignty*.

[25] Hashim Salamat, founding chairman of the MILF, died from illness in 2003.

Autonomous Region in Muslim Mindanao (ARMM)."[26] The agreement was signed in Manila on October 15, 2012. Under the principles laid out by the FAB, the prospective Bangsamoro government would expand the ARMM's powers, with the Framework emphasizing the "just and equitable" sharing of revenue from natural resources extracted and taxation levied in Mindanao. The succeeding months saw intense negotiations to hash out the details of the FAB. Four annexes, covering revenue and wealth sharing, political power sharing, and the postconflict disposition of combatants, detailed the specific modalities for organizing the prospective Bangsamoro government. The Annex on Transitional Arrangements and Modalities, Annex on Revenue Generation and Wealth Sharing, Annex on Power Sharing, and Annex on Normalization were signed on February 27, 2013, July 13, 2013, December 8, 2013, and January 25, 2014, respectively.[27] A separate FAB addendum was also signed, outlining the mechanisms for exploitation of resources in Bangsamoro waters.[28]

With some measure of consensus in place, the FAB led to the 2014 Comprehensive Agreement on the Bangsamoro, which "ends the armed hostilities between two parties" and "consolidates and affirms the understanding and commitment between the Government of the Philippines (GPH) and the Moro Islamic Liberation Front (MILF)" as written in prior agreements between the two parties. Implementing the CAB still requires its transformation into legislation—the Bangsamoro Basic Law (BBL).

As of this writing, the BBL has become indefinitely stalled in both chambers of the Philippine legislature, as a result of the Mamasapano Incident. This botched law enforcement raid, code-named Oplan Exodus, occurred in the municipality of Mamasapano, Maguindanao, on January 25, 2015. The raid was intended to kill or capture Zulkifli bin Hir, aka "Marwan," a Malaysian member of Jemaah Islamiyah. Marwan was shot and killed inside a community controlled by a MILF breakaway group, the Bangsamoro Islamic Freedom Movement (BIFM).[29] Marwan's hut was near a community hosting the MILF 105th Base Command. An ensuing clash between the PNP and MILF led to the deaths of 35 Philippine National Police (PNP) commandos. Nine other PNP were killed by the BIFM.[30] The heavy casualties inflicted by MILF on the PNP sparked public outrage. Politicians opposed to the BBL also used the incident to call into question the BBL's constitutionality and to lobby for junking it altogether.[31] Accusations of MILF duplicity were exacerbated further by insinuations

---

[26] GPH, "Framework Agreement on the Bangsamoro," *Official Gazette*, Oct. 15, 2014, www.gov.ph/2014/10/15/the-framework-agreement-on-the-bangsamoro.

[27] See GPH, "Framework Agreement on the Bangsamoro and Its Annexes," Oct. 15, 2012, *Official Gazette*, www.gov.ph/framework-agreement-on-the-bangsamoro-and-its-annexes/.

[28] "2012 Framework Agreement," Addendum on the Bangsamoro Waters and Zones of Joint Cooperation," signed Jan. 25, 2014.

[29] Media accounts would also refer to the armed BIFM members as "Bangsamoro Islamic Freedom Fighters," or BIFF. The hard-liners making up BIFM split from MILF over opposition to peace negotiations.

30 PNP Board of Inquiry (PNP-BOI), "The Mamasapano Report," Mar. 2015, http://pnp.gov.ph/portal/images/boimamasapano/boi_final.pdf.

[31] Lira Dalangin-Fernandez, "BBL 'Not Dead' but Won't Be Passed by March – Speaker," *Interaksyon*, Feb. 10, 2015, www.interaksyon.com/article/104779/bbl-not-dead-but-wont-be-passed-by-march---speaker; Maila Ager, "Cayetano Reveals Report of Threats vs Politicians Standing in Way of BBL Passage," Inquirer, Mar. 26, 2015, http://newsinfo.inquirer.net/681522/cayetano-reveals-report-of-threats-vs-politicians-standing-in-way-of-bbl-passage.

that the MILF had aided and abetted the wanted international terrorist Marwan.[32]

If the BBL ratification and the subsequent plebiscite are successful, the Bangsamoro government is expected to be in operation by 2016—significantly later than originally scheduled. But the Bangsamoro conflict is still open ended. As the Mamasapano incident demonstrated, even an existing cease-fire cannot be the sole guarantee for the rule of law. Cautious optimism is in order regarding the CAB's prospects for delivering on its promises, for high hopes have been dashed before. The 1996 FPA between MNLF and the Philippines, and its establishment of the ARMM, did not prevent MILF's emergence, and it did not stop Mindanao's recurrent violence.

Although it is too early to assess whether the peace settlement with MILF will resolve the core issues behind the Bangsamoro conflict, it is not too early to look closely at the CAB and the prospects for peace that it promises. CAB and BBL provisions concerning asymmetric relations between Manila and the Bangsamoro, decommissioning of MILF forces, and wealth sharing is intended to resolve the political economy of exclusion that feeds the Bangsamoro conflict. The asymmetrical relationship between the central government and the Bangsamoro government will effectively swing in the latter's favor—an important distinction from prior attempts to establish autonomy in Mindanao. In comparison, the ARMM had more limited mechanisms to exercise governance. The greater devolution of political authority and power that is expected from the BBL can counter the tendency toward governance in absentia, which derailed prior efforts at political autonomy and, ultimately, efforts to defuse the Bangsamoro conflict.

It is instructive to look into the experience of Misuari, the MNLF, and the ARMM, to see how unresolved power vacuums can undermine the post-CAB peace in Mindanao. After 1996, Misuari was elected Regional Governor of the ARMM, with 96 percent of the vote.[33] But it was a wasted opportunity. While Misuari and his closest allies were at the helm from 1996 to 2005, they failed to deliver the peace dividend they promised in their campaign. During Misuari's tenure, foreign donors and international development agencies poured massive assistance into the region, with disappointing results for the constituency. Misuari had the political clout to start dismantling the entrenched political interests and rent-seeking behavior of some traditional Muslim elites in Mindanao. But MNLF's nine-year rule of ARMM was marked by widespread complaints about Misuari's mismanagement of the regional government, along with allegations of pervasive corruption. Some ARMM civil servants even demanded to see their municipalities and provinces revert back to their original non-ARMM regions, which were dominated mostly by non-Muslim, even Christian, politicians. Misuari was present at the Office of the Regional Governor for only 187 days during his entire 1996-2001 term in office. This phenomenon of absenteeism creates a power vacuum and denies Filipino Muslims access to their government. "Remote-control administration" is, unfortunately, not limited to Misuari but extends to a large number of ARMM local mayors and governors who spend their time in "satellite offices" in neighboring non-ARMM cities.[34]

---

[32] Senate of the Philippines, "Committee Report on the Mamasapano Incident," *Philippine Star*, Mar. 17, 2015, www.philstar.com/headlines/2015/03/18/1434963/document-senate-panels-report-mamasapano-clash; PNP-BOI, "The Mamasapano Report."

[33] Tom Stern, *Nur Misuari: An Authorized Bibliography* (Mandaluyong City: Anvil, 2011), 124.

[34] The terms "remote-control administration" and "satellite offices" were a recurring feature in con-

Aside from issues of governance and exclusion, the CAB is expected to have a better chance of restraining MILF combatants than the 1996 FPA or the ARMM. Under the draft BBL's Article XI, a Bangsamoro police force will be organized for the "enforcement of maintenance and peace and order" in the Bangsamoro.[35] This is on top of provisions in the Annex on Normalization, which calls for the phased decommissioning of MILF weapons. The Normalization Annex prescribes the creation of an independent decommissioning body, composed of foreign and Philippine experts jointly nominated by the government and MILF. The most critical part of the annex is the explicit time frame set for decommissioning MILF weapons and forces by 2016. This provision was missing from the 1996 GPH-MNLF FPA. It was a policy failure with dire consequences: the existence of latent capability to conduct organized armed violence even after a peace settlement. It is a common but unfortunate sight to see MNLF members, ostensibly at peace with the government yet brandishing weapons and sporting MNLF insignia.

Failure to demobilize the MNLF led to armed violence in an apparent subversion, by Misuari loyalists, of firearms possession laws. In 2001, Misuari protested his legitimate ouster from the ARMM, by laying siege to the Cabatangan district in Zamboanga City, the Christian-majority urban center of Western Mindanao.[36] Misuari and his armed men were finally dislodged by a military operation that saw the MNLF founder fleeing behind the cover of civilian hostages. More than a decade later, in the 2013 Zamboanga City crisis, pro-Misuari gunmen assaulted the city center in a purportedly unavoidable escalation of a planned "peace rally" that, by all accounts, was a show of force.[37] The crisis was deemed the largest urban-warfare incident in Philippine history. Nearly 200 MNLF fighters were killed, and 10,000 homes were destroyed. The CAB is expected to make such dangerous flare-ups less likely.

Also, the CAB will provide increased sources of tax revenue, and a greater accrual of benefits to residents of the Bangsamoro, than the ARMM currently provides. Unlike the ARMM, which relies on block grants (Internal Revenue Allotment) from the national government, the Bangsamoro will be allowed greater leeway in legislating new taxation measures. A greater proportion of taxes collected from the Bangsamoro—an unprecedented 75 percent, compared to the 35 percent retained by the ARMM—will also be retained within the region. Another key distinction from prior efforts at Mindanao autonomy is the Bangsamoro's greater share of proceeds from natural resource exploitation: 100 percent for nonmetallic minerals and 75 percent for metallic minerals. Bangsamoro inland waters would also be subject to preferential arrangements.

The depth and breadth of resources available to the emergent Bangsamoro government are expected to reverse the high levels of economic deprivation in Mindanao. The seminal 2005 Philippine Human Development Report points to economic deprivation

---

versations with various AFP commanders in central and eastern Mindanao during the author's stint as a researcher-analyst with various units during 2007-11.

[35] OPAPP, "Draft Bangsamoro Basic Law," Sept. 16, 2014, www.opapp.gov.ph/resources/draft-bangsamoro-basic-law.

[36] Philippine Army, *The Cabatangan Crisis* (Taguig City: Office of the Deputy Chief of Staff for Operations, 2001).

[37] Joseph Franco, "The Zamboanga Standoff: The Role of the Nur Misuari Group," RSIS Commentaries, Sept. 17, 2013, www.rsis.edu.sg/rsis-publication/cens/2056-the-zamboanga-standoff-role-o/#.VRoljvmUc1Y.

as the most significant factor triggering the emergence of conflict in the Philippines (in both Mindanao and non-Mindanao provinces).[38] Economic deprivation, measured and observed through human development indicators (HDIs), is also recognized as the key trigger of organized armed violence by the Philippine military.[39] In Mindanao, the correlation of deprivation with conflict is apparent since nine of the 10 lowest-HDI provinces are in Mindanao.[40] Underdevelopment of the ARMM also manifests in measures other than HDI. For example, the most recent 2015 Labor Force Survey results, published by the National Statistics Office, reveal that the ARMM's labor force participation rate, at 55.8 percent, is the lowest of all regions in the Philippines. It is far below the national average of 63.8 percent.[41]

Assessing the Illicit Power Structure of MILF

The entire saga of peace negotiations with MILF reveals the importance of addressing the material issues behind the conflict. MILF is a purely materialistic organization, with the political economy and armed capability driving its strategy. Throughout its existence, MILF has functioned as an illicit power structure. Hashim Salamat and his successors have always postured themselves as a counterstate to the Republic of the Philippines and, thus, have organized MILF as a politico-military organization. Formally, MILF is the political structure that controls the Bangsamoro Islamic Armed Forces (BIAF). As an illicit power structure, the Moro Islamic Liberation Front has proved itself resilient and self-sustaining, with a resource base facilitated by its embeddedness within central Mindanao communities.

Before the 2000 All-Out War and the capture of Camp Abubakar, MILF maintained 46 camps across central Mindanao. At the time, MILF's operational posture was a semi-conventional force, with defined areas of operations (covering barangays, municipalities, and provinces in Mindanao) for each BIAF formation. Before the 2000 war, MILF had five divisions: one elite National Guard division and four field divisions, spread across Mindanao.[42] BIAF divisions are smaller than PAF infantry divisions and have less equipment. Limited manpower is a reflection of the community-dependent recruitment that the group relies on. MILF division commanders also have limited resources. Some MILF communities do not have sufficient agricultural surpluses to sustain a large armed formation. At its height, before the 2000 war, BIAF had 15,690 members—the rough equivalent of five AFP infantry brigades.[43] Unlike conventional state military forces,

---

[38] Human Development Network, "Philippine Human Development Report 2005," 2nd ed., 2005, http://hdr.undp.org/sites/default/files/philippines_2005_en.pdf.

[39] AFP, "Internal Peace and Security Plan 'Bayanihan,'" Jan. 1, 2011, www.army.mil.ph/ATR_Website/pdf_files/IPSP/IPSP%20Bayanihan.pdf.

[40] United Nations Development Programme, "Human Development Index Highlights Inequality, Slow Pace of Progress," July 29, 2013, www.ph.undp.org/content/philippines/en/home/presscenter/press-releases/2013/07/29/human-development-index-highlights-inequality-slow-pace-of-progress/. These provinces are Lanao del Sur (0.416), Zamboanga del Norte (0.384), Saranggani (0.371), Davao Oriental (0.356), Agusan del Sur (0.354), Zamboanga Sibugay (0.353), Tawi-Tawi (0.310), Maguindanao (0.300), and Sulu (0.266). Lanao, Tawi-tawi, Maguindanao, and Sulu are all ARMM provinces.

[41] The National Statistics Office published its most recent Labor Force Survey report in January 2015.

[42] Pobre, *In Assertion of Sovereignty*, 21-22.

[43] Ibid., 24.

MILF camps are not marked by actual physical boundaries (e.g., perimeter fencing) or physical camp infrastructure—with the exception of Camp Abubakar during its existence. Other MILF camps are more correctly thought of as agglomerations of *barangays* where the rebels wield influence.

After the capture of Camp Abubakar, the MILF Central Committee broke up its unwieldy field divisions and reorganized into 17 battalion-size "base commands," with even more decentralized command and control.[44] MILF base commanders became more focused on maintaining their influence over the villages they held, rather than engaging in Central Committee-ordered sorties against the AFP. The inward-looking nature of the base commands manifested in their preoccupation with acting as community-level adjudicators of land disputes and petty crime among Mindanao's MILF-supportive population. In short, it was a reversal from MILF's stated goal to build a counterstate. What emerged was a network of community armed groups whose political and economic interests are not fully aligned with MILF leadership.

The necessary use of agriculture to acquire financial resources has obliged MILF base commands to enter into unlikely partnerships. Central Mindanao MILF commands are typically clusters of farming communities. Thus, they sustain themselves partly through harnessing the agricultural potential of their encampments. In some cases, MILF base commands are compelled to interact and sell their harvests to local traders aligned with progovernment militias. Unlike the rugged archipelagic geography of western Mindanao (where the MNLF was rooted), central Mindanao features large tracts of irrigated paddies. This terrain made Camp Abubakar a prime MILF agricultural production site. Even after its capture by government forces, the patchwork of MILF communities covering the rice lands of Central Mindanao continued to yield harvests, which were sold to the traders in Mindanao's urban centers, such as Cotabato City.[45] MILF-supplied rice commanded preferential prices as sellers flaunted their links with the armed secessionist movement. This captive market for agricultural produce remains for the post-CAB MILF and provides its various base commands with a revenue stream.

MILF has also used less licit means to secure resources, exploiting kinship networks not only for legitimate political authority but also to divert government funds to MILF's purposes. During the 2000 war, approaches to Camp Abubakar were fortified by trenches and bunkers. It was later determined that the fortifications were made from concrete originally intended for irrigation projects but diverted by local politicians sympathetic to MILF.[46] These obstacles posed a serious challenge to military units spearheading the assault against Abubakar. The Battle of Matanog, which sought to dislodge 1,000 entrenched MILF, resulted in the biggest single-day loss of life for the Philippine military.[47]

Notwithstanding the charged discourse pitting MILF governance against Philippine governance, some MILF-affiliated clans have hedged their bets by having relatives

---

[44] Author interview with Rodolfo Garcia, chief of the GPH Peace Panel when the MOA-AD was junked, Quezon City, Dec. 2011.

[45] Office of the Army Chief of Staff for Operations (G3), Philippine Army (PA) "The Road to Abubakar," unpublished monograph, Fort Bonifacio, Taguig City, Philippines, 2001.

[46] Office of the Army Chief of Staff for Operations (G3), "Clearing the Narciso Ramos Highway," unpublished monograph, Fort Bonifacio, Taguig City, 2001.

[47] Nearly two dozen Army Scout Rangers and Philippine marines were killed. The debacle delayed the advance by a week. The bunkers were around eighteen inches thick and could withstand close artillery hits. See Pobre, *In Assertion of Sovereignty*.

on *both sides* of the conflict. The biographies of rogue MILF commanders involved in the 2008 clashes are an illustrative example of how MILF both subverts and contests the Philippine government's local administrative units. During the time of the clashes, Aleem Solaiman Pangalian of the 103rd Base Command had a nephew who was also the mayor of Lumbayanague, Lanao del Sur—the municipality serving as his unit is de facto headquarters. Further down the local government hierarchy, one of Pangalian's daughters was the village chief of a Lanao del Sur barangay that supported the AFP's counterinsurgency efforts.

The life of another MILF commander, Ameril Umbra Kato of the 105th Base Command, shows another angle to the role of kinship networks. Kato, who was also involved in the 2008 MOA-AD clashes, was from the outset known to be a firebrand within MILF. He and his notorious operations officer, Wahid Tondok, often figured in clan disputes involving the powerful pro-Manila Ampatuan clan.[48] Kato and the Ampatuan clan, a powerful Muslim political dynasty in Maguindanao province, had long been locked in a family feud over ownership of vast tracts of land. Clashes frequently erupted between the 105th BC and the Ampatuans' private militia despite the amicable relationship between the MILF Central Committee and the Ampatuans' principal political patron, then-President Gloria Macapagal-Arroyo. The attacks appear to be economically motivated, especially since clashes often erupted during the rice harvest. During harvests, Kato and his men would often venture from their redoubts to capture crops. This often led to escalation of violence, which drew in AFP units. The combined role of protracted clan disputes and stolen harvests in stoking community-level violence is often obscured. Thus, what starts as a criminal activity can easily be misperceived by the national media and Manila-based policymakers as sectarian conflict.[49]

Data on MILF's internal revenue from its supporters is even more obscure than data on revenue from illicit activities. The sparse data available, from the period before the 2000 war, concerns the amount of *zakat* (tax based on Islamic jurisprudence) levied on MILF-supporting peasants—only a small portion of the group's revenue stream. Philippine military intelligence sources estimate that zakat makes up only two percent of the MILF revenue stream.[50] And MILF's operating expenses are as opaque as its revenue. Although research exists on the political economy of the communist insurgency, no reliable open sources detail the financial status of MILF and the BIAF.[51]

Given zakat's meager contribution to MILF coffers, base commands needed less licit revenue schemes to complement their agriculture-derived financial resources. Each

---

[48] International Crisis Group, "The Collapse of Peace in Mindanao," Asia Policy Briefing no. 83, Oct. 23, 2008, www.crisisgroup.org/~/media/Files/asia/south-east-asia/philippines/b83_the_philippines__the_collapse_of_peace_in_mindanao.pdf.

[49] Joseph Franco, "Attacks in Mindanao: Overestimating the Bangsamoro Splinter Group," RSIS Commentaries, Aug. 13, 2013, www.rsis.edu.sg/rsis-publication/cens/2037-attacks-in-central-mindanao-o/#.VRpLTfmUc1Y; Chih Yuan Woon, "Reading the 'War on Terror' in the Philippines," Geopolitics 19, no. 3 (2014): 658.

[50] Author interview with assistant chief of staff for intelligence-Philippine Army, Fort Bonifacio, Taguig City, Dec. 2011.

[51] Alexander Magno, "The Insurgency that Would Not Go Away," in *Whither the Philippines in the 21st Century?* ed. Rodolfo Severino and Lorraine Carlos Salazar (Singapore: ISEAS, 2007). According to Magno's estimates, the communist New People's Army requires nearly US$(2007)90 million a year to fund its operations. Based on MILF's order of battle of around 11,000 fighters, its operating costs would be more than twice the NPA figure, at around US$200 million.

MILF field division (later base command) had a special operations group (SOG). A SOG was traditionally a few dozen operatives trained to carry out deniable attacks against Philippine security forces, using improvised explosive devices. In reality, though, MILF base commanders used their SOGs more often for profit-making activities. SOG operatives gained notoriety by threatening to bomb businesses or kidnap those who failed to pay a protection fee. The most prolific MILF SOG was the group attached to the now-defunct National Guard Division. This unit was responsible for the 2003 bombings of a major wharf and the international airport, both in Davao City, which killed dozens. Another form of extortion has been MILF's levy of "checkpoint fees" on motorists traveling roads in its areas of influence, such as the Narciso Ramos Highway, on the way to Camp Abubakar.[52]

Along with its use of coercion and illicit acquisition of resources, the MILF actively sought to project its image as a counterstate to the Philippine government. The narrative of an existing and sustainable Moro counterstate was intended not just to promote the idea of parity with Manila but also—and more importantly—to burnish MILF's reputation among its supporters. Thus, when Camp Abubakar still existed; it served not only as a headquarters but as the de facto capital of the idealized Bangsamoro state. Under Hashim Salamat, Abubakar was also intended to be the nucleus of an independent Bangsamoro state, covering seven municipalities across the provinces of Maguindanao and Lanao del Sur. The 10,000-hectare camp was dotted with structures that housed the trappings of the state, including Sharia courts and prisons intended for those found guilty of violating Islamic law. Also, MILF-sanctioned, home-based institutions offered Islamic banking. Retail stores were set up under cooperatives established by MILF commanders, and commanders studied at the Bedis Memorial Military Academy—MILF's response to the Philippine Military Academy. Bedis was also mandated to conduct basic training for rank-and-file recruits. There was even a BIAF Supply and Logistics Department, which housed MILF's limited firearms manufacturing capability. Weaponry crafted there included improvised B40 RPG rounds fashioned from 60mm mortar shells, crude single-shot grenade launchers (similar in form and function to the M79 40mm grenade launcher), and .50-caliber "sniper" rifles.[53]

Despite the trappings of statehood, MILF's power is, at root, a parochial organization with very localized sources of authority. In fact, research has revealed the organization's limited operational range. Use of geospatial methods shows that MILF's sorties were confined within a limited geography. Ninety-five percent of skirmishes with the AFP occurred within a 10-kilometer radius of MILF communities.[54] Even at their peak level of

---

[52] The levying of "toll fees" reportedly continues even after the fall of Abubakar, albeit in an ad hoc fashion, using mobile checkpoints established by local MILF commanders.

[53] In a 2009 author interview with members of the Philippine Navy Special Operations Group (NAVSOG) deployed in Lanao del Sur, it was pointed out that the "baby Barretts," or "Barit" (MILF allusion to the popular antimateriel rifle), crafted by MILF are durable enough to make only five shots. The rifles were also found to be effective only against area targets from a distance of 200 meters or less, rendering them undeserving of the "sniper" designation. NAVSOG operators surmised that the weapons were made for "show of force" activities.

[54] Patricio Abinales, "Sancho Panza: The Paradox of Muslim Separatism," in Severino and Salazar, *Whither the Philippines*. Based on the list of municipalities provided by Abinales, the author plotted the operating radii of various MILF units.

armed capability, MILF units based in Camp Abubakar could sortie out to a 15-kilometer radius at most. This geographic limitation is another manifestation of the embedded nature of the BIAF units in their communities.

Leadership positions in MILF are not objectively linked to military competence. Command and control of the BIAF is exercised mostly by self-taught soldiers, usually with backgrounds in religious instruction, rather than by actual graduates of Bedis. The notorious commanders Bravo, Kato, and Pangalian wielded authority and legitimacy over individual MILF fighters based on the cult of personality and their stature in the barangays.[55] Bravo, one of the parties deemed to have triggered the 2000 war, was a folk hero in the Lanao Provinces because of his Robin Hood-styled persona, aside from his religious credentials. Kato, on the other hand, is respected as an Islamic teacher, having received training in Saudi Arabia. Pangalian followed the mold of these rogue commanders, with Islamic learning obtained in Libya. Such charismatic MILF commanders led the charge against any attempts to engage in peace negotiations.

The exploits of these rogue MILF commanders highlight how local-level strongmen can subvert not only the government's institutions but also the broader revolutionary movement. Violence waged by Kato and Bravo was a marked departure from what the MILF leadership realized early on: that it was in the organization's best interests to scale down secessionist aspirations to the more feasible goal of attaining autonomy. Hashim Salamat and the MILF top echelons had acknowledged by 1997 that the 1976 Tripoli Agreement, earlier agreed to by MILF's predecessor, the MNLF, would be the upper limit of its negotiation goals.[56] In this sense, the 2000 war was a violent interregnum resulting from local tensions that boiled over into a total AFP-MILF conflict. But from 2001 to the present, MILF has repeatedly pledged its commitment to the peace process. Thus, while MILF's increasingly modest goals may appear as a recent development, this stance had actually been a key part of the group's strategic thought.[57] It was not just a last-minute concession made to the Philippine government to seal the deal, to pave the way for signing the CAB.

But MILF has never been a monolithic structure. The localized sources of power and the differing agendas of base commanders meant that opposition to the Central Committee's peace initiatives would be inevitable. Hawkish leaders such as Bravo and Kato periodically emerged to contest the Central Committee's pro-peace stance. The organization had dealt with these recidivists to violence, with mixed results. Bravo was apparently won over and has pledged his support to signing the Framework Agreement. Kato, on the other hand, resisted censure by the MILF Central Committee for his acts related to the botched MOA-AD. In 2011, he and some 300 followers from the 105th Base Command established a MILF splinter group, the Bangsamoro Islamic Freedom Movement (BIFM). BIFM purports to be the genuine voice of the Bangsamoro, aiming to use violence

---

[55] For discussion on the biographies of these three rogue commanders, see Pobre, *In Assertion of Sovereignty*.

[56] The Tripoli Agreement, signed Dec. 23, 1976, called for "establishment of Autonomy in the Southern Philippines within the realm of the sovereignty and territorial integrity of the Republic of the Philippines." OPAPP, "The 1976 Tripoli Agreement," www.opapp.gov.ph/resources/1976-tripoli-agreement.

[57] For discussion of MILF's strategic pragmatism, see Franco, "Violence and Peace Spoilers"; Franco, "Malaysia: Unsung Hero."

to achieve secession rather than adopt MILF's pro-autonomy stance. But its promised large-scale "war" fizzled out in less than a month.[58] On March 30, 2015, a month-long AFP offensive killed an estimated half of BIFM fighters.[59] The operation was a punitive action in response to BIFM's opportunistic involvement in the Mamasapano Incident.[60]

Aside from these notorious commanders' actions, some low-level (i.e., not pertaining to the GPH-MILF negotiations) violence has been ongoing among local MILF leaders, who are often embroiled in local-level disputes over land and resources, sometimes against other fellow secessionists. They can also be motivated by *rido*, or interclan disputes among Muslim families.[61] Examples abound of BIFM-MILF skirmishes.[62] MILF units have also clashed with the MNLF, local strongmen, and even other MILF groups.[63] The dynamics of intra-MILF infighting depend on the revenue streams involved—typically agriculturally based trade and extortion rackets. MILF unit commanders may engage in turf wars with each other as extortion/protection rackets overlap. Similar overlaps may also occur in arable zones that fall under the sphere of influence of two or more different MILF commanders.

Taken in this context, Kato's spoiling behavior, aimed at continuing the conflict, appears to be pragmatic—not resulting from an ideological opposition to the pro-peace MILF Central Committee. Allowing the conflict in Mindanao to fester gives Kato the milieu and polarizing discourse to mobilize the population in his private clan war with the Ampatuans. The continuation of conflict also prevents normalization of government functioning in Mindanao, allowing Kato and the BIFM to continue their revenue-generating schemes unhindered.

---

[58] Joseph Franco, "The 2012 Bangsamoro Framework Agreement: Lessons Learned," RSIS Commentaries, Oct. 9, 2012, www.rsis.edu.sg/rsis-publication/cens/1851-the-2012-bangsamoro-framework/#.VRqU8_mUc1Y.

[59] Carmela Fonbuena, "All-out Offensive vs BIFF Ends Today," *Rappler*, Mar. 30, 2015, www.rappler.com/nation/88362-afp-ends-offensive-vs-biff.

[60] BIFM, along with bands of armed residents, private militias, and ordinary bandits, joined the skirmish between MILF and the PNP. After MILF 105th Base Command forces withdrew from the main battle site, BIFM members went about delivering coup de grâce shots to the wounded police officers and carting away their weapons and equipment. See MILF Special Investigation Commission, "Report on the Mamasapano Incident," Mar. 24, 2015, www.gmanetwork.com/news/story/458026/news/nation/full-text-report-of-the-milf-on-the-mamasapano-clash.

[61] See Wilfredo Magno Torres III, ed., *Rido: Clan Feuding and Conflict Management in Mindanao* (Makati City: Asia Foundation, 2007); Simone Orendain, "Philippines Local Disputes Complicate Peace Talks with Separatists," Voice of America, June 26, 2011, www.voanews.com/content/philippines-local-disputes-complicate-peace-talks-with-separatists-124586334/141380.html.

[62] See Abigail Kwok, "1 Killed in Clash between 2 MILF Factions," InterAksyon, Nov. 20, 2011, www.interaksyon.com/article/17733/1-killed-in-clash-between-2-milf-factions; Jaime Laude, "Kato Commander Killed in Clash with MILF Rebels," *Philippine Star*, Nov. 21, 2011.

[63] Ali Macabalang, "MILF Members Figure in Separate Clashes in Maguindanao, Lanao del Sur," *Manila Bulletin*, Apr. 25, 2014.

International Involvement and Its Effects on MILF

The 2001 resumption of the GPH-MILF peace talks came with a plethora of cease-fire mechanisms. At the forefront is the Coordinating Committee on the Cessation of Hostilities (CCCH), which was originally tasked with verifying the location of MILF camps. It later transformed into a permanent monitoring body. Thirteen Local Monitoring Teams (LMTs), deployed across Mindanao, are the CCCH's operational arm. Each LMT has five members: two each from the GPH and MILF, and one from the religious sector, who is jointly chosen by the parties.

Unfortunately, the LMTs failed to hold the peace. In fact, months-long clashes erupted in Mindanao when a 2003 law enforcement operation in Buliok, North Cotabato, against the Pentagon kidnapping gang spilled over into MILF-controlled areas. The skirmishes demonstrated how the LMTs and the cease-fire mechanisms in general had limited clout to prevent ad hoc alliances between criminal gangs and low-level MILF commanders.[64]

Malaysia's conduct of back-channel talks and shuttle diplomacy in Kuala Lumpur needed to be translated into tangible actions at the operational and tactical levels. After all, what use were proclamations extolling the need to stay the course of crafting a negotiated political settlement if clashes continued to break out periodically in Mindanao? Cease-fire-related agreements needed to coincide with enhanced mechanisms to prevent clashes in MILF areas. This would entail establishing an independent body to investigate violent incidents and objectively discern whether they were the result of secessionist activities or ordinary banditry.

The Malaysia-led International Monitoring Team (IMT) grew from the OIC's proposal for a peacekeeping force. The first contingent, deployed on October 10, 2004, comprised 60 members, 46 of them Malaysians.[65] Other contingent members included military and civilian personnel from Japan, Indonesia, Libya, and other countries. IMT members were deployed near sites where AFP-MILF skirmished had occurred. While the Philippine military initially met them with ambivalence, it soon deemed them a trustworthy and indispensable partner for keeping the peace.[66] Indeed, after the IMT's deployment, clashes between BIAF and AFP units fell from 569 in one year to 10 the year after the IMT's deployment.[67]

Arguably, the 2008 MOA-AD clashes can be perceived as a failure of the IMT. But these clashes had more to do with the persistence of the local-level conflict triggers, such

---

[64] The Buliok complex operation started as a joint military and police operation to arrest the Pentagon kidnap-for-ransom gang. Gang leader Faisal Marohombsar sought refuge with relatives in North Cotabato who were MILF members. Marohombsar and his men sought to use the cover of the MILF-AFP cease-fire to evade pursuing government forces. See Philippine Center for Investigative Journalism, "The Many Lives of the Pentagon Gang," Public Eye 9, no. 1 (Jan.-Mar. 2003), http://pcij.org/imag/PublicEye/pentagon2.html.

[65] Ayesah Abubakar, "Keeping the Peace: The International Monitoring Team (IMT) in Mindanao," SEACSN, Jan.-June 2005, www.seacsn.usm.my/index.php?option=com_content&view=article&id=118:keeping-the-peace-the-international-monitoring-team-imt-in-mindanao&catid=39:content-january-june-2005.

[66] Franco, "Malaysia: Unsung Hero."

[67] OPAPP, "Zero GPH-MILF Armed Clashes since January 2012," June 20, 2012, www.opapp.gov.ph/milf/news/zero-gph-milf-armed-clashes-january-2012-leonen.

as land disputes. Moreover, the narrative of the IMT's failure was espoused primarily by actors who wanted to scuttle the peace negotiations: a peculiar convergence of interests of Muslim rebels such as Kato, on the one hand, and Christian politicians with private militias, on the other. The efficacy of the IMT and the AFP-MILF ceasefire shows most clearly in the fact that during the 2008 MOA-AD clashes, the MILF Central Committee was able to rein in the bulk of its forces. The IMT members were unfazed at being deployed close to the front lines during the AFP campaign against Kato, where some foreign observers were exposed to indirect fire.[68]

With the boost from the IMT came a renewed impetus for local stakeholders to be involved in monitoring the cease-fire. The LMTs, once derided by military commanders as ineffectual in stopping clashes, became most effective with the entry of the IMT. Local civil society actors emerged on the scene. These include the Mindanao People's Caucus's Bantay Ceasefire (Cease-fire Watch), which, starting in 2007, spearheaded IMT investigations in areas where IMT's mandate might be limited.[69] International civil society actors, such as the Center for Humanitarian Dialogue, Muhammadiyah, and the Asia Foundation, were also involved.[70] These actors added other avenues for informal dialogue between the Philippines and MILF. More importantly, they paved the way for a more inclusive consultation process for the negotiations' stakeholders. Again, resistance to the 2008 MOA-AD stemmed mainly from what was perceived as an opaque negotiation process. Having international and domestic civil society actors take the lead in community information drives and consultations assuaged both the central government's and the local Mindanao politicians' fears of a sellout of Philippine sovereignty. Even Christian politicians traditionally opposed to any form of Bangsamoro self-rule, who were indirect instigators of the skirmishes involving the MOA-AD, were convinced of the talks' newfound inclusiveness.[71]

Conclusions and Recommendations

Addressing the illicit power structure of the Moro Islamic Liberation Front was not immediately the intent of the internationalization strategy taken by the Philippine government. From the start, acceding to the MILF request to bring in Malaysia was a pragmatic short-term initiative to restart the stalled talks in 2001. But just as cease-fire mechanisms evolved from ad hoc affairs into developed institutions, so the limited international facilitation role evolved from a stopgap measure to a vital role in forming a more positive conception of peace. Once mechanisms such as the IMT were in place, the notion of "negative peace"—the simple absence of armed conflict—gave way to more developmental approaches. But it was an incremental process, from 2001 to the signing of the Comprehensive Agreement on the Bangsamoro in 2014.

---

[68] Author interview with Brig. Gen. Alejandro Estomo, Davao City, Aug. 2009.

[69] Kamarulzaman Askandar, "'Bantay Ceasefire': Grassroots Ceasefire Monitoring in Mindanao," SEACSN, Jan.-Mar. 2003, www.seacsn.usm.my/index.php?option=com_content&view=article&id=169:bantay-ceasefire-grassroots-ceasefire-monitoring-in-mindanao&catid=47:content-january-march-2003.

[70] See GPH, Comprehensive Agreement on the Bangsamoro, "Acknowledgment" section.

[71] "Piñol Wishes MILF Luck in Peace Talks," ABS-CBN News Online, Oct. 7, 2013, www.abs-cbnnews.com/nation/regions/10/07/12/pi%C3%B1ol-wishes-milf-luck-peace-talks.

The peace process is expected to move slowly despite the Aquino administration's stated goal to expedite it. The Mamasapano incident further set back the process, casting doubt on whether the draft Bangsamoro Basic Law will emerge unscathed in the Philippine legislature.

The protracted nature of the negotiations has posed significant challenges, primarily by allowing time for factions to emerge. Time has allowed rogue and recidivist MILF leaders, using the discourse of how the negotiations were going in circles, to consolidate their power bases. It also has allowed illicit revenue-generation activities to fester unchecked because they are marginalized as "low-priority" community-level, quality-of-life issues, distinct from the "high" strategic issue of resolving the Bangsamoro conflict. It is the combination of (a) the discourse surrounding ineffectual talks and (b) the entrenchment of illicit revenue streams that has driven the vicious cycle of continued conflict. Skirmishes and clashes between MILF forces and the Philippine military underscore how, during the initial stages, Manila seemed to have misdiagnosed the conflict as something that could be won through a military victory. Repeating the 2000 All-Out War would only inflame secessionist impulses, obscuring the political economy of conflict in Mindanao. Worse, a purely military solution would only serve to engender feelings of exclusion and worsen economic deprivation.

With the benefit of hindsight and the signed CAB, it is apparent that the strategy of including international observers to counter the eruption of localized conflict paid off. But it was not a perfect strategy. Since the start of Malaysian involvement in the peace process, a recurring discourse has run in Philippine media and policy circles regarding Kuala Lumpur's ulterior motives. Anti-Malaysian discourse is rooted in the historical but latent Philippine claim over Sabah (North Borneo). Setting aside the Sabah claim, the IMT was able to help keep the peace where it mattered: at the community level. The combination of (a) the legitimacy of international actors and (b) local knowledge created a balanced response, countering the adverse effect of central Mindanao's political economy in further entrenching the conflict.

Similarly, the balancing of justice and stabilization was a more recent development in the negotiations. Cease-fire mechanisms were the priority, with the more substantive aspects of post-peace agreement governance coming late into the picture. This, of course, was expected—violence must be brought down to a manageable level before any substantive talks could start. This pragmatic outlook was shared by both Manila and MILF, with the creation of even more novel stabilization arrangements, such as the Ad Hoc Joint Action Group. This group acted as a mechanism for the GPH and MILF to coordinate law enforcement operations by Philippine police, to prevent organized-crime groups from seeking refuge in rebel-influenced areas.

Both sides could have exercised greater political will to address the conflict. Manila could have emphasized the importance of nonmilitary approaches earlier. A more collaborative and comprehensive approach with both government and nongovernment stakeholders could have helped prevent the emergence of dysfunctional governance and power vacuums, which led to the periodic outbreak of skirmishes. A more resolute stance by the Philippine government against erring local elected officials would also have had to coincide with economic improvements and addressing culture of impunity and exclusion prevalent in Mindanao. This would have given more incentive for communities to come together in support of the peace negotiations.

MILF, for its part, could have helped expedite the peace process by exercising firmer control over its units, to prevent factionalization. The MILF Central Committee's limited command and control during the Mamasapano incident is a worrisome indicator of the leadership's loosening grip over its subordinate base commands. A revolutionary organization such as MILF is bound to exhibit heterogeneity among its ranks, but it is no excuse for the Central Committee's apparently hands-off approach toward rogue commanders. That MILF could rein in Bravo and Pangalian, preventing them from establishing a breakaway group like Kato's BIFM, shows that MILF can keep base commanders in line if it wants to.

The two steps forward, one step back progression of negotiations between the Philippines and MILF suggests that the sequence of events leading to successful peace negotiations is made up not of distinct steps but of repertoires of actions that merge and remerge once opportunities manifest. As the IMT moved from peace monitoring to actual demobilization of MILF fighters, so did the cease-fire committees move from merely validating locations of MILF camps to serving as the central fixture of GPH-MILF interactions outside the negotiations in Kuala Lumpur.

In addressing an illicit power structure such as MILF, three principles are key:

- First, local security must be provided to the communities. It cannot be stressed enough that to build a constituency for peace requires allowing spaces for discourse. Insecurity within communities promotes the possession of weapons. Weak rule of law prompts people to acquire firearms. In turn, illicit small arms facilitate the escalation of clan or personal disputes into armed confrontations. When such brush fires draw in Philippine security forces and MILF, this only reinforces prevailing notions of instability. Potential peace constituencies are therefore lost, which can lead to couching nonsectarian disputes in more divisive ideological/sectarian themes. What starts as a loose network of armed individuals, clans, or groups can be radicalized and recruited into the ranks of illicit power structures such as MILF. A still worse scenario is the crystallization of such disparate groups into a cohesive, extremist organization, as seen in the formation of the BIFM from a hard-line faction of MILF.

- Second, the material conditions that sustain conflict must never be overlooked. It is easy to cast conflict in Mindanao simplistically, as a dualistic interaction between secessionists and supporters of the Philippine state, or as a sectarian Muslim-Christian conflict. Policymakers tackling illicit power structures should not fall into the totalizing and polarizing discourse used by belligerents in the conflict. Tagging a community as "MILF-supporting," "MILF-influenced," or a "MILF camp," without adequate validation, can lead to a vicious cycle of pushing once-neutral parties into the arms of MILF. It must also be stressed that even from a military/counterinsurgency perspective, good strategy requires knowing the breadth and depth of the opponent's reality. Depth of knowledge is therefore relevant in envisioning and adapting to the complex postconflict environment.

- Finally, involving foreign or international actors is best done as soon as possible, to infuse new ideas and resources into intractable conflicts. Each conflict is unique, but the key to exploiting foreign involvement is to time it to the point when at least one of the belligerents has come to the realization that domestic resources are not enough. Caution and due diligence are naturally necessary, since a state would not want a third-party to create an irredentist impulse or inadvertently sanction a proxy war. In the MILF case, Malaysia's entry came with Manila's understanding that it was not in Kuala Lumpur's current national interest to stoke the Bangsamoro conflict. Had Malaysia arrived too soon on the scene, MILF would have misread this as supporting its secessionist goals. Manila would therefore have been compelled to take a heavy-handed approach to counter what would be perceived as Malaysian meddling. On the other hand, had Malaysian involvement been delayed, this would have signaled to MILF that Manila was opposed to having an international actor work out a political solution. In short, missing the right window for international involvement would have led to a more intransigent negotiating posture by both sides.

Ultimately, multiple stakeholders' efforts must be coordinated to address the challenges posed by an illicit power structure. States and state actors must realize that non-military initiatives have an indispensable role to play both during and after conflict periods. Governance by local elected officials must be exercised effectively and impartially, denying armed groups space to agitate communities and mobilize them for a protracted conflict. And international actors must be drawn in at the right time, playing the right roles to soften the negotiating positions of the belligerents. The Bangsamoro conflict, as waged by MILF, was the product of a dysfunctional state building process. Only an inclusive process could eliminate the illicit capture of political and economic resources that sustains the roots of violence in Mindanao.

Map No. 3902 Rev. 5 UNITED NATIONS
January 2004

Department for Peacekeeping Operations
Cartographic Section

8. Sierra Leone: The Revolutionary United Front

Ismail Rashid

Sierra Leone represents one of the most successful cases of UN peacekeeping and peacebuilding. More than 10 years after the declared cessation of its decade-long conflict in 2002, the country remained relatively stable.[1] Three successful national elections were held in 2002, 2007, and 2012. Two of them (2007 and 2012) resulted in the defeat of the theretofore incumbent Sierra Leone People's Party (SLPP) government by the opposition All People's Congress Party (APC). Thus, a key test for nascent democracy—the peaceful transition of power to an opposition party—albeit fraught with tension, occurred twice in the decade following one of the most brutal African civil wars in modern times. This is a major achievement that should not be overlooked.

The UN Mission in Sierra Leone (UNAMSIL), established in 1999, was a critical factor enabling stabilization. It started inauspiciously, however, with the peacekeepers initially failing to recognize the volatility of the situation, and especially the threat posed by the Revolutionary United Front (RUF). Even after signing the Lomé Peace Agreement in 1999, the RUF remained intent on seizing state power, unwilling to relinquish control over lucrative diamond areas, and hostile to international peacekeepers. Despite UNAMSIL's initial misreading of the nature of the threat, and other organizational and leadership stumbles, it was able to establish a well-resourced and robust force. Supported by the UK and several West African countries, it eventually dislodged the RUF from its captured territories. The Sierra Leone intelligence services, police, and army were then overhauled as part of a larger strategy of reconstituting state authority, building democratic institutions and practices, and reestablishing a functioning society and economy. All were the crucial elements for consolidating peace and ensuring stability, and the Sierra Leone experience has provided valuable insights about the challenge of confronting illicit power before, during, and after civil conflict and political transition.

It can also be argued that the lessons from Sierra Leone have done more to inform current international thinking on stabilization issues than any other conflict of the past twenty-five years. The Organization for Economic Cooperation and Development's Development Advisory Council (OECD DAC) leaned heavily on the UK and UN Department of Peacekeeping Operations (DPKO) experience when it drafted its *Handbook on Security System Reform* in 2005-6.[2] Also, the lessons from Sierra Leone have significantly informed subsequent DPKO guidelines and best practices; the ways in which war crimes tribunals are established and conducted continue to be heavily influenced by the Special Court for Sierra Leone; and, out of this conflict in particular, the Kimberley Process for regulating world trade in so-called blood diamonds was established.[3]

---

[1] The Sierra Leone civil war is generally defined as covering the decade between 1991 and 2002.

[2] OECD DAC, *OECD DAC Handbook on Security System Reform: Supporting Security and Justice*, Feb. 25, 2008, www.oecd.org/governance/governance-peace/conflictandfragility/oecddachandbookonsecuritysystemreformsupportingsecurityandjustice.htm.

[3] Kimberly Process, "The Kimberley Process (KP)," 2015, www.kimberleyprocess.com/en/about.

The Conflict and Its Causes

The Sierra Leone conflict defies the usual easy characterization of conflicts in Africa. It was not ethnic, religious, or ideological. Instead, its origins lay partly in Sierra Leone's precarious condition after gaining its independence from British colonial rule in 1961. During 1961-91, successive postcolonial governments, civilian and military, proved unable to establish policies or implement programs reflecting the collective interests and aspirations of their citizenry within a rapidly changing global system.[4] The government could not provide clean water, dependable electricity, decent education, or affordable health services.[5] Like other African governments, the APC regime could not fund its own government or national development programs, putting the country at the mercy of harsh World Bank and International Monetary Fund (IMF) structural adjustment programs.[6] For almost two decades, the APC repressed its opponents, conducted fraudulent elections, and fostered a culture of corruption and impunity.[7]

Student disaffection over this state of affairs and over repressive government actions gave rise to the RUF. Radical college student leaders helped recruit Foday Sankoh and other Sierra Leoneans for military training in Benghazi, Libya, for an armed revolution against the APC regime in 1987. By 1989, the students had completely abandoned their revolutionary project, but they had laid the foundation for a brutal civil war that would bleed Sierra Leone for a decade. Sankoh and two other Libyan trainees, Abu Kanu and Rashid Mansaray, would eventually link up with Charles Taylor, head of the National Patriotic Front of Liberia (NPFL), whom they had met in Libya in 1988. Taylor had started his own revolution against Samuel Doe's regime in Liberia in 1989. That war-ravaged nation gave Sankoh and his comrades a fertile recruiting ground and a launching pad for their war in Sierra Leone.

Taylor viewed the RUF as part of a greater panWest African "revolutionary" enterprise to replace repressive dictatorial governments.[8] The NPFL's and RUF's combined training, recruitment, and resource networks stretched across Sierra Leone, Liberia, Guinea, Côte D'Ivoire, Ghana, Burkina Faso, and Libya.[9] Taylor's support for the RUF in Sierra Leone reflected not only his pan-West African revolutionary ambitions but also his willingness to reciprocate for Sankoh and comrades' participation in the Liberian war. But as the 2004 Sierra Leone Truth and Reconciliation Commission (TRC) report later revealed, more cynical strategic considerations underpinned this support. Taylor wanted to disrupt the Economic Community of West African States (ECOWAS) Monitor-

---

[4] Adekeye Adebajo, *Building Peace in West Africa: Liberia, Sierra Leone and Guinea Bissau* (Boulder, CO: Lynne Rienner, 2002), 38-39.

[5] See Magbaily Fyle, ed., *The State and the Provision of Social Services in Sierra Leone since Independence, 1961-1991* (Dakar: CODESRIA, 1992); Earl Conteh Morgan and Mac Dixon-Fyle, *Sierra Leone at the End of the Twentieth Century: History, Politics, and Society* (Washington, DC: Peter Lang, 1999), 91-117.

[6] During the precivil war period, the APC governed the country from 1968 to 1992. Beginning in 1978, it was the sole political party in Sierra Leone to have legal status.

[7] Abdul Karim Koroma, *Sierra Leone: The Agony of a Nation* (Freetown, Sierra Leone: Andromeda, 1996), 242.

[8] Ibrahim Abdullah, "Bush Path to Destruction: The Origin and Character of the Revolutionary United Front / Sierra Leone," *Journal of Modern African Studies* 36, no. 2 (1998): 203-35.

[9] Ibid., 220.

ing Group's (ECOMOG's) deployment and its use of Sierra Leone as a base for Liberian operations. He also sought to counter anti-NPFL groups that coalesced in Sierra Leone and threatened his rapacious NPFL and mercenary troops' access to Sierra Leone's diamonds.[10] As the Commission noted, during the first year of the war, "NPFL commandos with a patent obsession for self-enrichment . . . choose to indulge themselves in looting and mining activities."[11]

It was Sankoh who announced (over BBC African Service) that the invasion of March 1991 represented the start of the armed phase of the liberation struggle to rid Sierra Leone of the APC government and establish "a just, democratic, and egalitarian society."[12] But it was Taylor who determined the operation's timing, strength, control, and direction. Of the roughly 2,000 troops who crossed into Sierra Leone, RUF combatants numbered only about 360. Charles Taylor's NPFL commandos numbered 1,600. Command of the field operations of the invasion was vested in two NPFL generals: Francis Mewon (Kailahun, Eastern Front) and Oliver Vandy (Pujehun, Southern Front). Materiel for the operation—trucks, 4x4 vehicles, AK-47s, and rocket-propelled grenades—came from Taylor. Taylor had pushed Sankoh, who was waiting for his own independent arms supplies from Libya, into invading Sierra Leone much earlier than Sankoh had planned. But although Taylor withdrew his NPFL troops from Sierra Leone after a year, his support and initial shaping operations had set the RUF on the path to becoming a lethal and intractable illicit power structure.

When it began in 1991, the NPFL/RUF invasion of Sierra Leone had two military objectives. First, the invaders wanted to control the main route to Koindu and establish a forward base, where they could await reinforcements from Liberia. Second, they aimed to capture the most significant military garrison in the eastern region: Moa Barracks, in Kailahun district. Not only were the barracks expected to yield weapons and ammunition, but their capture would deny the Sierra Leone army and ECOMOG support for their operations in eastern Sierra Leone, and in Liberia to the south and east. Possession of the base would also put the RUF in a position to capture Kenema, Sierra Leone's third-largest city.

The insurgents quickly overran major towns in Pujehun and Kailahun districts and were poised to attack Bo, Kenema, and Kono districts, but despite waves of victory during 1991-92, they ultimately failed in their military objectives. The 1,000 RUF and NPFL troops who attacked Moa Barracks in Daru, Kailahun district, found it heavily defended by Sierra Leonean and Guinean troops that had been quickly relocated from their ECOMOG contingent in Liberia. The RUF/NPFL's military plans were derailed, but the insurgents nonetheless caused a major shift in Sierra Leone's political leadership and, more significantly, in the nature of the war itself.

The fighting created the opportunity for a group of young Republic of Sierra Leone Military Forces (RSLMF) officers to depose the decrepit APC regime of Major General Momoh on April 29, 1992. The officers labeled their new administration the National Provisional Ruling Council (NPRC) and pledged to end the war quickly, reduce cor-

---

[10] Sierra Leone TRC, *Witness to Truth: Report of the Truth and Reconciliation Commission*, vol. 3A (Accra, Ghana: GPL Press, 2004), 118, www.sierra-leone.org/Other-Conflict/TRCVolume3A.pdf.

[11] Ibid., 142.

[12] Zubairu Wai, *Epistemologies of African Conflicts: Violence, Evolutionism, and the War in Sierra Leone* (New York: Palgrave Macmillan, 2012), 95.

ruption, and restore multiparty democracy. Headed by Captain Valentine Strasser, the NPRC received enthusiastic support from a disaffected population.[13] The NPRC quickly expanded the RSMLF from about 5,000 to 15,000 troops, drawing mostly from unemployed, poorly educated urban youth. Bolstered by the increased numbers and supported by Guinean soldiers and Nigerian airpower, the NPRC successfully flushed the RUF insurgents from nearly all their captured towns in the eastern and southern regions.

In response to its defeat by the NPRC in open warfare, the RUF resorted to guerrilla tactics, employing small, highly mobile units in stealth attacks and ambushes of military and civilian convoys. The RUF also engaged in extensive "false flag" operations, dressing combatants as RSMLF soldiers and targeting civilians, to discredit the Sierra Leone military. The RUF aimed to weaken the government by crippling its administrative and security apparatuses, and to shut down all major industrial and commercial activities and capture the diamondiferous regions. The RUF also wanted international attention and wanted to force the government to negotiate.

By late 1995, the RUF had achieved some of its aims after effectively shutting down the operations of Sierra Rutile in Bonthe district, and Sierra Leone Ore and Metal Company in Moyamba district. The RUF also overran the diamondiferous areas in the eastern and southern regions, securing access to a resource that would help it finance its military campaign and transform it into a transnational criminal enterprise.

The RUF guerrilla campaign produced a number of responses. First, a "sobel"—soldier by day, rebel by night—phenomenon emerged as RSLMF officers began collaborating with the RUF.[14] Sobel activities so discredited the RSLMF that the army became unwelcome in several areas of the country. This led to the establishment of Kamajors, a civil defense militia initially made up of hunters who sought to defend their communities from RUF and renegade soldiers. Over the course of the war, these militias, or Civil Defense Forces, cropped up all over the eastern, southern, and northern regions. Even Freetown had its own Civil Defense Unit.

The NPRC also contracted foreign mercenaries—Jersey-based Gurkha Security Services, and Executive Outcomes (EO), a South African mercenary outfit—to train the RSLMF, provide logistics, fight the war, and protect diamond areas. The Gurkha presence proved to be of limited military value, but EO personnel effectively supported government counterinsurgency operations against RUF bases around Freetown and in mining areas.[15]

RUF guerrilla successes forced the NPRC into peace negotiations. Sankoh flew to Abidjan, Côte d'Ivoire, and spent 10 months negotiating what would be the first of several peace accords with the NPRC government.

The Abidjan Peace Agreement, signed in November 1996, established a cease-fire and granted amnesty to RUF members.[16] It also called for the Disarmament, Demobiliza-

---

[13] Joseph A. Opala, "Ecstatic Renovation: Street Art Celebrating Sierra Leone's 1992 Revolution," African Affairs 93 (1994): 195-218.

[14] Although he was never formally tried, the Commission singled out Captain Tom Nyuma, NPRC deputy secretary of state for defense, as "foremost among the officers who put his personal interests ahead of his constitutional duties." Sierra Leone TRC, *Witness to Truth*, vol. 2, 52.

[15] EO's services, costing over $1 million a month, became difficult for the country to sustain and were terminated in 1997.

[16] Sierra Leone Web, "Peace Agreement between the Government of the Republic of Sierra Leone and the Revolutionary United Front of Sierra Leone (RUF/SL)," 2015, www.sierra-leone.org/abidjanaccord.html.

tion, and Reintegration (DDR) of armed combatants, transformation of the RUF into a political organization, and creation of a set of joint institutions to implement its provisions. Ultimately, the agreement failed because neither the government nor the RUF had the will or international support to implement it, and both sides repeatedly breached the cease-fire. Tellingly, Sankoh did not return from Côte d'Ivoire to Sierra Leone after signing the accord. Instead, he embarked on a tour of various African countries, ostensibly to raise financial support for the RUF's transformation into a legitimate political party.[17]

Clumsy handling of the army by the civilian government of Tejan Kabbah, elected in 1996, led to renewal of the conflict. The Kabbah administration had inherited an army that it did not trust, and that the public also distrusted because of its "sobel" activities and undisguised opposition to democratic elections and civilian rule. The mistrust was not unwarranted. On May 25, 1997, a group of disgruntled soldiers freed Major Johnny Paul Koroma, who been imprisoned for treasonable offenses, and staged a military coup, forcing the Kabbah government and thousands of Sierra Leoneans into exile in neighboring Guinea. The trigger for the coup was the government's attempt to downsize the army and reduce its rice rations, which had already been mostly siphoned off by senior officers, to the detriment of the lower ranks.[18]

The Armed Forces Redemption Council (AFRC), as the junta was called, naively invited the RUF to join it, form a "people's army," and end the war. Sankoh, in Nigeria awaiting charges of illegal possession of arms, endorsed the move, and the acting RUF commander, Sam Bockarie, moved into Freetown with thousands of mostly juvenile RUF commandos. The TRC later observed:

> This effort to end the war worked briefly in getting the RUF out of the bush but it was counter productive. It endorsed the assertion that the army was in connivance with the "rebels." This stiffened the people's resolve not to have anything to do with the new "people's army." All commercial enterprises closed shop; schools and offices remained closed for much of the nine months that the AFRC was in power. About eighty percent of the armed forces had forsworn their allegiance to the constitution and the elected government and joined the People's Army established by the AFRC.[19]

In response, the two international organizations that had underwritten the Abidjan Peace Agreement—ECOWAS and the United Nations—condemned the coup, and the United Nations imposed an arms embargo on the junta.

ECOMOG troops, supported by loyal RSLMF soldiers and Kamajors, eventually dislodged the renegade AFRC soldiers and their RUF allies from Freetown in February 1998, enabling the Kabbah government to return to power. But although ECOMOG secured some of the larger towns in the north and east, it could not decisively defeat the AFRC or RUF. Thus, the restoration of the Kabbah government belied the fact that the country had virtually returned to all-out war. After Kabbah announced the disbandment of the army, ECOMOG and Kamajors effectively became the principal defenders of the country's security. Vengeful recriminations by the Kabbah government, Kamajors,

---

[17] The Libyan Ambassador to Ghana, Mohamed Talibi, reportedly gave Sankoh $500,000 for humanitarian assistance and to help transform the RUF into a political movement. About $7,000 went to buying arms for the RUF; Sankoh spent the rest. See SayIt, "Charles Taylor Trial," April 14, 2010, http://charles-taylor.sayit.mysociety.org/hearing-14-april-2010/witness-dct-306-on-former-oath#s183095.

[18] Sierra Leone TRC, *Witness to Truth*, vol. 3A, 235.

[19] Ibid., 247.

and ordinary civilians against junta members and collaborators did not engender stability, and the RUF "vowed to make the country ungovernable."[20] For its part, the AFRC set its sights on retaking Freetown and reinstalling the military junta.

The AFRC engaged in mass recruitment and the abduction of civilians.[21] Eventually, it had 2,000 fighters armed with AK-47s, machine guns, rocket-propelled grenades, and heavy mortars. The AFRC blazed a trail of destruction and atrocities, forcing ECOMOG troops to pull back from major towns in the northern region. The RUF, with arms reportedly provided by Charles Taylor in Liberia and Blaise Campaoré in Burkina Faso, raced through Kono and Makeni and on to Freetown. The RUF then joined the AFRC, using thousands of civilians as human shields, in reentering Freetown on January 6, 1999.

The government and ECOMOG misjudged the scale and intensity of the attack, with horrendous consequences for the city.[22] According to the Commission, it ". . . quickly evolved into one of the most concentrated spates of human rights abuse and atrocities against civilians perpetrated by any group or groups during the entire history of the conflict."[23] ECOMOG, replenished by fresh battalions from Nigeria and supported by the Kamajors, succeeded in pushing AFRC and RUF combatants out of the city, but the retreating combatants left a horrific trail of mutilation, death, and destruction of government and private property. By May 1999, the financial and human cost of counterinsurgency operations had also become burdensome for Guinea, Ghana, and Nigeria, the main contributors to ECOMOG. Over 800 ECOMOG troops (mostly Nigerian) had been killed, and the operation was costing Nigeria $1 million a day.[24] Since it had become painfully obvious that the conflict could not be resolved by military force, international pressure mounted on the Kabbah government to recognize the RUF and negotiate with Sankoh.[25]

The Peace Settlement

The RUF's particular brand of militaristic orientation gave it four characteristics that made the peace process more difficult in subsequent negotiations than they had been in Abidjan.

- First, by 1999, leadership and power in the RUF had become concentrated in the hands of hard-core militarists. A credible political wing never developed within the organization.

- Second, because the RUF's main recruitment method had been kidnapping and coercion, some of its senior political cadre by this time, including those participating in the peace negotiations, were themselves abductees. For these men, whose

---

[20] Ibid., 297; Zubairu Wai, *Epistemologies of African Conflicts* (New York: Palgrave Macmillan, 2012), 110.

[21] Sierra Leone TRC, *Witness to Truth*, vol. 3A, 316.

[22] Jimmy Kandeh, "Subaltern Terror in Sierra Leone," in *Africa in Crisis: New Challenges and Possibilities*, ed. Tunde Zack-Williams, Diane Frost, and Alex Thompson (London: Pluto, 2002).

[23] Sierra Leone TRC, *Witness to Truth*, vol. 3A, 326.

[24] Ismail Rashid, "The Lomé Peace Negotiations," Accord 9 (2000): 26-33, www.c-r.org/sites/default/files/Accord%2009_5The%20Lome%20peace%20negotiations_2000_ENG.pdf.

[25] Lansana Gberie, *A Dirty War in West Africa: The RUF and the Destruction of Sierra Leone* (Bloomington, IN: Indiana Univ. Press, 2005), 157.

families had been murdered before their eyes and who were inducted into the RUF at an impressionable age through use of drugs and extreme violence, the RUF was all the family they knew, and they had nothing to go back to should the RUF cease to exist. They were invested in seeing it continue and prosper.

- Third, while the rank and file had become fiercely loyal to Sankoh and various field commanders, the RUF never developed the necessary organizational discipline or accountability mechanisms that would have prevented many of the egregious crimes committed during the war.

- Finally, it had become deeply wedded to criminal diamond- and arms-smuggling networks in Liberia and Guinea. Peace meant loss of profit, and they now had a long-term source of funding and a web of business interests throughout the illicit economy that made them a formidable force and an intractable foe.

Ironically, even though the RUF had been less successful militarily, it went into the peace negotiations in Lomé, Togo, in a stronger position than the renegade soldiers of the AFRC, who had spearheaded the deadliest and most disruptive military actions of the conflict's final phase.[26] The AFRC did not play a major part in the negotiations, nor was the signature of its leader, Johnny Paul Koroma, on the final document. The Kamajors, expecting that the government would adequately protect their interests, also did not play a significant role.

The Lomé agreement, which took two months to negotiate, built on the Abidjan Peace Agreement. Once again the parties agreed on a cease-fire and established a group, chaired by the UN Observer Mission in Sierra Leone (UNOMSIL), established in 1998, to monitor it. The parties reiterated the grant of an "absolute and free pardon," amnesty, and immunity from prosecution to Foday Sankoh and "combatants and collaborators" from all sides of the conflict.[27] They agreed on the need for ex-combatants from the RUF, Kamajors, and RSLMF to be disarmed, demobilized, and reintegrated into society and given the opportunity to be absorbed into a newly restructured and retrained national security force. A Commission for Consolidation for Peace and a committee of seven, under the chairmanship of ECOWAS, were tasked with implementing the treaty. The parties also quickly agreed on provisions for safeguarding humanitarian assistance and fostering human rights.[28]

Just as in the Abidjan negotiations, the thorniest issues in Lomé were power sharing and withdrawal of foreign troops. The RUF entered the Lomé negotiations in May 1999 determined to win at the negotiating table what it could not win on the ground: state power. For Sankoh and RUF, "power-sharing and transitional government meant sub-

---

[26] Virginia Page Fortna, *Does Peacekeeping Work? Shaping Belligerents' Choices after Civil War* (Princeton, NJ: Princeton Univ. Press, 2008), 69.

[27] Lomé Peace Agreement, Kroc Institute Peace Accords Matrix, 1999, https://peaceaccords.nd.edu/provision/amnesty-lom-peace-agreement. The UN special envoy, Francis Okelo, attached a disclaimer excluding egregious war crimes, genocide, and major violations of international humanitarian law from the amnesty provision. This proved important later for the Special Court that was set up to try those most responsible for war crimes in Sierra Leone.

[28] Rashid, "The Lomé Peace Negotiations," 26-33.

stantial control over the state apparatus."[29] The RUF demanded 11 ministerial positions, 11 parastatal positions, six ambassadorships, and the mayoral leadership of Freetown. It also demanded immediate withdrawal of ECOMOG troops from Sierra Leone.

Steadfastly arguing the need to protect the existing 1991 Constitution, the Kabbah government eventually conceded four ministerial and three deputy ministerial positions. In an act akin to posting the fox to guard the henhouse, Sankoh was made chairman of the Commission for the Management of Strategic Resources, National Reconstruction and Development (CMRRD), with the rank of a vice president. The Kabbah government had effectively conceded control over the diamond fields and other resources to him and RUF until 2002, when the next national elections were to be held. The RUF could not force ECOMOG out of the country, however. It had to accept that ECOMOG's forces would be beefed up, and its mandate reoriented to peacekeeping and supporting UN-OMSIL in the DDR process.

Despite the euphoria that greeted the signing of the Lomé Agreement, many questions remained about its implementation: How committed was the RUF to both the written provisions and the spirit of the document? And could the RUF delegates in Togo actually sell the treaty to the field commanders in Sierra Leone? Despite his public posturing, Sankoh's hold over RUF was uncertain, especially after two years in a Nigerian jail. The RUF's ability to transform itself from a military outfit into a credible political party also remained in question. The agreement, with its suggestions for creating a trust fund, training RUF members, and receiving support from the SLPP government, may have grossly overestimated RUF's capacity to change.

The RUF as an Illicit Power Structure

A Military Structure in Search of Power

During the Lomé peace negotiations, the international community remained deeply aware of the egregious crimes committed by various factions in the Sierra Leone civil war, but failed to realize the extent to which the RUF had become a militaristic criminal organization with connections to international crime syndicates. Even Reverend Jesse Jackson's impolitic comparison of Sankoh to Nelson Mandela could not obscure the reality that the RUF was not a credible political movement.[30] RUF leaders justified their atrocities, rapacious activities, and systematic destruction of the country largely on the general sense of public disaffection with the APC regime and, after its ouster, the poor performance of the NPRC and SLPP governments. Despite Foday Sankoh's many BBC interviews and incoherent "ideological" lectures to terrified communities overrun by the RUF, until 1995 many Sierra Leoneans did not know that the RUF even had political objectives.

The RUF's first significant revelation of its political project for Sierra Leone—a short statement, "About RUF" (Mar. 23, 1994), and a pamphlet, *Footpaths to Democracy* (1995)—appeared initially in the UK.[31] These documents contained an eclectic assemblage of ideas and slogans encompassing pan-Africanism, Qaddafi's *Green Book* socialism, libera-

---

[29] Ibid.
[30] Gberie, *A Dirty War*, 158.
[31] RUF, "Footpaths to Democracy," Sierra Leone Web, 2014, www.sierra-leone.org/AFRC-RUF/footpaths.html.

tion theology, and liberalism. The central theme of the documents was RUF's desire to "revolutionize" and violently "remake" state and society in Sierra Leone. Neither document contained a clear or coherent political program of how this was to be done, except through the endemic violence that Sierra Leoneans were experiencing at that time.

Within the motley group recruited from Sierra Leone, Côte d'Ivoire, Liberia, Ghana, and Nigeria and assembled in Liberia in early 1991, Sankoh had gained preeminence based on his age, military experience, and close relations with Charles Taylor. The top brass of the nascent RUF was drawn from a small group of militants trained in Libya and Burkina Faso. Some of them were experienced in combat, having fought as part of the "special forces" of West African mercenaries in the earlier NPFL campaign in Liberia, against Doe in 1989 and 1990. Ultimately, it was Sankoh, strongly supported by Taylor, who provided military direction and control.[32]

Once in Sierra Leone, the RUF attracted individuals and groups who had deep-seated grievances or felt alienated from the ruling APC regime and who bought into Sankoh's pronouncements about creating a just society. Others saw RUF and its war simply as an opportunity to loot or settle personal scores. In the mining areas of the eastern region, the RUF got much support from the illicit miners—mainly young school dropouts living by their wits. They joined the RUF because it protected them and gave them access to choice mining sites. Moreover, their everyday work culture as gangs under the control of a headman fit well with RUF fighting formations. During the guerrilla phase of the conflict, the kidnapping of civilians, especially young children, and their use as carriers, spies, sex slaves, and combatants became much more widespread and systematic. The use of illegal drugs to embolden RUF combatants also became rampant.

The RUF operated an effective chain of personal loyalties, from the juvenile soldiers in its "Small Boys Unit" to the battlefield commanders and right up to "Papay," as RUF members affectionately called Sankoh.[33] The RUF high command was composed of "battlefield combatants and other frontline operatives," with Sankoh at the helm.[34] The TRC later identified 21 people who were members of the high command at different times, but noted that it was "unrealistic to talk about a permanent hierarchy" and "difficult to discern any consistent and centralized vertical structure of leadership."[35] Nonetheless, RUF combatants' loyalty to Sankoh was particularly strong during 1994-96, when his control over the organization was unassailable. Many senior commanders who could have challenged his leadership had been killed earlier, in 1991-93. A RUF War Council, under the Chairmanship of S. Y. B. Rogers, a former civil servant, also met regularly to discuss the war and peace efforts. Its responsibilities were more political and administrative than martial, and as with the high command, its membership was unclear. The relationship between the two RUF structures is obscure, as is the extent to which Sankoh or his loyal commanders were answerable to the War Council.

---

[32] See Sierra Leone TRC, *Witness to Truth*, vol. 2, 48.

[33] The Small Boys Unit (SBU) was a group of children under the age of 15 who were forcibly recruited by the RUF as militants. Originally, the children were taken to carry ammo, food supplies, and equipment to the fighters, but as the war progressed, they were taken to special work camps where the boys were trained for war and the girls were made into sex slaves. Once they were sent out to fight, they plied the trades they were taught, and engaged in the murder of innocent civilians, including those close to their families.

[34] Sierra Leone TRC, *Witness to Truth*, vol. 2, 48.

[35] Ibid., 47.

As the RUF gained recognition and was drawn into peace negotiations, an internal rift began to open. The Abidjan Agreement had called for the organization to be transformed into a legitimate political party, but the RUF clearly lacked the capacity to do so, and the subsequent attempt by some RUF leaders to ensure the organization's participation in the political process created conflict between those who were supportive and those who were not. Those supporting the RUF's participation in the political process had little to show in concrete power gains or incentives from the Abidjan agreement; thus, they could not effectively sell the agreement to the die-hard militarists.

From 1997 until Sankoh's ouster in 2000, his grip over the RUF weakened even as his international profile rose. Following the Abidjan Peace Agreement in late 1996, the newly elected Kabbah government tried to help change the RUF's leadership. It colluded with Nigeria's military government to arrest Sankoh on arms-possession charges at Murtala Mohammed Airport in Lagos, on March 12, 1997. The Kabbah government then supported the attempt by some senior members of the RUF War Council—Fayia Musa, Ibrahim Deen-Jalloh, and Philip Palmer—who had been more supportive of the Abidjan Agreement, to replace Sankoh as leader. Their coup attempt failed after Sam Bockarie, whom Sankoh had anointed as interim leader, arrested the three leaders, Musa, Deen-Jalloh, and Palmer, effectively ending any opportunity for legitimization of the RUF under the Abidjan Accord.[36]

The RUF and Blood Diamonds

During Sankoh's absence, Sam Bockarie became the RUF's unchallenged leader. During his reign, there was no indication that the RUF War Council ever met or that battle group and battlefront commanders were called in for consultation and planning. Instead, on Sankoh's advice, Bockarie established strong ties with Charles Taylor, who had been elected president of Liberia, and sought his guidance. Taylor provided Bockarie and the RUF leadership with facilities and communications equipment in Monrovia, which he claimed were for facilitating the organization's peace efforts. In reality, these facilities became the base for the RUF (and AFRC) leadership to work jointly with Taylor in training and equipping RUF, planning attacks in Sierra Leone, and exploiting revenues from Sierra Leone diamonds to acquire weapons.[37]

Liberia's role as a hub of criminal activity is evident in the discrepancy between the records of its production of diamonds and the Antwerp figures of diamond imports from Liberia, vis-à-vis the two sets of figures for Sierra Leone in the same years. (See table 8.1.) Throughout the 1990s, Liberian diamond production averaged 100,000 to 150,000 carats per year, but imports recorded from the country ranged from 658,000 to 12,320,000 carats. Sierra Leone, with a greater production capacity of 8,500 to 347,000 carats, recorded exports to Antwerp ranging from 221,000 to 831,000 carats.[38]

---

[36] Musa later became a witness in the trial of Charles Taylor in 2010. SayIt, "Charles Taylor Trial," Apr. 16, 2010, http://charles-taylor.sayit.mysociety.org/hearing-16-april-2010/witness-dct-306-on-former-oath.

[37] Taylor's activities during this period were included in the criminal charges that would later be brought against him in the Special Court for Sierra Leone when, in 2011, he was convicted for aiding and abetting various crimes committed by the AFRC/RUF during 1996-2002.

[38] Ian Smillie, Lansana Gberie, and Ralph Hazleton, "The Heart of the Matter: Sierra Leone Diamonds

8.1 Diamond production and exports from Sierra Leone and Liberia to Antwerp, 1990-98 (thousands of carats)

		1990	1991	1992	1993	1994	1995	1996	1997	1998
Sierra Leone	Production	78	243	347	158	255	213	270	104	8.5
	Antwerp Imports	*221*	*534*	*831*	*344*	*526*	*455*	*566*	*803*	*770*
Liberia	Production	100	100	150	150	100	150	150	150	150
	Antwerp Imports	*5,523*	*658*	*1,909*	*5,006*	*3,268*	*10,677*	*12,320*	*5,803*	*2,558*

Source: Adapted from Smillie, Gberie, and Hazleton, "The Heart of the Matter," 32.

All the various armed factions in Sierra Leone benefited from the illegal sale of Sierra Leonean diamonds, but the RUF took the lion's share. In 1999-2001 alone, it siphoned off approximately $70 million of the $138 million worth of diamonds illegally exported from the country.[39] The trade in blood diamonds by the RUF (and AFRC) attracted "businessmen" from South Africa, Israel, Ukraine, and the UK who were willing to provide or transport arms and ammunition and, in some cases, hard drugs to the RUF. Diamond trafficking also paid for Ukrainian and Burkinabé mercenaries who contributed their military expertise. RUF leaders and Taylor took large commissions from some of these illegal transactions.[40]

By the time negotiations started in Lomé, the RUF had been battered by ECOMOG, but it remained highly militarized and criminalized. Sankoh's signature on the Lomé Agreement suggested that RUF had negotiable interests, encoded in the provisions of the agreement. This was only partly true, however. Those in the RUF who felt positive about the agreement started disarming and demobilizing. Those with nonnegotiable criminal interests could not be brought to the table. Even with Sankoh being offered the chairmanship of CMRRD, there was no way to negotiate with hard-core leaders such as Bockarie, who wanted to maintain long-term control of the diamond areas. Taylor, who played the role of peacemaker, also schemed to preserve his influence over the RUF, keep his access to profits from its trade in diamonds and arms, and continue to use members of the organization against his enemies in Liberia. Thus, his interests could not be publicly negotiated. By late 1999, these irreconcilable differences produced violent conflict between factions loyal to Bockarie and those loyal to Sankoh.

Owing to this intra-RUF schism over relinquishing control of the diamond fields, the Lomé Agreement began to unravel within six months. Despite the RUF's political gains and Sankoh's award of the potentially powerful chairmanship of the CMRRD,

---

and Human Security," report, Partnership Africa Canada, 2000, 32, http://cryptome.org/kimberly/kimberly-016.pdf.

[39] Diane Frost, *From Pit to Market: Politics and the Diamond Economy in Sierra Leone* (Rochester, NY: James Currey, 2012), 76.

[40] Ibid.

the RUF acted in a manner that clearly demonstrated a lack of commitment to the peace process. It refused to observe the cease-fire, remained on a belligerent war footing, and continued to attack government troops and international peacekeepers. It hampered the DDR process, stirring up impatience and disaffection among those combatants willing to disarm. Sankoh denounced the transformation of UNOMSIL, which was strictly an observer mission, to UNAMSIL, and the increase of its forces. Sankoh condemned the start of the deployment of UNAMSIL as illegal and inconsistent with the Lomé Agreement, which had mentioned only UNOMSIL and ECOMOG. According to UN reports, "Every effort made to explain the link between UNAMSIL and article XVI of the Lomé Agreement met with a pretense at understanding, only for UNAMSIL to be denounced again shortly thereafter."[41]

Most egregiously, on May 6, 2000, the RUF exploited the confusion created by the transition from ECOMOG to UNAMSIL, by kidnapping and disarming 498 UN peacekeepers who had been hastily deployed in Makeni, Magburaka, Kambia, and Kailahun.[42] The next day, the RUF shot down a UN helicopter. The UN peacekeeping mission in Sierra Leone tottered on the brink of collapse, and a wave of panic swept through Freetown amid rumors that the city was in danger of a third invasion from RUF and AFRC. The UK and other Western countries asked their citizens to leave the country. As the RUF increased its activities to oppose the peace process through violent means, all hope of transforming it to a licit organization whose fundamental interests might be reconcilable was gone.

The International Strategy and Its Impact

Saving the Mission

The crisis that UNAMSIL faced in May 2000 had little to do with the mandate it was given when established by UN Security Resolution 1270 (1999). UNSCR 1270 authorized a force of 6,000, later increased to 11,000 by Resolution 1289 (February 2000). But UNAMSIL was not only tasked to facilitate implementation of Lomé's key provisions and empowered to use force to defend itself. Its mandate was notable because it authorized UNAMSIL to protect civilians under imminent threat of physical violence. This was a more proactive style of UN peacekeeping, which the numbers did not support. The mismatch between capabilities and requirements stemmed largely from the United Nations' misreading of the operational environment in Sierra Leone, especially the RUF's intransigence, and an assumption that ECOMOG forces would remain in Sierra Leone to help maintain the peace. As Funmi Olonisakin, an expert in peace and security, put it, "In hindsight, the two assumptions of UNAMSIL's planners . . . that the RUF would abide by the terms of the Lomé Agreement, and that the UN could cope with challenges after ECOMOG's withdrawal—proved disastrously wrong."[43] As a result of its failure to fully

---

[41] "Report to 3rd Joint Implementation Committee meeting, May 13, 2000, 18," quoted in Shalini Chawla, "United Nations Mission in Sierra Leone: A Search for Peace," www.idsa-india.org/an-dec-00-10.html.

[42] Funmi Olonisakin, *Peacekeeping in Sierra Leone: The Story of UNAMSIL* (Boulder, CO: Lynne Rienner, 2008), 57-58.

[43] Ibid., 61.

grasp the scope of the challenge ahead, the UN Department of Peacekeeping Operations (DPKO) proposed deploying only 6,000 troops and selected the highly regarded Indian diplomat and soldier Major General Vijay Kumar Jetley as force commander. DPKO thought that selecting a non-West African commander might better project UNAMSIL's neutrality, but Jetley's credentials and UN neutrality meant little to RUF.

DPKO's assumption that ECOMOG would remain active in Sierra Leone was quickly invalidated by newly elected Nigerian President Olusegun Obasanjo's decision to withdraw Nigerian troops from the country. Though some Nigerian troops were later rehatted as UN peacekeepers, lack of coordination between UNAMSIL's deployment and ECOMOG's withdrawal created major problems that the RUF and the AFRC exploited. Newly arrived UN peacekeepers immediately found themselves in a volatile operating environment, with insufficient intelligence and no clear orders on how robustly they should respond to RUF threats and hostile actions. UNAMSIL's difficulties were further compounded by dissatisfaction over an Indian general's appointment as force commander. Despite having two Nigerians in the UNAMSIL top leadership—Oluyemi Adeniji as special representative of the secretary-general, and Brigadier General Mohammed Garba as deputy force commander—Nigerian officials felt that overall command of the troops should have been given to a West African and that West Africans should constitute the bulk of the peacekeeping force.[44]

This disaffection led to further squabbles over subordinate leadership positions, deployment strategies, and chain of command, with troops taking orders from their home countries rather than from the force commander. A May 2000 confidential internal report by Jetley exposed the deep divisions within the mission. Although never officially submitted to the UN Security Council (UNSC), the report was widely circulated among its members. Jetley's report highlighted the former ECOMOG troops' disaffection, suggested the RUF's preference for ECOMOG over UNAMSIL, and accused senior ECOMOG members of stealing Sierra Leonean diamonds. Allegations that its soldiers engaged in illicit trade with the RUF had long dogged ECOMOG. Nonetheless, the Jetley report sparked hostility from Nigerian diplomats and military officials, who called for his removal.[45]

The crisis of May 2000 produced two sets of responses that ultimately saved the Lomé Agreement and prevented the country from plunging into all-out conflict again. The first came in the form of short-term decisive military actions by the UK, the now progovernment (and reintegrated) AFRC, and Kamajors and loyalist segments of the Sierra Leone military.

As mentioned earlier, the AFRC had not taken an active part in the Lomé negotiations. Afterward, in an extraordinary series of statements, the AFRC's leadership declared, "We herald the dawn of a new era. The war has ended. The era of peace, forgiveness and reconciliation has come." The AFRC declared an end to its war against the government, announced its commitment to the peace accord, and begged the people's forgiveness for the wrongs committed by the AFRC and its members.[46]

---

[44] Ibid., 82. Olonisakin points out that it was not so much an issue of political sensitivity as of remuneration, since only the force commander was paid on a UN scale, whereas the deputy force commander retained his (much lower) regular salary.

[45] Ibid., 81-85.

[46] Sierra Leone Web, "Statement on the Historic Return to Freetown, Sierra Leone, of the Leaders of the

Their action stabilized the situation in Freetown, saved UNAMSIL, and crippled Sankoh's leadership over the RUF. In response to the RUF's abduction of UNAMSIL troops, UN Secretary-General Kofi Annan had asked the UK, U.S.A., and France to deploy a rapid-reaction force in Sierra Leone to save the mission. But it was the UK, acting independently of the United Nations, that ultimately intervened and changed the course of the conflict.[47]

The first UK intervention, code-named "Palliser," began on May 7, 2000, with the dual objectives of rescuing UK and other foreign nationals from the volatile situation in the country and saving the tottering peacekeeping mission. British troops quickly secured Lungi airport, repelled a RUF attack, and set up defensive positions in various parts of Freetown. The British evacuation of its and other nations' citizens during Palliser sent mixed signals. On the one hand, it underscored the fragility of the situation in the city. On the other, as President Kabbah conceded in his memoir, it raised public confidence and served as a deterrent to the RUF.[48]

Complementing Operation Palliser were equally decisive actions by civic groups and progovernment factions of the AFRC, led by Johnny Paul Koroma, against Sankoh. In reaction to the kidnapping of the UN troops, and rumors of another impending invasion of the city, hundreds of Sierra Leonian citizens and politicians marched to Sankoh's residence on May 6 and May 8, 2000, to preempt the invasion. With the knowledge and tacit support of government officials including Kabbah, Koroma and his AFRC loyalists in the military coordinated the arrest of prominent RUF officials in the city and orchestrated a violent retaliation against Sankoh's security guards. On May 8, Sankoh's RUF security detail, faced with hundreds of angry demonstrators, panicked and killed 21 protestors. The West Side Boys (a splinter group that had supported the AFRC), Kamajors, and other government operatives who had infiltrated the civic demonstration, then killed a number of Sankoh's RUF security guards. Sankoh escaped but was apprehended ten days later and detained without trial by the government of Sierra Leone, in accordance with emergency powers of detention.

The West Side Boys, even at their peak, numbered no more than 600. They were never believed to have any affiliation with the RUF and were known for their bizarre dress, extreme violence, and heavy drug use (mostly palm wine and heroin, purchased with illicit diamond revenues). At the time, they were portrayed as a criminal gang with no political purpose, enabled by the perpetual lawlessness and social breakdown of the country. A 2008 article in the *Journal of Modern African Studies* offered an alternative view of the West Side Boys as an effective military unit employing military and political techniques to achieve defined goals, but there is scant evidence to support that view. More likely, they are just another example of how illicit power structures arise opportunistically in a security and governance vacuum.

Alliance of the Revolutionary United Front of Sierra Leone and the Armed Forces Revolutionary Council," Oct. 3, 1999, www.sierra-leone.org/AFRC-RUF/AFRC-100399.html. See also "Position Statement of the Sierra Leone Army (SLA) and the Armed Forces Revolutionary Council (AFRC)," Sept. 18, 1999, www.sierra-leone.org/AFRC-RUF/AFRC-091899.html.

[47] Adekeye Adebajo and David Keen, "Sierra Leone," in *United Nations Interventionism, 1991-2004*, ed. Mats Berdal and Spyros Economides (Cambridge, UK: Cambridge Univ. Press, 2007), 246-73.

[48] Ahmad Tejan Kabbah, *Coming Back from the Brink in Sierra Leone* (Accra: EPP Book Services, 2010), 158.

The second UK intervention and show of force, Operation Barras, took place almost four months after Palliser. It was a much smaller operation, triggered by a monumental miscalculation by an errant faction of the West Side Boys. On August 25, 2000, they seized 11 UK troops on patrol in the strategic Okra Hills area and refused to release them through negotiations. The UK then launched Operation Barras on September 10, 2000, dismantling the base and freeing all hostages but one British officer, who died in the fighting. The operation eliminated the West Side Boys as a military threat to Freetown, effectively destroying the group. It also opened the flow of road traffic between Freetown and other regions of the country and helped ease the deployment of UNAMSIL troops to regions outside the city.

Defeating the RUF

Rather than merely reacting to breaches of the agreement with short-term military operations designed only to address the immediate security threat, the Sierra Leone government and the international community made their second round of responses to the May 2000 crisis based on a much deeper understanding of RUF as an illicit power structure. They now recognized that the RUF was not a political movement committed to peace, but rather a criminal organization driven by material greed. This realization motivated the Sierra Leone government to work closely with UK military experts to develop a strategic campaign plan that would either ensure total military defeat of the RUF or finally conclude the conflict on terms that would enable the peace process to succeed. The plan entailed severing RUF from its Liberian strategic center of gravity, particularly the planning and arms support provided by Taylor. It also aimed to restrict RUF profits from the illegal diamond trade through Liberia and Guinea. Crucially, the plan envisioned isolation and disruption of RUF's operational center of gravity by pushing for government control and investment in the diamondiferous areas. The plan, conducted in five major phases, included tactical coordination between the retrained Sierra Leone Army (SLA), Kamajors, Sierra Leone Police (SLP), and Guinean armed forces, with operational support, training, and equipment to be provided by the UK. Through its International Military Assistance Training Team (IMATT), the UK had started retraining units of the Sierra Leone armed forces in 1999 with the aim of making them a more democratically accountable, effective, and sustainable arm of the Sierra Leonean government.[49]

Even as the plan was being written in January 2001, various elements of its first two phases were already unfolding on the ground. The confidence building and disruption of RUF command centers envisaged in the first phase were already under way in late 2000, with a string of SLA/AFRC/Kamajor victories that cleared parts of the northern access to Lunsar and Makeni. RUF incursions into Guinea, over a trade dispute with Guinean soldiers and to support Taylor in his war against the Liberians United for Reconciliation and Democracy, triggered a massive Guinean offensive against the RUF, decimating its top leadership and effectively destroying its operational center of gravity in Sierra Leone by early 2001.[50] This broke the RUF's will to fight and weakened

---

[49] British Army, "The British Army in Africa," 2015, www.army.mod.uk/operations-deployments/22724.aspx.

[50] Lansana Gberie, "Destabilizing Guinea: Diamonds, Charles Taylor and the Potential for Wider Hu-

its ability to hold on to the diamond fields by force. It also impeded RUF's ability to be resupplied through Guinea.[51] Already cut off from its supply routes through Liberia, the RUF found it much harder to replenish weapons and ammunition.

While the RUF was being systematically defeated on the battlefield, authority within the organization was shifting from the more truculent leaders, such as Bockarie, to others, such as Issa Sesay, who were willing to work with UNAMSIL and the Sierra Leone government. By 2000, Bockarie had effectively relocated to Liberia, partly due to international pressure on Taylor to control him and also as a result of factional struggle within the RUF. In November 2000, RUF field commanders nominated Issa Sesay to take over from the incarcerated Sankoh. Nigerian President Obasanjo and Malian President Alpha Konaré persuaded Sankoh to accept the decision.

Sesay, a battlefront commander with considerable influence on RUF field combatants, displayed a genuine commitment to the peace process and implementation of the Lomé Agreement—a commitment thus far absent in RUF leadership. In Abuja, Nigeria, he signed a cease-fire with the government, agreeing to the unfettered deployment of UN peacekeepers across the country, the release of all arms that RUF had seized from UNAMSIL, and recommencement of the DDR process. Sesay later signed a second agreement in Abuja, with the government and in the presence of ECOWAS and UN officials, to relinquish RUF areas of control and allow conduction of the 2002 national elections. Transfer of control over the diamond fields to the government marked a significant milestone in the strategic campaign plan drawn by the Sierra Leone government and UK military advisers.

Consolidating UNAMSIL

The strategic campaign plan aimed to strengthen the government as well as UNAMSIL and its activities, but there is no indication that the UN peacekeeping force participated directly in its implementation. The United Nations did, however, undertake a series of complementary actions designed to strengthen UNAMSIL, diminish Liberian support for the rebels, and restrict the flow of blood diamonds from Sierra Leone and other conflict areas in Africa. The UNSC increased UNAMSIL's mandated troop strength from 6,000 in May 2000 to 17,500 by March 2001. UNSC Resolution 1313, adopted on August 4, 2000, set out a number of high-priority tasks and gave the mission teeth, authorizing it to "decisively counter the threat of RUF attack."[52] In response to West African demands for African leadership of the mission, Lieutenant General Daniel Opande of Kenya replaced Major General Jetley as force commander. Opande received broad acceptance and respect from the troops, enabling him to consolidate command and control within the UN contingent. He directed UNAMSIL efforts in the pacification and disarmament of RUF combatants in Kono, Kambia, and Makeni. The appointment of Major General

---

man Catastrophe," Occasional Paper no. 1, Partnership Africa Canada, Oct. 2001, www.pacweb.org/Documents/diamonds_KP/1_dastabilizing_Guinea_Eng_Oct2001.pdf.

[51] Sierra Leone TRC, *Witness to Truth*, vol. 3A, 460.

[52] UNSC, Resolution 13, S/RES/1313 (2000), 4 Aug. 2000, operative para. 3; UNSC, "Fifth Report of the Secretary-General on the United Nations Mission in Sierra Leone," July 31, 2000, http://daccess-dds-ny.un.org/doc/UNDOC/GEN/N00/554/71/PDF/N0055471.pdf?OpenElement.

Martin Agwai of Nigeria to replace Brigadier General Garba, who had been part of the controversy surrounding Jetley, also helped stabilize the mission.

UNAMSIL itself was reorganized with the addition of two deputy special representatives of the secretary-general to coordinate political and administrative affairs and governance and stabilization. The operating forces received better military equipment, maps, and communications capabilities. Perhaps most importantly of all, they developed better predeployment reconnaissance and information gathering and processing, which strengthened their ability to respond rapidly and appropriately to events on the ground. A significant contributor to this capacity was the improvement in the military information cell that UNOMSIL had established at UNAMSIL headquarters in the Mammy Yoko Hotel.

In neighboring Liberia, the international tide turned sharply against Charles Taylor, weakening his influence and ability to destabilize Sierra Leone. He was denounced for his military and criminal activities in Sierra Leone and Guinea. UNSC Resolution 1343 of March 7, 2001, strengthened the arms embargo against Liberia and banned its diamond exports. Thanks in part to an effective media campaign orchestrated by Global Witness and other nongovernmental organizations, the embargo on Liberia's diamond trade signaled a real shift by the international community against the traffic in conflict diamonds. The UN Security Council had already identified conflict diamonds as a major source of financing for armed conflict, especially in Angola, and UNSC 1295, passed in April 2000, had pushed for measures to curtail the circulation of such diamonds. The diamond industry responded by adopting a resolution and developing the Kimberley Process Certification Scheme to ensure that rough diamonds reaching the international marketplace were conflict free. The scheme, which came into operation in 2003 after the conflict in Sierra Leone had effectively ended, has faced criticism from some of its initial backers, including Global Witness and Partnership Africa Canada.[53] It has had some positive impact, however, on countries such as Sierra Leone, as evidenced by rising official exports of diamonds certified under the Kimberley Process. In 2012, Sierra Leone officially and legally exported 532,555 carats of diamonds, compared to 77,370 carats in 2000, when the mines were under RUF control.[54]

The events of 2000 also signaled a major attitudinal shift in the international community (United Nations, supported by the UK and the United States) toward more activist and robust enforcement of peace in Sierra Leone. RUF fighters participated in the DDR program, and by January 18, 2002, when the conflict was formally declared over, 24,352 RUF combatants had been disarmed. In addition to the general amnesty granted by the Lomé Agreement, combatants were further enticed with a mixture of small cash payments (300,000 Leones, or US$125) and various skills-training programs for participating in the DDR process. Many RUF members subsequently participated in the truth and reconciliation process.[55]

---

[53] See, for example, James Melik, "Diamonds: Does the Kimberley Process Work?" *BBC News*, June 28, 2010, www.bbc.com/news/10307046.

[54] See Bank of Sierra Leone, "Annual Report and Statement of Accounts 2001," www.bsl.gov.sl/pdf/Annual_Report_2001_complete.pdf; Bank of Sierra Leone, "Annual Report 2012," www.bsl.gov.sl/pdf/Annual_Report_2012.pdf.

[55] Article XXVI of the Lomé Agreement called for creation of a Truth and Reconciliation Commission

In 2002, the United Nations and the Sierra Leone government set up a hybrid special court that indicted 13 key leaders from RUF, Kamajors, and AFRC who were most responsible for war crimes and crimes against humanity during the latter phases of the war. Although there was a clear desire to end impunity and punish egregious violations of international humanitarian law, global fatigue was already setting in around long and costly international tribunals, such as those for Rwanda and the former Yugoslavia. In opting for a hybrid court with a limited number of indictments, the United Nations aimed for a less costly, "targeted and efficient" mechanism.[56] Despite this strategy, the court attracted criticism over its cost (which exceeded $200 million), mandate, and ultimate legacy. The court oversaw the successful conviction and incarceration of nine of the indictees, including Charles Taylor. Sankoh, former deputy defense minister Chief Samuel Hinga Norman[57], Bockarie, and Koroma were indicted but died either before being apprehended by the court or during the trial.

By the time the court started work in January 2002, the RUF was no longer an effective militia or criminal organization. It no longer controlled diamonds or any other national resources, and it had lost the capacity to derail the peace process through violence or intimidation. As envisioned in the Lomé Agreement, the remnants of RUF leadership organized themselves into a political party and participated in the 2002 national elections (the first since the conflict). Despite a lingering sense of insecurity in different parts of the country, the 2002 elections proceeded fairly well, and Kabbah was reelected in a landslide. The RUF Party (RUFP), led by Alimamy Pallo Bangura, received less than two percent of the vote, while Johnny Paul Koroma, heading the Peace and Liberation Party, received three percent. The elections were mostly peaceful, and RUFP accepted defeat without resorting to violence.[58]

UNAMSIL's Strategy for Security Sector Reform

The stable environment created by the extensive deployment of UN peacekeepers and their successful disarmament of the various armed groups enabled the Sierra Leone government, the UK, and other international partners to focus on longer-range projects of consolidating peace, extending government authority, and rebuilding the Sierra Leonean state, society, and economy. These projects also coincided with the goals envisaged in phases 35 of the strategic plan drawn up by the government and UK military advisers.

---

within three months of the signing of the agreement. Parliament enacted the Commission in 2000. It began in 2002 and, after collecting information and hearing testimony in 2003 and 2004, presented its final report in 2004. See Sierra Leone TRC, "Truth and Reconciliation Commission Report," 2004, Sierra Leone Web, www.sierra-leone.org/TRCDocuments.html.

[56] Michael Miklaucic, "Justice in the Balance: Taking Stock of International War Crimes Tribunals," in *Civilians and Modern War: Armed Conflict and the Ideology of Violence*, ed. Daniel Rothbart, Karina Korostelina, and Mohammed Cherkaoui (New York: Routledge, 2012).

[57] Samuel Hinga Norman was founder and leader of the Civil Defence Forces, commonly known as the Kamajors. Although the Kamajors supported the Kabbah government against the RUF, Hinga Norman was indicted by the Special Court for Sierra Leone for war crimes and crimes against humanity, and recruitment of child soldiers. He died on February 22, 2007, before he could be convicted, while undergoing medical treatment in Senegal.

[58] The RUFP did not participate in the 2007 elections, and in the recently concluded 2012 national elections, its presidential candidate, Eldred Collins, received around 13,000 votes, or 0.6 percent of the total.

The two major processes that international partners addressed with the Sierra Leone government were security sector reform and poverty reduction.[59] Both processes were grounded in the widely embraced paradigm that security and development are inextricably connected. Throughout the peacebuilding project in Sierra Leone, international and local stakeholders worked to maintain this connection.

Reforming the Sierra Leone Police

The Kabbah government recognized that restoring and professionalizing the Sierra Leone Police was a priority if long-term stability was to be achieved. With financial and technical support from the UK, the Sierra Leone government undertook an extensive security sector reform program involving the national intelligence services, police, army, and judiciary. For the police reform effort, it appointed Keith Biddle, a retired British senior police officer, as inspector general of the SLP. He would lead the restructuring and rebuilding of the police force from 1999 to 2003.

Biddle faced a monumental challenge. War, retirement, and desertion had reduced the SLP's numbers from a prewar high of 9,317 to around 6,600 in 1999. More than 900 serving police officers had been killed during the conflict, and many police stations had been destroyed. With the assistance of the Commonwealth Community Safety and Security Project, funded by the UK's Department for International Development, Biddle oversaw the rebuilding of police stations and barracks and the strengthening of the force to 9,500. He went beyond recruitment, overseeing extensive retraining and restructuring throughout the whole SLP.

To create opportunity and promote professionalism, Biddle reduced the number of ranks in the force from 22 to 10 and opened promotion to younger talented officers. Sixty senior officers were retrained at the Police Staff College at Bramshill, UK.[60] In a controversial move, the Special Security Division (SSD), the hated paramilitary police unit created by the repressive APC president, Siaka Stevens, was kept on, but it was also retrained and redesignated the Operational Support Division (OSD). Doing so allowed the SLP to hold on to experienced members of the force and prevented them from becoming a disaffected opposition. Another factor weighing in favor of retention was the minimal public outcry against the OSD, thanks to SSD officers' heroic defense of Freetown in 1999. UNAMSIL's Civilian Police component also helped by reinstating SLP presence throughout the country and restoring confidence in the force.

Biddle placed greater emphasis on accountability to civilian control and emphasized community policing, an approach known in Sierra Leone as "local needs policing."[61] From 2004, Local Policing Partnership Boards, which included citizens working voluntarily with the police on preventing crime, were set up at each police station. The Domestic Violence Unit was upgraded to a Family Support Unit (FSU), with a broader role in investigating domestic violence, sexual offenses, and cruelty against women and chil-

---

[59] Peter Albrecht and Paul Jackson, "Security System Transformation in Sierra Leone, 1997-2007," GFN-SSR and International Alert, 2009, 37, http://issat.dcaf.ch/content/download/33989/486204/file/Security%20System%20Transformation%20in%20Sierra%20Leone,%201997-2007.pdf.

[60] Ibid., 31-32.

[61] Bruce Baker, "Sierra Leone Police Reform: The Role of the UK Government" (paper, GRIPS State-Building Workshop: Organizing Police Forces in Post-Conflict Peace-Support Operations, Jan. 27-28, 2010).

dren. By 2007, 230 people manned 30 FSUs across the country. Overall public perception of the SLP improved. In particular, its professional performance during the 2002, 2007, and 2012 elections stands as a testament to the reform process's positive impact.[62]

In 2003, a Justice Sector Task Force was established to widen the narrow focus on the SLP into a sectorwide approach on issues of policing and justice. In 2005, the task force became the Justice Sector Development Program, which, in 2007, helped the government map out a Justice Sector Reform Strategy and Investment Plan. The plan's four main goals were "safer communities through strengthening police; better access to justice through improving quality of local courts and providing paralegal services; strengthened rule of law through addressing corruption and maladministration; and improved justice service delivery through improving the performance of justice institutions."[63] The UK's Department for International Development supported the strategy with a £30 million grant (roughly US$[2007]60 million) over three years. While there has been some progress, reforming the justice sector remains a major challenge due to the lack of skilled personnel, inadequate infrastructure, insufficient financial resources, and resistance from entrenched interests within the sector.

Rebuilding the Intelligence Service and the Sierra Leone Army

The outbreak and prolongation of the civil war and various military coups during 1991-99 reflected systemic and repeated failures in intelligence, which the Kabbah government, with the UK's assistance, sought to remedy. The government's ongoing ability to counter threats to security and stability would require a robust yet accountable intelligence architecture that could facilitate information gathering, transmission, and analysis from the districts, through the provinces, and on to the national level. This architecture was created during 1999-2002. The National Security Council, Office of National Security, and a host of other national and local structures were created to collect, process, and act on information important to national security. Parliament gave legal force to the reforms in the Sierra Leone National Security and Central Intelligence Act of 2002.[64]

Rebuilding the Sierra Leone armed forces proved to be the most expensive and time-consuming aspect of the UK-led security sector reform project. Nonetheless, it has been crucial to maintaining the peace and stability that Sierra Leone has enjoyed since the cessation of fighting in 2000 and the completion of UNAMSIL's mandate in 2005.

During 1997-99, Kabbah had seriously contemplated disbanding the SLA. Minister of Internal Affairs Charles Margai met with President Oscar Arias of Costa Rica in Arusha, Tanzania, in 1998 to learn about Costa Rica's experience without an army. But General Maxwell M. Khobe, whom Kabbah had appointed chief of defense staff, argued that it would be unwise to turn battle-tested veterans loose in a volatile environment with no army in place to either absorb or subdue them. He also pointed out that some of the Si-

---

[62] Kadi Fakondo, "Reforming and Building Capacity of the Sierra Leone Police, 1999–2007," in *Security Sector Reform in Sierra Leone 1997–2007: Views from the Front Line*, ed. Peter Albrecht and Paul Jackson (Geneva: DCAF, 2010), 161-170.

[63] Clare Castillejo, "Building Accountable Justice in Sierra Leone," FRIDE Working Paper no. 76, Jan. 2009, 3, http://fride.org/download/WP76_Building_Accountable_Eng_ene09.pdf.

[64] National Security and Central Intelligence Act 2002, Supplement to *Sierra Leone Gazette* vol. CXXXII, no. 42, July 4, 2002, www.sierra-leone.org/Laws/2002-10.pdf.

erra Leone soldiers had performed creditably during the war. Khobe recommended that instead of being disbanded, the army should be restructured, retrained, and reduced in size.[65] He died before the radical restructuring of the military began, but his recommendation was largely accepted, and the IMATT assumed the task of implementation.

Military reform and the DDR process were inextricably entwined. During the DDR process, nearly 2,500 former Kamajor, ARFC, and RUF combatants were screened, trained, and absorbed as part of the Military Reintegration Program.[66] Under Operation Cheetah, the Freetown military garrison was trained to become a rapid-reaction force with the capacity to deploy quickly and respond more effectively to security threats to the city. Under Operation Reassurance in 2006, retrained units were deployed in particularly vulnerable areas, especially border points. With revised and newly developed training manuals, veteran soldiers were retrained in how to function in a democratic environment.

From 2003 on, new recruits who were expected to be the core of the new breed of soldiers began training at Benguema Military Barracks.[67] As this new blood was being brought into the reconstituted military, older and long-serving officers were retired. By 2010, the army had been downsized from around 15,000 during the war to 10,500 in 2006. The strength of the army was further reduced to around 8,500 in 2012.[68] In the process, its command-and-control structures, administration, maintenance, personnel management systems, and training regimes were completely overhauled. The legacy of rape, abduction, sexual assault and other crimes against women has been an ongoing challenge that both the UK and the Sierra Leone government were determined to address in the SSR process. For its part, the military is still making serious efforts to recruit and train women and promote qualified individuals to command and leadership roles.[69]

The restructuring of the army followed reorganization of the Ministry of Defense. The IMATT led the effort. The team was funded and staffed largely by the UK, with support from Canadian, U.S., and West African military experts.[70] To restructure toward greater effectiveness and accountability, the Defense Ministry established two parallel organizations: the Joint Force Command, headed by the chief of the defense staff, and the Joint Support Command, headed by a civil servant as director general. Both were under the direction of the deputy minister of defense and the president. IMATT officers who oversaw the reorganization of the ministry felt that the joint command structure would minimize the ability of a single military officer to mobilize troops and other resources for

---

[65] Osman Gbla, "Security Sector Reform in Sierra Leone," in *Challenges to Security Sector Reform in the Horn of Africa*, ed. Len le Roux and Yemane Kidane, ISS Monograph Series no. 135, May 2007, 21, http://dspace.africaportal.org/jspui/bitstream/123456789/31144/1/M135FULL.pdf?1.

[66] Emmanuel Osho-Coker, "Governance and Security Sector Reform," in Albrecht and Jackson, *Security Sector Reform in Sierra Leone*, 109-18.

[67] Author interviews with national security officials, IMATT officers, and senior RSLAF military commanders, who preferred to remain anonymous, Aug. 2007, Freetown. Around 2003, IMATT also set up a new institution, the Horton Military Staff Academy, for training junior and senior officers.

[68] Alfred Nelson-Williams, "Restructuring the Republic of Sierra Leone Armed Forces," in Albrecht and Jackson, *Security Sector Reform in Sierra Leone*, 119-48.

[69] Ibid.

[70] The total number of IMATT officers has varied over the years. In 2007, it stood at around 105, of whom 88 were British, 11 Canadian, 3 American, 2 Nigerian, and 1 Jamaican. The tour of duty is normally six months, with some spending up to a year.

a coup. The restructuring of the ministry faced criticism, however. Joe Blell, then deputy minister of defense, regarded it as "far too complex" for Sierra Leoneans.[71] The creation of equivalencies in the grades and ranks of officials in the military and civilian wings of the ministry also caused resentment. The army top brass disliked seeing civilians with no military experience or extensive employment record receive the equivalent rank of colonels, brigadiers, and generals.[72]

The restructuring of the SLA has not resolved all its perennial problems. Although IMATT officers praised the "seamless" integration of the former combatants of the SLA, AFRC, RUF, and Kamajors into the new force, their training has been criticized as "hurried."[73] Military transport and equipment remains inadequate, and maintenance has been a recurrent problem.[74] Compensation, decent billeting, fuel, and promotions remain contentious issues as the Sierra Leone government struggles to finance an army of over 8,500 troops. The retirement of old soldiers created tensions during and after the reorganization process. There was fiscal pressure to reduce numbers fast. But a number of former senior military officials still had loyal followers, and for this reason, the IMATT in particular was worried about the possible consequences of putting too many former soldiers out on the streets. This led to a decision to make the retirements deliberately incremental despite financial concerns.

Pensions and payments to the families of officers killed and wounded added to fiscal tensions.[75] The Sierra Leone government was concerned about disbursing pension money before an independent verification committee could establish the amounts to be paid over and above salary, so it began to renege on its obligations to this group of pensioners. Ultimately, the UK's Agency for International Development made the interim payments—again, for security reasons. No one wanted disaffected former combatants on the streets, unable to support themselves and nursing a grievance against the government.

Still, these problems have not yet produced any serious security crisis.[76] The soldiers of the new army are aware of the public's deep mistrust and of how their involvement in politics and conflict in the 1990s alienated them from the public. In the 2007 and 2012 elections, the military remained neutral in its posture, and public perception has improved greatly.[77] Notably, because of the security reform program, Sierra Leone police and military officers have been deployed to support African Union and UN peacekeeping operations in Haiti, Sudan, Somalia, Lebanon, and Nepal.[78]

---

[71] Albrecht and Jackson, *Security Sector Transformation*, 99-100.

[72] Interviews with security officials, IMATT officers, and senior RSLAF military commanders, who preferred to remain anonymous, Aug. 2007, Freetown.

[73] Osman Gbla, "The Role of External Actors in Sierra Leone's Security Reform," in *When the State Fails: Studies on Interventions in the Sierra Leone Civil War*, ed. Tunde Zack-Williams (Uppsala: Pluto Press, 2012).

[74] Interviews with security officials, IMATT officers, and senior RSLAF military commanders, who preferred to remain anonymous, Aug. 2007, Freetown.

[75] Emmanuel B. Osho-Williams, "Governance and Security Sector Reform," in *Albrecht and Jackson, Security Sector Reform in Sierra Leone*, 109-18.

[76] These problems continue to plague the military. See Ken Josiah, "The Defender – Voice of the Institution or a Political Publication?" *Sierra Express Media*, May 8, 2012, www.sierraexpressmedia.com/?p=40126.

[77] Interviews with security officials, IMATT officers, and senior RSLAF military commanders, who preferred to remain anonymous, Aug. 2007, Freetown.

[78] See S. O. Williams, "A New Dawn in the RSLAF," *Sierra Express Media*, Mar. 26, 2012, www.sierraexpressmedia.com/?p=37362.

Recommendations and Conclusions

Recommendations

The restoration of peace and stability in Sierra Leone offers a number of valuable lessons regarding interrelationship between power structures, conflict resolution, and peacekeeping in war-ridden and volatile contexts.[79]

Analyze insurgents and illicit power structures properly. The Sierra Leone conflict and peace process underscored the importance of devoting more resources to understanding the nature of insurgent groups—their strategies and motivations, structure, and sources of support—so that national, regional, and international actors develop a coherent strategy for responding to them. Ignorance of RUF's true nature nearly scuttled the Lomé Agreement and the UNAMSIL mission. As late as 1999, the government of Sierra Leone, ECOMOG, the United Nations, and their supporters in the international community still clung to a set of expectations about RUF's transformation into a legitimate political actor—expectations that the RUF had neither the intent or the capacity to meet. The international community also misdiagnosed the RUF's entire motivation for being. The subsequent kidnapping of peacekeeping troops merely exposed what the international community should have long known: that the RUF had no regard for international law and human rights and that its objectives were fundamentally irreconcilable with the peace process.

Engage regional actors. The role of regional actors, especially ECOWAS and West African heads of state, was important in preventing the Sierra Leone interventions from failing completely, as happened in Somalia. ECOWAS leaders took the early initiative of trying to keep the peace when the Sierra Leone conflict was not high on the UN agenda. The personal diplomacy of the various leaders, including peacemakers such as Konan Bédié of Côte d'Ivoire, Gnassingbe Eyadema of Togo, and Alpha Konaré of Mali, were crucial in getting RUF and the Sierra Leone government to sign the two peace agreements in 1996 and 1999, and the two Abuja agreements in 2000 and 2001. Equally important was regional leaders' role in helping neutralize powerful illicit actors such as Charles Taylor and Blaise Compaoré, who had enabled and empowered the RUF at critical points in the conflict.

ECOMOG troops' performance was checkered, however, due to unclear mandates and insufficient resources, peacekeeping skills, and experience. And yet, despite the shortcomings, they helped contain the Sierra Leone war and twice saved the elected Kabbah government. ECOMOG troops' mixed performance up to 2000, and the exemplary performance of those rehatted under UNAMSIL afterward, showed what can be achieved when UN and regional organizations work together. The United Nations'

---

[79] See Clifford Bernath and Sayre Nyce, "A Peacekeeping Success: Lessons Learned from UNAMSIL," in *International Peacekeeping: The Yearbook of International Peacekeeping Operations*, vol. 8, ed. Harvey Langholtz, Boris Kondoch, and Alan Wells (Leiden, Netherlands: Brill, 2004), 119-42; Matthias Goldmann, "Sierra Leone: African Solutions to African Problems?" in *Max Planck Yearbook of United Nations Law 9* (July 2005): 475-515.

deployment of sufficient resources, peacekeeping expertise, and strong oversight contributed to the West African troops' effective performance. The UNAMSIL experience also highlights the need for the international community to support efforts by regional organizations.

Give robust mandates to international peacekeeping and enforcement. The UNAMSIL experience shows that in complex and prolonged conflicts involving an illicit power structure willing to use violence to oppose the peace process, international peacekeeping with inadequate intelligence and planning, limited mandates, and constrained personnel and logistical capacity cannot be effective. While speedy deployment is important, it can be disastrous without proper planning and preparation. The subsequent strengthening of UNAMSIL's enforcement mandate, provision of combat-capable troops and intelligence resources, and reorganization of the force proved crucial to transforming the volatile situation.

As the Report on the Panel of United Nations Peace Operations (Brahimi Report) suggested and as UNAMSIL's experience revealed, missions must have clear, achievable priorities and tasks, not only in peacekeeping and peacebuilding but also in peace enforcement.[80] The defeat of recalcitrant West Side Boys factions, effective DDR of former combatants, recovery of the diamondiferous areas, extension of government authority to all parts of the country, and support for democratic elections by the end of 2002 are examples what can be achieved by a UN Peace Operation that has clear and achievable priorities and is empowered by sufficient mandates and robust capabilities.

Implement quick-impact projects and sustained peacebuilding. Quick-impact projects—for example, reconstruction of basic infrastructure such as bridges, schools, community centers, health facilities, and even places of worship—also generated tremendous goodwill in Sierra Leone. These actions helped change UNAMSIL's image and public perception in a short time from one of derision to one of support, especially after the mission demonstrated its ability to confront violent threats against it.

Embrace the challenge of comprehensive security sector reform. The UNAMSIL experience also showed that for robust peacekeeping to succeed, it must be accompanied by sustained institution building across the security system. The mission became extensively involved in the process of strengthening democratic institutions and practices, civil society, human rights, and gender equity in society. Policing development was deliberately joined with the development of other justice sector institutions, and reform of the SLA and the intelligence services was accompanied by equally robust restructuring to ensure democratic civilian participation and control. DDR and SSR processes were linked to ensure that former combatants did not become disaffected spoilers of the reform process. Significantly, those who had suffered at the hands of security forces during the conflict—particularly women—were actively recruited to become part of the new order, whether as members of the security forces themselves or as part of community outreach and dialogue programs. For a mission confronted by an illicit power structure that may

---

[80] UN General Assembly, Security Council, "Report of the Panel on United Nations Special Operations," A/55/305 – S/2000/809, Aug. 21, 2000, www.unrol.org/files/brahimi%20report%20peacekeeping.pdf.

be violently opposed to the peace process, a dual strategy is essential: come prepared to defend the mandate and peace process by creating an environment conducive to peace, and then provide more attractive peaceful alternatives and institutions that can support long-term reconciliation and effectively manage the competition for wealth and power.

Enlist (and follow) leadership from the UNSC. The active response and leadership by the UN secretary-general, the UNSC, and, in particular, the UK after the May 2000 hostage-taking crisis saved UNAMSIL. As Sierra Leone's former colonial ruler and an influential permanent member of the UNSC, the UK provided advocacy and leverage that translated into robust support for peace negotiations, peacekeeping, peacebuilding, and peace enforcement in Sierra Leone. Its willingness to use military force to safeguard the UN peacekeeping mission sent a powerful signal to the AFRC and RUF insurgents.

Link DDR with rebuilding the security sector. The UK-sponsored security sector reform program, though implemented separately from UNAMSIL, complemented and strengthened the mission's overall goal of ensuring long-term stability. The program's implementation showed some positive effects of linking disarmament of combatants with absorption of some of them into an army that was being remodeled. This linkage fulfilled an important provision of the Lomé Agreement while rebuilding the capacity of the state's security forces.

Conclusions

The misreading of RUF's criminal character and militaristic intentions after it signed the Lomé Peace Agreement initially led the United Nations to misjudge the Sierra Leone situation and to adopt a strategy that ultimately jeopardized its peacekeeping mission. Decisive military actions by the UK and elements of the Sierra Leone military and Kamajors saved the mission from complete collapse. The strengthening of UNAMSIL's size and peace enforcement capacity, supported by a concerted plan of action to defeat and diminish RUF's operational capacity and cut it off from Charles Taylor, eventually stabilized the situation. Coupled with UK national commitment, a restructured UNAMSIL mission was able to complete the DDR program, extend government authority throughout the country, and support national elections in 2002. The TRC laid the foundations, imperfect instruments though they were, for postconflict national reconciliation and healing, and the Special Court underlined the international commitment to curtail impunity and to prevent and punish gross violations of international humanitarian law. The Kimberley Process constrained the illegal traffic in conflict diamonds, which had been one of the main enablers of the illicit economy, and the main revenue source for the RUF and its allies. With UN and UK support, Sierra Leone's intelligence and security forces were restructured, retrained, and equipped to anticipate and contain further conflict. Democratic control over the armed forces, though imperfect, was established.

Though defeated, the RUF has not disappeared. In accordance with the Lomé Agreement, it eventually became a political party that has participated peacefully in two national elections, polling around two percent of the vote in 2002 and less than one percent in 2012. Despite the United Nations' significant achievements in restoring peace and stability in Sierra Leone and helping rebuild state institutions, the picture is still far from rosy. Sierra Leone remains at the bottom of the UNDP Human Development Index. It continues to wrestle with deep social and economic challenges despite possessing tremendous natural resources and benefiting from more than two decades of IMF- and World Bank-supported poverty reduction assistance. National elections in 2007 and 2012 were conducted amid great anxiety and accompanied by widespread threats of violence.

Nonetheless, the achievements in restoring peace and security should not be understated. Coups and further violent conflict have been prevented so far. The three most recent Sierra Leonean governments have been stable, and various governmental institutions, though riddled with problems, function relatively well. In particular, the 2014-15 Ebola crisis highlighted significant gaps in the Sierra Leonean government's ability to manage a national crisis, but it should also be noted that at no time during the pandemic was the government in danger of collapse. That in itself indicates progress, and where there is progress, there is hope.

Map No. 4172 Rev.3 UNITED NATIONS
March 2008

Department of Field Support
Cartographic Section

9. Sri Lanka: State Response to the Liberation Tigers of Tamil Eelam as an Illicit Power Structure

Thomas A. Marks and
Lt. Gen. (Ret.) Tej Pratap Singh Brar

Until recently, Sri Lanka was the homeland of an illicit power structure unlike any other. The Liberation Tigers of Tamil Eelam (LTTE) was an insurgency that privileged terrorism as a method of action yet ultimately fielded land, air, and sea regular forces, rounded out by powerful special-operations and information capabilities. LTTE grew in capacity until it was capable of forcing the government to agree to a February 2002 cease-fire and the de facto existence of a Tamil state, or Tamil Eelam. But this victory of sorts produced a host of unforeseen consequences leading to the July 2006 resumption of hostilities. The result, in May 2009, was complete military defeat of the insurgency.

The Tamil Eelam case actually encompasses four distinct conflicts, generally referred to as Eelam I (1983-87), Eelam II (1990-95), Eelam III (1995-2002), and Eelam IV (2006-9). These dates are open to discussion given realities on the ground. The gap between Eelam I and II saw the interlude of the Indian Peace Keeping Force (IPKF), which clashed bitterly with LTTE. And the gap between Eelam III and IV saw the effective rule of the Tamil Eelam state in areas of the north and east. This was accompanied by an uneasy cease-fire. In fact, each of the Eelam conflicts involved periods of negotiation and cessation of hostilities, though all were problematic in implementation and intent (certainly on the part of LTTE). All involved foreign participation. Further complicating the picture, the IPKF years saw Sri Lanka fully committed to suppressing another insurgency on a wholly different front. This was JVP II, the second upsurge of the original Maoist Janatha Vimukthi Peramuna (JVP, or People's Liberation Front) uprising, which had erupted and was crushed in 1971 (JVP I). Total casualty figures for the Eelam insurgencies are subject to considerable disagreement but cannot be less than 120,000 dead.[1]

LTTE's end, when it came, was as spectacular as its three decades of existence. Having grown from a ragtag band of angry young men into an impressive guerrilla group, then to a full-fledged army, the self-proclaimed flag-bearer of Tamil nationalism found itself caught in the same position as the Confederacy in the 1861-65 American Civil War: outmobilized and outfought. Its sometime foreign supporters, notably its neighbor India, had deserted it, and even a pronounced global shift of attitudes on what was acceptable in warfighting could not turn outrage into tangible pressure on Colombo before the Tigers' end came. A force that at one point fielded as many as 35,000 combatants found its maneuvering space squeezed by the inexorable advance of government columns using punishing innovative tactics. A last stand on a narrow stretch of northeastern beach ended in annihilation, with considerable collateral casualties to civilians forced to accompany LTTE fighters as human shields.

---

[1] This estimate, even if accurate for the Eelam conflict, is surely off the mark when the JVP insurgencies are included. One expert, in fact, has noted that various sources put the number killed in JVP II alone at between 20,000 and 60,000, with 40,000 the most commonly cited figure. See Tom H. J. Hill, "The Deception of Victory: The JVP in Sri Lanka and the Long-Term Dynamics of Rebel Reintegration," *International Peacekeeping* 20, no. 3 (June 2013): 357-74.

LTTE itself admitted defeat on May 17, 2009, after basically all its major figures were killed in action. These included the near-mythical leader, Velupillai Prabhakaran, who had emerged in the late 1970s as the group's head and ruthlessly hung on to the position throughout the conflict. Ironically, the struggle has not yet been closed, for a shift in the political winds caused many governments, led by European nations and the United States, to turn on their Sri Lankan former partner. They joined cause-oriented groups in seeking sanction through international humanitarian and human rights law for what they saw as callous (and illegal) indifference to civilian casualties in the final period of struggle.

An outraged Sri Lanka became estranged from those democratic nations that it had the most in common with. So it reoriented its foreign policy to new regional forces, notably China. Even the recent January 2015 upset win by an opposition coalition headed by a former ruling party intimate, Maithripala Sirisena, is unlikely to result in a shift fully in the direction desired by those who seek to mandate that war be something other than what it has always been: barbarous and cruel.

Response to a "Terrorist" Threat

Sri Lanka's conflict did not end much differently from other historical instances of major combat. What sets it apart is the sheer savagery of the war that developed over three decades. Also significant was the complexity of the threat faced. Many governments labeled LTTE a terrorist organization. In fact, it was an insurgency in intent and methodology. It had, however, gone from using terrorism as a tool for mass mobilization to using it as the main element in its approach to achieving Tamil Eelam.

The problem for security forces everywhere is that early on, armed challenges to the government's writ appear much the same. A systemic response centered in use of force, to the near exclusion of other facets, may be inappropriate in counterterrorism, complicating the effort, but in counterinsurgency it can often be disastrous. Most commonly, abuse of the populace creates a new dynamic, which allows an operationally astute insurgent challenger of state power to mobilize additional support. This is precisely what occurred in Sri Lanka.

An Unlikely Setting for War

A less likely setting for conflict would be hard to imagine, for the West Virginia-size island was and still is a tropical paradise in its physical aspects. The human landscape, though, has been less Edenic. British colonialism (1815-1948) had left unresolved issues regarding the meaning of independence and societal composition.[2] The Buddhist, Sinhala-speaking majority—10,979,561 of 14,846,750 according to the 1981 census, or 73.95 percent—dominated the British-inspired parliamentary democracy. And yet, the principal minority group, overwhelmingly Hindu (with Christian pockets) and Tamil-speaking (1,886,872 or 12.71 percent), had maneuvered within the British imperial struc-

---

[2] See Harshan Kumarasingham, "The Jewel of the East Yet Has Its Flaws": The Deceptive Tranquility Surrounding Sri Lankan Independence," Heidelberg Papers in South Asian and Comparative Politics, Working Paper no. 72, June 2013, http://archiv.ub.uni-heidelberg.de/volltextserver/151-48/1/Heidelberg%20Papers_72_Kumarasingham.pdf.

ture to achieve a position of relative advantage in commerce and the professions. This inspired much resentment among the majority, which increasingly resorted to inequitable measures to improve its standing—for instance, by making Sinhala the language of the civil service.[3]

Two other Tamil-speaking populations inhabited the island: the 1,046,926 Muslims, known as "Moors" (7.05 percent), and the 818,656 Indian Tamils (5.51 percent)—the latter the remainder of a larger migrant population recruited in Tamil Nadu, India, by the British to work on the coffee (and, later, tea and rubber) plantations.[4] In this discussion, we refer to Sri Lankan Tamils simply as "Tamils," and those from India as "Indian Tamils" or "estate Tamils" (for they remain clustered on the plantations in the south). Moors are now generally called "Muslims." These smaller groups had their own parochial issues and did not generally participate in the increasingly raw political battle between the Tamils and the Sinhalese majority. The Tamil protest movement began with a demand for justice but moved increasingly from street action to protoinsurgency.

In the decades after achieving independence from Great Britain in 1948, Sri Lanka was remarkably unprepared to deal even with overt protest action, much less subversion and its challenges, whether terrorism or guerrilla action. Following the country's annexation by Great Britain in the three Kandyan Wars (1803–5, 1815, and 1817-18), its martial heritage had effectively ended.[5] In 1971, when JVP I occurred, the principal armed capacity of the state consisted of just 10,605 policemen, armed at best with the venerable .303 Lee Enfield rifle and scattered in small stations amid a population of 12.5 million.[6] The military was also small (the army numbered only 6,578 soldiers in five battalions) and indifferently equipped. These forces grew but little in the following decades, even as the population reached roughly 18 million.

Political efforts to improve the position of the Sinhala-speaking Buddhist majority increasingly clashed with the Tamil-minority efforts to retain theirs. Particularly resented by the Tamils were government efforts, carried out with international assistance, to open up unused lands in the north and east, through irrigation and resettlement, in areas traditionally regarded as Tamil homelands (although fully a third of all Tamils lived amid the majority).[7]

---

[3] See Thomas A. Marks, *Maoist Insurgency Since Vietnam* (London: Frank Cass, 1996), 174-252; Tej Pratap Brar, "Sri Lanka's Civil War" (paper presented at Radcliffe Institute for Advanced Study conference, "Postcolonial Wars: Current Perspectives on the Deferred Violence of Decolonialization," Oct. 30-31, 2008, Harvard Univ.

[4] See John Richardson, *Paradise Poisoned: Learning about Conflict, Terrorism and Development from Sri Lanka's Civil Wars* (Kandy, Sri Lanka: International Centre for Ethnic Studies, 2005), 441. On the communities mentioned, see Ilyas Ahmed H., "Estate Tamils of Sri Lanka: a Socio-Economic Review," International Journal of Sociology and Anthropology 6, no. 6 (June 2014): 184-91; Valentine Daniel, Charred Lullabies: Chapters in an Anthropology of Violence (Princeton, NJ: Princeton Univ. Press, 1996); Amer Ali, "The Genesis of the Muslim Community in Ceylon (Sri Lanka): A Historical Summary," Asian Studies 19 (Apr.-Dec. 1981), 65-82.

[5] See Geoffrey Powell, *The Kandyan Wars: The British Army in Ceylon 1803-18* (Barnsley, UK: Pen & Sword, 1973); Channa Wickremesekera, Kandy at War: Indigenous Military Resistance to European Expansion in Sri Lanka 1594-1818 (New Delhi: Manohar, 2004).

[6] See A. C. Alles, *Insurgency 1971*, 3rd ed. (Colombo: Mervyn Mendis, 1976).

[7] See Chelvadurai Manogaran, *Ethnic Conflict and Reconciliation in Sri Lanka* (Honolulu: Univ. of Hawaii Press, 1987), 78-114; Sumantra Bose, *Contested Lands: Israel-Palestine, Kashmir, Bosnia, Cyprus, and Sri Lanka* (Cambridge, MA: Harvard Univ. Press, 2007), 6-54.

Small groups of radical Tamil youth, influenced by Marxism, formed both at home and abroad. Their solution to their "oppression" was to call for "liberation," that is, the formation of a separate socialist or Marxist Tamil state, or *Tamil Eelam*. These radical youth numbered perhaps 200. Later, leaders sought to indoctrinate youth. This proved problematic because both socioeconomic-political grievances and the desire for revenge (in response to instances of state violence) lent themselves more readily to an embrace of communalism than to confusing Marxist-Leninist ideology.

It is noteworthy that Marxist-Leninist doctrine, as the prism through which the Eelam leadership interpreted societal realities (especially state violence), is simply absent from all major treatments of the conflict. This is curious given the extent to which the various groups, including LTTE, in their formative years embraced Marxism-Leninism for both vocabulary and analytical constructs.[8] Just where the tension between ideologically driven leadership and grievance-produced manpower would have led for the *Eelam* movement as a whole was never put to the test, since LTTE, even as it established its dominance, increasingly embraced communalism.

At this point, however, the Tamil people, whatever their plight, were not much interested in giving their support to aspiring revolutionaries. Whatever its flaws, Sri Lanka remained a functioning democracy. And without a mass base, the insurgents could do little more than plan future terrorist actions. Police and intelligence documents speak of small, isolated groups of a half-dozen or so would-be liberationists meeting in forest gatherings to plot their moves. The bombings and small-scale attacks they made on government supporters and police positions were irritating (though sometimes horrific) but dismissed as the logical consequence of radicalism.

There was a method to the upstart schemes, however. By 1975, contacts had been made with the Palestine Liberation Organization through its representatives in London. Shortly thereafter, Tamils began to train in the Middle East. At home, LTTE initiated its armed struggle with an April 7, 1978, ambush in which four members of a police party were killed and their weapons captured. This was followed by hit-and-run attacks that led Parliament to ban the "Liberation Tigers" on May 19, 1978.

Though the police bore the brunt of LTTE activities, the army was also committed early on. This was carried out through the normal procedures of parliament's voting to activate emergency law. The burden for implementation of precise dictates and prohibitions, modeled after those of the former British colonial power, fell to a postcolonial security apparatus inadequate to the task. By July 11, 1979, the government claimed that LTTE had killed 14 policemen. On that date, a state of emergency was declared in Jaffna and at the two airports in the Colombo vicinity. It was soon extended to the entire country and remained in force for 28 years (renewed at monthly intervals).[9]

---

[8] See, for example, the mimeographed publication by LTTE's eventual number two, Anton S. Balasingham, *On the Tamil National Question* (London: Polytechnic of the South Bank, 1978). On mobilization, see Bryan Pfaffenberger, "Ethnic Conflict and Youth Insurgency in Sri Lanka: The Social Origins of Tamil Separatism," in *Conflict and Peacemaking in Multiethnic Societies*, ed. Joseph V. Montville (New York: Lexington, 1991), 241-57; Siri T. Hettige, "Economic Policy, Changing Opportunities for Youth, and the Ethnic Conflict in Sri Lanka," in *Economy, Culture, and Civil War in Sri Lanka*, ed. Deborah Winslow and Michael D. Woost (Bloomington: Indiana Univ. Press, 2004), 115-30.

[9] The emergency was formally lifted on August 25, 2011. See Stephanie Nolen, "Sri Lanka Announces End of 28-Year State of Emergency," *Globe and Mail*, Aug. 25, 2011, www.theglobeandmail.com/news/world/sri-lanka-announces-end-of-28-year-state-of-emergency/article595949/.

A week later, Parliament passed the Prevention of Terrorism Act, which, though modeled after British legislation, contained a number of controversial provisions, such as the authority to detain for 18 months (in six renewable three-month increments), without trial, anyone suspected of activities connected with "terrorism." In the context of terrorism, murder and kidnapping were made punishable by life imprisonment. Members of the security forces acting within the scope of the Act were granted blanket immunity.[10]

Nevertheless, the situation continued to deteriorate. In Jaffna, Charles Anton, LTTE "military wing" commander, was killed in a firefight with Sri Lankan military on July 15, 1983. In retaliation, on July 23, an LTTE ambush left 13 soldiers dead. Their funeral in Colombo ignited widespread rioting and looting directed against Tamils. Elements of the political establishment had a hand in planning and leading the violence. At least 400 people were killed and 100,000 left homeless; another 200,000 to 250,000 fled to India. Police stood by, and in many cases, members of the armed forces participated in the violence.

Communalism Leads to Armed Reaction

This spasm of communal violence proved to be a critical turning point in the conflict, both traumatizing the Tamil community and providing LTTE with an influx of new manpower.[11] Thus, the ascendancy of radical leadership in the struggle for Tamil Eelam was complete.[12] Although more than three dozen different groups may have been active at one point, they were dominated by just five: Liberation Tigers of Tamil Eelam (LTTE); People's Liberation Organisation of Thamil Eelam; Tamil Eelam Liberation Organisation; Eelam People's Revolutionary Liberation Front; and Eelam Revolutionary Organisation. By ruthless application of terror against its rivals, LTTE emerged as the dominant force.[13] For funding, criminality (including apparent involvement in the drug trade) was quickly surpassed by donations (both actual and coerced), increasingly from Tamil Nadu (both private and public sources) but mainly from the Tamil diaspora.[14] For arms and equipment, groups also looked to India.

The groups existed within the larger strategic realities of the Cold War. Since 1977, Sri Lanka, under the United National Party (UNP) administration, was a Western-oriented democracy with a market economy. In contrast, neighboring India, closely linked to the Soviet Union, was a democracy with a socialist economic approach and a geostra-

---

[10] Government of Sri Lanka, "Prevention of Terrorism Act," July 20, 1979, www.sangam.org/FACTBOOK/PTA1979.htm; N. Manoharan, *Counterterrorism Legislation in Sri Lanka: Evaluating Efficacy* (Washington, DC: East-West Center, 2006).

[11] See Stanley J. Tambiah, *Leveling Crowds: Ethnonationalist Conflicts and Collective Violence in South Asia* (Berkeley: Univ. of California Press, 1996), 82-100.

[12] See Sixta Rinehart, *Volatile Social Movements and the Origins of Terrorism: The Radicalization of Change* (Boulder, CO: Lexington Books, 2013), 109-37.

[13] The single best treatment of LTTE is M. R. Narayan Swamy, Tigers of Lanka: From Boys to Guerrillas (Delhi: Konark, 1994). See also M. R. Narayan Swamy, *Inside an Elusive Mind: Prabhakaran – the First Profile of the World's Most Ruthless Guerrilla Leader* (Colombo: Vijitha Yapa, 2003).

[14] No single work serves as an authoritative source on funding of the Eelam groups (later LTTE alone). Although written well after the events discussed here, a useful reference is Anthony Davis, "Tamil Tiger International," *Jane's Intelligence Review* (Oct. 1996): 469-73. On support provided by Tamils in Canada, see Paul Kaihla, "Banker, Tiger, Soldier, Spy," *Maclean's*, Aug. 5, 1996, 28-32.

tegic view that called for thoroughgoing domination of its smaller South Asian neighbors. Apparently to gain information on developments concerning the Sri Lankan port of Trincomalee, which New Delhi feared that the West coveted as a base, Indian Prime Minister Indira Gandhi agreed in 1982 to a plan by the Research and Analysis Wing (RAW—India's equivalent of the CIA) to establish links with a number of Tamil terrorist organizations. India was not especially interested in the ideology of those who received its training. It sought to safeguard its regional position while calming aroused pro-Tamil communal passions within its own borders.[15]

Consequently, in Tamil Nadu, the Tamil-majority Indian state of 55 million directly across the narrow Palk Strait from Sri Lanka, an extensive network of bases was allowed to support the clandestine counterstate formed within Sri Lanka.[16] This enabled dramatic expansion of insurgent actions, and by the end of 1984, insurgent activity had grown to the point that it threatened government control of Tamil-majority areas in northern Sri Lanka. The security forces had increased in size and quality of weaponry, but a national concept of operations was lacking. The result was a steadily deteriorating situation and hundreds of dead, most of them civilians killed in terrorist acts.

The extent to which insurgent capabilities had developed was amply demonstrated in a well-coordinated attack on November 20, 1984, when a Tamil force of company size used overwhelming firepower and explosives to demolish the Chavakachcheri police station on the Jaffna peninsula (east of Jaffna City) and kill at least 27 policemen defending it. Ambushes on security forces continued, along with several large massacres of Sinhalese civilians living in areas deemed "traditional Tamil homelands" by the insurgents. Use of automatic weapons, mortars, and rocket-propelled grenades was reported.

It became clear to the authorities that security force capabilities needed a drastic upgrading—a task accomplished in remarkably short order. Oxford-educated Lalith Athulathmudalai, a possible successor to President Junius R. Jayewardene, was named head of a newly created (March 1984) Ministry of National Security, as well as deputy defense minister. (Jayewardene himself was defense minister.) This effectively placed control of the armed services and counterinsurgency operations under one man. Interservice coordination improved under a Joint Operations Center (JOC), formed February 11, 1985. Its commander, Cyril Ranatunga, a recalled veteran of the 1971 JVP I conflict and a former commander in Jaffna, was promoted from brigadier to lieutenant general.[17] New manpower, formations, and equipment resulted in better discipline and force disposition. To relieve pressure on the military, a new police field unit, Special Task Force (STF), was

---

[15] See Tom Marks, "India Is the Key to Peace in Sri Lanka," *Asian Wall Street Journal*, Sept. 1920, 1986, 8. This work involved access to numerous prisoners and captured documentation, supplemented by fieldwork in Tamil Nadu, where members of all groups were quite forthcoming concerning assistance they received from New Delhi and Tamil Nadu State (which was running its own foreign policy of sorts). It was rumored (but known only later) that RAW's station chief in Madras, K. V. Unnikrishnan, had been compromised by the CIA. For two years until his arrest, he reported on Indian support to LTTE. See Sandeep Unnithan, "Madras Café Brings Back Uncomfortable Memories of the CIA's Honey Trap," *India Today*, Aug. 29, 2013, http://indiatoday.intoday.in/story/madras-cafe-madras-honey-trap-john-abraham-cia-ltte-raw/1/304302.html.

[16] See G. Palanithurai and K. Mohanasundaram, *Dynamics of Tamil Nadu Politics in Sri Lankan Ethnicity* (New Delhi: Northern Book Centre, 1993).

[17] See Cyril Ranatunga, *Adventurous Journey: From Peace to War, Insurgency to Terrorism* (Colombo: Vijitha Yapa, 2009).

raised under the tutelage of former Special Air Services personnel, employed by KMS Ltd. STF took over primary responsibility for security in the Eastern Province in late 1984, freeing the army to concentrate on areas of the Northern Province (which included Jaffna).[18] The army fielded new special forces and commando units.

Nevertheless, the situation continued to worsen. Terrorism not only was destructive in its own terms but also incited further communal strife. Attacks on Muslims in April 1985, for example, sparked Muslim-Tamil riots and significant population displacement. Bombs were discovered in the capital even as attacks hit trains, buses, and other modes of transportation. On April 29, a parcel bomb damaged several buildings in the army headquarters complex in Colombo.

Then, on May 14, an outrage occurred beside which others paled. LTTE combatants disguised as security force personnel used a bus to enter one of Sri Lanka's most sacred shrines, the Sri Maha Bodhi, a bo (pipal) tree said to be the southern branch from the tree under which the Buddha attained enlightenment. Indiscriminately attacking worshippers, LTTE murdered some 180 pilgrims. All too predictably, communal riots followed. In the field, a quickening tempo of guerrilla attacks displayed rapidly growing insurgent numbers and capabilities.

India's covert role has already been discussed. In July-August 1985, it endeavored to be more constructive by hosting peace talks between all major Eelam groups, including the noninsurgent Tamil United Liberation Front and representatives of the Sri Lanka state, in Thimpu, the capital of Bhutan. At this point, it was already clear that LTTE was the most intransigent of the groups, and eventually its leadership had to be coerced by New Delhi to continue the discussions.

Although various principles were agreed on, LTTE's real intent was to escape the constraints being placed on it in Thimpu and return to its chosen course of action: armed struggle. And this it did.[19] Nineteen eighty-six began with attacks, massacres, and bombings in seemingly endless succession. On May 3, 1986, an Air Lanka flight from London to Colombo, continuing to the Maldives, was delayed in Colombo long enough that a bomb intended to explode in midflight detonated while the plane was still on the ground, killing 21 passengers and injuring 41.

Seeking a Way Forward

At this point, despite the substantial steps that had been taken toward peaceful resolution of the conflict, the situation was clearly out of control. The tactical changes in security had been reasonably effective, but the government response was hobbled by the state's inability to set forth a viable political solution within which stability operations could proceed. Focusing on "terrorism" rather than on an insurgency that used terrorism as but one of its weapons, Colombo ordered its military leaders to go after the militants and

---

[18] See Tom Marks, "Sri Lanka's Special Force: Professionalism in a Dirty War," *Soldier of Fortune* 13, no. 7 (July 1988): 32-39. For a negative assessment of the KMS role (and the UK's as well), see the highly skewed (but useful, in parts) Phil Miller, "Britain's Dirty War against the Tamil People – 1979-2009," International Human Rights Association, June 2014, www.tamilnet.com/img/publish/2014/07/britains_dirty_war.pdf.

[19] See Tamil Nation, "Conflict Resolution: Tamil EelamSri Lanka," 1998, http://tamilnation.co/conflictresolution/tamileelam/85thimpu/thimpu00.htm; P. Venkateshwar Rao, "Ethnic Conflict in Sri Lanka," *Asian Survey* 28, no. 4 (Apr. 1988): 419-36.

stamp out the violence. There was little movement toward political accommodation that would have isolated the insurgent hard core from the bulk of the movement.

At heart, the impasse stemmed from an unresolved debate on just what independent Sri Lanka was: a multiethnic nation-state or the last bastion of a religion that, at one point, had dominated much of South Asia: Buddhism. Tamils and other Sri Lankan minorities could participate as equals only in the former—a diverse, multifaith society. The latter concept of Sri Lankan society, though by no means the dominant choice among the socioeconomic-political elites—many of whom were trilingual and had schooled together in elite institutions (with English the lingua franca throughout the island)—gained greater currency as the Tamil response to state violence took on many of the chauvinistic aspects it purported to be struggling against. This was particularly the case with LTTE, which, even during its flirtation with Marxist-Leninism, was dominated by the chauvinism, if not outright racism, of Prabhakaran.

The result was that the struggle, which the government framed in the language of counterinsurgency and counterterrorism, was more accurately a clash of contending nationalisms, with an increasingly beleaguered element of the national elite seeking to champion the fluid boundaries that saw communities mix and intermarry.[20] To further complicate the situation, while language and community were at the core of each national conception, the Sinhalese essence was defined by a Buddhism that was also central to the resistance against colonialism. In contrast, Tamils had not only generally embraced the opportunities afforded by colonialism but were divided into the two communities discussed earlier: the indigenous Sri Lanka Tamils (further differentiated by region) and the Indian Tamils. Ideologically, whereas the Sinhalese increasingly used political Buddhism as a tool for mobilization, the Eelam movement was informed by either the secular ideology of Marxism or the raw emotions of communalism.[21] Both threads rejected Tamil society's traditional structures pertaining to caste and gender.[22]

President Jayewardene, an experienced politician, led the country from 1977 to 1989. Born in 1906, he was, in a sense, a representative from an earlier era. Seeking a way forward, he increasingly used his immediate family and a small circle of trusted associates to determine how best to proceed, and to assess who within the military leadership could best deal with the fluid situation. A strategic plan that Jayewardene opportunistically requested in mid-1986 from a visiting security consultant emphasized that military action must serve to implement a political solution through redress of grievances, area domination, increased international support, and astute diplomacy with India, as opposed to the defensive posture that dominated relations with New Delhi. The actual mechanics of implementation—particularly the tangible steps necessary to restore governmental

---

[20] For the Tamil dimension, see A. Jeyaratnam Wilson, *Sri Lankan Tamil Nationalism: Its Origins and Development in the 19th and 20th Centuries* (London: Hurst, 2000); Chelvadurai Manogaran and Bryan Pfaffenberger, eds., *The Sri Lankan Tamils: Ethnicity and Identity* (Boulder, CO: Westview, 1994). For the Sinhalese dimension, see Tessa Bartholomeusz, "First Among Equals: Buddhism and the Sri Lankan State," in Buddhism and Politics in Twentieth-Century Asia, ed. Ian Harris (New York: Continuum, 1999), 173-93.

[21] See Patrick Grant, *Buddhism and Ethnic Conflict in Sri Lanka* (Albany: State Univ. of New York Press, 2009); Tessa J. Bartholomeusz, *In Defense of Dharma: Just-War Ideology in Buddhist Sri Lanka* (New York: Routledge/Curzon, 2002). On the Marxist ideology, see Satchi Ponnambalam, *Sri Lanka: The National Question and the Tamil Liberation Struggle* (London: Zed Books, 1983).

[22] On Tamil communalism, see Thomas Marks, "People's War in Sri Lanka: Insurgency and Counterinsurgency," *Issues & Studies* 22, no. 8 (Aug. 1986): 63-100.

authority to all areas of the country—were quite straightforward and adhered closely to what was finally done successfully in 2006-9: successive domination of areas, population and resource control, and mobilization of those Tamils who were opposed to the Eelam group. In the end, political realities dictated that the military facets of response continue to dominate, whereas the politically necessary steps were not taken.[23]

Transformation of Threat

Although Colombo did not put together the necessary *national* campaign plan, it did come up with an approach for the military domination of insurgent-affected areas. By early 1987, pacification in the east and near north left only Jaffna as an insurgent stronghold. As the Tigers' position in the Jaffna peninsula collapsed, they became more fanatical. They adopted the suicide tactics normally associated with violent radical Islamist movements. Individual combatants were issued cyanide capsules so they could avoid capture. A "Black Tigers" commando was formed to carry out suicide attacks using individuals or vehicles. Debate continues over the precise inspiration for this shift, but the result was never in dispute: LTTE's violence became much more lethal.[24] Surprisingly, though, it was not these tactics, but India, that rescued LTTE.

When Sri Lankan forces launched Operation Liberation in May 1987 and appeared on the verge of delivering a knockout blow,[25] New Delhi, responding to domestic pressure, entered the conflict directly with the Indian Peace Keeping Force (IPKF), thus bringing to a conclusion the phase known as Eelam I. Sri Lankan forces returned to barracks, and India assumed responsibility for overseeing implementation of a yet-to-be-agreed-upon cessation of hostilities. After an initial honeymoon period, during which all Eelam groups but LTTE chose to align themselves with Indian expectations, hostilities began between IPKF and LTTE.[26]

Although the Indian presence was useful in a tactical sense—New Delhi was now bearing the burden and the casualties of fighting LTTE—it was strategically disastrous. It not only reinforced the nationalist aspects of the Eelam appeal among the Tamil base but also provoked a Sinhalese nationalist reaction in the south, which absorbed virtually all the attention of Sri Lankan security forces.[27]

As the Indians tried to deal with the Tamil insurgents, Sri Lanka was forced to move

---

23 See Thomas Marks, "Counterinsurgency and Operational Art," *Low Intensity Conflict & Law Enforcement* 13, no. 3 (Winter 2005): 168-211.

24 See R. Ramasubramanian, Suicide Terrorism in Sri Lanka (New Delhi: Institute of Peace and Conflict Studies, 2004); Mia Bloom, *Dying to Kill: The Allure of Suicide Terror* (New York: Columbia Univ. Press, 2005), 45-75.

25 See Channa Wickremesekara, "Operation Liberation: 25 Years On," *Groundviews*, May 28, 2012, http://groundviews.org/2012/05/28/operation-liberation-25-years-on/.

26 It is worth noting that this chapter's authors met as a consequence of the IPKF deployment, when both were billeted in Jaffna Fort: Brar as commanding officer of the IPKF battalion headquartered there, and Marks as a journalist embedded with the Sri Lankan partner battalion in the same location. Brar became a key interface with LTTE command personalities—a relationship that continued until the outbreak of hostilities. See Thomas Marks, "Sri Lankan Minefield: Gandhi's Troops Fail to Keep the Peace," *Soldier of Fortune 13*, no. 3 (March 1988): 36-45, 74-75; Thomas Marks, "Handling Snakes and Unfriendly Troops in Sri Lanka," *Honolulu Star-Bulletin*, Sept. 22, 1987, A-17.

27 See Rajan Hoole, Daya Somasundaram, K. Sritharan, and Rajani Thiranagama, *The Broken Palmyra: The Tamil Crisis in Sri Lanka – An Insider Account* (Claremont, CA: Sri Lanka Studies Institute, 1990).

troops south to deal with Sinhalese Maoists of the JVP. The group's 1971 insurgency had been crushed at a cost of some thousands dead and at least 16,500 youth detained. In this second effort to seize state power, the JVP gained influence far beyond its numbers by exploiting nationalist passions and using terrorism to murder those who did not comply with its demands. The industrial sector, thoroughly cowed by a spate of carefully selected assassinations, was functioning at a mere 20 percent capacity. Such economic paralysis, in turn, fed the JVP cause. Sri Lanka was staggering.

Reorganization of State Response

A change in leadership in Colombo, with Ranasinghe Premadasa replacing the retiring President Jayewardene, brought a government approach that turned the tide against the JVP. Crucial to this effort in the Sinhalese-speaking south was the employment of the area-domination techniques that had gradually become standard in dealing with the Tamil insurgency in the north. Particularly salient was the command-and-control structure that had evolved. This was implemented by an army that had become a more effective, powerful organization. Its 76 battalions were now deployed to areas where, among other things, they spoke the language of the inhabitants and had an excellent intelligence apparatus. It was these battalions that implemented the counterinsurgency effort. Administratively, Sri Lanka's nine provinces were already divided into 22 districts, each headed by a government agent (GA), who saw to it that services and programs were carried out. To deal with the insurgency, these GAs were paired with military coordinating officers (COs), responsible for the security effort in the district. Often, to simplify the chain of command, the CO would be the commander of a battalion assigned permanently to the district.

Only as the conflict progressed did the army place its battalions under numbered brigades—although these remained continually changing in composition—and its brigades under divisions. In theory, there was a brigade for each of Sri Lanka's nine provinces. These were grouped under three divisional headquarters, only two of which were operational at the time of the JVP insurgency, because the third was designated to cover the LTTE insurgent areas in the north. With IPK active there, the division was not active. Each brigade commander acted as chief CO for the province and reported to his area commander (who also commanded the division to which the brigade was assigned). Areas 1 and 2 divided the Sinhalese heartland into southern and northern sectors, respectively; Area 3 was the Tamil-populated zone under IPKF control and, thus, inactive.

This system of creating a grid using the administrative boundaries, implemented historically to good effect by many security forces (particularly the British), had the advantage of setting in place permanently assigned security personnel who could become thoroughly familiar with their areas. The COs and their local security forces could be assigned further assets, both military and civilian, as circumstances dictated. The COs controlled all security forces deployed in their districts. They were to work closely with the GAs to develop plans for the protection of normal civilian administrative and area development functions. For this work, they were aided by a permanent staff whose job was to know the area intimately. Intelligence assets remained assigned to the CO headquarters and guided the employment of operational personnel. They did not constantly rotate as combat units came and went.

At the head of the framework was the Joint Operations Center. But the JOC never really hit its stride as a coordinating body. Instead, it usurped actual command functions to such an extent that it *became* the military. The security service headquarters, especially the army's, were reduced to little more than administrative centers.

Although often lacking precise guidance from above, local military authorities nonetheless fashioned increasingly effective responses to the JVP insurgency. This was possible because the COs and operational commanders—older and wiser after their tours in the Tamil areas—proved quite capable of planning their own local campaigns. Decentralization, in a state lacking communications and oversight capabilities, led some individuals and units to dispense with the tedious business of legal process. Those suspected of subversion too often were simply imprisoned or killed. Under the combined authorized and unauthorized onslaught, the JVP collapsed.[28]

After ending this second Sinhalese Maoist insurgency, the security forces could return their attention to the Tamil campaign when India withdrew in January-March 1990 (after almost three years and casualties of 1,155 IPKF dead and 2,984 wounded). New Delhi's involvement remains highly controversial to date, with considerable disagreement concerning the achievements of its counterinsurgency effort.[29] Ultimately, relations between India and Sri Lanka were so strained that Sri Lanka appeared to be actually *assisting* the various Tamil insurgent groups in their resistance. At this point, the Indians knew it was time to leave. Ominously, it was a greatly strengthened LTTE that awaited Colombo in Eelam II.

Growth of LTTE Power

LTTE power grew during a round of post-IPKF negotiations, which the Tigers used to eliminate their Tamil insurgent rivals. The talks collapsed when LTTE demanded that police stations in Eastern Province be vacated, then massacred more than 300 policemen who had been ordered by their superiors to accept what turned out to be false LTTE guarantees of safety.[30] Widespread terrorism followed, and a leap from guerrilla to mobile warfare. The insurgents attacked in massed units, often of multiple battalion strength, supported by a variety of heavy weapons. Deaths numbered in the thousands, reaching a peak in July-August 1991 in a series of set-piece battles around Jaffna. The 25 days of fighting at Elephant Pass, the land bridge connecting the Jaffna peninsula with the rest of Sri Lanka, saw the first insurgent use of improvised armor (using bulldozer chassis and power train), supported by artillery and extensive concrete-reinforced siegeworks protected by thick concentrations of antiaircraft weapons. The battalion was in danger of being overrun when one of the LTTE armored bulldozers, followed by infantry, breached the perimeter, but the assault was turned back in fierce

---

[28] See Rohan Gunaratna, Sri Lanka: *A Lost Revolution? The Inside Story of the JVP* (Kandy, Sri Lanka: Institute of Fundamental Studies, 1990); C. A. Chandraprema, *Sri Lanka: The Years of Terror – The JVP Insurrection 1987-1989* (Colombo: Lake House Bookshop, 1991).

[29] See Rohan Gunaratna, *Indian Intervention in Sri Lanka: The Role of India's Intelligence Agencies* (Colombo: South Asian Network on Conflict Research, 1993). For the Indian perspective, see Shankar Bhaduri and Afsir Karim, *The Sri Lankan Crisis* (New Delhi: Lancer International, 1990).

[30] Some sources put the number of police murdered as high as 600.

hand-to-hand fighting. Relief came overland, through difficult terrain after landing on the eastern coast, but not before casualties on both sides totaled several thousand.[31]

Elsewhere, terrorist bombings and assassinations became routine. Even national leaders such as Rajiv Gandhi of India and Sri Lanka's President Premadasa fell to LTTE bomb attacks (on May 21, 1991, and May 1, 1993, respectively), along with numerous other important figures, such as Lalith Athulathmudali (April 1993) and members of the JOC upper echelons.[32] Heavy fighting in Jaffna in early 1994, as the security forces attempted to tighten their grip around Jaffna City, resulted in government casualties approaching those suffered by LTTE in the Elephant Pass action. The conflict had devolved into a tropical replay of World War I trench warfare.

Dingiri Banda Wijetunga, who had been prime minister since March 3, 1989, took over Premadasa's position as president on May 7, 1993. He would lead the country until November 12, 1994. Ironically, in Sri Lanka's mixed system, wherein the president dominates and, if his party controls Parliament, all but names the prime minister, Wijetunga had been selected for his "old school" grace and lack of further political ambitions. But he was experienced and well versed in the security situation. His preparation had included in-depth discussions, in mid-August 1991, with security experts who emphasized the imperative that armed action serve to facilitate a political program that addressed Tamil grievances and marginalization. Thus, he moved beyond mere return to the prewar status quo. Nevertheless, he could not reorient the counterinsurgency approach in his brief time in office.[33]

Only with the election of a coalition headed by the Sri Lanka Freedom Party (SLFP) in August 1994, followed by the November presidential victory of SLFP leader Chandrika Bandaranaike Kumaratunga, was politics again introduced into the debate on state response to the insurgent challenge. The SLFP sweep ended 17 years of UNP rule and led to a three-month cease-fire, during which Colombo sought to frame a solution acceptable to the warring sides. The effort came to an abrupt halt when LTTE again—as it had done in every previous instance—unilaterally ended the talks by a surprise attack on government forces. Eelam III had begun.

Significantly, the wave of assaults highlighted the degree to which LTTE had become a potent military threat. Its techniques included the use of underwater assets to destroy navy ships, as well as the introduction, somewhat later (April 1995), of surface-to-air missiles, which were eventually used to destroy five aircraft.[34] In the field, LTTE guerrilla

---

[31] For battlefield photos, see Thomas Marks, "Sri Lanka: Reform, Revolution or Ruin?" *Soldier of Fortune* 21, no. 6 (June 1996), 35-39. Forces in the camp numbered about 600, and attackers numbered in the thousands (a figure of 5,000 is often used). Lance Corporal Gamini Kularatne posthumously became the first recipient of Sri Lanka's highest award for gallantry, the Parama Weera Vibhushanaya, for his actions in assaulting the armored bulldozer that had broken through the defenses on July 14, 1991.

[32] Predictably, it is the Rajiv assassination that has exercised international attention. He was, after all, apparently on the verge of again becoming prime minister in the Indian election campaign, which created the opportunity for his targeting. See Rajeev Sharma, *Beyond the Tigers: Tracking Rajiv Gandhi's Assassination* (New Delhi: Kaveri Books, 2013). See also the fictional film The Terrorist, directed by Santosh Sivan (1994). The assassination (and a cameo for the Premadasa killing) serves as the backdrop for the film *Madras Cafe*, directed by Shoojit Sircar (2013), which, though widely acclaimed, was banned in Tamil Nadu for (ironically) its accurate portrayal of the Madras-supported LTTE.

[33] Author (Marks) interview with Dingiri Banda Wijetunga, Aug. 15, 1991, Colombo. Wijetunga was from the Kandy area, the heartland of Sinhalese nationalism.

[34] See Sri Lanka Ministry of Defence, *Humanitarian Operation Factual Analysis July 2006-May 2009* (Colombo: 2011), 21, www.defence.lk/news/20110801_Conf.pdf. The two aircraft lost on April 28 and 29, 1995,

formations fighting as light infantry regular military units proved capable of engaging with government forces on more or less even terms. What had begun as a campaign by terrorists had grown to main-force warfare (also termed mobile or maneuver warfare) augmented by terrorist and guerrilla action.

The State Tries Further Adaptation

These new circumstances demanded a review of the government's approach to the conflict. In mid-1995, therefore, a series of meetings was held to settle on a revised national strategy for ending the conflict. On the political side, as directed by President Kumaratunga, a plan was articulated that came close, in all but name, to abandoning the unitary state in favor of a federal system. Devolution of power to the provinces, several of which would likely be dominated by Tamil voters, would effectively allow the establishment of ethnolinguistic states, as in India's federal system. On the military side, as had President Jayewardene had done, President Kumaratunga kept the defense portfolio for herself. Meanwhile, she selected a trusted associate (reportedly her uncle), parliamentarian Anuruddha Ratwatte, as deputy minister and, hence, effectively minister. He had reached the rank of lieutenant colonel while a mobilized reservist and had military experience, but none at higher levels of command. This was to prove a key factor because the strategic review quickly became a fierce battle of opposing positions.

All participants in the debate basically agreed that for a political solution to be implemented, LTTE must be dealt with militarily. But there was considerable disagreement on the plan of operations. On one side were those who favored a *military-dominated* response—essentially a conventional assault on LTTE. Opposed were those who favored a *counterinsurgency* effort of systematically dominating areas, using force as the shield behind which restoration of government writ would occur. The first called for strike operations, the second for the classic "oil spot" approach—the systematic domination of areas, which were then linked in a steadily expanding flow. Essentially, it was this latter approach that had emerged during the Wijetunga presidency as the security forces' default position. It was not favored by Ratwatte, though, who sought something more decisive, in particular the liberation of Jaffna peninsula, which LTTE had held for a decade.[35]

Contextually, there were grounds for favoring such a direct approach. With the end of the Cold War, LTTE had quietly dropped all talk of Marxism, though it continued to portray itself as socialist. Its links with the Tamil diaspora had matured, but its rupture with New Delhi was complete. For its part, India, though still closely linked to Russia, had seen its Soviet patron collapse and cautiously reached out to establish more normal relations with the United States and other supporters of Colombo. No objections arose

---

in the vicinity of Palali Air Base in Jaffna, were Avro transports carrying soldiers on leave; ninety-seven died. In July 1995, an FMA IA 58 Pucará providing close air support was also shot down, its pilot lost. Several years later, on September 29, 1998, Lionair Flight 602, using an Antonov An-24RV, was downed, apparently by an LTTE surface-to-air missile, killing all fifty-five people aboard. Though the missile type has not been stipulated, as early as mid-1987, author Marks examined an SA-7 shoulder-fired missile manual (translated into Tamil) in an LTTE safe house in Jaffna.

[35] Author (Marks) interviews, including a July 1995 series of meetings with Anuruddha Ratwatte, Colombo.

when the United States agreed, in mid-1994, to begin a series of direct training missions conducted by special-operations elements.[36] (Washington would designate LTTE as a Foreign Terrorist Organization [FTO] on October 8, 1997.)[37] These training missions enhanced the already mature relations that Colombo enjoyed with the UK and India. Further military ties and assistance were developed with Pakistan and China, to a lesser extent with Israel (always a controversial proposition in Sri Lankan politics because of the views of the Sri Lankan Muslim population and because of the large number of expatriate workers employed in the Middle East). The upshot was that the military seemed in relatively good shape internally, with strong external linkages to provide a steady stream of assistance and material support.

Operation Riviresa (more clumsily in English, "Rays of Sunlight") was launched in October 1995 to retake Jaffna—a goal accomplished by December 2. Strong leadership overcame an array of personnel and operational difficulties, but the victory left the occupying forces in a perilous position, cut off by the extensive territory to the south and east that remained in LTTE hands. It had been a conventional response to an unconventional problem, executed successfully but "a bridge too far," leaving multiple brigades stranded in Jaffna, where they could be supplied only by sea or air. LTTE adroitly used a combination of main force and guerrilla units, together with special operations, to isolate exposed government units and then overrun them. These included headquarters elements, with even brigade and division headquarters being battered. In the rear area, LTTE detonated a suicide truck bomb in the financial heart of Colombo in February 1996, killing at least 75 and wounding more than 1,500.

A pressing need for further force development led Colombo to approach the U.S. firm Military Professional Resources Inc. (MPRI), which had impressed the Sri Lankan military with its apparent success in overseeing the modernization and training of the Croatians for the successful Croatian summer 1995 offensive (Operation Storm) against Serb-supported forces.[38] An assistance plan was developed with U.S. acquiescence, but this went no further than the proposal stage, when many in the Sri Lankan military higher command objected.

---

[36] Author (Marks) interviews with the assessment authors before their deployment, July 1994, Honolulu. As reflected in the Sri Lankan copy of the Special Operations Command Pacific assessment, dated July 20, 1994, U.S. involvement was focused on training and support functions.

[37] After moving quickly to designate LTTE as an FTO, Washington was much slower to ban its various fundraising fronts, such as the Tamil Rehabilitation Organization, named a specially designated global terrorist entity under Executive Order 13224 on November 15, 2007. The Maryland-based Tamil Foundation was not banned (under the same authority) until February 11, 2009.

[38] Author (Marks) interview with MPRI personnel, Apr. 12, 1996, Alexandria, VA.

Colombo Down for the Count

Much worse was to come when overextension of forces, and an inability to handle the complexities of main-force conventional operations, left the Sri Lankan military badly deployed. Disaster was not long in coming. On July 17, 1996, an estimated 3,000 to 4,000 LTTE combatants isolated and then overwhelmed an understrength brigade camp at Mullaitivu in the northeast, killing at least 1,520 members of the security forces. This exceeded the 1,454 total death toll for 1994 and shattered army morale. Desertion, already a problem, rapidly escalated even as the isolation of Jaffna tightened. The linkup effort, Operation Jayasikurui (Certain Victory), kicked off in May 1997 but quickly slowed to a crawl as LTTE repeatedly demonstrated the ability to use combinations of regular and irregular action to inflict crippling casualties on poorly deployed, numerically superior government forces. Stalemate followed.

LTTE, needing only to exist as a rump counterstate that mobilized its young for combat, had demonstrated the ability to construct mechanisms for human and fiscal resource generation that defied the coercive capacity of the state. Linkages extended abroad, from where virtually all funding came (US$20-30 million per year); and diasporic commercial activities enabled procurement of necessary weapons, ammunition, and supplies. Though the security forces could hold key positions and even dominate much of the east, they simply could not advance on the well-prepared, fortified LTTE positions in the north and northeast, which, in any case, were guarded by a veritable carpet of land mines.

Political disillusionment again followed and increased as LTTE continued to pull off spectacular actions: In 1998, a suicide bomber attacked the most sacred Buddhist shrine in the country, the Temple of the Tooth, in Kandy; Kumaratunga herself narrowly missed following Premadasa as a presidential assassination victim, surviving a 1999 LTTE bomb attack but losing an eye; the Elephant Pass camp, which had previously held out against superior numbers, fell in 2000; and in July 2001, a sapper attack on the international airport in Colombo destroyed 11 aircraft. Ratwatte, who, in the flush of victory after the recapture of Jaffna, had been made a full general by President Kumaratunga, was no longer in his position, having been replaced in 1999.[39]

It was not altogether surprising that in the December 2001 parliamentary elections, the UNP, led by Ranil Wickremasinghe, was returned to power by a shaken electorate. This left the political landscape badly fractured between the majority UNP and its leader, the prime minister, and the SLFP's Kumaratunga, still the powerful president in Sri Lanka's hybrid political system, which is similar to France's. That the two figures were longtime rivals with considerable personal animosity did not ease the situation.

Again, as at the end of the Cold War, changes in the international arena dealt a wild

---

[39] Precise reasons for Ratwatte's removal were unstated. Besides the operational disaster, he was implicated in a series of corruption scandals (still under investigation at the time of his death) and accused of death squad involvement (of which he was acquitted in January 2006). On the corruption charges, see Frederica Jansz, "The Crooked General," *Sunday Leader*, Sept. 1, 2002, www.thesundayleader.lk/archive/20020901/spotlight.htm; Frederica Jansz, "Anuruddha Ratwatte Corruption Case Re-Opened," Sunday Leader, July 18, 2010, www.thesundayleader.lk/2010/07/18/anuruddha-ratwatte-corruption-case-re-opened/. On the murder charges, see *BBC Sinhala*, "Ratwatte Acquitted on Murder Case," Jan. 20, 2006, www.bbc.com/sinhala/news/story/2006/01/060120_ratwatte.shtml.

card. The increased worldwide concern with terrorism, already a factor in the new millennium but becoming central after the 9/11 terrorist attacks in the United States, caused additional Western countries to proscribe LTTE and move against its fundraising activities on their soil. LTTE was already banned in the United States and India (it had been proscribed by New Delhi in 1992) when the UK announced its listing as a terrorist organization on February 28, 2001. This was an important step since the Tamil diaspora in the UK was larger than anywhere else except Malaysia.[40] Canada finally proscribed LTTE on April 14, 2006, and the next month, the entire European Union followed.[41]

When considering the role of the Tamil diaspora on the conflict, a distinction must be made between imperial legacy communities, such as the Tamils of Malaysia, who migrated there or were recruited in the service of the British empire, and more recent migrants produced, at least in part, by the war in Sri Lanka. It appears that no studies disaggregate these categories, but the available literature makes clear that large, active support communities for LTTE existed in the UK, the United States, South Africa, and Canada, with that of Canada being perhaps the leading source of funding.

Amid this growing shift of international sentiments, shortly after 9/11, in February 2002, for reasons that remain unclear, LTTE suddenly offered to negotiate with the new UNP government. The government accepted the offer, and an uneasy truce commenced. The cessation of hostilities was a mixed bag in that it exacerbated intra-Sinhalese community tensions while also failing to bring "peace."[42] LTTE used the restrictions on Sri Lankan security forces to move aggressively into Tamil areas where it had been excluded and to eliminate rival Tamil politicians. Throughout Tamil-populated areas, Tamil-language psychological operations continued to denounce the state. In October 2003, LTTE proposed an Interim Self-Governing Authority (ISGA), which would have pushed beyond de facto realities to make LTTE the legitimate power in the Northern and Eastern Provinces. This prompted a strong reaction in the increasingly restive Sinhalese-majority heartland in the south.[43]

Chandrika Kumaratunga watched uneasily and then, in early November 2003, asserted her power while Wickremasinghe was in Washington, meeting with U.S. President George W. Bush. Claiming that the UNP approach was threatening "the sovereignty of the state of Sri Lanka, its territorial integrity, and the security of the nation," she ousted the three UNP cabinet ministers most closely associated with the talks, dismissed Parliament, and ordered the army into Colombo's streets.

LTTE waited, but in the April 2004 parliamentary elections that resulted from talks

---

[40] See Kaihla, "Banker, Tiger, Soldier, Spy"; Nomi Morris, "The Canadian Connection: Sri Lanka Moves to Crush Tamil Rebels at Home and Abroad," *Maclean's*, Nov. 27, 1995, 28-29.

[41] Like the United States, after proscribing LTTE, Canada was slower to move against its front organizations. The important fundraising group World Tamil Movement, for example, was not banned until June 2008. In all cases, LTTE supporters vehemently opposed such proscription. On the worldview of diaspora members who championed LTTE as the authentic representative of the Tamil people, see Øivind Fuglerud, *Life on the Outside: The Tamil Diaspora and Long-Distance Nationalism* (London: Pluto Press, 1999).

[42] See G. H. Peiris, *Twilight of the Tigers: Peace Efforts and Power Struggles in Sri Lanka* (New Delhi: Oxford Univ. Press, 2009).

[43] See *BBC News*, "Full Text: Tamil Tiger Proposals," Nov. 1, 2003, http://news.bbc.co.uk/2/hi/south_asia/3232913.stm; Muttukrishna Sarvananthan, "Sri Lanka Interim Self-Governing Authority: A Critical Assessment," *Economic and Political Weekly* 38, no. 48 (Nov.29-Dec. 5, 2003): 5038-40, www.jstor.org/stable/4414338?seq=1#page_scan_tab_contents.

between the dueling Sinhalese parties, SLFP unexpectedly swept back into power at the head of a United People's Freedom Alliance. The Tigers withdrew from negotiations but did not renew active hostilities, for they were preoccupied with what had once seemed unthinkable: a split within the movement. After long chafing under LTTE's domination by northern Tamils, the eastern cadre, under the leadership of longtime LTTE stalwart Vinayagamoorthy Muralitharan (more commonly known as Colonel Karuna Amman), had finally revolted in March. Though they were crushed in intense fighting followed by a wholesale vetting and purge of eastern cadre and combatants, the fracture remained permanent.[44] The alienated eastern Tamils, represented by Karuna's Tamil Makkal Viduthalai Pulikal (TMVP, or Tamil People's Liberation Tigers), increasingly made common cause with the government. This would prove to be a key development.

As events on the ground strained the cease-fire, the devastating December 26, 2004, Indian Ocean tsunami left more than 35,000 dead in Sri Lanka. Tamil areas were hit particularly hard. International aid poured in, but the issue of how it should be distributed stripped the last fig leaf from the unspoken agreements that had given LTTE its Eelam. When LTTE demanded that aid be channeled through its own counterstate bureaucracy, with the original ISGA proposal taking on all the trappings of statehood, the strained cease-fire collapsed.[45]

The situation continued to deteriorate, although LTTE was careful not to move too aggressively. The "cease-fire" served as the ideal cover for eliminating anyone the group saw as standing in its way. This included even the Sri Lankan foreign minister, Lakshman Kadirgamar, an ethnic Tamil, assassinated in August 2005. Also murdered was Sarath Ambepitiya, the judge who had sentenced Prabhakaran in absentia to 200 years in jail for the 1996 bombing of Colombo, and literally hundreds of Tamil politicians and activists opposed to LTTE (as well as many who were simply misidentified). For whatever the rhetoric connected with the peace process, LTTE remained committed to Eelam. In his annual November 27 speech, delivered on LTTE Heroes Day, Prabhakaran, the "president and prime minister of Eelam" (as the Tamil media billed him), warned that LTTE intended to renew hostilities if the government made no tangible moves toward "peace."[46]

In what was seen at the time as merely a tactical error (though it ultimately proved fatal), LTTE ordered a boycott of a presidential election hastily held in November 2005 after a Supreme Court decision ruled that Chandrika Kumaratunga's presidential term had run its course. Hard-line Prime Minister Mahinda Rajapaksa eked out a narrow victory against Ranil Wickremasinghe on a 73 percent turnout.

---

[44] See D. B. S. Jeyaraj, "Tiger vs. Tiger: Tenth Anniversary of Revolt Led by Eastern LTTE Leader 'Col' Karuna," *Daily Mirror*, Apr. 12, 2014, www.dailymirror.lk/45822/tiger-vs-tiger-tenth-anniversary-of-revolt-led-by-eastern-ltte-leader-col-karuna; Ajit Kumar Singh, "Endgame in Sri Lanka," *Faultlines* 20 (2011): 131-70.

[45] See Zachariah Mampilly, "A Marriage of Inconvenience: Tsunami Aid and the Unraveling of the LTTE and the GoSL's Complex Dependency," Civil Wars 11, no. 3 (Sept. 2009): 302-20; Alan Keenan, "Building the Conflict Back Better: The Politics of Tsunami Relief and Reconstruction in Sri Lanka," in *Tsunami Recovery in Sri Lanka: Ethnic and Regional Dimensions*, ed. Dennis B. McGilvray and Michele R. Gamburd, (New York: Routledge, 2010), 17-39.

[46] See Kasun Ubayasiri, "An Illusive Leader's Annual Speech," *Tamil Nation*, 2006, http://tamilnation.co/ltte/vp/mahaveerar/06ubayasri.htm.

By all historical accounts, Wickremasinghe would have been the better option for LTTE's plans. But the Tigers' continued cease-fire violations, which dwarfed the government's in both number and scale, steeled the new Rajapaksa administration for what was to come. A string of prominent LTTE suicide attacks, including an attempt to kill the army head, Lieutenant General Sarath Fonseka, and a successful assassination of the army's number three, pushed the situation beyond redemption.

Last-gasp efforts by Norway, the lead facilitator of the attempted settlement, came to naught. Norway's role in the peace process became increasingly controversial as LTTE continued to escalate its provocations. Whatever may be said about Colombo's conduct, it did not begin to approach the wholesale brutality of the Tigers, whose actions were dominated by assassinations. That Norway and other international actors could not bring themselves to vigorously counter LTTE atrocities led in the end to the mediators' loss of legitimacy.[47]

As fighting became more general, suicide attacks hit targets even in the deep south, such as Galle. By August 2006, Sri Lanka was again at war, in Eelam IV.

Transformation of Response

What followed was unlike what had gone before. The crushing of LTTE, often touted as a victory for counterinsurgency, was, in reality, the end of a civil war between a state and a rival counterstate. What ended LTTE's three decades of struggle was an operational clash of arms akin to the American Civil War in its ferocity, albeit distinct in tactics and societal features. What occurred was a signal illustration of military adaptation executed in concert with national mobilization, on the government's part, while LTTE proved unable to do the same. Examined more strategically and theoretically, the vanquishing of LTTE as an illicit power structure, and the postwar conflict it unleashed, serve to illustrate the profound changes that globalization has brought about in everything from the way that insurgency is waged to what is permissible in response.

Precisely what occurred operationally is easier to describe than to explain. For even after five years, considerable disagreement continues regarding just who initiated key aspects of the military's strategic adaptation to the operational situation, and just who was responsible for a series of astute tactical decisions in the field. The basis for renewed combat obviously lay in national mobilization. This was brought about, first, by the powerful *sangha's* (Buddhist clergy's) appeals to what effectively was holy war, and, second, by the government, with Mahinda Rajapaksa as president and his army veteran brother, Gotabhaya, as defense secretary, marshaling the financial support and determination necessary to rearm and reequip an expanded military.[48]

---

[47] See Gunnar Sørbø, Jonathan Goodhand, Bart Klem, Ada Elisabeth Nissen, and Hilde Selbervik, Pawns of Peace: *Evaluation of Norwegian Peace Efforts in Sri Lanka, 1997-2009* (Oslo: Norad, 2011), www.oecd.org/countries/srilanka/49035074.pdf; Kristine Höglund and Isak Svensson, "Mediating between Tigers and Lions: Norwegian Peace Diplomacy in Sri Lanka's Civil War," in *War and Peace in Transition: Changing Roles of External Actors*, ed. Karin Aggestam and Annika Björkdahl (Lund, Sweden: Nordic Academic Press, 2009), 147-69; Maria Groeneveld-Savisaar and Siniša Vuković, "Terror, Muscle, and Negotiation: Failure of Multiparty Mediation in Sri Lanka," in *Engaging Extremists: Trade-Offs, Timing, and Diplomacy*, ed. I. William Zartman and Guy Olivier Faure (Washington, DC: USIP Press, 2011), 105-35.

[48] See C. A. Chandraprema, *Gōta's War: The Crushing of Tamil Tiger Terrorism in Sri Lanka* (self-published, 2012 [available at Amazon.com]). For a discussion of mobilization of manpower (Sri Lanka never had to

In the field, General (following promotion) Fonseka insisted on a free hand that allowed him to field the best overall combat leadership yet. Innovative deployment of entire battalions as squad-size units or smaller, schooled in light infantry (i.e., commando) tactics and able to call in supporting fire, dramatically multiplied the defensive demands for an LTTE now struggling to defend its pseudo nation-state. Its governance, though innovative in some respects, had remained grounded in coercion, which dampened the enthusiasm of a populace being asked to mobilize in defense of Eelam. Indeed, one of the most contradictory aspects of the entire conflict was that throughout, a substantial proportion of Tamils, as well as nearly the entire Indian Tamil and Tamil-speaking Muslim populations, remained within government-controlled areas.[49]

First steps to seal off the battlespace and strangle LTTE's supply lines came with a successful high-seas campaign that hunted down and destroyed LTTE's oceangoing merchant navy. Simultaneously, development of high-speed coastal craft and tactics succeeded in neutralizing LTTE's hitherto formidable swarm of maritime suicide craft. The air force, though faced (even in the final phase of the struggle) with LTTE suicide efforts to attack Colombo, used overhead imagery and ground patrol coordination of targeting to eliminate the insurgent air arm.

On the ground, the actual conduct of reducing LTTE's counterstate followed the geographic plan that had been laid out originally in the 1985 planning documents. Seizure of the Eastern Province by July 2007, with help of the defecting eastern Tamil elements of the TMVP (perhaps a majority of its most effective combatants), allowed converging columns to draw an ever tighter noose around LTTE forces trapped in the northeast coastal area. This happened even as the first provincial elections were held to foster legitimacy for political reincorporation of previously LTTE-held areas.[50] TMVP, registered as a political party affiliated with the ruling coalition, emerged dominant in the March 2008 elections for local councils, and in the provincial elections in May. A split between Karuna and his deputy, Sivanesathurai Chandrakanthan, resulted in the latter's becoming the first elected chief minister of Eastern Province. Karuna later became a deputy minister in the government, and vice president of the ruling SLFP.

In the west, Mannar District fell by August 2008, and government forces were then able to move east to link up with military and TMVP elements in Eastern Province. Other forces cleared Jaffna peninsula and pushed south. The LTTE administrative center of Kilinochchi was abandoned and fell to the government in early January 2009. By early 2009, the remaining LTTE combatants, with perhaps 30,000 civilian hostages being used

---

resort to a draft), see Michele Ruth Gamburd, "The Economics of Enlisting: A Village View of Armed Service," in Winslow and Woost, *Economy, Culture, and Civil War in Sri Lanka*, 151-67.

[49] Author's (Marks) road counts of vehicles and individuals, and examination of the relevant logs kept at major government checkpoints ringing LTTE-held areas, consistently revealed flight away from Tamil Eelam and toward government-held areas—thus, toward relative safety. Efforts to conduct longitudinal studies on such internally-displaced-persons populations achieved mixed success. See H. L. Seneviratne and Maria Stavropoulou, "Sri Lanka's Vicious Circle of Displacement," in *The Forsaken People: Case Studies of the Internally Displaced*, ed. Roberta Cohen and Francis M. Deng (Washington, DC: Brookings Institution Press, 1998), 359-98.

[50] See Cathrine Brun and Nicholas Van Hear, "Shifting between the Local and Transnational: Space, Power and Politics in War-torn Sri Lanka," in *Trysts with Democracy: Political Practice in South Asia*, ed. Stig Madsen, Kenneth Bo Nielsen, and Uwe Skoda (London: Anthem Press, 2011), 239-60.

as human shields, were trapped in the coastal area of the Nanthi Kadal Lagoon, north of Mullaitivu. There, five divisions—a force normally described as fielding some 50,000 combatants—crushed them by mid-May 2009. In all aspects except the innovative tactics used by the Sri Lankan infantry, the 2006-9 endgame of the conflict had been major combat as might be seen on any conventional battlefield, featuring everything from heavy artillery to rocket launchers, extensive minefields, and suicide attacks.[51]

This last point highlights that the final years, coming as they did at the end of three decades of ever more vicious conflict that progressively brutalized all facets of Sri Lankan life, most resembled the island battles of the Second World War's Pacific Theater, especially the battle for Okinawa, which, like Sri Lanka, was heavily populated.[52] It was this reality that increasingly galvanized human rights advocacy groups, whose voices grew shriller as the end became ever more "like Berlin." When Colombo refused to heed calls from advocacy groups and certain Western governments, among them the United States and the UK, to allow some form of humanitarian intervention, advocacy gave way to outright opposition and siding with the defeated insurgents. This posture continues today.

In this respect—an external network of interested parties endeavoring to exert pressure directly on strategic choice—the Sri Lankan case transcends the mere "facts on the ground." The tangible conflict, horrific though it was, nevertheless was fought by a democracy that adhered throughout to the rule of law (albeit with very sharp elbows). That major combat places the rule of law under severe strain is a reality that Americans should readily recognize, particularly given the trajectory of American warfighting since Sherman's March to the Sea during the Civil War. There appear to be no credible sources claiming that Sherman gratuitously inflicted harm on the innocent, but few sources dispute his intense determination to embrace the very horror of war for the purpose of bringing it to a conclusion—a stance that delivered victory, however flawed it might be. This was the position that Sri Lanka found itself in. The war simply had to end if the country was to survive.

Lessons in an Era of Illicit Power Structures

It is challenging, after the short breathing space of five years, to draw lessons from this most vexing case of an illicit power structure challenging a licit power structure that erred.

LTTE was an insurgency that struggled to transcend its origins as a traditional rebellion in order to leverage the new possibilities in a post-Cold War world. This it did, both physically and virtually, integrally linking its struggle to regional and global Tamil

---

[51] See Ivan Welch, "Infantry Innovations in Insurgencies: Sri Lanka's Experience," *Infantry* (MayJune 2013): 28-31; *Daily FT*, "General Sarath Fonseka Reveals Untold Story of Eelam War IV," Mar. 10, 2015, www.ft.lk/2015/03/10/general-sarath-fonseka-reveals-untold-story-of-eelam-war-iv/; K. M. de Silva, Sri Lanka and the Defeat of the LTTE (New Delhi: Penguin Books India, 2012); Ahmed S. Hashim, *When Counterinsurgency Wins: Sri Lanka's Defeat of the Tamil Tigers* (Philadelphia: Univ. of Pennsylvania Press, 2013).

[52] See S. E. Selvadurai and M. L. R. Smith, "Black Tigers, Bronze Lotus: The Evolution and Dynamics of Sri Lanka's Strategies of Dirty War," *Studies in Conflict & Terrorism* 36, no. 7 (2013): 547-72; M. L. R. Smith and Sophie Roberts, "War in the Gray: Exploring the Concept of Dirty War," *Studies in Conflict & Terrorism* 31, no. 5 (2008): 377-98.

communities—the Tamils of southern India and the Sri Lankan Tamil global diaspora, respectively—in such a way that it could retain the strategic advantage long enough to achieve its goal of Eelam. In the process, it became almost legendary for its melding of commitment to destruction with its imagery of a new world emerging. With its suicide bombers and the cyanide capsules worn by its combatants—many of both being women—it set the Sri Lankan state back on its heels time after time.[53] Meanwhile, the dictatorial Eelam world it created was hailed for giving a people dignity and freedom, not only driving off the communal Sinhalese oppressors but also, in the process, shattering Tamil bonds of caste and gender inequity.[54]

The conflict waged by the state, which began as ineffective counterinsurgency and gradually grew to equally ineffective civil warfighting, illustrated another set of lessons. At each stage in the conflict, Sri Lanka struggled to comprehend just what it was involved in—and came up short. Initially, it treated protoinsurgency as emerging terrorism, thus emphasizing kinetic response when it should have been addressing the roots of conflict. Later, having mastered counterinsurgency's martial facets, it neglected the necessity of a *holistic* response, resulting in India's intervention. In the post-Indian context, the emergence of hybrid war—the blending of irregular and regular warfare with criminality and even (in its attempts to use chlorine gas in shells at one point) "WMD (weapons of mass destruction) warfare"—was mistaken for conventional conflict, resulting in devastating government defeats and LTTE's temporary victory. Finally, in the renewed 2006-9 fighting, a new civil-military team engaged in the functional equivalent of national mobilization and delivered a virtuoso display of integrating strategic, operational, and tactical levels of combat to deliver a knockout punch.

LTTE's end, when it came, had all the characteristics of the Second World War's denouement in Berlin or the ashes of Japan's incinerated cities. Colombo, ecstatic over its triumph, simply could not comprehend that it had again missed the bigger picture: the fundamental shift to an age of "new war" (more recently termed "hybrid warfare" by the world's militaries), in which powerful advocacy groups sought to make impossible, both practically and conceptually, the "total war" of past eras. It was just such a total war that Sri Lanka had fought. Its warfighting adaptation had been almost completely in the application of kinetic power, without the reforms in human rights and legal components necessary to engage in combat within what has become a global fishbowl. Colombo's strategists were quite ignorant of (and certainly unprepared for) the corresponding growth of new global norms, notably "R2P" (the responsibility to protect) and the right to intervene, together with the accompanying demands of what has been termed "the liberal peace." Indeed, it would be difficult to understate the mounting intensity with which both state and nonstate actors sought to slow, even end, Colombo's final push toward LTTE's annihilation, or the resulting sense of betrayal that Colombo ultimately felt toward the international community.[55]

In the events outlined above, a pathway led from the world of traditional war to what

---

[53] See Tamara Herath, *Women in Terrorism: Case of the LTTE* (New Delhi: Sage, 2012).

[54] See N. Malathy, *A Fleeting Moment in my Country: The Last Years of the LTTE De-Facto State* (Atlanta: Clear Day Books, 2012).

[55] See Damien Kingsbury, Sri Lanka and the *Responsibility to Protect: Politics, Ethnicity and Genocide* (New York: Routledge, 2012); David Lewis, "The Failure of a Liberal Peace: Sri Lanka's Counter-insurgency in Global Perspective," *Conflict, Security & Development* 10, no. 5 (Nov. 2010): 647-71.

has been called "new war," "postmodern war," "postheroic war," or even (though from a different theoretical angle) "fourth-generation warfare."[56] Regardless of terminology, the heart of the strategic matter was that in the post-Cold War global arena, use of force was to be legitimate, discriminating, and secondary to more compelling concerns (e.g., human security). It could be argued that this describes the strategic (and even tactical) requirements of counterinsurgency.

But counterinsurgency balances its kinetic and nonkinetic facets as required for successful mobilization to the extent necessary for victory, whereas advocates of the new approach to warfare see the use of kinetics as itself a symptom of a larger failure.[57] To use force to resolve the issue at hand—in this case, a drive for separatism—was to forfeit legitimacy. To add to this the bloodshed and destruction inherent in total war was to cross into criminality, which is precisely what very vocal and active voices asserted in demanding legal actions against the victors following the May 2009 obliteration of Eelam.[58]

Indeed, if any one characteristic may be seen as central to postmodern war, it is the supremacy of framing and narrative over the tangible imperatives of war. And it was in this area that Sri Lanka found itself thoroughly on the defensive. Colombo's frame of "victory" was all but overwhelmed by a shrill countering frame of "repression," and Colombo's narrative trumpeting a triumph over terrorism was all but swamped by a rival narrative of communal repression and barbarism. Warfare, as traditionally waged, found itself struggling to deal with *lawfare:* attempts to use new international norms and the law to force cessation of hostilities, intervention by external actors (state and nonstate), and prosecution of key government figures.[59] Matters were not eased by what can only be described as the shrill moralizing of both state (particularly the United States and the UK) and nonstate (particularly international human rights groups) critics.

And yet, given the astonishing level of brutality and suffering that Sri Lanka had endured for three decades, its wounded attitude was quite comprehensible, as were the realities that emerged from the major combat that ended only with LTTE's surrender.

A globalized world has so empowered *netwar* at the geostrategic-legal level of international relations that it all but compels the waging of conflict in the intangible rather than tangible dimension.[60] *Facts on the ground* count for far less than *facts in the mind,*

---

[56] See Mats Berdal, "The 'New Wars' Thesis Revisited," in *The Changing Character of War*, ed. Hew Strachan and Sibylle Scheipers (New York: Oxford Univ. Press, 2011), 109-33; Paul Richards, "New War: An Ethnographic Approach," in *No Peace No War: An Anthropology of Contemporary Armed Conflicts*, ed. Paul Richards (Athens, OH: Ohio Univ. Press, 2005), 1-21.

[57] See M. L. R. Smith and David Martin Jones, *The Political Impossibility of Modern Counterinsurgency* (New York: Columbia Univ. Press, 2015).

[58] See, for example, International Crisis Group (ICG), "War Crimes in Sri Lanka," ICG Asia Report no. 191, May 17, 2010; Frances Harrison, *Still Counting the Dead: Survivors of Sri Lanka's Hidden War* (London: Portobello Books, 2012).

[59] "Lawfare" is most often applied to the efforts of substate actors, both legal and illegal, to use the law as a weapon to impose their will on others; hence the play on "warfare." A growing body of discussion and scholarship is available on the topic, including a blog (jointly sponsored by the Lawfare Institute and Brookings) that adopts the more expansive definition: the use of law as a weapon of war (irrespective of user). See Lawfare Institute and Brookings, "Hard National Security Choices," 2015, www.lawfareblog.com/; Tom Marks, "Lawfare's Role in Irregular Conflict," *inFocus* 4, no. 2 (Summer 2010): 12-14.

[60] See John Arquilla and David Ronfeldt, The Advent of Netwar (Santa Monica, CA: RAND, 1996); David Ronfeldt, John Arquilla, Graham E. Fuller, and Melissa Fuller, *The Zapatista Social Netwar in Chiapas* (Santa Monica, CA: RAND, 1998).

never mind whether those "facts" are true or later prove false. Seeing is no longer believing. Indeed, believing has become seeing, with disabling pressure from a networked world directed against the party judged to be "in the wrong," that is, the party judged to have forfeited legitimacy.

If we imagine the Chiapas conflict, which inspired the emergence of the *netwar* concept, ending not in retreat by the Mexican state but in elimination of the Zapatista challenge, Mexico would be in a position not so different from that occupied now by Sri Lanka. It has secured its desired end state of an indivisible Lanka, the land of the Buddha, through achieving the objective of LTTE's destruction. But its ways (which included not only material but also psychological national mobilization) have been found wanting. Communal chauvinism, goes the critique, provided the fuel that allowed an overhauled war machine to "win," and democracy itself was collateral damage, along with justice.[61] In such an assessment, the reality of an illicit power structure that had done as much as any in the post-World War II era to earn the label "evil" becomes irrelevant.

This, too, may be seen as emblematic of the new age of war. Ultimately, the conflict morphed into one of dueling narratives on the fundamental merits or demerits of Sri Lanka's democratic, market polity. In such a battle, the increasingly problematic and despicable nature of LTTE's decision making and actions was irrelevant, as if the very intensity of Colombo's transgression in "winning ugly" revealed much about Colombo's structural and moral inadequacy, and rather less concerning LTTE's evil agency. It is in examining this process that we can draw lessons.

---

[61] See Gordon Weiss, *The Cage: The Fight for Sri Lanka and the Last Days of the Tamil Tigers* (London: Bodley Head, 2011).

10. It Takes a Thief to Catch a Thief: Illicit Power and the Intelligence Challenge

Michelle Hughes

Viewing the current chaos in the Middle East through a 2015 lens, it seems almost quaint that only a few short years ago, within certain defense circles, serious discussion and debate about the theories behind Islamic jihad was seen as a distraction. Not that people were not interested or that jihadists' motivations were deemed unimportant. Interest was high, and no responsible military commander would ever conclude that understanding the enemy was not a priority. But at the same time, there was an underlying sense of a limit to how much academic discussion was constructive while in the midst of prosecuting two wars. Was jihadist motivation really a military problem to solve? And how deep must one dive into philosophy or legal theory to address the immediate security imperatives in Afghanistan and Iraq?

So it should have come as no surprise that when a Department of Defense concept development team at U.S. Joint Forces Command (JFCOM) started writing its Unified Action Handbook Series in 2009, there was a bit of a struggle to include, in a subchapter about dealing with illicit and informal power structures, a brief summary titled "The Fiqh of Jihad."[1]

"The Fiqh of Jihad" was the product of research by Pakistani political scientist Nasra Hassan, into why jihadists believe as they do, and how those beliefs affect Western efforts to resolve conflict and improve stability in the Muslim world. Here are Hassan's findings:

1. For the jihadis, the Holy Quran is divine, and their practices are firmly supported in the Book.

2. There is a general feeling that the concept of "nation states" is against the Book because sovereignty belongs to Allah alone.

3. Under Sharia, the only state entity is the caliphate; thus, man-made borders are irrelevant, and national laws made within those borders do not count.

4. Allah is the sole source of law, and duties are divinely ordained. Rights flow only from following duties. Thus, man-made laws conferring rights do not count.

The author gratefully acknowledges Clifford Aims's contributions to this chapter. And special thanks to Lieutenant Colonel Jody Vittori, U.S. Air Force (ret.), formerly assistant professor / Air Force Chair, College of International Security Affairs, National Defense University, who contributed significantly to the section on intelligence collection.

[1] The term *"fiqh"* refers to the various processes of Islamic legal thought and reasoning, including the decisions arrived at through them. The discussion of jihadi *fiqh* in this chapter was derived from a presentation by the former director of the UN Information Service in Austria. See Nasra Hassan, "The Fiqh of Jihad," paper presented at the Salzburg Global Forum on International and Islamic Law, "Finding Common Ground," (Oct. 29, 2008). A slightly longer version of Hassan's findings and conclusions was previously published by Michelle Hughes, David S. Gordon, et al., in the *Handbook for Military Support to Rule of Law and Security Sector Reform* (Washington, DC: JFCOM, 2011), www.dtic.mil/doctrine/doctrine/jwfc/ruleoflaw_hbk.pdf.

5. International law is not justice; it is arbitrary and exclusionary.

6. Territory must be conquered, reconquered, or liberated through jihad. The means determine the nature of the entity that emerges. Thus, elections do not confer legitimacy, nor do treaties.

7. Jihad must be permanent.

8. In the absence of a caliph, local leaders can issue a call to jihad.

The *Handbook* authors argued that "The Fiqh of Jihad," while perhaps not authoritative, nonetheless represented the type of analysis that planners and policymakers needed to undertake if they were to fully comprehend how U.S. stabilization and reconstruction—particularly in the areas of law enforcement, governance, and the rule of law—would be perceived by power brokers (and their constituents) in the battlespace. This, in turn, would help both planners and operators more accurately mitigate risk and recognize opportunity. "Just because it [does not] make sense to us," the authors argued, "[does not] mean it does not make sense to *them*. And we have to understand and accept, not dismiss, alternative perspectives on power and authority."[2]

But senior military leaders wanted to see something less "ambiguous." This was all very interesting, they said, but all it did was confirm how irrational the jihadists were. What was really needed, they believed, was more of a "template" that, when filled in with collected intelligence, would enable its users to sort and classify any organization into one of a set number of discrete categories. Tools, tactics, and procedures could then be applied accordingly.

At its core, the argument reflected two different ideas about what readers of the *Handbook* needed. Did they need a specific concept that could be universally applied, or did they need something that would prompt them to think in a different way about what they might encounter in the battlespace? What was the use, leadership wanted to know, of presenting a set of interconnected ideas that raised numerous possibilities but did not lead to clear answers and did not represent a repeatable, generalizable process showing input and output, and cause and effect?

"The Fiqh of Jihad" eventually made it into the *Handbook*, but only as an illustration of what is meant by "motivation," which was a characteristic included in a template for understanding illicit power structures. The type of thinking it illustrated, and the way that such thought could be manifest throughout countries dealing with the problem of Islamist extremism and violence, was not further developed. It was not target oriented, and it simply did not fit well into intelligence or assessment models at the time.[3]

A recurring theme throughout the cases in this book is the challenge that (usually) Western interveners face when they try to understand the operating environment of a conflict, and the roles and risks that illicit power structures present within that environment. The problem is steeped in complexity. A single power structure can arise out of several quite different motivations. Illicit power structures are opportunistic. They re-

---

[2] Author's notes and records of command guidance given during the 2009-11 drafting process for the *Handbook for Military Support to Rule of Law and Security Sector Reform* (2011).
[3] Ibid.

flect political, cultural, religious, economic, criminal, and ideological grievances, alliances, and relationships. They thrive wherever they find gaps in governance, institutional capacity, and political will. They often operate in the gray areas between legitimate and illegitimate, moving back and forth across the spectrum of legitimacy in response to political and security imperatives, economic opportunities, and social need. They are a moving target. Thus, they present a cognitive nightmare for strategists, policymakers, military leaders, and law enforcement, all of whom have distinct information requirements that must be met if they are to mitigate illicit power's pernicious effects on peace and stability.

The intelligence community has struggled to respond to the challenge, primarily because it is looking for a different threat. Unlike an external threat that seeks to impose a new order (political, social, or religious) from without, many illicit power structures work from within the existing order and wish to maintain the status quo in order to further enrich a small, select portion of the population. And unlike conventional security forces with transparent structures and predictable modalities, illicit power structures do not fit neat templates used to generate information requirements, or set-piece frameworks for analysis.

To be accurate and predictive, analysis of illicit power must be multidimensional and—it is worth stressing—empathetic. If the intelligence focus is limited to a single aspect or dimension of a power structure, the resulting conclusions and, more importantly, the courses of action derived from them can lead to ineffective or even counterproductive responses and effects. When intelligence analysis is filtered through the lens of our organizations, functions, governance, and philosophies, it will reflect a distorted view of *us*, rather than an accurate view of *them*. In short, traditional intelligence gathering literally does not see the problem. Often, as a result, intelligence analysis can actually provide cover for illicit actors by focusing the consumer on the wrong problem.

Had the members of a NATO special-operations task force facilitating an Afghan-led rule of law shura in Northern Afghanistan in 2012 been asked whether they understood the problem of power in their operational environment, they likely would have concluded that however deep their own information gathering and analysis, it had not been enough to support this particular task.

The rule of law shura was the brainchild of Major General Sadat, the chief of antiterrorism prosecutions in the Office of the Afghan Attorney General. Frustrated by his inability to bring high-profile cases before provincial courts, Sadat was meeting throughout the country with provincial and district leaders to identify problems and negotiate solutions. The special-ops task forces in the provinces had been given the task of helping their Afghan counterparts make the transition from Coalition-led military operations designed to kill or capture Taliban operatives, to Afghan law enforcement-led evidence-based operations. Supporting Major General Sadat's initiative was a great idea that Afghans and Coalition alike agreed should have happened sooner.

The first part of the shura was relatively predictable and easy for the task force members to follow. Afghan prosecutors bemoaned the lack of solid investigative standards in the police; police commanders complained about corruption in the courts, and the lack of good intelligence; the governor and subgovernors demanded greater security and more say in setting the priority of efforts by all concerned—in other words, the usual litany of complaints. But the conversation took an unexpected turn when local elders spoke up to describe their difficulty in communicating government objectives and authority to their people. They said that their people did not understand how the government (and the internationals) could enter their villages and just announce a set of rights and obligations that had never existed before.

The elders explained that their people were uneducated. Many of the villagers had heard the terms "democracy" and "human rights" for the first time only recently, through Western communications. Based on their inexperience, the villagers had processed those terms to mean that the government had no authority to detain them unless they, the individuals, voted to allow it. Applied to arrests and detentions, democracy meant, therefore, a "right to release" because, after all, were not individual human rights more important than government rights? That was what international forces and development specialists had told them. Besides, the only true law came from Allah, anyway.

The elders further explained, the net effect was that whenever a member of the village was arrested on insurgency-related charges, the villagers demanded that the individual be released and his crimes be addressed through their own local process. This was why they were not cooperating with the formal prosecutorial investigations. No one disputed that the detainee had committed a crime; it was the governmental process and authority that they did not understand.

The Afghan leaders at the rule of law shura seemed to suffer no confusion over the logic and instead debated ways to work around these troublesome attitudes. But after the shura, the task force members struggled to process what they had heard. All of them had studied Afghan culture before their deployment. They had received countless briefings on the Taliban, warlordism, and the hierarchical lay-down of the numerous known power structures they would confront in the battlespace. Many had studied Islam so they would recognize difficult cultural issues they might encounter, and had already completed multiple deployments. Several were civilian police officers in their home countries, chosen for this deployment because of their experience with the rule of law. But none had looked at how *informal* power and authority was conferred, and no one could describe how an isolated, uneducated Afghan population perceived government authority and the law.

In fact, the Coalition operators felt as though they had fallen through the looking glass. To them, this whole crazy village idea about democracy and authority was completely illogical. It showed what happens when you are dealing with people who have no formal education, and it was just one more example of how nothing in Afghanistan made any sense. But it *did* make sense—to the Afghans. In many ways, their belief system on authority mirrored the *fiqh* of jihad, which the Taliban and the warlords—both disruptive illicit power structures in the region—knew and understood. The warlords and the Taliban had been able to communicate jihadist thinking effectively to exploit local perceptions and undermine the legitimate exercise of government authority by Afghan law enforcement. The Coalition task force, meanwhile, continued to be baffled. It was clear that even as they prepared for the end of their mission, they were still trying to understand the environment in which both warlords and Taliban fighters continued to operate with impunity.[4]

This chapter acknowledges how difficult it is to understand the problem of power,

---

4 Author's personal observations and conclusions from fieldwork and interviews with U.S. and Swedish special-operations task force members in Afghanistan, Oct.-Nov. 2012.

particularly in places that we do not control. As the cases and vignettes throughout this book illustrate, failure even to recognize the presence of illicit power structures, let alone grasp how they arise and thrive, is a central challenge to confronting impunity. This chapter looks at some of the issues and lays out a basic sequence and structure for identifying critical information requirements. It discusses common attributes of illicit power structures, provides a start point for understanding them, and discusses fundamental questions that should be asked and answered—in the initial planning phase of an operation and throughout execution. Finally, we will look at some examples of how to find relevant information and piece it together in a way that works. Ultimately, there are no silver bullets or neat templates. The challenges are real. But with flexible, adaptive, creative approaches, the challenge is surmountable.

Analyzing the Environment: Why this first?

It might seem counterintuitive to look at the environment that surrounds an illicit power structure before looking at the structure itself. But as this volume's case studies show time and again, power brokers and power structures arose out of conditions that created opportunities and grievances, concentrated authority or resources in the hands of bad actors who otherwise would not have gotten them, or exposed gaps in sovereign capacity to check power or control behavior. Into the vacuum, illicit power emerges both to exploit and to create violence and instability. For this reason, it is critical to conduct a *vulnerability assessment* of the operational environment, focused specifically on inhibiting the rise of illicit power.

Table 10.1 Vulnerabilities that Allowed Illicit Power Structures to Emerge

Case Study	Impact	Vulnerabilities that Enabled the IPS to Emerge
Afghanistan: Traffickers and Truckers	Illicit economy; obstruction of peace process	Corruption; lack of oversight and accountability (theirs and ours); nonfunctioning licit economy
Afghanistan: Criminal Patronage Networks	Capture of state institutions	Corruption; lack of oversight, accountability, and political will; inattention to justice capacity building
Liberia: Durable Illicit Power Structures	Capture of state institutions; illicit economy; state's inability to maintain legitimate monopoly of force	Entrenched system of political and personal patronage and uncontrolled illicit economic activity; perversion of justice and security systems
Odessa Network: Russia/Ukraine Weapons Trafficking	Capture of state institutions and critical transportation infrastructure, thus limiting options to cut off flow of weapons to conflict zones and bad actors	Absence of legitimate control over commercial and investment activity; inattention to economic development; corruption, patronage, and unchecked use of violence to gain economic control

Table 10.1 Vulnerabilities that Allowed Illicit Power Structures to Emerge (cont.)

Case Study	Impact	Vulnerabilities that Enabled the IPS to Emerge
Timor-Leste: Security Sector Reform	Failure of reconstructed security institutions; restoration of partisan power bases	Unresolved political power sharing exacerbated by external, regional interests and interference; identity-based patronage
Iraq: Jaish al-Mahdi	Perception of lawlessness and insecurity that permitted competing illicit groups to take matters into their own hands; rise of uncontrolled militias and economic competition for resources and infrastructure	Ambiguity between licit and illicit power; presence of strong, entrenched criminal organizations; unanticipated and misjudged political factionalism; presence of newly demobilized security forces
Colombia: Fuerzas Armadas Revolucionarias Colombianas (FARC)	Endemic violence; human rights violations; emergence of drug trafficking organizations, and consolidation of ungoverned spaces by illicit, antigovernment actors	Tradition of rebellion and guerrilla warfare, combined with absence of government control; proliferation of nonstate security actors working hand in hand with unaccountable government forces
Haiti: Gangs of Cité Soleil	Co-optation of security, justice, and law enforcement; rampant human rights violations; unchecked violence, predation, and criminality; unending cycles of poverty and violence; mass migration and humanitarian disaster	Governmental corruption and extreme incompetence; formal and informal security forces as tools of regime repression and control
Philippines: Moro Islamic Liberation Front	Sustained religious and ethnic strife and competition for resources and recognition; uncontrolled banditry and armed aggression	Disfunctional, nonrepresentational statebuilding; unequal economic rights and minority protection
Sri Lanka: Liberation Tigers of Tamil Eelam (LTTE)	Uncontrolled violence and insecurity; fraudulent elections; discredited government; withdrawal of international support; ungoverned territory and, ultimately, civil war	Government oppression; caste and gender inequity; rampant human rights violations; no accountability of security forces or state-sponsored nonstate actors
Sierra Leone: Revolutionary United Front (RUF)	Fraudulent elections; capture of state resources; crippled and discredited government and government security institutions; uncontrolled violence and insecurity	Governmental ineptitude; repression; corruption; unchecked illicit economic activity

The vulnerability assessment goes beyond mere situational awareness. Viewed one way, an illicit power structure begins in much the same way as a burglar casing a neighborhood for potential victims. The burglar notes who has a dog or an alarm system, who is consistently away from home during specific hours, and who has expensive cars and, by extrapolation, valuables that will yield the greatest return for the risk. Illicit actors do the same: continually looking for exploitable weaknesses in the aftermath of war and transition. Ineffective governance, poor accountability, high levels of corruption with weak enforcement mechanisms, incompetent or inexperienced security forces, unaccountable donor assistance, political infighting, poor communications between population and government, unresolved grievances, poverty—all these conditions create an environment ripe for consolidation of power by bad actors.

It is often said that "it takes a thief to catch a thief," but it is not as if we had not done this type of empathetic analysis before. In their 2012 book *The Endgame*, Michael Gordon and Bernard Trainor provide several examples in which smart, well-intentioned people tried to look at power from the perspective of those who presented the greatest risk of exercising it illicitly—only to be shut down or ignored by policymakers and senior leaders. In one account, a "red team" of military officers and civilian analysts conducted a classified review of America's military strategy in Iraq, a year and a half before the 2007 "surge."[5] Examining the extant counterinsurgency strategy, the red team's report concluded that the effects of a strategy, which was already perceived to be failing, on bad actors' ability to hinder regional peace and stability were enormous. "The insurgency is resilient and capable of regenerating itself," the report noted. "The Coalition and the Iraqi government have not been able to suppress the insurgency in an enduring way," and it might be emboldened if it thought that the main American goal was to disengage from Iraq. The insurgents would then be "less likely to cut the political deals that would be needed to shore up the new Iraq." The report went on to conclude that Iraqis who had stood by the Americans might also lose confidence in their ally: "Fears of abandonment might lead the Iraqis to hedge their bets by developing greater reliance on Iran." If the transition to self-reliance takes place before the defeat of the insurgency, the Iraqi government and the insurgents could seek external support from neighboring states (e.g., Syria and Iran) in order to fight on, potentially leading to "civil war along the lines of the one in Afghanistan in the 1990s."[6]

Red-teaming is a doctrinal element of operational design, and what this particular red team did in 2005 was exactly the sort of analysis that needs to be done when thinking about illicit power structures.[7] By looking at conditions, possibilities, and risks from the perspective of the potential power brokers and their allies, it may be possible to preclude or at least limit their emergence as enemies of peace. By not looking at these issues, we almost guarantee that opportunities to limit the rise of impunity will be lost.

Most current intelligence frameworks do a good job of laying out the essential issues within any operating environment, and these can also be used effectively to guide the analysis of vulnerability to illicit power, *provided that intelligence collectors and analysts*

---

[5] Michael Gordon and Bernard Trainor, *The Endgame: The Inside Story of the Struggle for Iraq, from George W. Bush to Barack Obama* (New York: Vintage Books, 2012), Kindle ed. location 3248-3269.
[6] Ibid., Kindle location 3263.
[7] JP 5-0 at III-6.

look at the problem specifically through the lens of illicit power. When illicit power is not viewed this way, critical warnings and indicators are consistently and predictably overlooked, underemphasized, or misunderstood. Certainly, analyst bias or inexperience plays a role, but so does the attempt to shoehorn information into prescribed categories that, although relevant, were not developed specifically to address what we know about the particular problem.

To illustrate, a common construct used by U.S. military analysts for both analyzing and visualizing the environment is PMESII—the acronym for "political, military, economic, social, infrastructure, and information" systems.[8] But as comprehensive as PMESII may seem, there are questions about vulnerability to illicit power that need to be asked but that do not fit neatly within the framework. And even where they appear to have a category, their inclusion is not part of conventional application. So, for example, a variation on PMESII that begins to uncover overlooked issues might include the following:

Governance (instead of "Political"). Does the host nation government have the political, technical, and institutional capacity to control its territory, protect its sovereignty, and uphold its laws? If not (which is likely), then who or what is most likely to fill the gap? Political will aside, does the sovereign government even have the specialized oversight and audit, investigative, prosecutorial, and judicial competence to prevent, detect, and punish complex crimes such as corruption and graft?

Military. To what extent are security institutions influenced by nonstate power brokers and informal power structures? Do the security institutions have bureaucratic and procedural checks and balances that can inoculate them against corruption, graft, undue influence, and abuse of power? What is the history of accountability within the security apparatus? What discipline and accountability mechanisms have traditionally worked?

Economic (with the added dimension of "Economic Infrastructure"). What is the extent and nature of the relationship between the licit and illicit economies before, during, and after the conflict? How might it evolve, and why? If there was illegal trade in licit goods before the conflict, in what ways will trade networks merge or expand to include illicit goods such as weapons and drugs? In what ways can the existing licit trade networks be co-opted for illicit purposes? What law enforcement vacuums will allow illicit networks to expand?

Commercial (as a separate dimension of "Economic"). What are the normal business relationships between people, organizations, governments, and political factions? When a reconstruction contract is let, for example, who benefits, and why? What are the business interests of key political figures, and how are they likely to influence strategic decisions and the consolidation of governmental power?

---

[8] According to U.S. military doctrine, analyzing the adversary's PMESII systems can lead to identifying key nodes, links, and vulnerabilities, which can then be targeted kinetically or nonkinetically to achieve desired effects. JP 5-0 at III-8.

Social. Do illicit power structures and their networks and activities generate positive benefits for the population? For stabilization? For society? How will those benefits be replicated by licit authorities if they are taken away? On the other hand, what might be the benefit to the illicit power structure if legitimate authority does not step in to control them?

Cultural (as a separate dimension of "Social"). What are the cultural attitudes toward power, patronage, and authority? How are they conferred? What keeps them going? How is legitimacy or illegitimacy determined? Is there any expectation of reciprocity between the people and those in power over them? If so, what forms does it take?

Information. Is there a culture of transparency? How are public records assembled, maintained, and accessed? What are their vulnerabilities to manipulation? How competent are the media to report on politics, power, and crime? Does the legal system support free speech, civic participation in governmental oversight, and a free press?[9]

Hundreds of variations on these themes can be used to guide the acquisition of relevant information and to assess both the likelihood and the potential effects of an illicit power structure emerging as a threat to stabilization. What matters most is not which framework is used, but that the analysis is done specifically to address the potential risks from illicit power. This is a critical first step.

Analyzing Emerging Illicit Power Structures: Who are we dealing with, and what do they want?

As illicit power begins to emerge, the analysis acquires another layer: the taxonomy of the power structure itself. Because power is multidimensional, the analysis must be multidimensional as well. If the focus is limited to a single attribute or dimension of a power structure, the resulting conclusions—and, more importantly, courses of action—can lead to ineffective or even counterproductive responses and effects.

In almost all cases, illicit power has four defining attributes: (1) objectives, (2) motivations, (3) organizational structure, and (4) behavior. Each attribute should be viewed as a critical information requirement, and each must be addressed as a vital aspect of any effort to neutralize the negative impact of an illicit organization.

---

[9] This has been a particularly difficult area for Westerners to evaluate accurately and illustrates the problem of "mirroring" when doing assessments. Our tendency is to evaluate media capacity by measuring technical skills and intellectual capital within the host nation's media, the extent to which civil society is present, and whether freedom of the press is enshrined in national law. Often overlooked are the ways that state and media collaborate to create the *perception* of what is real and what is not. Russia's manipulation of the entire field of journalism, from print to television, to social media, is a prime example. Everything looks good on the surface, but the underlying reportage is carefully controlled and subtly, consistently skewed so that neither the population nor the international community ever gets the real story behind the version presented.

Objectives: What do they hope to achieve?

Illicit power structures are generally distinguished by the type of objective they pursue. They seek either (a) destruction of the international order or its component parts—including individual states—and their replacement with an alternative order; or (b) redistribution of power, position, and resources within the existing state, in the interest of the members of the illicit power structure.

Objectives can evolve and change, and *both may coexist within the same organization* at different levels or even within the same level, even though this may appear illogical at the time. The case study in Chapter 2, on the Jaish al-Mahdi in Iraq, illustrates this phenomenon. Its authors, Phil Williams and Dan Bisbee, describe the changes in the organization's motivations, principal players, and constituent parts over the course of the Iraq War, in response to the situation on the ground, political choices made by both the Iraqi government and the international community, and challenges and opportunities presented by each. In this case, both objectives did indeed coexist within the JAM leadership at various times, but U.S. decision makers either could not, or chose not to, recognize them. That was unfortunate because where both objectives are present, they create opportunities to exploit the internal inconsistency and weaken or even neutralize the organization.

When the ideology, intentions, and objectives of a power structure are fundamentally at odds with the underlying principles of a rule-based system of sovereign states, then any accommodation satisfactory to both state power and illicit power is unlikely. For example, it is hard to imagine how groups seeking to do away with the state system altogether—such as those promoting a global proletarian revolution or a global Islamic caliphate—will accept any *long-term* outcome based on the modern system of individual sovereign states. It is also difficult to imagine that an organization with an unwavering commitment to destroy the state in its current form can be accommodated within a process of political compromise with the ruling state authority.

Those opposed to the existing system who will not compromise within the context of that system can be described as "intransigent," "revolutionary," or "absolutist." But this does not mean that no accommodation is possible with the power structure's more moderate factions—which could neutralize its more extremist members' ability to carry out their intransigent agenda. And meanwhile, of course, the absolutists may be targeted individually.

The peace settlement with the RUF in Sierra Leone is an example of how this two-track approach can inform not only the peace agreement itself but also subsequent institution building and community reconciliation.[10] In the case of the RUF, intransigents and absolutists were identified and isolated from their (reconcilable) former colleagues. Some were filtered into a judicial process. The rest faded into obscurity. They ceased to matter once their power base ceased to exist.

Illicit power structures that do not fundamentally oppose the principle of the state but, rather, wish to achieve a better position within the state can be classified as having *limited* objectives. They may wish to reform the state in the larger public interest (real or

---

[10] See chapter 8 of this volume: "Sierra Leone: The Revolutionary United Front," by Ismail Rashid.

perceived) or exploit the state to pursue narrow parochial interests, as is often the case with those who seek to capture state institutions for personal profit, power, or ideology. For obvious reasons, the tactics and strategies for countering them will vary. But the important thing is to recognize that no successful or enduring power structure will remain static. Thus, it is necessary to constantly reevaluate illicit power structures in order to ensure that we are not misreading their fundamental goals and objectives as we try to counter or neutralize the threats they present, and as they react to our attempts.

Motivation: What makes them behave as they do?

In war and transition, a common mistake is to assume that belligerents want conflict resolution, peace, or cessation of hostilities or have broader social aspirations such as reconstruction, stabilization, democratization, or development.[11] In fact, some belligerents—often in the form of illicit power structures—benefit substantially from conflict and therefore protect their material interests by fueling and sustaining the conflict that keeps them in business. These are often referred to as "conflict entrepreneurs." Their motivation is the same as any other entrepreneur's: profit. It is well documented that during the Balkan Wars of the 1990s, the criminal elements of all three major ethnic groups involved made extensive fortunes during the fighting and that the peace established in late 1996 cut into their profits dramatically. Similarly, Gretchen Peters, in her study of Afghanistan's Pashtun trucking networks, points out that as organized crime and corruption became the norm, and profits from drugs and weapons increased because of continued insecurity, the prospects for peace and reconciliation decreased accordingly. This phenomenon is not restricted to nonstate actors. At the same time that the Taliban was profiting from the ongoing conflict in Afghanistan, Afghan political leaders were, too.[12] As conflict entrepreneurs become more entrenched and diversified, the overlap between illicit and licit activities, whether economic, social, or political, increases. And so do the complexity and the disincentives for peace.

Theories hold that greed is the fundamental motivation behind all illicit power structures and that without a nexus between illicit power and illicit wealth, there is no illicit power structure, but something else instead—some other kind of spoiler or belligerent that, while certainly not good, is otherwise undefined. This oversimplifies the analysis and is not supported by the cases. Identity politics remain a critical motivator, as do injustice and economic or political deprivation. The latter motivators are referred to as *grievance* or *need* based. The existence of organizations such as al-Qaeda and Hezbollah in the Middle East, the FARC in Colombia, the Interahamwe in Rwanda, and, looking back, the Bolsheviks in Russia confirms the continued emergence of organizations motivated to conflict by belief, creed, or need, as opposed to mere greed.

In fact, it is most likely that illicit power structures are motivated by a combination of factors. These factors may be inconsistent within the group and can rapidly change because of internal evolution or external pressures. Generally, however, one factor will be the key motivator that drives those in leadership positions to direct the behavior of the overall organization. It is important to be realistic in determining what that pre-

---

[11] David Keen, "The Economic Functions of Violence in Civil Wars," *Adelphi Papers 38*, no. 320 (1998).

[12] See chapter 5 of this volume: "Traffickers and Truckers," by Gretchen Peters.

dominant factor is, yet be vigilant in continually reassessing all assumptions about what motivates illicit power.

Organization: How are they structured?

Organizational structure represents one of the most basic intelligence requirements. Its primacy is evident in the emphasis on command and control, leadership, equipment, and supporting infrastructure that is so often heard when planning with military and law enforcement. A fuller assessment also requires that key leaders be identified, along with their roles, responsibilities, and personalities. This last factor is particularly important because many illicit networks function under charismatic leaders with authoritarian styles, and organizational behavior becomes an extension of the leadership style.

Organizational analysis is critically important; this much is self-evident. But the intelligence and information challenge is our tendency to look for structures that mirror our own concepts of efficiency, control, authority, and autonomy. Because of this bias in both collection and analysis, we have difficulty recognizing and understanding structures that may be less hierarchical and more relationship based and organic than we are used to.

In an effort to create an organizing principle to collate information around, researchers have often classified the basic distinction between illicit power organizations as hierarchical versus networked.[13] While there are numerous varieties of each, this core distinction goes to the very essence of the organization, determining its patterns of communication, command, and control.

Table 10.2 Basic Organizational Structures

Hierarchical Organizations	Networked Organizations
• Well-defined chain of command • Clear lines of authority • Enduring vertical subordinate-superior relationships, decision making, and communication patterns	• Less defined chain of command • Decentralized decision making • Decentralized channels of communication • More adaptive, rapid, and flexible • Less predictable

The way an illicit power structure is organized reveals critical information about its propensities, strengths, and vulnerabilities. Rigidly hierarchical organizations can be hobbled by disruptions in the chain of command, but negotiations with hierarchical organizations are likely to follow a more predictable course than negotiations with decentralized networks. Hierarchical organizations are also generally thought to be vulnerable to "decapitation"—the removal or isolation of the highest level of leadership. An example of this is the weakening of the Peruvian Maoist insurgent group Sendero Luminoso after the capture and incarceration of its leader, Abimael Guzmán, in 1992.

---

[13] For a discussion of hierarchical and networked organizations, see John Arquilla and David Ronfeldt, *The Advent of Netwar* (Santa Monica, CA: RAND, 1996).

For years following Guzmán's capture, it was generally believed that Sendero Luminoso had been completely neutralized. But surviving factions later turned to narcotics trafficking and, during the first decade of the twenty-first century, engaged in sporadic violence, primarily against government security forces, in an effort to force a reconciliation process. Because the Peruvian government considered the resurgent violence manageable, however, no negotiations occurred.

Co-opting or otherwise alienating mid-level commanders from the high-level commanders can be effective in diluting the impact of hierarchical organizations. So can alienating the organization from the general population. This is a modification of modern counterinsurgency strategy, which states that in a COIN environment, the emphasis must be on "the populace, host nation, and insurgents." External support is considered a subset of popular support for insurgents.[14] Our analysis elevates external support, making it a key consideration in its own right. Networked organizations may have more difficulty reaching categorical and final decisions, and they can be compromised if their communications network is disrupted, but they are also harder to map with any certainty and may require equally networked responses.

In 1996, researchers from the Rand Corporation developed a theory known as Netwar.[15] Netwar emphasized the use of advanced communications and information technology as a basis for neutralizing the operations of highly networked organizations. The idea included the concept of *swarming*, which features striking "from all directions at a particular point or points, by means of a sustainable 'pulsing' of force and/or fire, close-in as well as from stand-off positions." This represented a technical solution targeting the specific organization of networks.

Networked organizations may also be weakened by disrupting network communications, picking them off one node at a time until the network no longer possesses critical mass to exert significant effects, or by eliminating critical, less redundant nodes, such as bomb makers, ideologues, technicians, and financiers. Recent trends in U.S. and allied approaches to countering extremism in the Middle East reflect this thinking. Of the nine core elements of the 2014 U.S. strategy to counter ISIS, three were focused on attacking critical enablers.[16]

Behavior: How do they operate?

Illicit power structures have idiosyncratic organizational cultures and dominant behaviors. Because objectives, motivations, and organization are frequently invisible to outsiders, the behaviors—a power structure's "modality"—are what we see. As a result, modality, or behavior, is usually the primary lens through which outsiders assess an illicit or informal power structure. An organization's capacity or propensity for engag-

---

[14] Dept. of the Army, *FM 3-24 Counterinsurgency Field Manual*, Dec. 2006, 79, 104, http://usacac.army.mil/cac2/Repository/Materials/COIN-FM3-24.pdf.

[15] See generally, John Arquilla and David Ronfeldt, *Networks and Netwars: The Future of Terror, Crime, and Militancy* (Santa Monica, CA: RAND, 2001).

[16] White House, "FACT SHEET: Strategy to Counter the Islamic State of Iraq and the Levant (ISIL)," Sept. 10, 2014, www.whitehouse.gov/the-press-office/2014/09/10/fact-sheet-strategy-counter-islamic-state-iraq-and-levant-isil.

ing in violent behavior influences its modality. Modality, in the sense of an illicit power structure's efforts to exercise influence and power, can usually be classified into three types: (a) persuasion, (b) coercion, and (c) inducement.

Although the three types of behavior will most frequently be used in combination, one is usually prevalent. Thus, if an organization's primary modality is persuasion, we may not be terribly concerned, since some degree of nonviolent persuasion is acceptable political behavior. Unfortunately, coercion and inducement are the means more typically used by illicit power structures to exert influence on their environment. These behaviors are inherently corrupt and often violent. They perpetuate insecurity, subvert the legitimate exercise of governmental power and authority, and undermine stability, development, and sustainable peace. The willingness of an organization to move from nonviolent persuasion to more venal behaviors, or to tolerate coercion and violent inducement from its members or associates, is often the signal that the organization has turned away from the government or that its objectives have changed.

Any response to an illicit power structure should be conditioned by the objectives and motivations of that structure. But in the near term, understanding the dominant *behavior* will likely be the key to finding the most effective immediate counteraction.

Analyzing the Enabling Environment

Understanding the operational environment is considered essential if the commander is to be able to visualize the problem and the desired end state and devise an operational approach that turns current conditions into a desired end.[17] U.S. military doctrine describes "operational environment" thus:

> The composite of the conditions, circumstances, and influences that affect the employment of capabilities and bear on the decisions of the commander. It encompasses physical areas and factors of the air, land, maritime, and space domains and the information environment (which includes cyberspace). Included within these areas are the adversary, friendly, and neutral actors that are relevant to a specific joint operation. Understanding the operational environment helps the JFC to better identify the problem; anticipate potential outcomes; and understand the results of various friendly, adversary, and neutral actions and how these actions affect achieving the military end state.[18]

Illicit power is part of that operational environment. But in considering how to address the problem of illicit power structures from their own perspective, it can be more useful to think of their operational environment as an "enabling environment." This more accurately captures the opportunistic nature of most illicit power. As the cases demonstrate, illicit power structures fill gaps in licit frameworks and activities and take advantage of failures in leadership, accountability, and institutional capacity. But they do not usually create anything new. Instead, they tend to feed off existing systems, networks, relationships, and infrastructure. Even hardcore intransigents such as the RUF in Sierra Leone relied on existing commercial networks for their ability to traffic

---

[17] Ibid.

[18] Joint Chiefs of Staff, *Joint Operation Planning (JP 5-0)*, Aug. 11, 2011, III-8, www.dtic.mil/doctrine/new_pubs/jp5_0.pdf.

in diamonds. And the weapons dealers in the Odessa Network built their entire business model on the back of the licit shipping and transportation industries and the legal frameworks that govern them. As a third layer of analysis, merging an understanding of the enabling environment with the vulnerability assessment, and with the taxonomy of the power structure itself, creates a more complete picture. It is this layer that facilitates development of realistic, actionable recommendations.

What does this enabling environment look like? Doctrinal publications contain exhaustive (and exhausting) lists of characteristics, factors, and acronym-inspiring elements. But looking across multiple examples of illicit power types, we can come up with a more simplified model that provides a start point for asking the right questions and searching for the right data without being unnecessarily formulaic:

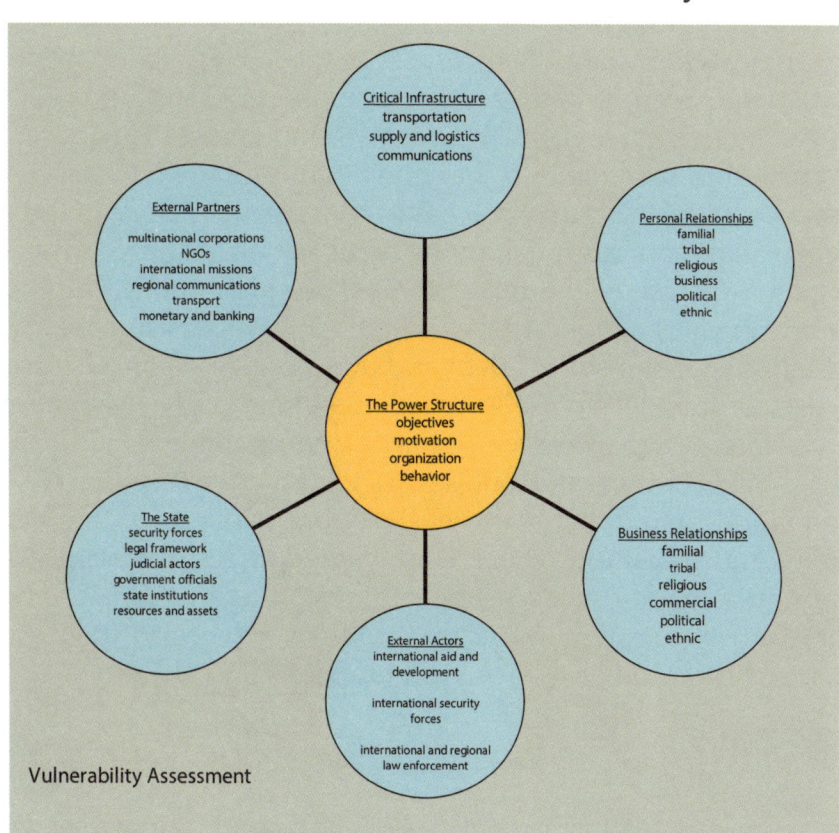

Figure 10.1 Illicit Power Structures: A Framework for Analysis and Understanding

This model shows how the world looks (generally) from the perspective of a would-be illicit power broker or structure. It merges the three different types of analysis that are necessary for us to develop an accurate picture of the operational environment. The vulnerability assessment that was done during planning (and updated as conditions change) represents the conditions that allow the power structure to emerge. The emergent illicit power structure is shown in the middle, described by the four main characteristics as noted above. Finally, the illicit power structure's enabling environment is represented by the spheres that surround it. Each sphere represents a general category of influencers and enablers that can be properly understood only if viewed in the context of the vulnerabilities in the operational environment. This model is not prescriptive.

Rather, it is a guide that helps us ask the right questions, focus our data collection, and more accurately develop strategies to counter the threat.

Economic Enablers

As experience shows us, there are priorities. Business relationships are vitally important to illicit power analysis. The nexus between illicit power and economic interests is undeniable. Even ideologically motivated organizations such as the FARC eventually networked with drug trafficking organizations. Less creed-based organizations rely on a web of contacts, contracts (written and unwritten, legal and illegal), and profit-motivated organizations and individuals to support their enterprise. They leverage licit economic structures and transactions, creating financial, political, and personal risk for those who never had any intention of getting involved in illicit activity, and profit motive for those who do not mind getting involved.

During the war in Afghanistan, a task force was specifically set up to look at contracting by the International Security Assistance Force (ISAF) Coalition in an effort to root out corruption. As the task force members began to look across functional areas to find the threads of corruption, they realized that understanding the business relationships—in this case, within the trucking industry—gave them a truer picture of the political environment in Afghanistan than did much of the other network analysis they were seeing. They began to gain a picture of the relationships, not just among contractors but also between power brokers at the highest levels of the Afghan government. By further tracing the contractual relationships of Afghan and other regional contractors, they quickly realized that every contract let by the international community was underpinned by a web of commercial interests that crossed political, ethnic, and even national boundaries.

At every level, businessmen and politicians were perverting the ISAF contracting system to siphon money and material out of the support structure that was essential to security operations and transition. Those who wanted to resist were nearly powerless. They were caught in a web of enforceable commercial agreements. Not only was the personal cost of integrity too high, but if they could not perform their obligations, the legal and financial costs would also have been too high.[19]

Governmental Enablers

The role of state actors and the institutions, assets, and legal framework within the state is another priority area that is underestimated and not well understood. We discussed the importance of the state in the vulnerability assessment, but the state as an enabler has many dimensions that must be reexamined against the specific threat posed by emergent power. Governance matters, and as with the illicit power structure itself, we must understand key governmental leaders, their direct and indirect involvement, and their incentives. In her book *Thieves of State,* journalist and Carnegie Foundation fellow Sarah Chayes describes how, at the extreme end of the spectrum, the state itself becomes a

---

[19] Conclusions presented by Gert Berthold, a former team member on Project 2010, at the NATO Building Integrity Expert Consultation (unclassified), National Defense Univ., Dec. 3-4, 2014.

criminal organization and a full participant in illicit activities.[20] Our case studies support her research. As we have seen, illicit power structures almost never function without some level of complicity by various state individuals and institutions.

In other instances, only certain security forces, state institutions, or rule of law actors will have been co-opted. Or they may simply be incapable of doing their job. Once identified, they can often be replaced or retrained, or an acceptable substitute capability can found. In Kenya, for example, a failing train-and-equip mission to help the Kenyan Navy interdict extremists coming out of Somalia was turned around simply by changing out the law enforcement partner that the Navy relied on for the law enforcement component of its concept of operations (CONOPS). The partner in the original CONOPS was the Kenyan National Police. When they proved unequal to the task (due to corruption and ineptitude), they were replaced by members of the Kenyan Wildlife Service, who both had the jurisdictional authority to make arrests in Kenyan territorial waters and were competent to perform the necessary maritime investigations.[21]

State assets and resources, when poorly or corruptly managed, also become tools in the hands of an illicit organization. And where the legal framework itself contains loopholes and exceptions, illicit actors will fully exploit the gaps.

External and Internal Enablers

External actors are almost always enablers. Regional and multinational corporations and infrastructure provide a backbone for illicit activities, especially trafficking, transportation, communications, and support. Multinational businesses can provide a cloak of legitimacy, and the legal structures that protect their licit operations can be manipulated to also protect their illicit partners. Other regional state actors provide direct and indirect support and sanctuary and impede law enforcement and cross-border operations. International missions inadvertently enable illicit power by limiting their own mandates in ways that constrain their security forces and development partners, not allowing them to address root causes of conflict or preventing them from requiring stricter accountability or law enforcement as a condition of aid. The influx of foreign aid that internationals bring to the environment provides incentives for graft and corruption and distorts normal patterns of commerce and trade, creating disincentives for long-term accountability and reform.[22]

Although not listed specifically in the model above, the population itself is also an enabler. The state and the illicit power structure may interact, but they do so within the context of the people. Dissatisfied or intimidated populations provide support for illicit power structures and the patronage networks that sustain them. While in many

---

[20] Sarah Chayes, *Thieves of State: Why Corruption Threatens Global Security* (New York: Norton, 2015).

[21] Author's observations from a joint security-sector reform assessment she conducted with Coalition Joint Task Force—Horn of Africa, Oct. 2006, Manda Bay, Kenya (unclassified).

[22] On external states using warlords as proxies to advance their foreign-policy interests, see Kimberly Marten, *Warlords: Strong-Arm Brokers in Weak States* (Ithaca, NY: Cornell Univ. Press, 2012). On external support to insurgencies and terrorist organizations, see Daniel Byman, *Deadly Connections: States that Sponsor Terrorism* (Cambridge, UK: Cambridge Univ. Press, 2005); Daniel Byman et al., *Trends in Outside Support for Insurgent Movements* (Santa Monica, CA: RAND, 2001), www.rand.org/content/dam/rand/pubs/monograph_reports/2007/MR1405.pdf.

ways the populace may be seen as a passive victim, it must also be seen as comprising a dynamic set of actors in their own right.[23] For example, typically, when conditions become intolerable for the people, they revolt—not against the illicit power structure, but against the government. There are some exceptions, such as in Mexico, where the citizenry in some areas took up arms against drug traffickers, or Afghanistan, where villagers have occasionally revolted against the Taliban. Depending on where one sits, this can be argued as reassertion of the rule of law in the absence of state action. But it can also be vigilantism, and eventually, the government will have to step in and retake control. Otherwise, it becomes just another opportunity for illicit power to emerge.

As a final note on the analysis, it is important to remember that the framework described here, and the three levels within it, are neither static nor linear. There will be significant overlap between the areas of analysis. Some state actors, for example, may be members of the illicit power structure that the state should theoretically be fighting, but because of political expediency or government reforms, they may find themselves on the other side, challenging the same illicit power structures that they themselves are a part of. The populace may selectively provide support or withdraw it from either the illicit power structures or the government. Likewise, external actors may provide significant on-site support, financing, or resourcing for illicit power brokers one day, and withdraw that support the next, causing an external support structure to collapse. All are dynamic entities, whose interactions must be evaluated, then reevaluated continually as conditions change.

Overcoming the Collection Challenge

Intelligence agencies—particularly military intelligence agencies (who are likely to be prevalent in conflict zones)—generally do not collect the types of business-, population-, and governance-centered intelligence required to understand how an illicit power structure arises and thrives. To win the support of the population, the legitimacy of the government is key to strategic success. In such situations, however, intelligence collection rarely focuses on the types of information required to understand the roots of the problem.[24]

Barriers and Impediments to More Comprehensive Collection

Western intelligence has become very good at tracking command and control, particularly in hierarchical organizations, and understanding tools, tactics, and procedures.

---

[23] Few academic studies rigorously assess the role of the population with insurgent or other illicit power structures. Two excellent ones are Stathis Kalyvas, *The Logic of Violence in Civil War* (Cambridge, UK: Cambridge Univ. Press, 2006); and Jeremy Weinstein, *Inside Rebellion: The Politics of Insurgent Violence* (Cambridge, UK: Cambridge Univ. Press, 2007). But many other authors have written on the role of the population vis-à-vis insurgent forces. See, for instance, David Kilcullen, *The Accidental Guerrilla: Fighting Small Wars in the Midst of a Big One* (Oxford, UK: Oxford Univ. Press, 2011).

[24] For an in-depth examination of this issue as it relates to the insurgency in Afghanistan, see Michael Flynn, Matt Pottinger, and Paul Batchelor, "Fixing Intel: A Blueprint for Making Intelligence Relevant in Afghanistan," CNAS, Jan. 2010, www.cnas.org/files/documents/publications/AfghanIntel_Flynn_Jan2010_code507_voices.pdf.

And we are getting better at following the money, although, as the Joumaa Web case (Chapter 12) points out, we are less good at achieving enduring impact through financial intelligence-based operations. Once we leave our own borders, we understand even less about local or national personalities, patronage networks, ties to criminality, the roles of external actors and external security forces, and the unwritten "rules of the game" for exercising power and influence within a particular society or culture. The information exists but is largely invisible to more traditional, threat-oriented collectors during wartime.

Part of the problem is that this information is accessed through very different channels. For example, governance specialists may generally understand the workings of both formal and informal government. But those specialists often reside on an embassy political staff or in a nongovernmental (NGO) or intergovernmental organization (IGO) rather than inside an intelligence agency or law enforcement organization. Information on criminal networks may be held by various law enforcement, financial, or regulatory bodies. But it may be subject to limited distribution because of legal constraints such as privacy and espionage laws, rules of evidence, criminal procedure codes, and intelligence and law enforcement oversight mechanisms. The news media may have extensive information networks and nuanced contextual understanding but, for obvious reasons, will be reluctant to share sources and methods with intelligence organizations and personnel.

The intelligence community also may not be familiar with alternative, nontraditional information sources in a given area. It may be unwilling or unable to tap into open source networks. There may be barriers between actual languages or between the lexicons used in different fields. Senior leaders often hesitate to use information based on sources of uncertain reliability or that fall outside standard information security channels. Indeed, intelligence professionals and military leaders often view unclassified information as "second class." Distribution of information is also difficult. For example, because of legal and (more often) "cultural" differences between military forces and law enforcement, their intelligence collection efforts will remain stovepiped to a degree, even where interagency coordination efforts are working well. Other information sources, such as IGOs and NGO networks, may not be accessible at all.

Also, intelligence collection rarely focuses on the types of information that define business relationships. Contractual relationships, domestic and international regulatory compliance (real or feigned), and custom and practice in how business is conducted within the operational environment require a different perspective from the normal security-oriented collection effort. To be truly effective, both collectors and analysts must not only understand how illicit activity works; they also have to understand how *licit* activities are conducted. Because illicit networks often live off licit systems, only by understanding how legitimate activity is conducted can we detect the anomalies. But collection against licit actors is often constrained by legal, policy, and privacy concerns. This means that much of the collection depends on the use of open sources, public records, and public communications.

Is it intelligence, or is it evidence?

Another factor complicating the challenge of intelligence collection against illicit power is that the problem is fundamentally one of law enforcement. Thus, traditional collection activities may be effective in identifying security threats and facilitating military targeting. But they are not as helpful in the transition from military-led security operations, under the law of armed conflict during wartime, to evidence-based law enforcement operations under the host nation's domestic law in the transition to peace. Information and intelligence has to be usable to effectively prosecute members of an illicit power structure in both foreign and domestic courts. The process by which information is collected and handled becomes as relevant to the effort's success as the information itself.

The Public Domain

What makes illicit power structures so interesting from a detection standpoint is how frequently they hide in plain sight. Business entities are registered, regulated, and taxed. Their assets (real and personal property and profits) are often publicly declared. Their websites describe what they purport to do, who their leaders and staff are, who they work with, and what services they claim to provide. Both the Odessa Network and Joumaa Web case studies illustrate the tremendous amount of information about an illicit power structure's enabling environment that can be found in the public domain. In the case of the Odessa Network, the entire network analysis was completed using only publicly available information. But to access it, we have to know what we are looking for. A key factor in C4ADS's success in collating that publicly available information was that the C4ADS analysts knew their topic—they knew what questions to ask and where to go for answers. They also spoke the language and knew generally how things worked on the ground. Their techniques were closer to those used by good investigative reporters than those used by intelligence services.

For more traditional intelligence organizations, this can be a real challenge because this type of problem is not their core skill. To close the information gap, new sources of information collection, processing, and distribution may be required.[25] Collection plans may need to include a process for regularly interviewing specialists: those who work in various government ministries; advisory or liaison teams who work with local elites, aid agencies, NGOs, and multinational corporations; and so on. Local and external news media reporting should also be reviewed, both as a source of information and for possible disinformation. And, of course, social media have become another critical source.

Institutional capacity building within the host nation also comes into play. The host nation government is, after all, our primary transition partner in most instances. But this means that our capacity building activities have to include robust technical development, within the host nation's law rather than our own. Effective intelligence-led policing, for example, requires specialized training for police *and* prosecutors so they know how such operations can be technically managed and properly applied within the existing legal system, and so they know what is required in order to bring evidence collected through such investigations to trial. Their training must be not only complementary but

---

[25] Flynn, Pottinger, and Batchelor, "Fixing Intel."

also collaborative. To be effective, they have to be able to work a case as a team. Judges may require specialized training and continuing professional education so they can try complex cases against illicit organizations.

Anyone operating in a domestic legal environment will need to have a solid understanding of existing laws and codes of criminal procedure. Where reform is required, legislative engagement with the host nation needs to be carefully thought through, and conducted with continuity of effort. It must also be coordinated with related institutional reform. Intelligence requirements also have to be considered when designing institutional reform strategies and programs. Each institution has its own unique set of information requirements and intelligence needs. But too often, international missions look only at the obvious ones: first, the need for a national intelligence service to protect the country from external threats; and second, a law enforcement intelligence capacity to address threats to national security. The information and intelligence requirements for other institutions to conduct their business securely, competently, and accountably falls to a distant third.

Resources for Intelligence Professionals

This section focuses on some newer resources that may not be well known to intelligence analysts. While military doctrine and civilian agencies' handbooks are important, resources outside standard intelligence channels can also be useful in confronting the collection challenge. In some instances, outside organizations have created specific guides or regularly produce information that is helpful in mapping illicit power structures and their enablers. These resources are unclassified and easily accessible to security professionals.

Appropriate Engagement with NGOs

Nongovernmental organizations also routinely publish reports and analysis that provide useful insights and alternative perspectives. NGO workers in the field often have an excellent understanding of the societies where they work; some NGOs even provide references and other sources specifically geared toward security professionals. The most notable is Transparency International's Defence and Security Programme (TI-DSP).[26] At one time, TI trained NATO troops for operations in Afghanistan. Its handbooks not only outline best practices for undertaking stabilization and peacekeeping in highly corrupt, criminalized societies but also explain how to conduct accountability- and anticorruption-focused training for security forces. They outline methods for doing country and regional assessments of the levels of corruption within security forces and defense ministries. Their work can be used to guide collection and analysis on one of the most critical vulnerabilities in the operational environment, and one of the key enablers of illicit power.[27] TI specifically addresses intelligence considerations and provides guid-

---

[26] For a library of its publications, see TI UK, "TI-DSP Reports," 2015, www.ti-defence.org/publications/dsp-pubs.html.

[27] See TI, "Corruption as a Threat to Stability and Peace," 2014, http://ti-defence.org/publications/dsp-pubs/257-corruption-threat-stability-peace.html. This compilation of case studies from Afghanistan, Kosovo, and West Africa demonstrates how corruption hinders stability in postconflict states and under-

ance on criminal patronage networks, narcotics trafficking networks, and land mafias, as well as such groups' use and abuse of foreign aid and military contracts. For military practitioners, Transparency International summarizes the existing military doctrine and guidance related to corruption and criminal patronage networks.[28]

Another excellent source is the Asia Foundation's handbook on how political settlements between elites work in conflict-prone societies. [29] Addressing the problem of conflict entrepreneurs, the Asia Foundation illustrates how, in many cases, the states that provide the most conducive environment for illicit power structures function under what the Asia Foundation terms a "political settlement." Elites, both legal and illicit, band together to settle large-scale violence by essentially dividing the "spoils" the country has to offer. Actors in conflict situations create political settlements to limit the violence and disburse the rents to various power brokers. The Asia Foundation's handbook is especially noteworthy because it gives practitioners practical advice on how to map the various power-broker networks that keep these actors in power, as well as strategies for marginalizing them and bringing reformist actors more to the fore.

The United Nations is also a great resource for intelligence and rule of law practitioners. The UN Office of Drugs and Crime website provides a wide variety of assessments, assessment tools, and handbooks that are useful for understanding various criminal networks, markets, and banking and logistics systems that illicit power structures use to finance and resource themselves.[30] The UN Department of Peacekeeping Operations contains national and regional assessments of current and past peace operations and is a trove of material on best practices for institutional development, including comprehensive assessment frameworks that can also guide intelligence collection and analysis.

Many tools for national security professionals are cowritten by military leaders or others well steeped in military issues. For example, the United States Institute of Peace has developed a number of handbooks for practitioners.[31] It also provides fact sheets, studies, and reports on specific countries and groups within those countries. The number and range of think tanks is ever expanding, but many have been around for decades and have established a solid reputation. These organizations publish everything from short, one-page fact sheets to longer, book-length works.[32]

Some think tanks and NGOs focus on specific topics that are particularly useful in confronting the problem of illicit power. The Fund for Peace publishes the yearly "Fragile States Index" as well as a variety of reports on various transnational threats.[33] Global

---

mines peace worldwide, and offers solutions for international organizations seeking to mitigate the risk.

[28] See, for example, TI Germany, "Corruption Threats to Stabilisation Missions and Defence Capacity Building," Robert Bosch Stiftung, Feb. 2015, www.dropbox.com/s/dlpaoxbpde8obsj/150206%20Policy%20Paper%20Stabilisation%20and%20Capacity_Building%20Transparency%202015.pdf.

[29] Thomas Parks and William Cole, "Political Settlements: Implications for International Development Policy and Practice," Asia Foundation Occasional Paper no. 2, July 2010, http://asiafoundation.org/resources/pdfs/PoliticalSettlementsFINAL.pdf.

[30] UNODC, "UNODC ENews," June 23, 2015, www.unodc.org/unodc/enewsunodc/index.html.

[31] USIP, "Articles and Publications," 2015, www.usip.org/publications.

[32] See, for example, Council on Foreign Relations, Brookings Institution, Carnegie Endowment for International Peace, Center for Strategic and International Security, and International Institute for Strategic Studies.

[33] Fund for Peace, "Fragile States Index 2015," http://global.fundforpeace.org/.

Witness publishes reports on how natural resources help a wide array of illicit actors fund and resource conflict in many locations worldwide. The Stockholm International Peace Research Institute (SIPRI) provides information on armaments and smuggling networks as well as group and regional analysis.

Leveraging Social Network Analysis

Illicit power structure linkages can be mapped and understood using social network analysis. This is one of the principal tools for understanding the organizational structures and their patronage networks. But many intelligence analysis personnel have no background in this type of analysis. Fortunately, there are books, YouTube videos, and free MOOCs (for "massive open online courses") that provide a basic understanding of how to conduct social network analysis. Doctrinally, *FM 3-24 Counterinsurgency Field Manual*'s Appendix B provides a basic glossary of key terms on social network analysis, as well as other analytical tools. And the British *Joint Doctrine Publication 3-40 Security and Stabilisation: The Military Contribution* provides additional considerations for key players who should be mapped, as well as some additional analytical and technical tools for such mapping. Specialized software packages can greatly help analysts conducting network analysis, but whiteboards, chart paper, and other low-tech means can also be tremendously effective. Network analysis will have to include not only the power brokers themselves but also their families, protégées, economic interests (e.g., front companies, banks), and criminal interests.

Using information from certain sources does require some delicacy. NGOs are highly sensitive to being openly exploited for information purposes, but often willingly share their expertise and experience. Outside organizations have political sensitivities that need to be respected, and security considerations as well. Their ability to operate in hostile environments (including politically hostile environments where their legal risk is significant and real) depends largely on their being perceived as neutral. They are increasingly vulnerable to accusations of colluding with a hostile entity or, worse, spies. As the recent spate of kidnappings and murders in the Middle East and the Horn of Africa has shown, this is a risk both to organizations and to their personnel.[34] Thus, formal and informal information-sharing agreements and rules of engagement for working with different organizations, even on an ad hoc basis, may have to be developed. Risks and benefits must be carefully weighed.

Conclusion: The "So What" for National Security Professionals

Even armed with nontraditional resources, deeper, more comprehensive background study, and a more empathetic approach, security professionals will always find illicit power elusive, tenacious, and persistent. But does it "take a thief to catch a thief," or can licit actors understand the illicit without having to step over the line themselves? Going back to our burglar-casing-a-neighborhood analogy may be instructive.

---

[34] Scott Feil, "Laying the Foundation: Enhancing Security Capabilities," in *Winning the Peace: An American Strategy for Post-Conflict Reconstruction*, ed. Robert C. Orr (Washington, DC: Center for Strategic and International Studies, 2004), 53.

First, it is an obvious point but worth remembering that for all we know about catching and prosecuting thieves, thefts occur everywhere, every day. Effective policing reduces the risk to a manageable level and makes successful prosecution more likely, but it does not eliminate the problem of theft any more than we can eliminate illicit power. As discussed earlier, bad actors are always waiting in the wings of chaos and transition, looking for opportunities to emerge. The goal of greater understanding is more effective containment—keeping the problem at a manageable level and ensuring that illicit power does not preclude sustainable peace. So it is interesting to note that the techniques for managing the burglary problem are strikingly analogous to the approach we advocate here for managing illicit power.

The Center for Problem-Oriented Policing (POP), a nonprofit organization that draws on practitioners, academics, and researchers to codify best practices in community policing, emphasizes the importance of understanding the operational environment from the thief's perspective (our "vulnerability assessment"): "Analyzing the local problem carefully will help you design a more effective response strategy. Your main emphasis should be on understanding the environmental settings in which the thefts occur in your suburban residential communities, and identifying those people in your community who can help change those settings."[35]

As thefts occur (or as illicit power emerges), the task shifts from the environment to the thief (our "taxonomy"): "The most important first step must be the collection of relevant data. It is only through the systematic collection of information concerning characteristics of location, times, and methods used by offenders that a clear picture of the problem will emerge." Understanding the problem from the thief's perspective is critical. "Interviewing offenders as to their motivation and methods can help police develop new approaches to the problem, and to determine which efforts they have employed are effective and which are not. The San Diego police used this method and determined that the perpetrators in their area typically focused on apartment complexes, worked in pairs and traveled to the area by car from several miles away."[36]

The focus then shifts to the enabling environment, with the purpose of developing realistic and practical courses of action: "Gathering detailed information from offenders can reveal the type of offender and suggest proper courses of action. . . . For example, if your analysis reveals that professionals are stealing vehicles for stripping or resale, investigation can focus on identifying suitable locations for the thieves to carry out such an operation."[37]

The POP approach makes sense for countering neighborhood theft, and on paper, our methodology retrospectively supports the lessons from case studies in the book. But will it really make a difference in our effort to understand and confront the problem of illicit power? Would it have helped the special-operations force (SOF) in Afghanistan? With peace negotiations in Sierra Leone? With capacity building in Timor Leste?

The answer is probably yes. By the time of the conversation with the elders regarding detainees, special-ops forces in Afghanistan had been doing superb work for almost ten

---

[35] Center for Problem-Oriented Policing, "Thefts of and from Cars on Residential Streets and Driveways," Guide no. 46, 2007, www.popcenter.org/problems/residential_car_theft/.
[36] Ibid.
[37] Ibid.

years in the province where the rule of law shura took place. But their focus had been on military security operations, not civilian-led evidence-based operations. That is, they had been dealing with a different set of issues and objectives. The local population's perceptions of law enforcement authority, while certainly of interest, had not been a primary concern. For the task force, attitudes about such lofty ideals as democratization and human rights were the purview of others, such as NGOs and civilian members of the provincial reconstruction teams. But even if the NGOs and civilians were asking the right questions, they were not asking them from the perspective of security and detection of illicit power, and they likely did not understand the significance of these attitudes to the SOF initiative. As many of the SOF team members said later, had they understood the villagers' [mis]perceptions about democracy, human rights, and governmental authority, they would have approached the problem of evidence-based operations differently. They would likely have spent more time facilitating the type of dialogue that took place during the Afghans' rule of law shura. They would have worked more closely with their Afghan partners to describe "democracy" in the Afghan context, and their communications strategy with the community would have been different in both its substance and its delivery.

There is no magic formula for getting the information and intelligence picture right, but ultimately, it comes down to knowing how to ask the right questions, at the right time, of the right people, in the right places. Those accustomed to standard enemy-focused intelligence collection and analysis may resist the expanded scope of information needed to detect, understand, and counter illicit power. In conflict and political transition especially, information-gathering resources are already spread thin; access to public information may be severely constrained. Both analysts and collectors must contend with the competing requirements of day-to-day operations over long-term, contextual understanding. But at the same time, leaders, policymakers, planners, and commanders need to be reminded that parallel information-gathering networks and ready sources of information already exist, and that learning how to work collaboratively with them will be not an extra requirement, but a true force multiplier. And as the cases in this volume demonstrate, a deeper understanding of the risks from illicit power, illicit power structures, and their enabling environment will enable more focused and effective response.

11. Weapons Trafficking and the Odessa Network: How One Small Think Tank was Able to Unpack One Very Big Problem, and the Lessons It Teaches Us

David E. A. Johnson

> Oh, what a tangled web we weave, when first we practice to deceive!
> —Sir Walter Scott, *Marmion*

Today's security environment is very different from the one we faced even two decades ago. Asymmetric and hybrid threats take advantage of the industrial-age doctrine, law, and organizational and geographic boundaries that we created to fight the World Wars and manage the Cold War. Informal and illicit power structures embed themselves in licit systems to promote political violence and gain control of vital resources. Persistent security challenges, such as arms trafficking and environmental crime, feed the local conflict economies and empower the illicit networks. This expansion of mission scope provides a daunting challenge for the intelligence community. But the arrival of the information age also provides the tools necessary to understand and influence these power structures. Using these tools, a C4ADS report, *The Odessa Network*, has mapped the intricate web of entities and processes used to execute a series of illicit arms transfers.

C4ADS (www.c4ads.org), winner of the Google Chairman's 2014 "New Digital Age" Grant, is a tiny Washington, DC-based, nonpartisan, nonprofit organization dedicated to enhancing global security.[1] Leveraging unique people, unique data, and disruptive emerging technology, it does quantifiable investigative reporting on illicit networks, addressing complex threat systems that cross regional and governmental department boundaries.

In 2010, the C4ADS leadership felt that a bureaucracy-bound, sensor-driven defense and intelligence establishment, designed never to lose sight of the Japanese Fleet again, was floundering in its attempts to understand the new threats and take advantage of the opportunities inherent in the new information paradigm.[2] The C4ADS leadership team further realized that traditional academia, think tank, and contractor models could not accomplish this, either.

To address this gap, they adopted an approach advocated by one of their fellows, Dr. John W. Bodnar, in his book *Warning Analysis for the Information Age*.[3] They selected and trained a few ambitious young international-relations student-analysts who spoke more than one language fluently, had lived abroad for at least a year, and were capable of quantifiable analysis for the core team. They modified the "starfish and spider" organizational concept, creating a flat hierarchy with greater end-

---

[1] Reuters, "Google Executive Chairman Eric Schmidt Names 10 Recipients for the 'New Digital Age' Grants," press release, Mar. 10, 2014, www.reuters.com/article/2014/03/10/idUSnMKW16JQpa+1cc+MKW20140310.

[2] David E. A. Johnson and Newton Howard, "Network Intelligence: An Emerging Discipline" (proceedings, Intelligence and Security Informatics Conference, Istanbul, Aug. 22-24, 2012), http://ieeexplore.ieee.org/xpl/login.jsp?tp=&arnumber=6298848&url=http%3A%2F%2Fieeexplore.ieee.org%2Fiel5%2F6298753%2F6298809%2F06298848.

[3] John W Bodnar, *Warning Analysis for the Information Age: Rethinking the Intelligence Process* (Washington, DC: Joint Military Intelligence College, 2003)..

point access to enhance situational awareness and still execute effective projects.[4] This change—along with the special access to sources, and the unusual freedom to focus on the problem set, that being a nonprofit organization with a donor-based business model permits—enabled rapid change and innovation. The C4ADS leadership team partnered with Google and Palantir; downloaded FOCA, NodeXL, QGIS, and other free software tools; and set out to explore the boundaries of open source intelligence.

In the winter of 2012, the chief operating officer, Farley Mesko, conceived a flagship project to examine the hypothesis that Russia was illicitly resupplying Syria's Assad regime with arms. He teamed up with analyst and Russian-language laureate Tom Wallace. In the process of their investigation, they successfully unraveled a complex web of activities that included licit and illicit organizations, multiple governments, and a social network cutting across four continents. This report had a huge effect. Pulitzer Prizewinning journalist Joby Warrick described *The Odessa Network* in a *Washington Post* article.[5] The president of Lithuania tweeted about the report. The Greek Navy seized two ships identified in the report, which were carrying 39 tons of arms. Military Sealift Command canceled large contracts with an identified shipping company. A Russian private military contractor identified in the report was found working in Syria, resulting in a change to Russian law. And the specter of corruption and Russian control raised in the report may have contributed to the downfall of the Yanukovych government in Ukraine.

The report used a wide range of open source local media reports in Arabic, Spanish, Russian, and Ukrainian; court cases and contract disputes from Russia, Ukraine, and the United States; incorporation documents made available as exhibits in those cases; and business directories from Europe, the Middle East, Africa, and elsewhere.[6] *The Odessa Network* demonstrated the utility of combining open source data, foreign language and regional expertise, and cutting-edge technology in answering complex research questions, thereby bridging the gap between tactical data and strategic insight.

The best policy-level insights are derived from tactical-level data that has been traced upward into abstraction. Deep investigation into weapons export provided political insights into the inner workings of Putin's Russia: "Abstract themes find detailed expression in the Odessa Network: reassertion of State control over strategic assets, keeping regime stakeholders loyal through sanctioned corruption, "power vertical" relationships, and fusion of public and private entities."[7] All data in the report was open source or commercially purchased. The report examined both licit and illicit weapons transfers. The authors used the term "illicit" to describe transfers perceived as contrary to international norms, not to imply violation of any international laws or agreements.

The Odessa Network is not any better or worse than other informal power structures, but its study provided some interesting lessons. In the process of discovering the details

---

[4] Ori Brafman and Rod Beckstrom, *The Starfish and the Spider: The Unstoppable Power of Leaderless Organizations* (New York: Penguin Books, 2006).

[5] Joby Warrick, "Ukrainian Port Eyed as Analysts Seek Syria's Arms Source, *Washington Post*, Sept. 7, 2013, www.washingtonpost.com/world/national-security/ukrainian-port-eyed-as-analysts-seek-syrias-arms-source/2013/09/07/f61b0082-1710-11e3-a2ec-b47e45e6f8ef_story.html.

[6] Tom Wallace and Farley Mesko, *The Odessa Network*, C4ADS, Sept. 2013, 10, www.globalinitiative.net/download/arms-trafficking/arms(2)/C4ads%20-%20The%20Odessa%20Network%20Mapping%20facilitators%20of%20Russian%20and%20Ukranian%20Arms%20Transfers%20-%20Sept%202013.pdf.

[7] Ibid., 11.

of Russian arms flows to Syria, the C4ADS team learned a great deal about bringing illicit power structures to light. First, they identified the C4ADS Postulates for Discovery of Illicit Networks:

1. All illicit networks touch licit networks somewhere.

2. To have an illicit capability, one must have a licit capability.

3. Human networks do not behave like mechanical networks.[8]

Using these postulates effectively will require that the government official or contractor be comfortable thinking outside the usual areas of expertise and authority. The postulates imply that publicly available data about licit activities can enable us to observe threat systems. Moreover, it is vital that the analyst understand how these commercial and consumer systems function.

As a nonprofit doing investigative reporting, C4ADS conducts legal review of its work and provides caveats with its products so that readers do not misinterpret the fully sourced, supported findings of the reports. C4ADS does this to protect those enablers of illicit networks who may be unwittingly complicit and not doing anything illegal or immoral. This is as opposed to facilitators, who knowingly support perpetrators of illicit activity through their own licit and illicit activity. Finally, the postulates imply that simple interdiction and action by any single entity or agency cannot have a useful long-term effect on the threat system. Because these networks are embedded in stable, licit systems, a sustained impact requires better understanding of illicit networks across regional and governmental department boundaries, and longitudinal study of the results of action against them. Next, the project identified some shared characteristics of illicit networks: local government acquiescence, key individual connectors, and leverage of the licit financial system.

Most importantly to our study, C4ADS also discovered a replicable process for gathering and analyzing open source information across a number of analytical methods and tools. This process highlights the need for integration across intelligence disciplines and staff functions. Other C4ADS efforts include reports on Somali Piracy (for the World Bank), on insurgency and contraband trafficking in Mali (cited by RAND), and on global ivory poaching and wildlife crime.[9] Each of these works has further reinforced the lessons taken from *The Odessa Network* project.

Use the Correct Tool for the Task

C4ADS's approach to the challenge of illuminating the complex illicit power structure involved in suspected Russian state-sponsored arms trafficking required unique people, unique data, emerging technology, and a top-down analytical direction. These were applied to create useful, credible data sets, and insightful, supported analysis.

---

[8] David E. A. Johnson, "Transforming Ideas into Operations" (PSOTEW WG3 presentation, National Defense University, Apr. 14, 2015).

[9] Michael Shurkin, and Stephanie Pezard, Achieving Peace in *Northern Mali: Past Agreements, Local Conflicts, and the Prospects for a Durable Settlement* (Santa Monica, CA: RAND, 2015).

Unique People

The C4ADS organization was very small and flat, consisting of a board of directors, an executive director, analyst leaders, analysts, interns, and fellows. The leadership selected project-specific analysts. They were generally recent undergraduate or graduate international-relations students from top-tier universities, who had lived at least a year in their region. They had open contacts with foreign entities, were ambitious, and were well connected, through family, peers, and professors, even up to the highest levels of industry and government. Experience was not a selection criterion. As Frederick the Great once noted, "A mule who has carried a pack for ten campaigns under Prince Eugene will be no better a tactician for it."[10] In fact, time spent isolated in a bureaucracy and learning local processes may be counterproductive. Before selection, interns attended a three-month unpaid internship and training program. The leadership encouraged analysts to bypass the "chain of command" and contact anyone in or outside the organization, regardless of title, who could assist with their problem. The young analyst was expected to develop the project, build the funding proposal, lead research, conduct on-the-ground interviews, prepare the report, coordinate with media, publish, and brief conferences and officials. One of the analysts on the Odessa Network report briefed the U.S. vice president's national security staff. While this approach may not be fully replicable in government, academic, or corporate structures, partnering with such teams is possible.

Unique Data

C4ADS used a definition of "open source" not used by the U.S. government. For C4ADS, open source data included anything legally commercially purchasable or freely available in the public domain. This included *general media, expert interview, commercial and public databases, and gray literature.*

General media contains print, video, audio, and online sources (including the current and archived websites and social media of identified organizations and individuals). For event data, C4ADS avoided U.S. outlets and preferred to examine international and local sources in the original language first. The analyst, depending on the question, may develop favorite sources based on a personal understanding of source history, bias, and credibility.

Expert interview includes not only individuals with a reputation and publications, but also anyone with the access and placement to provide useful, credible data or contacts.

Commercial and public databases contain a wide variety of data sets: telecommunications metadata, online feeds, satellite imagery, ship and aircraft transponder records, geospatial data, full-spectrum emissions data, court cases, property ownership, corporate registries, and financial data.

Finally, *gray literature* is the unpublished and published research of academic and commercial entities. Thus, open source includes access to elements of *all* the intelligence disciplines.

While the authors did travel to Ukraine for supplementary information, most of the work was done in the United States. By data foraging for readily available information and not data mining through restricted areas, the authors were able to develop

---

[10] Jay Luvaas, ed., trans., *Frederick the Great on the Art of War* (New York: Free Press, 1966), 47.

the product without doing anything even approaching espionage. Moreover, with all data appropriately source tagged and caveated, they were ready to build a strong argument. C4ADS holds itself to the standard of nonpartisan and credible conclusions drawn from the available data by a reasonable person. By leveraging all the open source components, the Odessa Network analysts could fill information gaps in even data-poor environments to produce data advantage and credibility.

Emerging Technology

Once the data is gathered, the analyst must apply qualitative and quantitative analytical techniques to create insight. Over 10 years, C4ADS has explored techniques ranging from dynamic, multimodal social network analysis with geospatial fusion, to simple descriptive statistics. The authors used a donated Palantir platform that included many of these tools. Another use of the term "open source" describes a business model for essentially free software. As a nonprofit, C4ADS readily adopts and tests these tools and uses donated software and hardware packages. The C4ADS team chose their bins, coded, and structured data manually. Because they were originally trained in CASOS-ORA and QGIS, the transition to Palantir was not difficult. The authors tested a number of open source and commercial data scrapers, including FOCA, and, of course, used Excel from Microsoft Office. As planners at CJSOTF-AP in 2003 in Baghdad proved, it is even possible to build a predictive analysis engine at no cost, from Microsoft Access and Falcon View.[11] The freedom to choose the tools and emerging technologies to search and process everything from Facebook photos and imagery metadata to transponder locations, Russian court documents, and corporate registries enabled the analytical team to find a way around obstacles.

Hypothesis-Driven Analysis

The C4ADS Postulates can help in identifying activities and even phenomena to observe. But without understanding the research question, it is difficult to select the appropriate analytical approach. Traditional sensor-driven approaches assume an understanding of the threat process model, and hope—mostly in vain—that data patterns will "speak" to the analyst. The analyst is frequently in danger of data overload and, most importantly, cannot define why a pattern change is important. Hope is not an analytical process. The alternative is top-down, hypothesis-driven analysis that seeks to define, instead of assume, a useful threat process model and provides data credibility and advantage. Bodnar indicates that both approaches are useful.[12] But few organizations have integrated hypothesis-driven methods effectively into their arsenal. C4ADS took a hypothesis-driven approach to the project.

The overall hybrid adaptive process is cyclic and consists of four phases: *hypothesis development, data foraging, sense making,* and *argumentation and testing. Hypothesis development* relies heavily on the context and the analyst's mental flexibility, general knowledge, personal history, and time spent on the problem. This requires an analyst with broad

---

[11] Author was chief of plans and current plans, CJSOTF-AP, when Maj. Ace Campbell and Maj. Larry Fauconet, two U.S. Army Reserve officers, developed this tool.

[12] Bodnar, *Warning Analysis for the Information Age.*

exposure to the enablers being examined. The initial research question has a tentative answer (hypothesis) that generates questions about the threat process. These questions are used to drive research.

Data foraging requires intuition and choices concerning the sources and binning of structured, unstructured, coded, uncoded, qualitative, and quantitative data.

Sense making leverages the use of appropriate analytical tools and descriptive statistics to provide evidence that supports understanding of a portion of the threat process model. The Odessa authors took particular care not to fixate on any particular tool or source, since these are entirely secondary to answering the question being asked.

The *argumentation and testing* phase builds the argument that supports the hypothesis. This argument must be structured to answer the question asked, in a way that convinces an informed audience. Moreover, at each phase there are formal and informal tests, such as analysis of competing hypotheses, that can result in reforming the hypothesis.[13] While the team continually adapted the process, this general approach was at the root of the C4ADS Odessa Network Project.

Create a Threat Process Model

At the start of the Syrian Civil War, C4ADS was looking for a project to demonstrate the power of data-driven open source analysis. Tom Wallace, a national Russian scholar laureate at the University of Michigan and a Wolcott fellow at the George Washington University, asked himself, if the Assad regime was depleting its stocks of arms, how was it being resupplied? His logical hypothesis was that Syria's longtime ally Russia was rearming the Assad regime despite international resolutions. Further, he wondered, if this was so, how would Russia do this, and how could he prove it? Farley Mesko immediately began to apply the analytical concepts used at C4ADS to support this initial hypothesis and set of research questions.

Build the Data Set

The beginning and end of the abductive inferential intelligence cycle is to create and confirm a useful threat process model.[14] The first step for C4ADS was to conduct general research into the descriptions of abstract activities surrounding arms transfers in existing literature. To do this, the team created a data set of Russian and Ukrainian weapons transfers and the ports, companies, and ships used to facilitate them: "The critical assumption was that transporting arms shipments to sensitive foreign customers requires a great deal of trust between contractors and the government suppliers. Therefore, suppliers are likely to replicate the use of companies, ships and patterns of behavior found in previous weapons exports."[15] The data set covered 12 years and 22 recipient countries, some licit and documented, others undocumented. Each shipment event had a dozen attributes, which could be used for correlation and analysis modes.

---

[13] CIA, Center for the Study of Intelligence, "A Tradecraft Primer: Structured Analytic Techniques for Improving Intelligence Analysis," Mar. 2009, https://www.cia.gov/library/center-for-the-study-of-intelligence/csi-publications/books-and-monographs/Tradecraft%20Primer-apr09.pdf.

[14] E. L. Waltz, *Information Warfare: Principles and Operations* (Norwood, MA: Artech House, 1998).

[15] Wallace and Mesko, The Odessa Network, 9.

Countries and shipments were determined by comparing lists of recipient countries to values of arms transfers from the Stockholm International Peace Research Institute (SIPRI) Arms Transfers and Military Expenditures databases, searching for international media coverage of shipment events (for those that were of sufficient size or nature to warrant international media coverage), and native-language local media coverage of the arrivals of shipments. This C4ADS data set drew heavily on unconventional open data sources such as photos posted by local Venezuelan military bloggers or Cambodian activists, of ships unloading weapons. Ship or owner names were determined from pictures or from searching for unique International Maritime Organization ID (IMOID) identifiers.[16]

For example, some data on one shipment was obtained by locating several pictures on Phoenix Shipping's website showing various types of military equipment—including tanks and armored personnel carriers (APCs)—being loaded onto one or more unidentified ships. The tanks and APCs appeared to be loaded on the same ship. EXIF data revealed that the pictures were from October 15, 2003. The team cross-referenced SIPRI arms transfer databases for tanks and APCs delivered from Russia or Ukraine to a foreign customer for whom maritime transport would be necessary (i.e., not a landlocked neighbor such as Uzbekistan or Kazakhstan).[17]

"Out of sample" events are also analytically useful because they show the involvement of weapons transporters in moving other interesting cargoes. One event identified a ship that was also used for smuggling cigarettes into the UK.[18]

Confirm Data Set Usefulness

The next step is to determine whether the data set is useful. First, we need to determine whether the sampled data set is representative of a larger population under study. Because Russia exports to many more countries than the sample size in the report, the sample size at first appeared unrepresentative. But when the data was controlled for only those countries that would receive shipments by sea, based on geography or the type of weapons transferred, the set was acceptably representative.

Figure 11.1

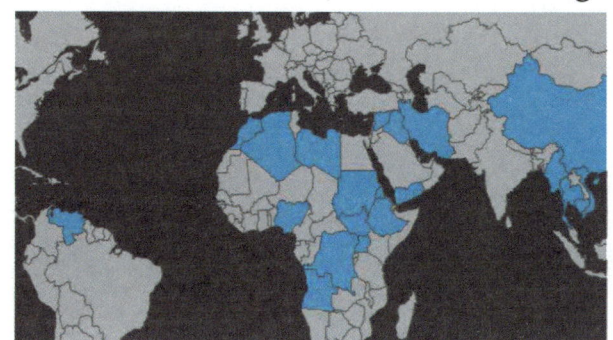

Countries in our data set   *All possible candidate countries*

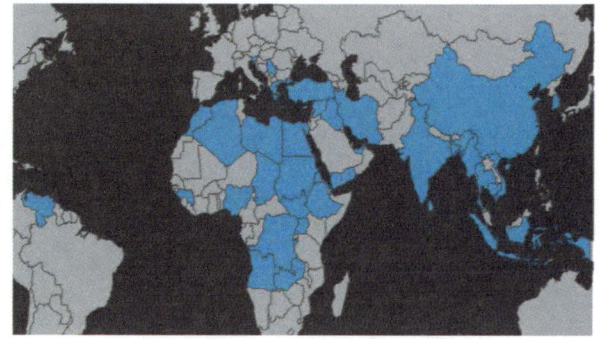

Source: Wallace and Mesko, *The Odessa Network*, 31.

---

[16] Ibid., 12.
[17] Ibid., 14.
[18] Ibid., 19.

Build the Model

The data set was also examined for insights. First, the analysts looked at entities involved. This appeared to indicate direct government ownership and control of the weapons being transferred. The team cross-referenced weapons being sold with acknowledged state arms transfers. Only one event in the data set was clearly not under official control. The team also examined the possibility of selection bias. Heavy weapons of the type likely to be transferred by sea are easier for governments to monitor and likely harder to sell illicitly than small arms and light weapons (SALW). While no SALW incidents were detected, this remains a possibility.

In the case of Russia, there is some ambiguity with the term "government control," which implies a unity of purpose among state agencies, and a distinction between private and public sectors, which may not exist in chaotic and corrupt political systems. The team determined that the involvement of official state weapons export agencies was the screening criterion to identify government control. With this definition, there is little evidence that private "lords of war" continue to sell major military hardware in the Putin era. In fact, examination of this alternate hypothesis uncovered draconian punishment for those who violated this norm. The Putin coalition of *siloviki* took back control of a large portion of the economy from the oligarchs after their heyday in the 1990s. And now "the facilitators of weapons exports are a critical element of domestic politics."[19] As the bridge between foreign customers and Russian producers, facilitators create the profits that are redistributed to regime stakeholders. This led the team to expect a strong connection between high-ranking Russian officials and shipment facilitators.

A small number of companies, based mostly in Ukraine and the EU, facilitate a high percentage of Russian and Ukrainian weapons exports. The chart and table below depict the salient roles played by some of these firms. This narrowing of the field led to some key research subquestions.

---

[19] Ibid., 33.

Figure 11.2 Shipment counts

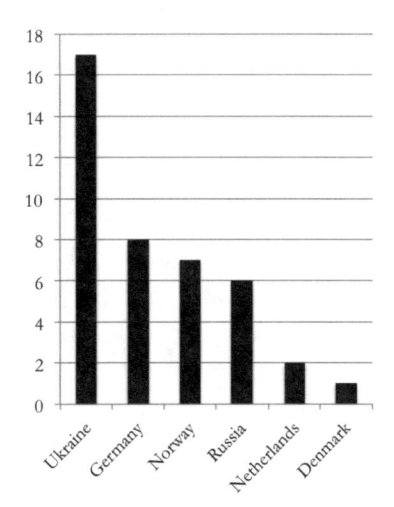

Company	Country	Count
Kaalbye	Ukraine	10
Eide	Norway	7
Phoenix Trans-Servis	Ukraine	4
Briese	Germany	4
Westberg	Russia	2
Beluga	Germany	2
Spliethoff	Netherlands	2
Nortrop	Ukraine	1
Ukrainian Danube	Ukraine	1
Barwil Ukraine	Ukraine	1
FEMCO	Russia	1
North-Western	Russia	1
AnRussTrans	Russia	1
Balchart	Russia	1
Eckhof	Germany	1
Held Bereederungs	Germany	1
Clipper Group	Denmark	1

Source: Wallace and Mesko, *The Odessa Network*, 34.

Subquestions we answer in our discussion of the report's findings: Who are these companies and their key personnel? What qualities enable them to carry out so many arms shipments? Are these firms connected? How? Do some firms have a particular specialty?[20]

Second, the team examined location data and discovered that most shipments originated at the Ukrainian port of Oktyabrsk. This insight drove more subquestions: What is special about Oktyabrsk? What are the mechanisms for controlling and safeguarding the weapons? Since Oktyabrsk is in Ukraine, how does Russia ensure that its interests are met? These and other questions are answered in the discussion later in this chapter.

Examining the 12 or so properties of the events led not only to these insights and research subquestions, but also to insights into the more abstract activities that enabled the transfer of arms. The team grouped these into six geographic and functional clusters.

Figure 11.3 Odessa Network by location and function

Location	Entitiy	Function
Moscow and Kiev	goverment agencies	ownership of weapons
Odessa, Ukraine	shipping companies	AZ logistics integration
EU and Russia	shipping companies	specialized shipping services
Oktyabrsk, Ukraine	port and port authority	weapons loading
Africa and Middle East	private security companies	sensitive cargo protection
Latvia	banks	financial service

Source: Wallace and Mesko, *The Odessa Network*, 9.

---

20 Ibid., 33.

Test the Model: Syria

Since the team had now developed a threat process model for Russian arms exports, the next challenge was to apply it to a particular case for confirmation and refinement. Using the model as background knowledge, the team obtained Automatic Identification System (AIS) transponder data for all ports in Syria, Russia, and Ukraine from January1, 2012, to June 30, 2013.[21]

The team constructed a nearly complete log of commercial maritime traffic entering or leaving ports in these countries over the given time frame, complete with exact date, time, and location. Using the unique IMOID identifier for each ship, they cross-referenced the port call data set against ship registry records, allowing them to determine ship owner, manager, tonnage, flag, and so on. Given the byzantine financial and organizational arrangements used by the Odessa Network (including the use of shell, holding, and management companies that obscured ownership and control), this entailed significant investigative research. This was accomplished by using Palantir to integrate diverse data sets drawn from sources such as Ukrainian court records, SWIFT transaction receipts, Russian business directories, international shipping registries, and more. Once this combined database was complete, the team searched it for the "signature" of Former Soviet Union (FSU) arms shipments discovered through our analysis: ships owned or operated by companies with a track record of transporting Russian and Ukrainian weapons, transiting from Oktyabrsk, St. Petersburg, or Kaliningrad to Syrian ports such as Tartus, al-Ladhiqiya, and Baniyas.

The team added one more possible selection criterion: "AIS discrepancy." Ships can turn off their transponders, broadcast a false name or IMOID or Maritime Mobile Service Identity, or even "spoof" their signals to make it appear as if the ship were in an entirely different location or were an entirely different ship. Ships carrying Russian weapons to Syria, such as the MV *Katsman*, have turned off their AIS transponders, and Iranian vessels routinely spoof their signals.[22] The International Maritime Organization publishes lists of ships that are detected with these discrepancies, which were included in the database. Putting all this together, the team identified shipment events that match patterns of ownership and behavior seen in past Russian weapons shipments.

While it is impossible, using only open source and commercial data, to say exactly what cargo was contained in each of the above shipments, the key finding is that in 2012 and 2013, many ships from the Odessa Network left from known ports of origin for Russian weapons shipments and went directly to Syria or embarked on voyages that make sense only if large portions of their movements were obscured. The evidence (e.g., Syrian port calls by Odessa Network-linked ships, AIS discrepancies coinciding with known Russian seaborne arms shipments) presents a strong circumstantial case that these ships and companies are moving weapons or other sensitive cargo to the Assad regime. Bolstering this evidence is the fact that most of the interdicted and publicly reported Russian arms shipments were also carried out by members of this network.[23]

Mesko and Wallace had to address the alternate hypothesis that Syrian companies, instead of the Odessa Network, did the shipping. An assumption of their paper was that when the Russian or Ukrainian government sells weapons abroad, it (not the purchaser) coordinates the transportation of weapons, most often through the Odessa Network.

---

[21] Ibid., 67.
[22] Ibid., 67, fns. 344, 345.
[23] Ibid., 68.

While they believed this was the case for the vast majority of arms transfers, it was plausible that in some cases, the purchasing country itself would handle logistics. The two are not mutually exclusive. It could be that both Russia or Ukraine and the purchasing country handle different subsections of an overall arms transfer. For example, Kremlin-linked FSU facilitators might handle the highest-value (in both the military and financial senses) weapons and systems, while "native" facilitators from the purchasing country handle the lower-value weapons.

Interestingly, however, ships owned or operated by companies based in Syria or common intermediary countries (e.g., Lebanon and Egypt) made up a higher percentage of 2012 traffic at Oktyabrsk compared to other, busier Ukrainian ports, such as Odessa. The percentage of Syrian traffic at Oktyabrsk was significantly higher even compared to Nikolaev, just a few kilometers up the Bug River. Also, bulk grain carriers are, in fact, ideal for shipping large quantities of small arms; for example, court documents from the 2007 U.S. trial of legendary Syrian arms dealer Monzer al-Kassar show that he intended to use the grain carrier MV *Anastasia* to move thousands of assault rifles and grenades to the FARC in Colombia. It is, of course, possible that no correlation exists between an abnormally high percentage of Syrian ships loading cargo at a port that is the epicenter of Russian arms exports and then traveling to Tartus, and the continued flow of Russian weapons into Syria.[24]

The application of our understanding of the Odessa Network confirmed aspects of our model, brought to light additional potential indicators of illicit arms traffic, and identified alternative and competing hypotheses for exploration. The end result was a useful mapping of the Russian arms trade, in particular the Odessa Network.

The Odessa Network: An Informal Power Structure

We have thus far avoided, as much as possible, the details of the report, which is readily available online at no cost. But some minimal detail is necessary to describe the utility of the report and its approach.

When we compared the number of known weapons shipments carried out by the Odessa Network facilitators and enablers to those carried out by unconnected facilitators, we could see that this network was responsible for the vast bulk of seaborne arms transfers in the data set. Thus, we can use common equipment, procedures, and locations used by these facilitators as a detectable "signature" of Russian and Ukrainian arms shipments.[25]

This signature includes logistics integration in Odessa, government connections in Moscow and Kiev, the Oktyabrsk port of origin, Russian and EU specialized services, African private security companies, and Eastern European financial services. What remained was for the team to further "pull the thread" and identify the key entities and their relationship to other facilitators and enablers. This produced the kind of detail that can enable effective interdiction, both kinetic and nonkinetic, depending on licit or illicit status of the target, strategy, available tools, and long-term objectives.[26]

---

[24] Ibid., 69.
[25] Ibid., 66.
[26] Sean F. Everton, *Disrupting Dark Networks* (New York: Cambridge Univ. Press, 2012).

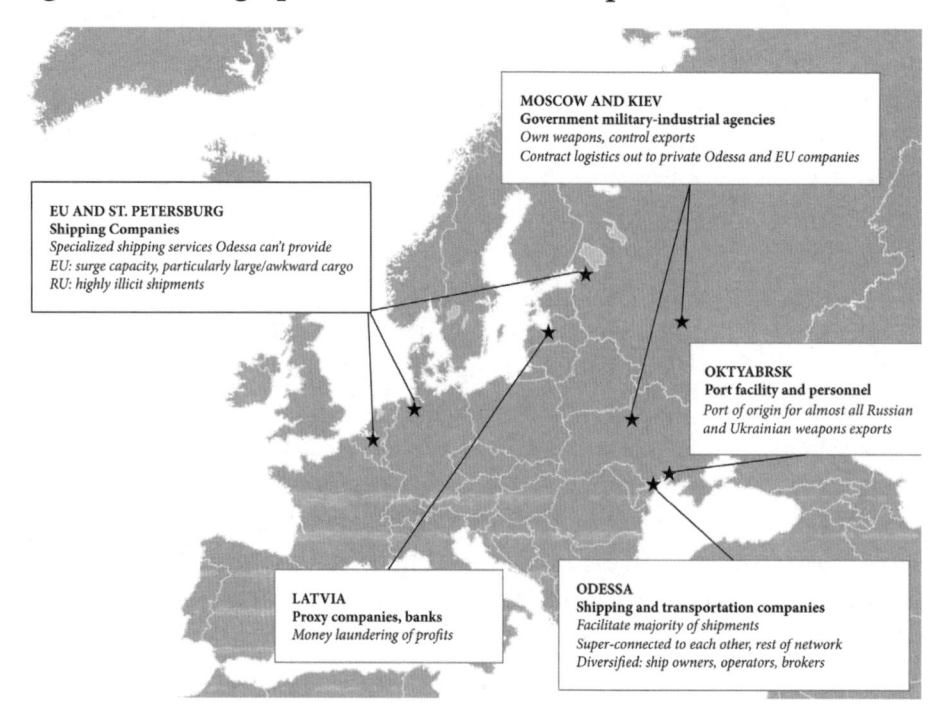

Figure 11.4 Geographic and functional map of the Odessa Network

EU AND ST. PETERSBURG
Shipping Companies
Specialized shipping services Odessa can't provide
EU: surge capacity, particularly large/awkward cargo
RU: highly illicit shipments

MOSCOW AND KIEV
Government military-industrial agencies
Own weapons, control exports
Contract logistics out to private Odessa and EU companies

OKTYABRSK
Port facility and personnel
Port of origin for almost all Russian and Ukrainian weapons exports

LATVIA
Proxy companies, banks
Money laundering of profits

ODESSA
Shipping and transportation companies
Facilitate majority of shipments
Super-connected to each other, rest of network
Diversified: ship owners, operators, brokers

Source: Wallace and Mesko, *The Odessa Network*, 36.

End-to-End Logistics

One of the authors' critical insights, from which the report gets its name, was that a small group of firms based in Odessa seemed to be at the heart of the largest number of weapons exports in the data set, providing end-to-end logistics integration. This group, led by a large shipping conglomerate, was built on a bewildering set of connections between companies, family, former employees, schoolmates, and friends. Link and network analysis outlined these connections in stark relief. This network benefited from business diversification, external connectedness, and government connections.

Business diversification is a part of normal business strategies that include vertical and horizontal integration, used to expand market segment share, reduce supply chain costs, and offset local market risk. Most businesses generally have relationships with a wide variety of trusted subcontractors, and employees change jobs within the same industry and spin off other companies and suppliers. In the case of the Odessa Network, the side effect was to ensure that the network could execute sensitive contracts without routinely relying on services outside its control. And as Mesko and Wallace point out, handling expensive and sensitive cargoes bound for embargoed conflict zones or U.S. strategic competitors encouraged complexity in order to obscure the effort and reduce the chance of outside interference.[27]

In 1996, Igor Urbansky and Boris Kogan founded Kaalbye Group in Odessa. Weak governance, stockpiles of former Soviet weaponry, and foreign demand made Odessa a hub of international arms shipping in the 1990s. One trafficker, Leonid Minin, was a

---
[27] Wallace and Mesko, *The Odessa Network*, 36.

major broker of arms to Charles Taylor in Liberia, the RUF in Sierra Leone, and the Ivory Coast, and he even tried to sell an aircraft carrier to Turkey. While most shippers have a specific market niche, Kaalbye Group had a wide variety of subsidiaries, including Kaalbye Oil Services, Ukrainian Maritime Agency, Kaalbye Marine Service, and Kaalbye Shipping Cyprus. These companies provided crewing, chartering, freight forwarding, container and bulk shipping, project and heavy-lift work, and luxury yacht services.[28]

Shared personnel and connections across the industry indicate potential ties to a variety of external brokers and services, such as Phoenix Trans-Servis Shipping. Phoenix is a freight forwarding company specializing in defense-related cargoes. It had strong ties to government officials in Ukraine, and a work history with Kaalbye and EU shipping firms. Beyond Phoenix, Kaalbye shared personnel and connections with a number of other companies through the Varvarenko family. Finally, Kaalbye frequently worked with companies such as Briese, a well-known German shipping company, and Tomex Team, led by Vadim Alperin.[29]

Alperin is a highly connected Ukrainian whose company managed the Kaalbye-owned MV *Faina*, which was seized by Somali pirates. His company was selected by Syrian businessman Youssef Hares, head of the Hares Group and partner to a number of FSU oligarchs. Hares recently leased a ship for trade with Libya.[30]

Figure 11.5 Dual-listed personnel

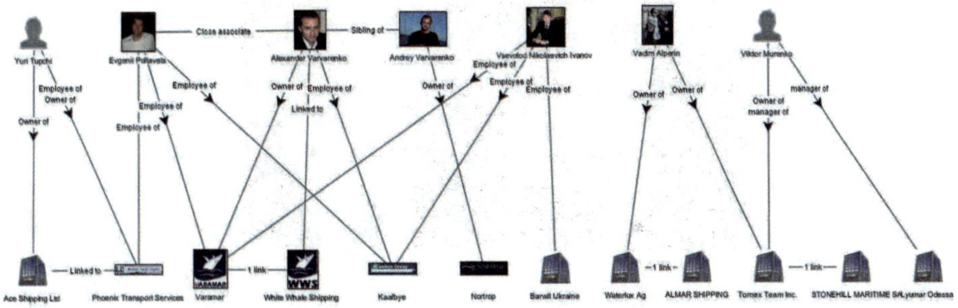

Source: Wallace and Mesko, *The Odessa Network*, 39.

As the report indicates, this network was capable of taking advantage of any business opportunity, shipping both licit and illicit cargoes. While it was not within the report's purview to define any action by the network or supporting services as illegal or immoral, the report maps the links to the event data set and provides sources that support the authors' reasonable conclusions. Connections to key government officials in Moscow and Kiev were essential to making the weapons export line of business possible.

---

[28] Ibid., 39, incl. fns. 143, 144.
[29] Ibid., 41.
[30] Ibid., 42.

Government Connections

The Odessa Network firms were not independent arms merchants. They were logistics contractors for the Russian and Ukrainian governments. State agencies such as Rosoboronexport and Ukrspetsexport owned the weapons and brokered almost all foreign sales. The Odessa Network companies played a critical role in making these arms transfers happen, but they did so only on behalf of powerful customers in Moscow and Kiev.[31] The key assumption was that must be persistent links and contractual relationships between the Odessa Network and government officials.

Igor Levitin. Phoenix Trans-Servis is an Odessa shipping firm that has brokered multiple weapons shipments and openly advertises its connections to Russian and Ukrainian defense industrial concerns. One former Phoenix employee is Igor Levitin, who served as Russia's minister of transportation during 2004–12. He currently works as a personal adviser to Vladimir Putin and has served on the boards of directors of Sheremetyevo Airport and Aeroflot.[32]

Figure 11.6 Levitin and Putin

Igor Urbansky. Igor Urbansky, the founder of Kaalbye, served as deputy minister of transport in Ukraine from 2006 to roughly 2009. He enjoys extensive contacts among the Ukrainian defense establishment and the Party of Regions clique. Urbansky served as a captain in the Soviet cargo fleet before starting his first business, Evas, in Odessa in 1992. He went on to found Kaalbye Shipping. Urbansky was involved in the sale of Kh-55 cruise missiles to Iran and China in 2000 (in cooperation with corrupt Ukrainian and Russian intelligence officials) and in shipping military equipment to Angola in 2001. He was closely linked to military and intelligence figures in Ukraine. For example, in 2007, he personally paid for then-Minister of the Interior Vasily Tsushko, with whom he served in the Odessa Socialist Party, to be flown to Germany for medical treatment. The sophistication and quantity of weapons in deals personally attended to by Urbansky (e.g., cruise missiles) leaves little doubt of his connections to defense officials.[33]

---

[31] Ibid., 43.
[32] Ibid., 43, incl. fns. 190, 191.
[33] Ibid., 44, incl. fns. 199, 203, 204.

Boris Kogan. Boris Kogan was one of the cofounders of Kaalbye Shipping Ukraine and serves as a director and senior manager of the company. He was a close business partner of Kaalbye cofounder Igor Urbansky. Kogan was also on the board of directors of a Russian company, RT-Logistika. RT-Logistika is 51 percent owned by Russian Technologies, the enormous state holding firm (headed by Putin's former KGB colleague and close ally Sergey Chemezov), which owns a variety of industrial companies, including Rosoboronexport and much of the Russian defense industry.

RT-Logistika is deeply involved in transporting weapons. For example, in October 2012, it arranged a cargo plane to transport sensory equipment for Syrian Pantsir SAM complexes from Moscow to Damascus, which the Turkish Air Force intercepted. Kogan was the only RT-Logistika board member who was not a senior Russian defense-industrial figure. Kogan personally knew and worked with some of the most senior defense-industrial figures in the Russian government. The company linking them, RT-Logistika, actively moved Russian weapons to Syria. The other company Kogan was involved with (Kaalbye) was by far the most frequent facilitator of Russian and Ukrainian weapons shipments.[34]

As shown in Figure 11.7, almost all the major Odessa-based facilitators were connected to organs of Russian and Ukrainian state power through personal connections.

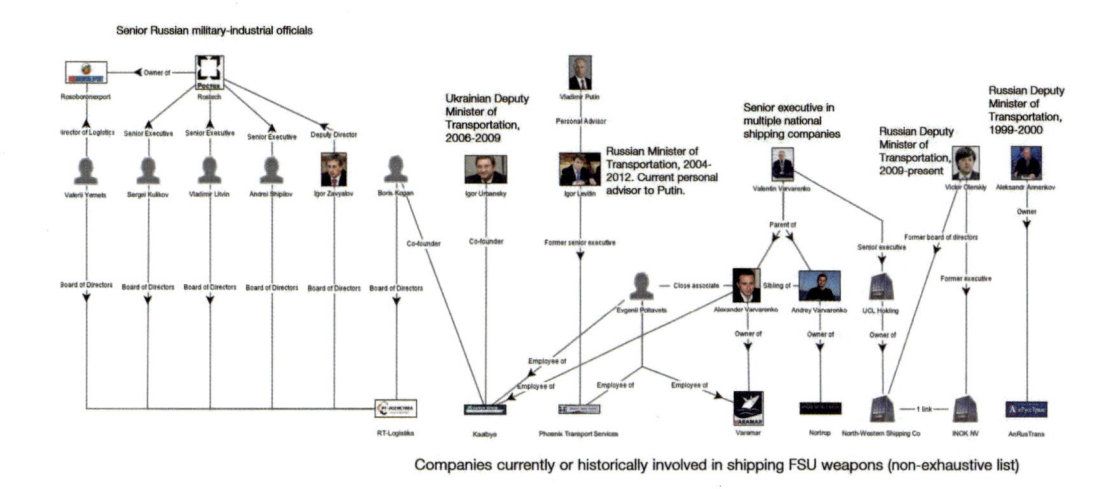

Figure 11.7 Political connections of companies in the Odessa Network

Source: Wallace and Mesko, *The Odessa Network*, 49.

In addition to these connections, the port of Oktyabrsk in Ukraine stood out as the port of origin for the largest number of events in the data set.

Key Locations

The port of Oktyabrsk was the point of origin for almost all Russian and Ukrainian weapons exports in our data set. Located in the city of Nikolaev in Ukraine and specially

---
[34] Ibid., 45, incl. fns. 205, 206, 207.

built by the Soviet Union to ship weapons, Oktyabrsk possesses a number of qualities making it well suited for arms exports: advantageous geography, specialized equipment, transportation infrastructure to major FSU defense-industrial plants, and more. Russian state weapons export agencies maintain offices and personnel in Oktyabrsk. Moscow exerts significant control over the port despite its technically being in Ukraine. (The port's owner and operators had close ties to the Russian military and the Kremlin.)[35]

The economics of maritime transport make the time and distance savings of leaving from Oktyabrsk very significant. From a shipper's perspective, every extra mile and day of transit eats away profit — more fuel costs, more crew costs, and more time spent fulfilling a contract, when profitability often depends on fitting as many voyages as possible into a year.

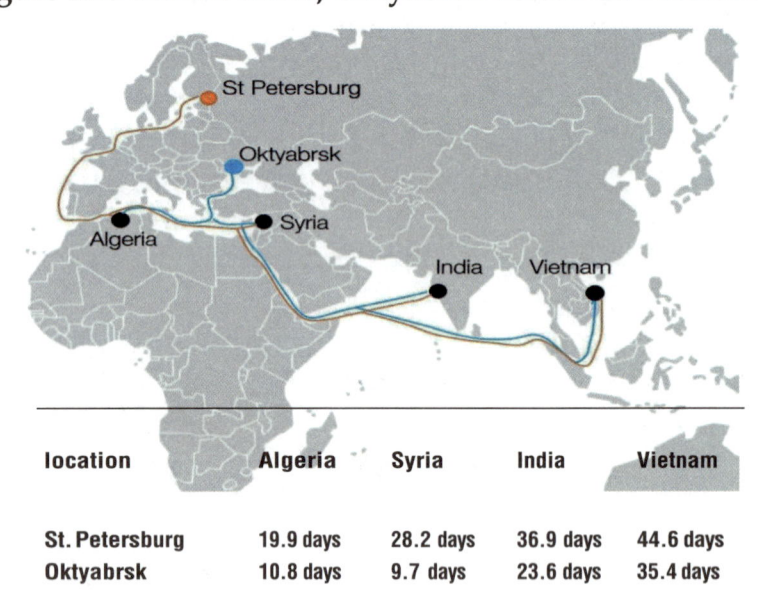

Figure 11.8 Transit times, Oktyabrsk versus St. Petersburg

location	Algeria	Syria	India	Vietnam
St. Petersburg	19.9 days	28.2 days	36.9 days	44.6 days
Oktyabrsk	10.8 days	9.7 days	23.6 days	35.4 days

Source: Wallace and Mesko, *The Odessa Network*, 50.

Port manager Andrey Yegorov was born in Sochi (now part of Russia), served as a submarine commander in the Russian Navy Black Sea Fleet until 2000, and achieved the rank of captain. He graduated from multiple elite Soviet/Russian military academies and did not receive Ukrainian citizenship until 2000. Yegorov was reported to be a client of the Russian-Ukrainian oligarch Vadim Novinsky, the supposed owner of Oktyabrsk.[36]

The report did not identify any weapons shipments by Russian companies to Syria that did not originate in St. Petersburg or Kaliningrad. The C4ADS team suggested, as the reason for this anomaly, that perhaps both international pressure and a desire for a greater degree of control outweighed the purely economic considerations.[37] This indicates that special circumstances may lead to using companies outside the Odessa Network to transport arms.

---

[35] Ibid., 50.
[36] Ibid., 52, incl. fns. 233, 234.
[37] Ibid., 53.

Specialized Shipping Services

The Odessa Network could not provide all shipping services. Companies that provided recognized unique services were better suited to transport certain categories of arms due to size, port infrastructure, or political sensitivity.

Warship and submarine transportation: Eide Marine Services (Norway). Eide's distinction is functional, not political: its barge ships allow it to move large military cargoes that few others can. Eide does not need access to Odessa's political connections to find customers: given that it has moved military cargoes not just for Russia but also for the Canadian, U.S., and Swedish navies, it is fair to say that it is a service that sells itself. Thus, while Eide is an important logistics contractor for moving Russian weapons, it was not truly a part of the Odessa Network.[38]

Heavy-lift transportation: Briese, BBC, and others. A German firm, Briese Schiffahrts GmbH and Co. KG, is another transporter of Russian and Ukrainian weapons. Based in Leer, Briese is one of the largest shipping companies in Germany and among the largest heavy-lift shippers in the world, with a fleet of over 140 ships. Heavy-lift ships typically carry their own cranes and thus can move large, heavy, and unusually shaped cargo such as tanks and artillery. (Bulk carriers and normal cargo ships cannot.) Briese and other EU heavy-lift firms had an important functional role in the Odessa Network: heavy-weapons shipments to countries with poor infrastructure. The report outlined potential connections to a Kaalbye Shipping shell company, Primetransport Ltd. Briese's corporate structure highlights the complexities of this analysis.[39]

Figure 11.9 Briese corporate structure

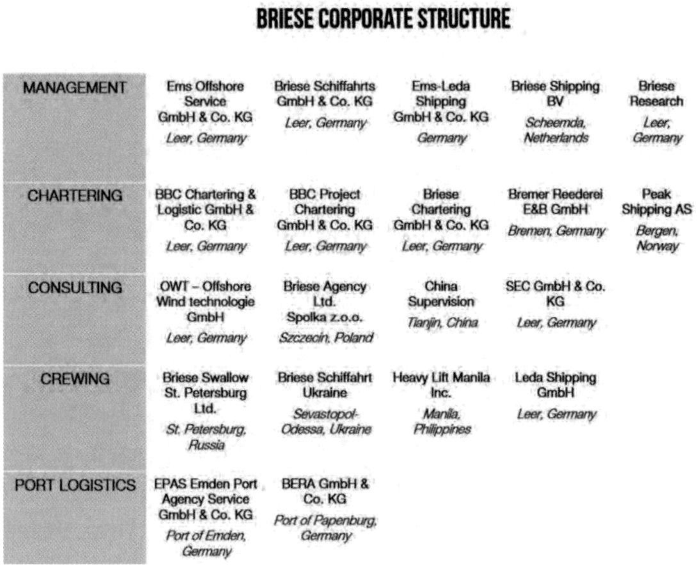

MANAGEMENT	Ems Offshore Service GmbH & Co. KG, Leer, Germany	Briese Schiffahrts GmbH & Co. KG, Leer, Germany	Ems-Leda Shipping GmbH & Co. KG, Germany	Briese Shipping BV, Scheemda, Netherlands	Briese Research, Leer, Germany
CHARTERING	BBC Chartering & Logistic GmbH & Co. KG, Leer, Germany	BBC Project Chartering GmbH & Co. KG, Leer, Germany	Briese Chartering GmbH & Co. KG, Leer, Germany	Bremer Reederei E&B GmbH, Bremen, Germany	Peak Shipping AS, Bergen, Norway
CONSULTING	OWT – Offshore Wind technologie GmbH, Leer, Germany	Briese Agency Ltd. Spolka z.o.o., Szczecin, Poland	China Supervision, Tianjin, China	SEC GmbH & Co. KG, Leer, Germany	
CREWING	Briese Swallow St. Petersburg Ltd., St. Petersburg, Russia	Briese Schiffahrt Ukraine, Sevastopol-Odessa, Ukraine	Heavy Lift Manila Inc., Manila, Philippines	Leda Shipping GmbH, Leer, Germany	
PORT LOGISTICS	EPAS Emden Port Agency Service GmbH & Co. KG, Port of Emden, Germany	BERA GmbH & Co. KG, Port of Papenburg, Germany			

Source: Wallace and Mesko, *The Odessa Network*, 56.

---

[38] Ibid., 54, incl. fns. 248, 249, 250.
[39] Ibid., 55.

Highly sensitive shipments: Balchart, Westberg, FEMCO, et al. Russian shipping companies appeared to play a small but significant role in the maritime export of FSU weapons. The data set showed only four arms transfers facilitated by Russian companies: three to Syria, and one to an unknown customer of sufficient interest that the U.S. State Department formally complained about the transfer. The C4ADS team found no weapons transfers facilitated by any Russian shipping company that were not highly illicit, whereas the Ukrainian and EU elements of the Odessa Network had plenty of licit shipments (e.g., to Vietnam and Venezuela). The second common feature among Russian companies shipping weapons was deep involvement by the Russian government. The authors postulated two tentative hypotheses: One, Russian shipping companies were more likely than non-Russian companies to be contracted by state weapons export agencies for illicit shipments. Two, they had particularly strong connections to the Russian government, including through the Odessa Network.[40]

Private Security

Odessa shipping companies appeared to own or employ multiple private maritime security companies (PMSCs), including Moran Security Group, Muse Professional Group, Helicon Security, Changsuk Security Group, and Al Mina Security Group. These companies' business model revolved around staffing ships passing through dangerous areas (particularly the Gulf of Aden and the Gulf of Guinea) with heavily armed FSU military veterans, who provided protection from pirates. Though hijacking is a threat for any shipowner, it is particularly troubling when the cargo is sensitive military equipment, as happened in 2008 when Somali pirates hijacked the MV *Faina*, a Kaalbye ship carrying Ukrainian weapons to South Sudan.[41]

Foreign observers, local media, and on-the-ground contacts have reported a growing number of Russian and Ukrainian private military security companies and arms dealers operating in war zones such as Somalia and the Democratic Republic of the Congo. FSU entities offer a wide range of services, acting as a one-stop shop for governments and militant groups alike to purchase weapons, mercenary services, trained pilots, and so on. Muse was active in Somalia, where a Somalia Monitoring Group report listed it as guarding ships entering and leaving Bosaso Port. Muse also worked with the Yemeni coast guard to contract services out to the highest bidder.[42] All these entities shared a need to transfer cash.

Financial Services

Russian and Ukrainian weapons are big business, both for the governments that export them and for the Odessa Network companies contracted as transporters. Russia exported over $17.6 billion in weapons in 2012, and Ukraine exported $1.3 billion worth in the same year. The Odessa Network company leaders facilitating these weapons flows

---

[40] Ibid., 58.
[41] Ibid., 60, incl. fn. 286.
[42] Ibid., 61, incl. fns. 306, 307.

earned significant profits. There was evidence that some of the Odessa Network companies employed Eastern European banks known for, or accused of, money laundering, and a series of Panamanian companies run by Eastern European nationals who acted as proxy directors. The analytical team found salient connections and "pulled the thread" to see what unraveled.[43]

The Odessa Network operated in a region of the world where financial crime is endemic, large-scale, and persistent. Russian and Ukrainian arms exports occurred in this context of massive and systemic financial evasion. There are functional reasons to launder money in arms shipping, particularly in concealing illicit arms transfers. Even licit transfers are sensitive to early disclosure, and many agencies are adept at "following the money" once they know where to look.[44] Illicit redistribution of national wealth to regime stakeholders was how United Russia and the Party of Regions maintained stability: they made collaborating with the state more profitable than challenging it. Money laundering also plays an important role in Russian foreign policy. A small but well-documented example is the Sluzhba Vneshney Razvedki's (Foreign Intelligence Service's) funding the election of a pro-Russian candidate in a Latvian mayoral election. The funds for this type of operation are not licit.[45]

Some FSU money laundering relies on proxy directors and shell companies. In this scheme, companies recruit people with no business qualifications to lend their names as directors to businesses located in offshore havens such as Panama, allowing the "real" owners to conduct business in virtual anonymity. One shell company can then be named a director of yet another new company, creating a daisy chain of shell companies facilitating anonymous financial transactions and obscuring ownership.[46]

The international banking system can be used to facilitate laundering and transfer of illicit income, even with reasonable compliance measures in place. A well-known bank in one country may have a correspondent relationship with a bank in another country, which shares an unknown relationship with a sanctioned bank in a third country, allowing ill-gotten gains to be accessed in the original country from a reputable financial-services provider. Banks and other financial-services providers expend significant effort seeking to avoid compliance challenges, and their due-diligence efforts can act as a form of early warning. Government agencies (while recognizing banks' need for legal certainty before sharing information) may want to investigate why a financial-services provider decides to sever a correspondent relationship. Even the most well-run and licit banks cannot share information with government agencies unless they are first immunized against legal action by customers, foreign governments, and even other government agencies due to privacy regulations. When they have such certainty, they could warn government regulators of the compliance risks that caused them to terminate a relationship with another financial services provider. The ability to prevent illicit gains from being transferred thru licit correspondent relationships is especially important since financial relationships are the glue that binds the network and its facilitators and enablers together.

---

[43] Ibid., 62, incl. fns. 308, 309.
[44] Ibid., 62, incl. fn. 317.
[45] Ibid., 63, fn. 320.
[46] Ibid., fn. 322.

Conclusion: The Need for a Comprehensive Approach

From the study of the Odessa Network, C4ADS identified postulates that enable the discovery of other illicit power structures. The research process is replicable and produces credible, useful insights. Finally, the insights provided in the Odessa Network report improve our understanding of the challenge of informal or illicit power.

The team derived the C4ADS postulates for discovery of illicit networks by examining the agents within the Odessa Network.

1. All illicit networks touch licit networks somewhere. The Odessa Network had a web of perpetrators conducting illicit activity, facilitators knowingly supporting that activity through licit and illicit activity of their own, and enablers unwittingly conducting licit activity that supported the perpetrators' and facilitators' actions. There is nothing inherently wrong with relationships forged in school, government service, business, or even prison. And yet, these bonds can create the trust necessary to bind an illicit or informal power structure.

2. To have an illicit capability, one must have a licit capability. In the Odessa Network case, the standard business motivations, processes, and assets involved in licit transportation of goods were mirrored and sometimes co-opted for illicit transport.

3. Human networks do not behave like mechanical networks. Human networks are resilient and cannot be broken. They are influenced and disrupted, but because they are biological and have unbounded connections to licit entities and activities, disruption is more difficult than with a minimum connected cut problem, such as efficient Allied interdiction of German rail networks during the Second World War.[47]

C4ADS created the right tool for conducting the kind of research required to understand an illicit power structure. The team then asked a research question that was relevant and built an event data set across a large enough time span to provide useful volume and variety. They examined the event properties for correlations, commonalities, and insight using descriptive statistics, geospatial analysis, and social network analysis. These insights were clustered into activity and location functions to describe a threat process model. The model was tested and further refined against a salient, related, current research subquestion. Throughout the process, alternative and competing hypotheses were examined for each of the relationships observed. C4ADS then presented the model in publication as a signature for a particular illicit power structure. This process, a variation on net-chain analysis, avoids the cultural-mirroring challenges that generally come from sensor-driven approaches that assume known functions. C4ADS has continued to refine the approach, build a larger data set of entities and relationships, train successive intern/analyst classes, and build best known practices with more interesting and useful open source and commercial databases in each new report.

---

[47] Alexander Gutfraind, Los Alamos National Laboratory, "New Models of Interdiction in Networked Systems," *MORS Phalanx* 44, no.2 (June 2011): 25-27. http://mmsengineering.ca/sasha-web/docs/gutfraind.phalanx.scan.pdf.

The Odessa Network highlights three key aspects of an illicit or informal power structure: government connection, key individual connectors, and financial services support. First, the network must have a relationship with local government. This can be a function of official sanction, poor governance, or official corruption. Next, key individuals act as super bridges connecting perpetrators, facilitators, and enablers, because illicit activity travels at the speed of trust. Finally, illicit activity generates significant income to sustain itself and requires the use of licit financial systems and regular business processes to hide, launder, and transfer resources.

Influencing or disrupting these informal and illicit power structures requires a sustained and comprehensive approach. Because of the unbounded nature of a human network, no single agency can have lasting impact. In *Disrupting Dark Networks,* Sean Everton provides a strategic matrix that includes kinetic and nonkinetic activities targeting the basic functional elements of an illicit power structure.[48] To take the strategic framework approach a step further will require a lead agency to identify outside agencies with the capacity, capability, and authority to act on the licit and illicit components, and provide those agencies with the information and motivation to address their portion of the problem set. Profit-motivated facilitators or enablers may be influenced through both kinetic and nonkinetic approaches. This will require analytical capability that recognizes more than geographic areas of responsibility, influence, and interest. For example, the Department of Defense would have to gather intelligence on domains of influence and interest, such as media, business, or finance, during times of peace. Informal and illicit power structures leverage these licit systems, thereby influencing diplomatic, informational, military, economic, and political elements of national power.

After publication of *The Odessa Network,* C4ADS was sued in U.S. court by a Ukrainian shipping conglomerate and, more recently, an Eastern European bank. The prestigious Washington law firms these companies chose as enablers, containing well-known American lawyers and politicians, have represented other affiliated known and allegedly illicit international clients. In this case, under the Washington, DC, Anti-Strategic Lawsuits Against Public Participation statute, the decision, with prejudice, in favor of C4ADS set several precedents and resulted in yet more illicit activity being read into the public record.[49] Rather than having the intended chilling effect, this effort at "lawfare" has highlighted just what a threat today's open source intelligence is to these illicit power structures. Once chided in intelligence circles as being static, irrelevant background information that produced no data advantage and could easily be deception, open source is fast becoming the main source of credible and useful intelligence. While Naval analysts have long thrown away the classified red book in favor of *Jane's Fighting Ships,* the use of now readily available and searchable public records, appropriate data sets and analytical approaches, and emerging data management tools throws the classified intelligence community on its ear, making old priorities, techniques, and access

---

[48] Sean F. Everton, *Disrupting Dark Networks* (New York: Cambridge Univ. Press, 2012).

[49] Superior Court District of Columbia, *Center for Advanced Defense Studies v. Kaalbye Shipping International et al.,* 2014 CA 002273, Apr. 7, 2015, http://dcslapplaw.com/files/2015/04/C4ADS_Superior_Court_Opinion.pdf.

controls obsolete and even counterproductive.[50] We should, however, expect bureaucratic resistance to real reform. Cramming effective responses to this paradigm shift into the same old organizational and personnel frameworks would not be easy, since our own networked power structures are also resilient. But the intelligence community can emphasize exploitation of open source data over classified approaches, integrate hypothesis-driven analytical methodologies, and support a network of strategic partnerships with innovative nonprofit organizations.

---

[50] Author interview with Capt. (Ret.) Peter O'Brien, USN, former director of fleet intelligence, DIA, Mar. 2015, Washington, DC; HIS, *Jane's Fighting Ships,* Aug. 13, 2015, https://www.ihs.com/products/janes-fighting-ships.html.

12. Financial Tools and Sanctions: Following the Money and the Joumaa Web

Robert ("J. R.") McBrien

In 2011, two units of the U.S. Treasury Department, acting on DEA-led investigations, imposed closely aligned financial countermeasures to expose, disrupt, and incapacitate a major international drug trafficking and money laundering network linked to the terrorist organization Hezbollah. The network, headed by Lebanese national Ayman Joumaa, had exploited the U.S. and international financial systems to launder hundreds of millions of dollars from narcotics trafficking and other criminal activities in Europe and the Middle East, moving the illicit moneys through West Africa and back into Lebanon. The U.S. dollars moving through this scheme included the proceeds of car sales between the United States and West Africa, transacted to conceal the money's criminal origins. A significant portion of this illicit money moved through courier and security networks controlled by Hezbollah or affiliated individuals.

The Lebanese Canadian Bank (LCB), its subsidiaries, and five Lebanese money exchange houses participated with Joumaa's network in a complex money laundering scheme of global breadth. LCB not only facilitated the Joumaa network's drug trafficking activities by laundering hundreds of millions of dollars through its accounts but also provided financial channels for Hezbollah, which drew financial support from the network's criminal enterprises. For ease of discussion, we refer to this web of illicit networks and actors as the "Joumaa Web."

This chapter explores the concept of *counter threat finance*, also known as "follow the money," and chronicles the real-life application of its core tools against a transnational network of illicit actors. The tools include economic sanctions, special anti-money laundering measures, civil forfeitures, and criminal investigations and charges. Their sustained application has resulted in criminal indictments, denial of assets and of access to the legitimate financial system, and exposure and disruption of the players' criminal activities.

This chapter introduces the tools and provides some perspective on their use against the transnational criminal and threat networks that constitute illicit power structures.[1] We do not address the investigative techniques and information collection practices that support use of sanctions and traditional criminal justice tools. Nor do we examine the military and other national security means that may be employed as elements of a broader, whole-of-government approach to countering the financial challenge of transnational threats. (Covering those aspects of counter threat finance would be a book-length enterprise in itself.)

As this case study shows, following the money is not a speedy process. The measures described, from the first actions in early 2011 to the July 2014 Kingpin Act designation, span three and a half years. Factor in the underlying DEA investigations that began in 2007, and this counter threat finance case has spanned more than seven years. The judicial proceedings have continued over three years, and some still have not reached final

---

The author and editors extend their sincerest gratitude to Dane Shelly and Josh Meservey for their research and analysis on the seizing and freezing of assets held by Muammar Qaddafi and associates.

[1] See especially, *Department of Defense Counternarcotics and Global Threats Strategy*, Apr. 27, 2011, 4.

resolution. For example, Ayman Joumaa, though indicted as well as designated, remains a fugitive.

This chapter seeks to improve whole-of-government collaboration and cross-fertilization of disciplines by making the military and other national security elements more familiar with the core players using follow-the-money measures in a real-life campaign against a transnational criminal web. The lessons and insights are intended to illuminate planning factors, expectations, and risks for planners and strategists alike. Critically, the case study does not tout counter threat finance as a silver bullet. But it does demonstrate how one key element, "follow the money," achieves a specific subset of results within the complex of whole-of-government tools for combating networked illicit actors.

The Money Trail

Since the 9/11 attacks, "follow the money" has become a central (though not always well understood) tenet of anti-money laundering and counter threat finance.[2] This principle contemplates a set of distinct but interrelated tools that government can use against the individuals, entities, organizations, and networks making up the phenomena that have come to be known as *transnational threats*.

Transnational Threats

These dangers to U.S. national security have two essential properties that, together, make them a prime target for a particular set of counter threat financial tools: they contain an element of criminality, and their participants have an underlying need for money. Consequently, effective countermeasures against them must include disrupting their nexus with the funding that keeps them in business.

With few exceptions, the illicit power structures discussed throughout the volume have this element in common: attacking their finances is an essential element of disrupting and degrading the threat. Whenever possible, we must disrupt or dismantle the sources, facilitators, and channels of financing that sustain transnational threat networks. That is, we must *follow the money*. But what does this really mean in practice?

The Agencies Involved

The principal agencies in counter threat finance are law enforcement and regulatory, and the authorities that they use are embedded in our criminal and civil justice systems and in administrative law. They are fundamentally non-Defense Department and non-military. The lead agencies in this study are the Justice Department—principally, the

---

[2] See, for example, Department of Defense Directive "DoD Counter Threat Finance (CTF) Policy," DoDD 5205.14, Aug. 19, 2010, Incorporating Change 1, Nov. 16, 2012, at Glossary, Part II, Definitions: "DoD CTF activities and capabilities . . . to deny, disrupt, destroy, or defeat finance systems and networks that negatively affect U.S. interests. . . . DoD CTF counters financing used to engage in terrorist activities and illicit networks that traffic narcotics, WMD, [IEDs], other weapons, persons, precursor chemicals, and related activities that support an adversary's ability to negatively affect U.S. interests." This DoD definition should be distinguished from that of the Financial Action Task Force (FATF) definition of "AML/CFT." See FATF, "Glossary," 2015, www.fatf-gafi.org/pages/glossary/.

Drug Enforcement Administration (DEA) and four United States Attorneys' offices—and the Treasury Department, through the Office of Foreign Assets Control (OFAC) and the Financial Crimes Enforcement Network (FinCEN).

These lead agencies work in close cooperation with multiple other agencies—federal, local, and foreign. Many of the actions involved have been coordinated through interagency task forces, particularly the Justice Department's Special Operations Division, which is headed by DEA. Drug trafficking and related money laundering investigations by DEA have been the foundation for most of the government countermeasures against this web of illicit actors. Nevertheless, in the mosaic of investigations and actions that have been under way since 2007, coordination through a single, integrated operating center has not been possible or, indeed, necessary. Inevitably, command and control in this multidisciplinary exercise of authorities is imperfect. But the mixing of authorities has worked synergistically, and the interagency participants continue to learn from it.

The Tool Kit for Countering Illicit Finance

Nearly every federal criminal investigation now includes inquiry into possible money laundering and other finance-related offenses. And along with criminal prosecution, criminal or civil forfeiture of assets is another possible option. When a national security dimension involving foreign threats is added, economic sanctions and special anti-money laundering measures become major parts of the follow-the-money arsenal. To disrupt the financial structures that keep the transnational threat operating, four established authorities are dominant: *financial (economic) sanctions, Section 311 actions under the PATRIOT Act, criminal and civil forfeitures, and criminal prosecutions.*

Sanctions

Whenever foreign governments or nonstate foreign adversaries—which include transnational illicit networks—are found to present a threat to the U.S. national security, foreign policy, or economy, economic sanctions are often first in the arsenal of legal instruments used to attack these foreign adversaries and their financial resources.[3] Sanctions directed against nonstate actors, such as international drug cartels, foreign terrorist groups, and other transnational threat networks, are often termed "financial sanctions"—not only because they focus on the financing of the drug trade, terrorism, and other threat activities but also because they help protect the U.S. financial system from manipulation and misuse by illicit actors.[4]

Sanctions against Governments. The use of economic sanctions has grown as an instrument of national power directed against foreign governments and ruling regimes that have adversarial relationships with the United States. These include several regimes and governments that are subjects of this book's case studies. Indeed, since 1987, when sanc-

---

[3] Michael Miklaucic and Jacqueline Brewer, eds., *Convergence: Illicit Networks and National Security in the Age of Globalization* (Washington, DC: National Defense Univ. Press, 2013).

[4] See, for example, U.S. Treasury Dept., "Executive Orders," May 28, 2015, www.treasury.gov/resource-center/sanctions/Pages/eolinks.aspx.

tions were imposed against Panama under the Noriega regime, through March 2015, economic sanctions have been employed in international emergencies in more than 30 national security situations.[5] Some of these emergencies, most prominently Iran, involve multiple Executive Orders and statutes. In many instances, U.S. sanctions have been complemented by international sanctions involving, for example, the United Nations, the European Union, and the Organization of African Unity. This multilateralizing of sanctions has not been confined to the well-known international sanctions against al-Qaeda and Iran, and those related to Russia and Ukraine. For example, many in the international community joined in the sanctions against Qaddafi's Libya and Bashar al-Assad's government in Syria.

Sanctions against nonstate actors. Economic or financial sanctions against nonstate actors have gained wider recognition since the United States imposed sanctions against al-Qaeda and other foreign terrorist groups following the attacks of September 11, 2001.[6] But this innovative use of targeted sanctions against nonstate actors actually began in 1995, with two innovative Executive Orders: E.O. 12947, against terrorists threatening the Middle East peace process, and E.O. 12978, against drug cartels centered in Colombia.[7] Those two Executive Orders focused the power of economic sanctions through tailored Specially Designated Nationals (SDN) programs designed and implemented by the U.S. Treasury Department's Office of Foreign Assets Control (OFAC). OFAC is the lead office responsible for implementing economic and trade sanctions based on U.S. foreign policy and national security goals.

Sanctions directed against nonstate actors—often referred to in the post-9/11 world as "targeted sanctions"—are focused on particular individuals, companies, and groups of nonstate actors that commonly operate as illicit networks. These include the transnational threat networks discussed in this chapter. Targeted sanctions are also employed against specific entities and individuals involved with sanctioned foreign governments and illicit regimes, but the number and frequency of designations against nonstate actors such as transnational threat networks are much greater than designations related to governments and illicit regimes.

Imposition of sanctions depends heavily on the use of OFAC's Specially Designated Nationals List (SDN list). The SDN list further defines and identifies the individuals and entities that make up the sanctioned foreign government or sanctioned transnational threat network. In most programs, this includes entities and individuals that are controlled by the principal targets of the sanctions, act for them, or provide services or support. This is where the front companies, middlemen, facilitators, financiers, and penetrators of the legitimate economy are identified and exposed to the financial and commercial world and to the public at large. This is also the means through which sus-

---

[5] Ibid.

[6] E.O. 13224, *Blocking Property and Prohibiting Transactions with Persons Who Commit, Threaten to Commit, or Support Terrorism*, Sept. 23, 2001, 66 Fed. Reg. 49079, www.treasury.gov/resource-center/sanctions/Documents/13224.pdf.

[7] E.O. 12947, *Prohibiting Transactions with Terrorists Who Threaten to Disrupt the Middle East Peace Process*, Jan. 23, 1995, 60 Fed. Reg. 5079, www.treasury.gov/resource-center/sanctions/Documents/12947.pdf; E.O. 12978, *Blocking Assets and Prohibiting Transactions with Significant Narcotics Traffickers*, Oct. 21, 1995, 60 Fed. Reg. 54579, www.treasury.gov/resource-center/sanctions/Documents/12978.pdf.

tained and adaptable sanctions pressure is maintained on the threat network. Sanctions designations are now a key instrument in the set of financial countermeasures that the U.S. government employs against both established and emerging transnational threats. Following the U.S. lead, sanctions designations on a multilateral scale have developed significantly since the terrorist attacks of 9/11.

The most recent new program of nonstate sanctions against a transnational threat began in July 2011 with Executive Order 13581, issued to deal with transnational criminal organizations (TCOs).[8] That Executive Order was the lead-off measure directed by the national Strategy to Combat Transnational Organized Crime.[9] The sanctions imposed by E.O. 13581 were directed by the TOC strategy; and they address, in part, the strategy's policy objectives. The most specific policy objectives are worth restating in full because they reflect what may be viewed as a policy concept for dealing with the broader set of transnational threats:

"Break the economic power of transnational criminal networks and protect strategic markets and the U.S. financial system from TOC penetration and abuse."[10]

"Defeat transnational criminal networks that pose the greatest threat to national security, by targeting their infrastructures, depriving them of their enabling means, and preventing the criminal facilitation of terrorist activities."[11]

The use of sanctions, particularly targeted sanctions designations, against nonstate actors was introduced in 1995, and those innovative early Executive Orders had objectives much like those of the TOC E.O. They are the foundation on which the subsequent sanctions programs against nonstate adversaries—including the Foreign Narcotics Kingpin Designation Act ("Kingpin Act"), which has been center stage in the actions against the Joumaa Web—have been modeled.

Although the policy framework of the TOC Executive Order helps us understand the use of economic sanctions against the Joumaa Web, this particular E.O. has not been one of the sanctions tools employed against the multiple networks involved. The sanctions tools used against the Joumaa Web—more than 180 designations under the Kingpin Act and one under E.O. 13224 (on global terrorists)—already existed and had been used many times before the TOC Executive Order was issued. Indeed, at the time of this writing, more than 1,700 Kingpin Act designations have occurred since June 2000.

Seizing and freezing. A misconception with economic sanctions is that they are principally concerned with, and measured by, the value of the assets that are blocked (frozen). Although blocking funds and other property and, thus, actually immobilizing or freezing economic wealth is a powerful intended consequence of sanctions, it is not the

---

[8] E.O. 13581, *Blocking Property of Transnational Criminal Organizations*, July 24, 2011, 76 Fed. Reg. 44757, www.treasury.gov/resource-center/sanctions/Documents/13581.pdf.

[9] White House, "Strategy to Combat Transnational Organized Crime: Addressing Converging Threats to National Security," July 2011, www.whitehouse.gov/sites/default/files/Strategy_to_Combat_Transnational_Organized_Crime_July_2011.pdf. See especially key policy objective number 3, "Strategy," 20.

[10] Ibid., 14.

[11] Ibid.

principal instrument of impact. More often, it is merely part of a larger strategy of *denial and disruption.*

For example, on February 25, 2011, President Obama signed Executive Order 13566, which froze the assets of Colonel Muammar Qaddafi, his family members, and the government-owned enterprises that he was using to fund his regime. According to the *Washington Post,* banks around the world froze over $30 billion of Qaddafi's assets within 72 hours of the signing of the Executive Order. Thus began multiple rounds of sanctions on the Libyan government, by which the international community achieved several objectives. First, the international freeze of Qaddafi's assets held abroad deprived him of most of his foreign currency reserves. Meanwhile, targeted sanctions on the Libyan oil sector made earnings from exports impossible. Thus deprived of his foreign holdings and prevented from earning new income, Qaddafi had trouble buying tribal support or hiring mercenaries. Also, targeted sanctions on the energy sector and Libyan ports prevented him from importing fuel for his trucks and tanks. Thus, the freezing of his foreign assets was a key link in the combined U.S., EU, and UN sanctions that helped starve Qaddafi of the funds and fuel he needed to stay in power. That was a major success for sanctions, but only a tactical success in the geopolitical sense. Most obviously, achieving those U.S. and international sanctions objectives has not led to a desired geopolitical end state of a stable, nonrepressive, and nonthreatening Libyan government. The case of Libya demonstrates that sanctions and their targeted economic effects are no substitute for a comprehensive, whole-of-government, preferably multilateral national security-foreign policy strategy. This means planning well beyond sanctions.

Denial and disruption. Denial and disruption, against nonstate adversaries especially but also against hostile regimes and governments, are essential levers in achieving the goals of degrading, delegitimizing, and incapacitating the target. This is particularly important in sanctions programs targeting nonstate actors, because the forms and expectations of the leverage gained through sanctions are different from those with foreign state/regime sanctions.

Sanctions imposed against a foreign state or ruling regime are only rarely undertaken with the express goal of regime change. (Those levied against the Qaddafi regime in Libya illustrate a rare and, unfortunately, tragic instance of sanctions being used successfully to help weaken a government enough that it toppled. The goals of having a stable, nonrepressive, and nonthreatening Libyan government and a largely intact country failed even though the sanctions were a tactical success.) More commonly, sanctions against governments/regimes are intended to achieve leverage that leads to a desirable change in behavior (e.g., abandoning a nuclear weapons program) or a cessation in an activity (e.g., stopping human rights abuses, eliminating terrorist safe havens).

The results are not always successful, however, even when many governments join in. For example, as of this writing, Bashar al-Assad continues the war on his own people despite sanctions against the Syrian government, and strong international sanctions have not yet persuaded Iran's supreme leader to shutter that nation's nuclear program. Although a deal with Iran was reached in July 2015, it remains to be seen whether or how Iran will follow through on its reluctant commitments. These are just two of the most recent examples of sanctions failing to persuade bad actors to give up their most cherished goals.

In many cases, multiple changes in behavior or cessations of activities are the desired end state. Even though the international and domestic political dynamics are more complex for foreign states and ruling regimes than for nonstate actors, concessions leading to the elimination or reduction of sanctions may be a more realistic possible outcome with state actors than nonstate actors.

Cessation of the "casus belli" is much less likely to be relevant to criminals, extremists, their financiers and facilitators, and other participants in their networks than to a governing regime. The hybrid criminal-extremist-facilitator networks and their key individual players (for the purposes of this discussion, *transnational threat networks*) are more likely to reduce or cease illicit activity or change bad behavior because they have been degraded and incapacitated than because they have been persuaded through the leverage brought by sanctions. Transnational threat networks and their specific participants are a new and malignant manifestation of an increasingly interconnected and mobile world. But most of the threat elements are still common criminal activities by traditional bad actors, albeit with greater scope, complexity, and sophistication than before. Consequently, the realistic desired end state in cases of sanctions against transnational threat networks is not to bring the bad actors into alignment with international norms of nation-state behavior. Rather, we seek a combination of complementary objectives against some of the targeted networks: deter, contain, or minimize the threat; disrupt, degrade, and delegitimize their financial and commercial backbones; impede, halt, and reverse their penetration of the legitimate economy. Ultimately, the optimal result is to incapacitate the transnational threat network. The more probable result is to *mitigate the severity* of the threat.

Sanctions are a powerful mechanism for combating nonstate threats and their evasive schemes for illicit finance. But sanctions alone are not enough. While they are particularly appropriate (and perhaps uniquely adaptable) in the transnational threat environment, it is their synergistic use with other tools—and not exclusively financial tools—that provides the best opportunities for effective countermeasures against transnational threat networks. By looking at the other follow-the-money tools, both new and old, we will see how an evolving whole-of-government approach can attack transnational threat networks with "official networks" instead of in the isolation of bureaucratic silos.

Section 311 Actions

In the aftermath of the al-Qaeda terrorist attacks of 9/11, Congress enacted the USA PATRIOT Act.[12] Among the act's many provisions is Section 311.[13] Section 311 (31 U.S.C. 5318A) gives the treasury secretary authority to find that a foreign jurisdiction, financial institution, class of transaction, or type of account is "of primary money laundering concern."

The authority for Section 311 has been delegated to the director of Treasury's Financial Crimes Enforcement Network ("FinCEN"). Under Section 311, domestic financial institutions and financial agencies (e.g., banks) can be required to take "special measures" against an entity identified by Treasury as being of primary money laundering

---

[12] USA PATRIOT Act, Pub. L., Oct. 26, 2001, 107-56.

[13] USA PATRIOT Act, Section 311, 31 U.S.C. section 5318A, www.sec.gov/about/offices/ocie/aml/patriotact2001.pdf.

concern. The range of options can be adapted to protect the U.S. financial system from specific money laundering and terrorist financing risks.

The most potent of the special measures is Special Measure 5: prohibitions on the opening or maintenance of any correspondent or payable-through accounts of the entity or jurisdiction identified as "of primary money laundering concern."[14] According to the FFIEC BSA/AML Examination Manual, "a 'correspondent account' is an account established by a [U.S.] bank for a foreign bank to receive deposits from, or to make payments or other disbursements on behalf of the foreign bank, or to handle other financial transactions related to the foreign bank." These transactions generally are done on behalf of the bank's customers, thus enabling international transactions. Payable-through accounts (PTAs) are distinct from correspondent accounts because a PTA permits the foreign bank's customers to access funds in the U.S. bank's account of the foreign bank. With a PTA, the individual customers are basically subaccount holders with direct access to the bank. For a foreign financial institution, this determination can be a catastrophic disruption of its ability to operate, as well as a reputational disaster.

It is important to understand that the "special measures" of Section 311 do not involve directly either the freezing of assets, such as could occur with an OFAC designation, or the seizing and forfeiting of assets by the Justice Department (discussed below). They do, however, lay the groundwork for the Justice Department and U.S. Attorneys' offices to undertake civil forfeiture proceedings and, in some cases, arrive at forfeiture settlements, as has occurred with LCB and other financial institutions involved in the Joumaa Web.

Since the inception of Section 311, "special measures" have been taken against four foreign jurisdictions and 19 foreign financial institutions. Five of those 311 actions against financial institutions are currently active, and of those, three are associated with the Joumaa Web. One of Treasury's more recent Section 311 actions was also the first use of the authority against nonbank financial institutions: Rmeiti Exchange and Halawi Exchange. Both were based in Lebanon, and both facilitated money laundering by Hezbollah and by Colombian and other drug trafficking and money laundering networks. We discuss these two entities and their networks in greater depth below. And using them, we can explore the synergistic relationship between Section 311, sanctions designations, and more traditional law enforcement-centered follow-the-money authorities.

Seizure and Forfeiture

Among nonsanction financial countermeasures, the most prominent are those that have been used the longest: criminal investigations and prosecutions, and their companions, seizure and forfeiture. Indictments for money laundering activities became increasingly important over the latter half of the twentieth century. It is important to bear in mind that these money laundering cases commonly involve other traditional substantive criminal charges, which constitute the predicate offenses (Specified Unlawful Activities, or SUAs) that are necessary for the money laundering counts in the indictments.[15]

---

[14] Ibid., 117; Bank Secrecy Act/Anti-Money Laundering Examination Manual, Federal Financial Institutions Examination Council (FFIEC), Apr. 29, 2010, 1830185, 198-200.

[15] 18 U.S.C. section 1956(c) (7).

The forfeiture authorities under federal law are key instruments in the tool kit, and while this is not a primer in asset forfeiture, some basics are worth mentioning. Federal forfeiture can be criminal or noncriminal. Criminal forfeiture occurs in the context of a federal criminal prosecution of a defendant. The property used or derived from the alleged crime (e.g., bank accounts, securities, real estate, boats, automobiles) is charged in the indictment along with the individuals or entities who are the defendants. It is a decision for the jury to find the property forfeitable or not. An affirmative finding by the jury results in the court issuing an order of forfeiture in addition to the sentences for convicted defendants.

There are also two noncriminal forfeiture processes. One is civil judicial forfeiture. This is an action brought against the property itself and adjudicated in court or through settlement. The owner of the property need not be charged with a crime. The other is administrative forfeiture against the property itself and does not involve the courts. There is separate, specific statutory authority for administrative forfeiture; and it includes monetary instruments.[16] When seizure and forfeiture proceedings are completed successfully for the federal government, title to the forfeited assets (e.g., cash, bank accounts, securities, real property, vehicles) passes to the U.S. government.

Criminal Prosecution

As seen in the cases described below, the main goal from the criminal justice system perspective is for the criminal investigations to result in indictments, arrests, and criminal prosecutions and convictions if the evidence developed is sufficient.[17] But that does not exclude cooperation where "criminal, civil, regulatory, and/or agency administrative parallel . . . proceedings" may also arise.[18]

The criminal investigations of the Joumaa Web have led to indictments of at least three different offenders for drug trafficking and money laundering. But while those actions are fundamental aspects of the U.S. criminal justice system and integral to confronting global illicit actors, they are unlikely, by themselves, to project U.S. power against foreign threat networks that reach around the world. Nor are they sufficient to safeguard the U.S. financial system and channels of global commerce from such networks. The inherent limitations on jurisdictional reach, on the types of evidence and nature of the defendants, on capture and extradition, and on the transnational nature of threat finance mean that other measures are also needed to bring a unity of effort against actors whose pursuits not only are criminal but also present a threat to U.S. national security.

---

[16] U.S. Justice Dept., "Types of Federal Forfeiture," Mar. 9, 2015, www.justice.gov/afp/types-federal-forfeiture.

[17] U.S. Justice Dept., U.S. Attorneys Organization and Functions Manual, "27. Coordination of Parallel Criminal, Civil, Regulatory, and Administrative Proceedings," Memorandum of Jan. 30, 2012, www.justice.gov/usao/eousa/foia_reading_room/usam/title1/doj00027.htm.

[18] Ibid.

Important Distinctions between CTF Measures

It is important to recognize the major distinctions between these instruments of counter threat finance. The traditional law enforcement approach may involve arresting suspects, seizing assets and contraband, putting individual violators in prison, and causing violators' assets to be forfeited to the United States. These processes occur daily as a core aspect of our criminal justice system and are a major impediment to those criminals and networks that fit into the broader realm of transnational threats to our national security and foreign policy.

In the case of blocking (or freezing) assets under sanctions programs, title to the blocked assets does not pass to the U.S. government; it remains with the owner. But *control* of the blocked assets resides with the government. Their use is denied to the target of the sanctions as well as to all others. Beyond that denial of specific blocked assets is the broader impact of the exposure and isolation of the sanctioned parties. This includes denial of access to the U.S. financial system, and disruption, degradation, and delegitimization of the designated entities and individuals.

Although Section 311 actions require an administrative process to find that their target is "of primary money laundering concern," they have the benefit of not requiring a prior declaration of a national emergency, as is necessary with economic sanctions programs enforced by OFAC. It is important to understand that the "special measures" of Section 311 do not involve directly either the freezing of assets, such as would likely occur with an OFAC designation, or the seizing and forfeiting of assets by the Justice Department. But as in the case of the Lebanese Canadian Bank (LCB, discussed in more detail below), the use of Section 311 can ultimately result in a substantial civil forfeiture as well as termination of the entity.

CTF "Follow the Money" Tools in Action

How do CTF operations work in practice? What does a successful, synergistic operation look like? Who is involved, what resources are required, and how long does it take? The answers to these questions vary widely from case to case, but closer examination of the Joumaa Web operation is instructive and provides a healthy dose of reality as we strive to confront impunity within an increasingly internationalized system of banking, finance, and transnational crime.

The following timeline for the Joumaa Web operation does not begin with the DEA investigations that started in 2007. But the DEA operations are relevant and must be kept in mind when thinking about planning horizons from an operational perspective. For the purposes of this study, we can begin with the 18 separate events that started in January 2011 as a set of Kingpin designations, against a single drug trafficking and money laundering organization that was believed to be associated with Hezbollah. Nearly four years later, a multiplicity of discrete actions—investigations, sanctions designations, special measures, indictments, and forfeiture activities—had collectively exposed the Joumaa Web, a complex network of illicit organizations and individuals reaching around the globe. By July 2014, the known linkages included other crime-terror hybrids, drug traffickers, money launderers, and terror financiers in more than 25 countries. This network encompassed everything from illegal drug, arms, and human trafficking to money laundering and an ostensibly licit traffic in used cars.

Setting the Stage

To understand the events as they unfold, it is worth taking a moment to briefly examine common characteristics of the web of organizations, entities, and individuals that form this vast, intricate transnational threat network. The most publicized view centers the web on one entity, the Lebanese Canadian Bank, and LCB's involvement with Hezbollah. An alternative view centers it on one individual: Ayman Joumaa. Given the complex relationships of the participants, the scope and diversity of their activities, and the routes and geography involved, neither viewpoint is entirely accurate. We are looking at a web of connections with multiple elements, including Lebanon, South America, Africa, used cars, drug cartels, money laundering, players with dual nationalities, hundreds of millions of dollars in bulk cash transfers and structured wire transactions, and, woven throughout, Hezbollah.

Chronology of Follow-the-Money Actions

An actual blow-by-blow account of legal measures taken against the Joumaa Web will paint a clearer picture of how whole-of-government coordination works to disrupt, degrade, and delegitimize a transnational threat.

January 2011. Kingpin Act designation action. The first public action against the Joumaa drug trafficking and money laundering organization was OFAC's designation of Colombian/Lebanese national Ayman Saied Joumaa, his organization, nine other individuals, and 19 entities as specially designated narcotics traffickers (SDNTs) under the Kingpin Act.[19] One immediate effect of this set of designations was the blocking (asset freezing) of more than 300 used cars awaiting export from the United States.[20] Interagency collaboration and coordination were publicly underscored by the Drug Enforcement Administration's announcement that OFAC's designation of the Joumaa network was based on "an ongoing DEA investigation."[21]

February 2011. PATRIOT Act 311 action. Two weeks after the Joumaa network designation, the Treasury Department's Financial Crimes Enforcement Network (FinCEN) announced the identification, under Section 311 of the PATRIOT Act, of LCB as a financial institution of primary money laundering concern. Treasury's press announcement and the FinCEN report in support of its 311 action linked LCB to the Joumaa DT/MLO and to the global terrorist organization Hezbollah. DEA's administrator spoke out in

---

[19] U.S. Treasury Dept., "Treasury Targets Major Lebanese-Based Drug Trafficking and Money Laundering Network."

[20] U.S. v. LCB, et al., Verified Amended Complaint, 11 Civ. 9186 (PAE), SDNY, Oct. 26, 2012, para. 5, and Schedule A, para. V.

[21] DEA, "DEA Investigation Leads to OFAC Designation of Lebanese-Based Drug Trafficking and Money Laundering Network."

support of the 311 action, as she had done earlier with OFAC's designation of the Joumaa network, further underscoring Hezbollah's connection to LCB, Joumaa, drug trafficking, and money laundering.[22]

November 2011. Drug trafficking and money laundering indictment of Joumaa. The interagency actions against Ayman Joumaa and his organization expanded when the U.S. Attorney in Alexandria, Virginia, indicted Joumaa on drug trafficking and money laundering charges.[23] This time, the connections were to Colombian drug traffickers and to the brutally violent Mexican drug trafficking organization los Zetas. President Obama had previously added the Zetas to the Kingpin Act list in April 2009, and on July 24, 2011, the group had been identified as a TCO in the Annex to Executive Order 13581 on Transnational Organized Crime.[24]

December 2011. OFAC Kingpin Act designations of the Cheaitelly-Khansa DT/MLO. The second related OFAC designation action involving networking organizations identified Jorge Fadlallah Cheaitelly and Mohamad Zouheir El Khansa, along with nine other individuals and 28 entities involved in an international drug trafficking and money laundering network based in Colombia and Panama, with a global span and major connections in Lebanon and Hong Kong.[25]

December 2011. Forfeiture and civil money laundering complaint. On December 15, 2011, the U.S. Attorney for the Southern District of New York filed a forfeiture and civil money laundering complaint seeking nearly $450 million in civil money laundering penalties against LCB and six other entities that had been designated by OFAC or subjected to the 311 action by FinCEN, and the forfeiture of "property in or traceable to the money laundering offenses" of the LCB group and thirty used-car purchasers involved with the Joumaa-LCB networks.[26]

June 2012. More OFAC designations under the Kingpin Act and the E.O. on terrorism: the Harb Network. In June 2012, OFAC designated four individuals and three entities under the Kingpin Act, and an additional connected individual under the terrorism Executive Order. The "Harb Network" was a Colombia- and Venezuela-based organization that laundered money for the Joumaa network through the Lebanese financial sector. The designations expanded the maze of connections to Joumaa and Hezbollah and high-

---

[22] U.S. Treasury Dept., "Treasury Identifies Lebanese Canadian Bank Sal as a 'Primary Money Laundering Concern,'" Treasury Press Center, Feb. 10, 2011, www.treasury.gov/press-center/press-releases/Pages/tg1057.aspx. Note that the Treasury press release contains links to FinCEN's contemporaneous finding against LCB and to its Notice of Proposed Rule Making.

[23] U.S. Justice Dept., "U.S. Charges Alleged Lebanese Drug Kingpin with Laundering Drug Proceeds for Mexican and Colombian Drug Cartels," press release, U.S. Attorney's Office, Eastern District of Virginia (VAED), Dec. 13, 2011; U.S. v. Ayman Joumaa, U.S. District Court (EDVA), Nov. 23, 2011, www.vaed.uscourts.gov.

[24] EO 13581.

[25] U.S. Treasury Dept., "Treasury Targets Key Panama-based Money Laundering Operation."

[26] *U.S. v. LCB, et al.*, Forfeiture and Civil Money Laundering Complaint and Post-Complaint Restraining Order, Case 1:11-cv-09186-RJH, Docs. 1 and 2, SDNY, Dec. 15, 2011.

lighted the international linkages between illicit actors in Colombia, Venezuela, and Lebanon.[27]

October 2012. Amended forfeiture and civil money laundering complaint. This complaint was filed in the Southern District of New York, against LCB, the other six entities named in the original complaint, and four used-car buyers.[28] In the ongoing process of the government perfecting its evidence against LCB and the other persons involved, the amended forfeiture complaint was a crucial step toward an eventual June 2013 settlement agreement with LCB, and forfeiture.

February 2013. OFAC Kingpin Act designation of José Linares Castillo. This Colombian cocaine boss had ties to the narco-terrorist group FARC. In naming Linares Castillo under the Kingpin Act, Treasury unraveled another strand in the drug trafficking and money laundering web.[29]

February 2013. Drug trafficking and money laundering indictment of Rodriguez Vasquez associates. On February 14, 2013, the U.S. Attorney for the Eastern District of Texas indicted 17 individuals on drug trafficking and money laundering charges. The indictment was unsealed on August 5, 2013, and Colombian authorities arrested several of those named. In February 2014, OFAC designated seven of the accused under authority of the Kingpin Act. The sanctions designation action would further extend the reach of U.S. authorities against these foreign drug traffickers and money launderers.[30]

April 2013. Another Section 311 action by FinCEN. The pressure exerted against the Joumaa Web and its interconnections grew when FinCEN named two Lebanese exchange houses, Kassem Rmeiti & Co. for Exchange and Halawi Exchange Co., as foreign financial institutions of primary money laundering concern. The 311 action against the two exchange houses was a countermeasure to their role in filling the gap for the Joumaa network's money laundering after the 311 action against LCB. This was the first use of Section 311 against a nonbank financial institution.[31]

---

[27] U.S. Treasury Dept., "Treasury Targets Major Money Laundering Network Linked to Drug Trafficker Ayman Joumaa and a Key Hizballah Supporter in South America," Treasury Press Center, June 27, 2012, www.treasury.gov/press-center/press-releases/Pages/tg1624.aspx.

[28] *U.S. v. LCB, et al.*, Verified Amended Complaint 11 Civ. 9186 (PAE), USDC SDNY, Oct. 26, 2012.

[29] U.S. Treasury Dept., "Treasury Designates Head of Aviation Drug Smuggling Operation: Action Targets Drug Trafficking Organization Tied to the FARC," Treasury Press Center, Feb. 20, 2013, www.treasury.gov/press-center/press-releases/Pages/tg1857.aspx.

[30] U.S. Justice Dept., "Seventeen Colombians Indicted in Eastern District of Texas Drug Conspiracy," press release, Aug. 26, 2013, www.justice.gov/usao/txe/News/2013/edtx-colombia-vasquez082613.html; U.S. Treasury Dept., "Treasury Designates Colombian Narcotics Trafficker," Treasury Press Center, Feb. 19, 2014, www.treasury.gov/press-center/press-releases/Pages/jl2295.aspx; *U.S. v. Fernain Rodriguez-Vasquez, et al.*, No. 4:13-cr-00038-MAC-ALM, Doc. 38, filed 2/27/13, U.S. District Court, Eastern District of Texas.

[31] U.S. Treasury Dept., "Treasury Identifies Kassem Rmeiti & Co. for Exchange and Halawi Exchange Co. as Financial Institutions of 'Primary Money Laundering Concern,'" www.treasury.gov/press-center/press-releases/Pages/jl1908.aspx; U.S. Treasury Dept., "Notice of Finding that Kassem Rmeiti & Co. for Exchange Is a Financial Institution of Primary Money Laundering Concern"; U.S. Treasury Dept., "Notice of Finding that Halawi Exchange Co. Is a Financial Institution of Primary Money Laundering Concern."

June 2013. $102 million settlement. On June 25, 2013, one of the paths of CTF actions against the Joumaa Web's drug and money laundering connections achieved a much-desired outcome when a settlement was reached requiring LCB to forfeit $102 million to the U.S. government. The U.S. Attorney for Manhattan (SDNY) praised DEA's New York OCDETF, a multiagency unit, for its "outstanding work" on the investigation.[32]

July 2013. Double Actions: OFAC designates, and the U.S. Attorney in Miami indicts, persons in another money laundering network connected to Joumaa. OFAC made 31 more designations under the Kingpin Act. It named nine individuals and 22 entities involved in yet another network (Guberek Network) of international drug-money launderers with connections to Ayman Joumaa and José Linares Castillo.[33] That same day, the U.S. Attorney for the Southern District of Florida indicted four key individuals in the network for their alleged participation in an international money laundering conspiracy investigated by the DEA.[34] The government actions against the Joumaa Web and its visible dimensions continued to expand.

October 2013. OFAC designates Spain-based associates of the Guberek Network. The CTF net affecting the Joumaa Web expanded again when OFAC used the Kingpin Act to designate two individuals (spouses) and the five entities they controlled in Spain and Peru, for their role as a "major node in the Guberek money laundering network, which provides a significant pipeline for illicit narcotics proceeds to flow from Europe to Colombia."[35]

February 2014. OFAC designates a Colombia-based drug trafficker and his network. Actions against participants in the Joumaa Web expanded again with the February 2014 designation of a significant Colombian drug trafficker, seven associates, and five entities. This group, the Fernain Rodriguez Vasquez network, was identified with connections to

---

[32] U.S. Justice Dept., "Manhattan U.S. Attorney Announces $102 Million Settlement of Civil Forfeiture and Money Laundering Claims Against Lebanese Canadian Bank," SDNY press release, June 25, 2013, www.justice.gov/usao/nys/pressreleases/June13/LCBSettlementPR.php; *U.S. v. Lebanese Canadian Bank SAL, et al., All Assets of Lebanese Canadian Bank SAL or Assets Traceable Thereto, et al.,* Stipulation and Order of Settlement, 11 Civ. 9186 (PAE), SDNY, June 25, 2013.

[33] U.S. Treasury Dept., "Treasury Targets Major Money Laundering Network Operating out of Colombia," Treasury Press Center, July 9, 2013, www.treasury.gov/press-center/press-releases/Pages/jl2002.aspx.

[34] U.S. Justice Dept., "Four Colombian Nationals Charged in International Drug Money Laundering Conspiracy," press release, July 9, 2013, www.justice.gov/usao/fls/PressReleases/130709-01.html; *U.S. v. Solorzano-Lozano, Grimberg Ravinovcz, Guberek-Grimberg and Ceballos-Bueno,* 1:13-cr-20497-MGC U.S. District Court, FLSD, July 9, 2013.

[35] According to the Treasury Department, Isaac Perez Guberek Ravinovicz, a Colombian national, and his son, Henry Guberek Grimberg, a dual Colombian and Israeli national, led a money laundering network based in Bogotá, Colombia, that laundered narcotics proceeds for numerous drug trafficking organizations, including organizations based in Colombia. When announcing the Kingpin designation against the Gubarek Grimburg network, Treasury officials specifically noted that Ayman Joumaa and José Evaristo Linares Castillo, who was designated in February 2013, were known to have laundered their drug proceeds through this money laundering network. U.S. Treasury Dept., "Treasury Targets Major Money Laundering Network Operating Out of Colombia"; U.S. Treasury Dept., "Treasury Targets Spanish Cell of Guberek Money Laundering Network," Treasury Press Center, Oct. 29, 2013, www.treasury.gov/press-center/press-releases/Pages/jl2193.aspx.

the Lebanese-Colombian Cheaitelly network (itself designated on December 29, 2011), the Colombian narco-terrorist group FARC, Mexico's los Zetas DTO, and the Sinaloa Cartel.[36] The designation further exposed the expansive transnational connections and illicit activities among the actors in the Joumaa Web, and the U.S. government was able to impose sanctions that would cut the foreign nonstate actors off from transactions and commerce with the United States.

May 2014. OFAC Kingpin Act designation. Panama connection of the Cheaitelly DT/MLO. Treasury's OFAC designated eight more individuals and 20 entities, centered in Panama and operating internationally, tied to the Cheaitelly DT/MLO designated in December 2011.[37]

June 2014. OFAC Kingpin Act designation: la Oficina de Envigado. Another link in the web of networks appeared with the designation of the Medellín-based DTO la Oficina de Envigado. La Oficina was cited for its recent support to the Sinaloa Cartel, and its interconnection with Colombia's AUC narco-terrorist organization and AUC's rural paramilitary successors.[38] The link to the Joumaa Web became more apparent when OFAC designated the Mejía Salazar DMLO, which had connections with la Oficina and Ayman Joumaa's network.

July 2014. OFAC Kingpin Act designation: Medellín-based Mejía Salazar drug money laundering organization. Treasury's OFAC highlighted the close money laundering connections between the Mejía Salazar DMLO, la Oficina, and Ayman Joumaa's international networks.[39]

Analysis of the Joumaa Web Operations

As of the date of this case study, the web of organizations, entities, and individuals that form this vast, intricate transnational threat network can be summarized in the following table:

---

[36] U.S. Treasury Dept., "Treasury Designates Colombian Narcotics Trafficker – Action Targets Individuals and Entities Tied to the FARC and Mexican Cartels," Treasury Press Center, Feb. 19, 2014, www.treasury.gov/press-center/press-releases/Pages/jl2295.aspx.

[37] U.S. Treasury Dept., "Treasury Targets Major Money Laundering Network Operation Based in Panama," Treasury Press Center, May 14, 2014, www.treasury.gov/press-center/press-releases/Pages/jl2397.aspx.

[38] U.S. Treasury Dept., "Treasury Designates Colombian Organized Crime Group La Oficina de Envigado."

[39] U.S. Treasury Dept., "Treasury Designates a Medellin, Colombia-based Drug Money Laundering Network."

Essential Elements of the Joumaa Web

- Lebanese banks, money exchange houses, and money launderers
- African banking and commercial subsidiaries
- drug and money laundering networks centered in South and Central America
- drugs moving from South America to West Africa and on to Europe and the Middle East
- close family members in banks and drug trafficking / money laundering organizations
- dual-nationality participants, including several members of the Lebanese diaspora
- trade-based money laundering (TBML) involving used cars and Asian commercial goods
- bulk cash transfers and structured electronic wire transfers
- multiple U.S.-based and West African car dealerships
- regular transfers involving hundreds of millions of dollars
- Hezbollah
- other groups with a terror-crime nexus

This far-flung array of terrorist and other criminal enterprises included not only the Ayman Joumaa drug trafficking/money laundering organization (DT/MLO), but also the LCB, five Lebanese money exchange houses, and the terrorist organization Hezbollah. Also identified and exposed during this time frame were seven other drug trafficking and money laundering networks, the Mexican drug trafficking organizations los Zetas (also classified as a TCO) and the Sinaloa Cartel, the Colombian narco-terrorist organization FARC, the Colombian drug trafficking crime group known as la Oficina de Envigado (including remnants of the narco-terrorist Autodefensas Unidas de Colombia [AUC]), and the Medellín drug money laundering network of Pedro Claver Mejía Salazar.[40]

To expose these elements and degrade their capabilities, federal government agencies had to employ legal mechanisms from the full range of CTF tools. Specifically, the federal government took the following actions against the Joumaa Web:

1. Ten separate but interrelated OFAC designation actions under the Kingpin Act, against eight DT/MLOs, resulting in 183 designations (117 entities and 66 individuals);

---

[40] U.S. Treasury Dept., "Treasury Designates Colombian Organized Crime Group La Oficina de Envigado for Role in International Narcotics Trafficking," June 26, 2014, www.treasury.gov/press-center/press-releases/Pages/jl2441.aspx; U.S. Treasury Dept., "Treasury Designates a Medellin, Colombia-based Drug Money Laundering Network with Ties to La Oficina de Envigado and Ayman Saied Joumaa," Treasury Press Center, July I, 2014. www.treasury.gov/press-center/press-releases/Pages/tg1035.aspx.

2. Two separate but interconnected PATRIOT Act Section 311 determinations by FinCEN that three Lebanon-based financial institutions and their subsidiaries are of primary money laundering concern;

3. Three separate federal indictments of individual leaders of two of the designated DT/MLOs (Ayman Joumaa, indicted in the Eastern District of Virginia; four key figures of the Guberek DMLO, indicted in the Southern District of Florida; and drug trafficking and money laundering associates of the Rodriguez Vasquez organization, indicted in the Eastern District of Texas);[41]

4. The $102 million settlement, in Manhattan, of the SDNY's civil forfeiture-money laundering complaint against LCB, and the $720,000 forfeiture-money by one of the Lebanese money exchange houses (Ayash Exchange) designated by OFAC as part of the Joumaa DT/MLO;

5. Other Hezbollah network connections exposed in 15 additional OFAC designation actions, most of them under the global terrorism Executive Order, and one additional 311 action against a Lebanese-originated bank located in Cyprus.

Federal agencies also relied on the traditional law enforcement tools that were being used by domestic and international law enforcement partners throughout the investigation, in multiple jurisdictions at home and abroad. How they did so, and the lessons learned from these events, bears further analysis.

2011: from designation through indictment. The series of government actions that occurred from January to December 2011 had a major impact on the Joumaa Web and the U.S. government's ability to degrade the network. Following OFAC's designation of Ayman Joumaa under the Kingpin Act, sanctions included the designations of nine other individuals, including three of Joumaa's brothers, and 19 entities located in Lebanon, Panama, Colombia, Benin, and the Republic of the Congo. All were alleged to be connected to Joumaa's DT/MLO.[42]

FinCEN found LCB to be a financial institution of primary money laundering concern because Joumaa's network was moving illegal drugs from South America to Europe and the Middle East via West Africa and laundering hundreds of millions of dollars monthly through accounts held at LCB. The activities involved trade-based money laundering of consumer goods throughout the world and included multiple used-car dealerships in the United States and West Africa. The finding for the Patriot Act 311 action in February 2011 also exposed the terrorist organization Hezbollah's links to LCB and Joumaa and to the international narcotics trafficking and money laundering network that LCB

---

[41] A third federal indictment occurred in the Eastern District of Texas on February 14, 2013, and was unsealed and announced in August 2013. U.S. Justice Dept., "Seventeen Colombians Indicted in Eastern District of Texas Drug Conspiracy," U.S. Attorney's Office, Eastern District of Texas, press release, Aug. 26, 2013, www.justice.gov/usao/txe/News/2013/edtx-colombia-vasquez082613.html.

[42] U.S. Treasury Dept., "Treasury Targets Major Lebanese-Based Drug Trafficking and Money Laundering Network," Treasury Press Center, Jan. 26, 2011, www.treasury.gov/press-center/press-releases/Pages/tg1035.aspx.

was facilitating. At the time of the OFAC designation action, Joumaa's network was described as a "complex money laundering scheme moving hundreds of millions of dollars of illicitly derived proceeds through businesses operated by him and his associates."[43] DEA's administrator remarked, "These are not legitimate businesses. These are illegal enterprises that fuel the drug trade and its violence and corruption. As we continue to follow the money trail, we starve these traffickers of their assets and eventually put their criminal networks out of business."[44]

In terms of countermeasures against Joumaa's network, the designations of the other individuals and entities were as significant as Joumaa's own designation because they exposed these individuals and entities publicly, disrupted their ability to conduct transactions involving the U.S. financial system, and cut them off from trade with persons subject to U.S. jurisdiction. This further constrained the ability of Joumaa's network to engage in money laundering and drug trafficking and complicated its connections with others in the Joumaa Web.

Expanded operations as the Joumaa Web is exposed. Joumaa's indictment in November 2011 was a pivotal moment, and events unfolded rapidly thereafter. OFAC continued its designations. A series of nine additional designations occurred during the next three and a half years. The first of those, occurring in December 2011, was against the international DT/MLO of Jorge Fadlallah Cheaitelly ("JFC") and Mohamad Zouheir El Khansa. The Treasury Department described their network as a "Panama-based money laundering operation linked to Mexican and Colombian drug cartels." It was also linked to Lebanon and Hong Kong. That designation action covered JFC and El Khansa along with nine other individuals, including three of JFC's siblings. Among the other individuals designated under the Kingpin Act was Ali Mohamad Saleh of Colombia, who appeared again in a subsequent designation further linking Hezbollah to the Joumaa Web. Twenty-eight businesses of the JFC network were designated. Seventeen—at least three of them money exchange businesses—were in Panama. Among them was a Cheaitelly-controlled business in the Colon Free Zone, linked to Ayman Joumaa and his brothers.[45] Among all the designation actions eventually taken against the Joumaa Web, this was the largest group of entities to be designated at one time.

Meanwhile, shortly after Joumaa was indicted in the Eastern District of Virginia, and nearly simultaneously with the first JFC network designations, the U.S. Attorney in Manhattan filed the civil forfeiture and money laundering complaint against LCB and others involved in Joumaa's Hezbollah-connected DT/MLO. The federal complaint also included the two Lebanese money exchange businesses Ellissa Holding and the Ayash Exchange, along with other front companies that OFAC had designated as part of Joumaa's network. At that point in the stream of ongoing CTF actions, all the follow-the-money tools were in play. The June 2013 settlement of the civil forfeiture case against

---

[43] Ibid.

[44] DEA, "DEA Investigation Leads to OFAC Designation of Lebanese-based Drug Trafficking and Money Laundering Network," DEA Public Affairs, Jan. 26, 2011, www.justice.gov/dea/divisions/hq/2011/hq012611.shtml.

[45] U.S. Treasury Dept., "Treasury Targets Key Panama-based Money Laundering Operation Linked to Mexican and Colombian Drug Cartels," Treasury Press Center, Dec. 29, 2011, www.treasury.gov/press-center/press-releases/Pages/tg1390.aspx.

LCB was a major victory for the government, and forfeiture from Ayash Exchange followed within the month.[46]

Together with the indictment of Ayman Joumaa, and the civil forfeiture and money laundering lawsuit brought against LCB the following month, the blend of sanctions designations, Section 311 actions, criminal indictments, and civil forfeitures had exposed the linkages among eight DT/MLOs and their connections to Hezbollah as well as to the Sinaloa Cartel, los Zetas, FARC, the Envigado organized crime group in Colombia, and other groups and individuals. From that point to the present, the actions illuminated the vast reach and complexity of connections within the Ayman Joumaa network and the broader Joumaa Web.

Impact and implications as operations progress. Two months before the LCB settlement and forfeiture, Treasury's FinCEN identified two Lebanese exchange houses, Rmeiti Exchange and Halawi Exchange, as foreign financial institutions of primary money laundering concern under Section 311. This was the first use of Section 311 against a nonbank financial institution, and it enabled the government to effectively cut off the Rmeiti and Halawi Exchanges from the U.S. financial system. David Cohen, Treasury's undersecretary for terrorism and financial intelligence at the time, explained that after the 311 action against LCB, "the Joumaa narcotics network turned to Rmeiti Exchange and Halawi Exchange to handle its money laundering needs."[47] LCB was out of the game, but substitutes had taken its place, handling the network's money laundering, including the TBML schemes involving American used cars and Asian consumer goods.[48] Now many of the same actions that had neutralized LCB were being employed against its successors.

The network continued to be exposed and disrupted. Using additional 311 actions, FinCEN's finding on the Rmeiti and Halawi exchange houses connected the dots further by exposing that two individuals associated with Joumaa and designated by OFAC under the Kingpin Act in June 2012, Abbas Hussein Harb and his partner, Ibrahim Chedli, "regularly coordinated and executed financial transactions . . . that were processed through the Halawi Exchange."[49] (The Harb network is discussed below.) The Rmeiti Exchange was described as working with the Halawi Exchange in money laundering and also engaging in trade-based money laundering for the Kingpin-designated Ellissa Exchange and its designated owner, Ali Mohammed Kharroubi.[50] Both Halawi Exchange and Rmeiti Exchange were also cited for facilitating or promoting money laundering for Hezbollah.[51]

---

[46] U.S. Justice Dept., "Manhattan U.S. Attorney Announces $102 Million Settlement of Civil Forfeiture and Money Laundering Claims against Lebanese Canadian Bank," June 25, 2013. www.justice.gov/usao/nys/pressreleases/June13/LCBSettlementPR.php.

[47] U.S. Treasury Dept., "Treasury Identifies Kassem Rmeiti & Co. for Exchange and Halawi Exchange Co. as Financial Institutions of 'Primary Money Laundering Concern,'" www.treasury.gov/press-center/press-releases/Pages/jl1908.aspx.

[48] Ibid.

[49] U.S. Treasury Dept., "Notice of Finding that Halawi Exchange Co. Is a Financial Institution of Primary Money Laundering Concern," 78 Fed. Reg. 24596, Apr. 25, 2013, www.gpo.gov/fdsys/pkg/FR-2013-04-25/pdf/2013-09785.pdf.

[50] Ibid.

[51] Ibid.

The June 2012 sanctions action against the Harb money laundering organization imposed Kingpin Act designations on Lebanese-Venezuelan individuals and business entities closely allied with Joumaa. That action also included the terrorism designation, under E.O. 13224, of Ali Mohamad Saleh, the Colombian national who had been designated in December 2011, under the Kingpin Act, as part of the JFC network. Ali Mohamad Saleh was described as a "key Hezbollah facilitator" who had raised funds for Hezbollah and coordinated transfers of money from Colombia, via Venezuela, to Hezbollah in Lebanon.[52] Again the Hezbollah connection was prominent.

More fronts and individuals were linked to the Joumaa Web in July 2013 when OFAC made 31 more designations under the Kingpin Act. Nine individuals and 22 entities were involved in the Colombia-based Guberek international drug money laundering network (DMLO). The Guberek network, tied to both Ayman Joumaa and Colombian cocaine boss José Evaristo Linares Castillo, was connected into the web with Hezbollah and the FARC.

On May 14, 2014, 20 additional entities and eight more individuals connected to JFC were designated. Fourteen of those entities were in Panama and included several shell companies used for money laundering. Again relatives of JFC were among those designated, along with the Panamanian attorney involved with the JFC network's fronts and money laundering. This designation expanded the total designations against JFC's network to 48 entities, of which 31 were located in Panama.[53]

Either Joumaa or his close associate Cheaitelly was directly implicated in all but one Kingpin Act designation of Joumaa Web actors occurring from February 2011 to July 2014. The exception, the June 2014 Kingpin Act designation of the Colombian organized crime group la Oficina de Envigado, was later connected to Joumaa and the Joumaa Web via the designation of the Pedro Claver Mejía Salazar network in July 2014.

More intersections. An additional aspect of this set of CTF actions was disclosed in the SDNY's amended civil forfeiture and money laundering complaint against LCB, filed in late October 2012.[54] The complaint set forth detailed information about LCB's connections to, and actions on behalf of, Hezbollah, including its connections to Hezbollah networks and financial channels that were the focus of four earlier OFAC designations from September 2006 to December 2010.[55] Those links included eight designated individuals, nine entities, and multiple subsidiaries on several different continents. They added four

---

[52] U.S. Treasury Dept., "Treasury Targets Major Money Laundering Network Linked to Drug Trafficker Ayman Joumaa and a Key Hizballah Supporter in South America."

[53] U.S. Treasury Dept., "Treasury Targets Major Money Laundering Network Operation Based in Panama.

[54] *U.S. v. LCB, et al.,* Amended Forfeiture and Civil Money Laundering Complaint, 11 Civ. 9186 (PAE), U.S. District Court, SDNY, Oct. 26, 2012.

[55] Ibid., 2833; U.S. Treasury Dept., "Treasury Designation Targets Hizballah's Bank," Treasury Press Center, Sept. 7, 2006, www.treasury.gov/press-center/press-releases/Pages/hp83.aspx; U.S. Treasury Dept., "Twin Treasury Actions Take Aim at Hizballah's Support Network," Treasury Press Center, July 24, 2007, www.treasury.gov/press-center/press-releases/Pages/hp503.aspx; U.S. Treasury Dept., "Treasury Targets Hizballah Network in Africa," Treasury Press Center, May 27, 2009. www.treasury.gov/press-center/press-releases/Page s/tg149.aspx; U.S. Treasury Dept., "Treasury Targets Hizballah Financial Network," Treasury Press Center, Dec. 9, 2010, www.treasury.gov/press-center/press-releases/Pages/tg997.aspx.

African countries, one Caribbean island, and Argentina, Brazil, and Paraguay, the three countries of South America's Tri-Border Area, to the more than 20 countries already identified in the web of networks and connections making up the Joumaa Web transnational threat complex. Those four counterterrorism designations not only intersect with the follow-the-money measures against the Joumaa Web, but also fit within a larger group of 15 mainly counterterrorism sanctions designations and one Section 311 action that were directed against Hezbollah's network of supporters and facilitators over eleven years (June 2004-February 2015). These other financial actions against Hezbollah and affiliates occurred independently of, but in the same time frame as, the Joumaa Web actions.

Observations and Implications: What Planners Need to Know

Civilian Lead and the Requirement for a Whole-of-Government Approach

The mutually reinforcing counter threat finance approach used in the Joumaa Web cases presents a valuable real-world example of a whole-of-government effort to counter complex illicit threats. The Joumaa Web cases also demonstrate that CTF actions do not operate in an exclusive "CTF stovepipe." Quite the contrary; they operate in a dynamic environment that includes and relies on traditional law enforcement methods and authorities. The innovation that successful CTF illustrates is not found in the special administrative authorities and designations themselves, but rather in how they are orchestrated synergistically across disciplines. The Joumaa Web cases combined criminal investigations, indictments, and forfeiture proceedings. They also leveraged OFAC's and FinCEN's ability to coordinate the inherent authorities of multiple U.S. government departments, bureaus, and agencies. Thus, the principle lesson is that tracking the complex financial networks that enable illicit power structures to succeed requires a sustained, well-resourced interagency effort with a legally empowered civilian lead.

For military planners, the importance of civilian lead cannot be overstated. CTF represents a balancing of security resources and shared responsibilities between defense, law enforcement, intelligence, and diplomacy. Defense policy emphasizes the primacy of multiagency CTF capabilities. In accordance with current directives, DOD's primary contributions include (a) incorporation of CTF operations into joint campaign plans; and (b) collection, analysis, and interagency sharing of signal, financial, and human intelligence supporting the necessary targeting and interdiction to attack or block financial lines of communication and disrupt networks.[56]

These activities must be carefully coordinated to ensure that any military support to CTF does not compromise the web of criminal and civil actions and objectives that civilian agencies are pursuing in legal jurisdictions around the United States. Where overseas operations occur, the lines of command authority between geographic and functional combatant commands can create additional coordination challenges, par-

---

[56] Joint Chiefs of Staff, *Commander's Handbook for Counter Threat Finance*, Sept. 13, 2011, https://publicintelligence.net/ufouo-joint-chiefs-of-staff-commanders-handbook-for-counter-threat-finance/; U.S. Defense Dept., "DoD Directive 5205.14, Aug. 19, 2010, www.dtic.mil/whs/directives/corres/pdf/520514p.pdf.

ticularly when they do not align with similar lines of authority used by the lead civilian agencies. Each side of the equation, military and civilian, has its own set of security-related objectives that are furthered by CTF. Determining the means for reaching those objectives, and whose objectives take primacy, will always be a challenge. Recognizing the challenges, and establishing the mechanisms to align operations, targets, objectives, and authorities so that turf conflicts do not arise, is a critical component of strategic planning and operational design.

Timing and Synchronization

Timelines and planning horizons are critical planning factor. It is important to recall that the Joumaa Web operations spanned more than seven years, and three of the four federal court actions that resulted are still not resolved. CTF is an effective enabler in the fight against illicit power, but it is not a quick fix. Targets and opportunities take years of intelligence and investigation to develop. Because effective illicit finance networks are complex networks, the cases against them tend to be unusually long and difficult to prove. Cross-border law enforcement operations require significant diplomatic engagement. Powerful international economic and banking interests come into play. Each of these factors adds not just to the difficulty of CTF but also to the time required.

Also, follow-the-money operations are not linear. The juxtaposition of the government's actions under different authorities is not always a clear sequence of cause-and-effect actions. For example, the first step in the actions against the Joumaa Web was the designation of Ayman Joumaa and others in his drug trafficking/money laundering network. That sanctions designation action preceded Joumaa's criminal indictment by 11 months. Although some of the evidence concerning these two CTF measures is similar, the measures are not interdependent. (Joumaa's indictment reported the designation of his network as factual background, but the designation did not itself give rise to the charges in that indictment.) Interestingly, if Joumaa committed specific acts after his Kingpin Act designation, those might be used as evidence of violations of the criminal or civil provisions of the Kingpin Act. Those criminal penalties range up to 10 years' imprisonment for most violators, and up to 30 years when a corporate official is involved, along with large fines).[57]

Often, a gap occurs between the public announcement of an indictment and later designations directed against the same persons or others in the illicit network. In the Joumaa Web series of actions, the U.S. Attorney announced the indictment of Fernain Rodriguez Vasquez and others in August 2013; and the OFAC designations occurred in February 2014. But in many other Kingpin Act designation cases, the indictments and designations occur simultaneously or nearly so. The indictments, in Miami, of key individuals in the Guberek Grimburg network (another Joumaa Web participant) were announced on the same day that OFAC sanctions designations were issued against the Guberek Grimburg network. In any of these situations, the designations, and their public disclosure, are coordinated and managed between OFAC, the criminal investigative agencies, the Justice Department, and others in a manner that will protect investiga-

---

[57] Foreign Narcotics Kingpin Designation Act, 21 U.S.C. 1901-1908.

tive and prosecutorial efforts. When indictments, designations, and 311 actions are used against actors in the same threat networks, convictions on the criminal charges are not necessary for the use of either designations or 311 actions.

Complexity

Time is not the only planning factor that cannot be precisely predicted. The degree to which CTF investigation will expand in both geographic scope and operational complexity can be only guessed at when an investigation begins. This is not "mission creep." It is the reality of how illicit organizations' financial networks function. It is highly likely that when the Joumaa Web operations began, no one anticipated how broad a network would come to light. As time and investigation went on, the Joumaa Web exposed intersections between transnational drug trafficking and organized-crime enterprises, polycrime money launderers and threat finance facilitators, global terrorist organizations, and separate bad actors whose connections arise mainly through similar financial and facilitation needs. As outlined in the preceding sections, the Joumaa Web eventually included some 25 countries on five continents; more than 18 separate networks, organizations, and entities; and countless individuals.

Daniel Glaser, the U.S. Treasury Department's assistant secretary for terrorism financing, pointed out this unanticipated complexity during his remarks before the 2013 Annual Arab Banking Conference in Beirut: "The facts of LCB demonstrate that what at first appears to be a criminal money laundering scheme might actually have broader implications. In the case of LCB, Hizballah benefitted from a global narcotics trafficking and money laundering network. This should not be surprising given Hizballah's involvement in a wide range of illicit activities."[58] In hindsight, Assistant Secretary Glaser was correct: it should not have been surprising. But hindsight can sometimes encourage foresight. As we move forward in our efforts to counter ever-evolving and increasingly networked illicit power structures, this is one lesson from the Joumaa Web that we need to take with us: however complex we believe the illicit finance network to be when the investigation effort begins, it will likely prove far more complex and far-reaching before the investigation ends.

Interagency Coordination and the Government CTF Players

Public actions dealing with the Joumaa Web of networks are the result of close coordination and cooperation among multiple agencies. DEA investigations were the basis for the OFAC designations, the Section 311 actions by FinCEN, the criminal indictments, the forfeitures achieved by U.S. Attorneys' offices, and the global exposure of this multifaceted web of transnational threat networks. Many other agencies have been involved, in addition to the lead agencies described above. They include, for example, Homeland Security's Immigration and Customs Enforcement (ICE), Customs and Border Protection, the Bureau of Alcohol, Tobacco, and Firearms, the FBI, IRS Criminal Investigations,

---

[58] U.S. Treasury Dept., "Remarks of Assistant Secretary Glaser on 'Protecting the Lebanese Financial Sector from Illicit Finance' at the 2013 Annual Arab Banking Conference," Treasury Press Center, Nov. 15, 2013, www.treasury.gov/press-center/press-releases/Pages/jl2219.aspx.

the multiagency New York Organized Crime and Drug Enforcement Task Force, the New York Police Department, and a foreign law enforcement agency, the Colombian National Police.

It is also important to note that for designations under the Kingpin Act, the statute requires significant interagency coordination. Thus, in the OFAC designations of the Joumaa Web networks, the entities and individuals were listed after formal consultation by Treasury with DOJ, FBI, DEA, DHS, CIA, National Intelligence, Defense, and State. This formal process took place in addition to the day-to-day working collaboration between those agencies (for example, OFAC and DEA) with direct interest in the specific persons to be listed.

Options and Effects

On the positive side, the federal actions in the Joumaa Web cases also demonstrate that there is no single path to disrupting the money trail and denying assets and financial access. Indeed, there are many options, thereby enabling a sustained campaign on several fronts. Applying multiple elements of counter threat finance against the various key players of an illicit-threat network should include some of the "fixers, super fixers, and shadow facilitators" who provide money laundering, threat financing services, and other enabling activities.[59] These should be pursued as aggressively, and on as broad a scale, as the drug kingpins, high-value terrorists, and other key actors. Both the secret and the merely compromised channels of commerce, aided by their middlemen, are vital supply lines feeding the entire web of transnational illicit enterprises. And CTF measures directed against these smaller middlemen, fixers, and facilitators can be as damaging to transnational threat networks as efforts against the high-value leaders, with far less political cost or operational risk.

Conclusion

General Stanley McChrystal wrote, "It takes a network to defeat a network."[60] He may have been referring to countering insurgents when he said it, but the principle applies equally to counter threat finance.

The finances and assets of transnational threat networks and their participants present a unique spectrum of vulnerabilities and opportunities for a whole-of-government approach to disrupting, containing, or even eliminating those networks and their participants. But counter threat finance alone is not enough. As the cross-disciplinary approach of CTF matures, the set of direct activities that could be used to thwart substantive offenses (e.g., drug and human trafficking) and gravely transcendent perils (e.g., terrorist possession of weapons of mass destruction) should not be isolated from CTF. On the contrary, nonfinancial investigative and direct operational methods ("action tools") need more integration with CTF measures. Ultimately, neither CTF measures nor other investigative, intelligence, and action tools should operate in stovepipes. The universe of transnational threats is too diverse, too adaptable, and too innovative for either isolated or short-term countermeasures. Synergy will always be vital to success.

---

[59] Miklaucic and Brewer, *Convergence*, ch. 5.
[60] Stanley A. McChrystal, "Becoming the Enemy," Foreign Policy, Mar.-Apr. 2011.

13. Recruitment and Radicalization: The Role of Social Media and New Technology

Maeghin Alarid

As parts of the Middle East imploded following the euphoria of the Arab Spring, some Americans shocked friends, family, policymakers, and pundits alike by leaving home to join the Islamic State of Iraq and the Levant (ISIL) and other terrorist organizations operating in the chaos of Syria, Iraq, and Libya. By May 2015, hundreds of U.S. citizens had joined the fight on the side of the extremists, and many others had been apprehended in the attempt. Even greater numbers of recruits left Europe, making their way across the Mediterranean, or overland through Turkey, to join the self-declared caliphate in its latest call to jihad. In 2014 in the UK alone, conservative estimates pegged the numbers of radicalized individuals leaving to join ISIL at five per week.[1] By 2015, the numbers were generally believed to have climbed significantly.

In their efforts at both radicalization and recruitment, terrorists, militias, and other illicit organizations have used social media in a calculated strategy that confounds many in the West. As a CNN article recently concluded, "Violent extremists like the self-proclaimed Islamic State of Iraq and Syria, or ISIS, have become increasingly sophisticated at creating dense, global networks of support online, networks that are helping these groups run virtual circles around governments and communities."[2]

Terrorist groups have good reason to use social media, whose popularity suits them in many ways. In 2015, the Internet is fast overtaking conventional forms of media such as books, magazines, and television to become the leading research and entertainment platform.[3] Social media outlets allow them to present themselves as just another part of mainstream news. Most social media platforms are easy to use and cost little or nothing. With them, terrorists can tailor their message to narrow audience niches, enlisting the help of the virtual world to enter the homes of millions of people.[4]

This chapter details how illicit organizations, including international terrorist groups, use the Internet and social media to radicalize and recruit individuals online and carry out attacks. Focusing specifically on al-Qaeda and the Islamic State of Iraq and the Levant (ISIL), we also examine how and why terrorists target women for recruitment—a particularly disturbing trend. Finally, we discuss how the West can turn the tables on the terrorists, and outline a basic approach to deterrence, using the same social media platforms and techniques as those favored by the terrorists themselves.

---

[1] Ben Brumfield, "Officials: 3 Denver Girls Played Hooky from School and Tried to Join ISIS," CNN News, Oct. 22, 2014, www.cnn.com/2014/10/22/us/colorado-teens-syria-odyssey/.

[2] Quintan Wiktorowicz and Shahed Amanullah, "How Tech Can Fight Extremism," CNN News, Feb. 17, 2015, www.cnn.com/2015/02/16/opinion/wiktorowicz-tech-fighting-extremism/

[3] Gabriel Weimann, "Social Media's Appeal to Terrorists," *Insite Blog on Terrorism and Extremism*, Oct, 3, 2014, http://news.siteintelgroup.com/blog/index.php/entry/295-social-media's-appeal-to-terrorists.

[4] Ibid.

How Online Radicalization Happens

Contrary to popular belief, many people who are radicalized online are not devout Muslims. Quite the contrary; some do not consider themselves very religious at all. It is also easy to categorize anyone recruited online as gullible, but this is simply not the case. In fact, many people who become caught up in online propaganda did not seek it out—it found them. The demographics vary widely. They may be barely or highly educated, young or old, male or female. Even financial status is no indicator. Online radicalization occurs in all economic classes. It reaches those of lower economic means and those who are financially stable.

The common denominator seems to be that everyone who is radicalized and recruited online feels sympathetic toward that group's cause, and people who feel there is "something missing" from their lives appear to be more susceptible than others. Radicalization is more widespread where conditions of inequality and political frustration prevail. It often takes root in people who sympathize with the plight of the oppressed and wish to show their solidarity. Radicalized men and women alike often feel despair, humiliation, and outrage over injustice and perceive few options for influencing change.[5] One brief moment of intense emotion evoked in them while they watch a YouTube video of innocent victims in Africa or the Middle East can be all it takes to spark their interest.

Once someone is mobilized, next steps vary. Some begin to research the causes that various extremist groups are fighting for. This leads to their discovery of the radical groups—and, more troublingly, to the groups' discovery of *them*. The Internet makes it easy to be found. A candidate for recruitment may come to the group's attention by making a financial donation, downloading extremist propaganda, entering a jihadi chat room, or visiting radical pages on Facebook. In today's environment, we see numerous examples of the radicalization process, from interest to recruitment, through execution of an actual mission, happening entirely online.

Radicalizing and recruiting online has great advantages over the traditional (and riskier) public communications. Terrorist groups can reach out to an incalculably vast audience. With no travel required, cost is minimal, no logistics or transportation support is needed, and the odds of detection are low. And the newly radicalized need not necessarily pack up and head for the Middle East—jihadi groups encourage attacks at home to avoid the risk of infiltration while traveling.

The threshold for engaging in cyber jihad is markedly lower than for someone who gives up a familiar, comfortable life to travel to an actual battle zone and risk death or capture. If the notion of online activism as a proper, respectable, and sufficient form of jihad wins wide acceptance within radical circles, we can expect ever-increasing efforts in online propaganda and cyber attacks. This could further inspire yet more individuals, facilitating both radicalization and recruitment, and lead to a new cycle of attacks.

From al-Qaeda to ISIL: An Increasingly Sophisticated Approach

The use of social media for recruitment and radicalization did not just suddenly emerge with ISIL. Even before 9/11, al-Qaeda recognized the value in harnessing the "new" media. It used the Internet to broadcast its message worldwide and prospect for recruits.

---

[5] Ibid.

It used the Internet for operational purposes, too. Much of the planning for the 9/11 attacks happened using online platforms. The attackers communicated with one another through the Internet and researched their targets online.[6]

Over time, terrorist groups have adapted their efforts with the changing battlefield landscape. In an undated letter to Taliban leader Mullah Omar that was quoted in a 2014 *International Security* article, al-Qaeda's leader, Osama bin Laden, observed that "90 percent of the preparation for war is effective use of the media." Al-Qaeda has long advocated *ghazwa ma'lumatiyya* ("information operations") and *harb electroniyya* ("electronic warfare").

The article's authors pointed out that bin Laden's successor, Ayman al-Zawahiri, shared this view. He spoke of the "jihad of the spear" and the "jihad of the *bayan*" (message, declaration). He considered the latter more important and praised the "knights of the media jihad," the "clandestine mujahideen," who are conducting it.[7] Al-Zawahiri practiced what he preached, making his declarations known in several videos and magazines posted online.

Social Media Platforms: Recruiting and Building Communities of Terrorist Practice

Social media differ from traditional media in several fundamental ways. They enable any terrorist group to reach a huge audience and circulate its message worldwide, and they provide a way to ensure that the group's propaganda lives forever online. And social media are democratic in the sense that they enable anyone to publish or access information online.[8] Social media are often used to incite fear in the general public and deliver threats (using tweets as one method), create a sense of community, radicalize others, romanticize Sharia law and the Islamic State, and offer travel advice and logistics for recruits.

As the Critical Incident Analysis Group (CIAG) points out, the Internet allows for vital dialogue between extremist ideas and inquisitive minds to take place in a virtual setting where infiltration is difficult: "For the post-Iraq (post-2003) generation especially, Internet chat rooms are now supplementing and replacing mosques, community centres and coffee shops as venues for recruitment."[9]

Nearly everyone in business today, licit or illicit, understands the importance of creating online communities. Even small mom-and-pop shops that were reluctant to provide Wi-Fi to their customers or build a website to market their business know that if they want to succeed and grow their customer base, an online community is a great way to start. And higher educational institutions realize that to stay relevant, they must routinely offer online courses and degree programs. According to Don Hinchcliffe of ZDNet, "Online communities are seen as a way to organize people and accomplish work

---

[6] Critical Incident Analysis Group (CIAG), "NETworked Radicalization: A Counter-Strategy," 2007, http://cchs.gwu.edu/sites/cchs.gwu.edu/files/downloads/HSPI_Report_11.pdf.

[7] Jerry Mark Long and Alex Wilner, "Delegitimizing al-Qaida: Defeating an 'Army Whose Men Love Death,'" *International Security* 39, no. 1 (Summer 2014).

[8] Gabriel Weimann, "New Terrorism and New Media," Wilson Center, 2014, www.wilsoncenter.org/sites/default/files/STIP_140501_new_terrorism_F.pdf.

[9] CIAG, "NETworked Radicalization."

in a collaborative manner, across geographic and demographic boundaries. The world is beginning to understand that online communities aren't just for socializing; they are also for getting things done."[10]

Illicit power structures, and terrorist organizations in particular, understand this as well as anyone. As this study of al-Qaeda and ISIL shows, such groups have adapted the same platforms that we all commonly use and rely on. They use them to foster intraorganizational communication, radicalize the vulnerable, mobilize supporters, and recruit people from all over the world. And over time, they have become increasingly effective and increasingly efficient. Here are some key ways that they are leveraging the cyber universe to their own ends.

Open Source Journals and Publications

Cyber magazines have become a signature communications platform for both al-Qaeda and the Islamic State. *Inspire,* the al-Qaeda in the Arabian Peninsula (AQAP) online magazine, widely available on the Internet as early as 2010, was published in English to reach a vast Western audience. True to its name, the magazine succeeded in inspiring others to take up the cause. Several editions of *Inspire* provided lengthy, detailed instructions on how to plan and execute bomb attacks, and lone individuals and groups acting on behalf of AQAP made or attempted terrorist attacks after downloading its material. Most notably, Tamerlan and Dzhokhar Tsarnaev, the brothers who committed the terrorist bombings at the 2013 Boston Marathon, used homemade bombs made from ordinary pressure cookers, using a recipe they obtained from *Inspire*.[11]

The magazine also routinely uses past terrorist attacks to illustrate which methods work well against Western forces, and warns followers about which methods to avoid. Articles detail specific attacks, such as Umar Farouk Abdulmutallab's unsuccessful Christmas Day bombing attempt aboard a transatlantic flight bound for Detroit. The magazine uses such examples to help future jihadists accomplish their objective while avoiding capture and incarceration.

Like AQAP, ISIL publishes an online magazine: *Dabiq*. The magazine's name was carefully chosen. Dabiq is a small Syrian town close to the Turkish border, and the prophetic location where Muslims battle infidels.[12] The production quality is impressive, and the articles contain adroitly wrought arguments legitimizing ISIL's actions against the West. It is illustrated with skillfully edited images of world leaders and regional events, depicting the Islamic State as rational and moral, and anyone who disagrees with its philosophy as venal and self-serving.

As of March 2015, at least nine issues of *Dabiq* had been published online.[13] The mag-

---

[10] Dion Hinchcliffe, "Ten Leading Platforms for Creating Online Communities," Sept. 4, 2008, www.zdnet.com/article/ten-leading-platforms-for-creating-online-communities/.

[11] Azmat Khan, "The Magazine that 'Inspired' the Boston Bombers," *Frontline*, Apr. 30, 2013. www.pbs.org/wgbh/pages/frontline/iraq-war-on-terror/topsecretamerica/the-magazine-that-inspired-the-boston-bombers/.

[12] *BBC News*, "Why Islamic State Chose Town of Dabiq for Propaganda," Nov. 17, 2014, www.bbc.com/news/world-middle-east-30083303.

[13] Clarion Project, "The Islamic State's (ISIS, ISIL) Magazine," Sept. 10, 2014, www.clarionproject.org/news/islamic-state-isis-isil-propaganda-magazine-dabiq#.

azine, printed in English and several other languages, features articles on such topics as holy war, the importance of community, and unity within the ranks. And, of course, no issue would be complete without several colorfully written articles boasting of ISIL victories on the battlefield.[14] The National Consortium for the Study of Terrorism and Responses to Terrorism (START) compared the AQAP and ISIL publications. It found *Dabiq* notable "because the emphasis is on maintaining a strong media brand and disseminating highly ideologically-congruent propaganda to promote radicalism among distant operatives, sympathizers, and foreign fighters whereas *Inspire* had a strong focus on training."[15]

ISIL's online recruitment tactics do not rely solely on *Dabiq*. The magazine is only one method the group uses to reach its wide audience. Early in its existence, ISIL understood the value of using social media and video game technology to radicalize as many individuals as possible, specifically targeting those who are young and computer savvy. ISIL's aim is to reach every corner of the earth and gain recruits to expand the self-declared caliphate. Its use of new technology and social media is unprecedented and unlike anything seen before in a terrorist group. Its marketing campaign is truly impressive, and it is happening on a massive scale.

"Liking" Jihad

By January 2014, the average age of Facebook users was 30 years, with almost half of all users logging in daily. As of April 2015, the online social network had roughly 1.3 billion active accounts every month, and 54.2 million individual and group pages.[16] Terrorist groups understand that if they want to reach out to the younger generation, this is an excellent vehicle. A 2010 report by the Department of Homeland Security listed several ways that terrorist groups use Facebook, and we have seen both al-Qaeda and ISIL use them all:

- "as a way to share operational and tactical information, such as bomb recipes, AK-47 maintenance and use, tactical shooting, etc.;

- as a gateway to extremist sites and other online radical content by linking on Facebook group pages and in discussion forums;

- as a media outlet for terrorist propaganda and extremist ideological messaging;

- as a wealth of information for remote reconnaissance for targeting purposes."[17]

---

[14] *Counter Jihad Report*, "Islamic State Magazine Asks, 'Did You Think We Were Joking?'" May 26, 2015, http://counterjihadreport.com/tag/they-plot-and-allah-plots/.

[15] START, "The Islamic State of Iraq and the Levant: Branding, Leadership Culture and Lethal Attraction," report, Univ. of Maryland, Nov. 2014, www.start.umd.edu/pubs/START_ISIL%20Branding%20Leadership%20Culture%20and%20Lethal%20Attraction_Ligon_Nov2014.pdf.

[16] Statistic Brain Research Institute, "Facebook Statistics," Apr. 14, 2015, www.statisticbrain.com/facebook-statistics/.

[17] Jana Winter, "Al Qaeda Looks to Make New 'Friends'—on Facebook," *Fox News*, Jan. 9, 2010, www.foxnews.com/tech/2010/12/09/facebook-friends-terror.

This list will not remain static. As Facebook and similar online social networks evolve, seemingly infinite avenues for promoting jihadist messages, disseminating threats to enemies, and infiltrating homes, families, and safe havens will emerge. The only limits on terrorists' use of this popular social media site are the bounds of human imagination.

Tweeting Terrorism

Twitter is another Internet social network that terrorist groups are using to their advantage. With it, they can get their propaganda out in almost real time. But the real targets of the tweets are not necessarily the recruiting base. Rather, the targets increasingly are Western institutions, news media, and anyone else who will react to a well-timed terrorist tweet. Terrorist groups use Twitter to put out fake news stories, entice followers, and win sympathy. There have been many instances of mainstream media mistaking terrorist-originated tweets as legitimate sources of breaking news. One particularly notable incident occurred in April 2013, when the Associated Press Twitter account was hacked and an ominous message posted: "Breaking: Two Explosions in the White House and Barack Obama Is Injured."[18] Within minutes of this false Twitter report, stocks on Wall Street plummeted.

In an effort to prevent itself from becoming a platform of choice for illicit goals and objectives, Twitter has suspended many accounts, but to no avail. The terrorists have proved nimbler than the administrators, and their sites generate an almost immediate response as soon as they appear. For example, when the Syrian extremist organization al-Nusra Front's Twitter accounts were shut down, the group had opened up a new account that had more than 20,000 followers within 24 hours.[19]

Getting It on Video

YouTube is another social medium that terrorists are using effectively. Their online recruitment videos even have sound tracks of slickly produced hip-hop music. According to a 2007 Associated Press article, this immensely popular platform receives "tens of thousands of new videos daily, and users watch over a hundred million per day, making content difficult to monitor."[20] According to statistics provided by YouTube, 300 hours of videos are uploaded to the site *every minute*.[21] Because of the sheer volume of videos added to the site each day, YouTube struggles with the Herculean task of filtering terrorist content. The site displays a "promotes terrorism" tag underneath all videos and relies on viewers to use this tag to flag inappropriate content for removal. Even so, the site cannot keep up with the onslaught of new content.[22] "Though authorities have identified

---

[18] Adam Shell, "Stocks Gyrate Wildly after Fake Terror Tweet," *USA Today*, Apr. 23, 2013, www.usatoday.com/story/money/markets/2013/04/23/stocks-gyrate-wildly-after-fake-terror-tweet/2107089/.

[19] Weimann, "New Terrorism and New Media."

[20] Tariq Panja, "Militant Islamic Groups Turn to YouTube," Associated Press, Feb. 11, 2007, http://www.nbcnews.com/id/17107489/#.VYr-FE3JCUk.

[21] YouTube, "Statistics," Oct. 10, 2014, www.youtube.com/yt/press/statistics.html.

[22] *NBC News*, "Ads Shown Before YouTube ISIS Videos Catch Companies Off-Guard," Mar. 10, 2015, www.nbcnews.com/storyline/isis-terror/ads-shown-isis-videos-youtube-catch-companies-guard-n320946.

a number of jihadist propaganda pieces—some of them viewed by thousands—these are often replaced almost as soon as they are removed."[23]

Removal comes at a cost, however, for those who are trying to counter radicalization and recruitment. For the intelligence community in particular, it is vitally important to analyze the tradeoff between the damage that jihadist videos can do and the potential intelligence that they may divulge. The complex debate whether to let these terrorist groups have a voice online is ongoing.[24] It may be possible for companies such as YouTube (as well as Google, Twitter, and Facebook) to use algorithms to detect violent language and content, much as they use them to ferret out illegal content involving children, and prevent these videos from being posted. This would deny the jihadists a useful medium, but what impact will it have on the intelligence community, which has gotten valuable inside information from such videos? To further complicate the dilemma, terrorists are keenly aware that their postings and videos are being monitored; thus, the actual intelligence value to be gained from their YouTube activity is questionable.

Radicalization and Video Game Technology

Terrorist groups worldwide have made no secret of their attempts to appeal to a younger generation. Various articles written as early as 2006 discuss how websites created by terrorist groups were meant to draw in "a computer savvy, media-saturated, video game-addicted generation."[25] One early jihadist website featured a video game titled "Quest for Bush," in which players "fight Americans and proceed to different levels including 'Jihad Growing Up' and 'Americans' Hell.'"[26] The game was released in 2006 by the Global Islamic Media Front, a radical organization with ties to al-Qaeda. According to one article, its "players are prompted to advance through six missions against soldiers who look like Bush, followed by a seventh mission against a character that looks like the president that takes place in a desert-like region. During the game, jihadist songs are played in the background."[27]

Quest for Bush was not the first, and certainly not the last, of its genre. But in 2006, it was the most extreme addition to a small but growing list of Islamic extremist video games, monitored by the Defense Department and much blogged about in gaming circles. Some are free, others not. But as the *Washington Post* describes them, they all champion issues "from an Islamic perspective, in stark contrast to many Western-made games that generally cast Muslims and Arabs as the bad guys. Furthermore, they underscore a brewing game-design war between East and West, a simmering tension over who is writing (and rewriting) history."[28]

---

[23] Tariq Panja, "Militant Islamic Groups Turn to YouTube," *Washington Post*, Feb. 11, 2007.

[24] Yigal Carmon and Steven Stalinsky, "Terrorist Use of U.S. Social Media Is a National Security Threat," *Forbes*, Jan. 30, 2015, www.forbes.com/sites/realspin/2015/01/30/terrorist-use-of-u-s-social-media-is-a-national-security-threat/2/.

[25] Bruce Hoffman, "The Use of the Internet by Islamic Extremists," Testimony before the House Permanent Select Committee on Intelligence, May 4, 2006, RAND, www.rand.org/content/dam/rand/pubs/testimonies/2006/RAND_CT262-1.pdf.

[26] Jose Antonio Vargas, "Way Radical, Dude," *Washington Post*, Oct. 9, 2006, www.washingtonpost.com/wp-dyn/content/article/2006/10/08/AR2006100800931.html.

[27] CNN, "Web Video Game Aim: 'Kill' Bush Characters," *CNN World News*, Sept. 18, 2006, www.cnn.com/2006/WORLD/meast/09/18/bush.game/index.html.

[28] Vargas, "Way Radical, Dude."

Terrorist groups have been using video game technology in their recruitment efforts since early in the first decade of this century. Recently, though, ISIL's efforts have reached unprecedented levels in scale and capabilities. Adapting the most popular video game of 2012, Grand Theft Auto, it created its own modifications so that players can role-play as members of ISIL engaged in combat.[29] Specific modifications made to the game include ISIL militants killing law enforcement officers and attacking military convoys with explosives, and ISIL snipers shooting American soldiers.[30]

Perhaps, in using video game development as a recruiting tactic, ISIL has taken a page from the U.S. military playbook. The U.S. Army published the video game America's Army in 2002, to be used as a recruitment video to reach the younger generation.[31] America's Army has been wildly successful and was listed as one of the top ten video games in the world in 2002-8.[32] Although a direct link has not been established, it is certainly in the realm of possibility that ISIL was aware of this marketing campaign by the U.S. Army, saw how successful the video game was at gaining recruits, and appropriated the idea to increase its own ranks.

Terrorist groups worldwide learn tactics from one another as well as from their sworn enemies. Video game technology is just one more way for them to reach a large audience. If militaries have been using video games not only to recruit soldiers but also to train them, as Professor Corey Mead's book *War Play* suggests, then why would adversarial groups not use this technology in the same way?[33] And indeed, they are.

Using Specialized Technology to Increase Reach and Effectiveness

Terrorists use many innovative methods to communicate online. The CIAG report states:

- Terrorists can draft an email message and save it as a draft rather than sending it, so that anyone with access to that email account can log in and read the message. Known as "dead drops," these communications are less subject to interception [than an email that has been sent].

- Terrorists can post training manuals online or even hack into a legitimate website and hide training materials "deep in seemingly innocuous subdirectories of the legitimate site," a process known as "parasiting."[34]

- Terrorists can conduct research on potential targets online, where both text and imagery, including satellite photography, is frequently available. Google Earth, for instance, has been used to target British soldiers in Iraq with increasing accuracy.

---

[29] START, "The Islamic State of Iraq and the Levant."

[30] *INQUISITR*, "ISIS Uses 'Grand Theft Auto' Mock Up to Recruit and Boost Morale," Sept. 19, 2014, www.inquisitr.com/1486558/isis-uses-grand-theft-auto-mock-up-to-recruit-and-boost-morale/.

[31] Brian Kennedy, "Uncle Sam Wants You (To Play This Game)," *New York Times*, July 11, 2002.

[32] Corey Mead, "Military Recruiters Have Gone Too Far," *TIME*, Sept. 17, 2013.

[33] Corey Mead, *War Play: Video Games and the Future of Armed Conflict* (Boston: Houghton Mifflin Harcourt, 2013).

[34] Technical Analysis Group, "Examining the Cyber Capabilities of Islamic Terrorist Groups," Dartmouth College, Nov. 2003, www.ists.dartmouth.edu/library/164.pdf, quoted in CIAG, "NETworked Radicalization."

- Terrorists can appeal anonymously for donations of financial or other support via websites.[35]

Another way that terrorists use social networking sites is with a method known as *narrowcasting*. According to Gabriel Weimann, "Narrowcasting aims messages at specific segments of the public defined by values, preferences, demographic attributes, or subscription. An online page, video, or chat's name, images, appeals, and information are tailored to match the profile of a particular social group."[36] Terrorist groups view online profiles and even user history and then target online pages, videos, or extremist chat rooms to match a particular individual. Using these methods of deduction, terrorist groups winnow their target audience by age, gender, and historical preferences. This technique is a classic Marketing 101 tool, revealing which sites users visit most often and what kinds of products they view online, and thus guiding marketers in what products to promote to them. Terrorist groups view people's profiles and decide whom to target and how best to approach each individual. "An online page, video, or chat's name, images, appeals, and information are tailored to match the profile of a particular social group. These methods enable terrorists to target youth especially."[37]

Understanding the many ways that the Internet can be used to their advantage is vital for terrorist organizations. "Social media also has numerous technical advantages for terrorists: sharing, uploading or downloading files and videos no longer requires access to computers or cyber-savvy members capable of using sophisticated computers and advanced programs. Using smart phones and social media platforms allows simple, free and fast access to all."[38]

It is vital that policymakers monitor the various methods by which terrorist groups use the Internet—and specifically social media. These groups are constantly trying to keep a step ahead of the game in their radicalization and recruitment efforts, while contriving new ways to attack their enemies.

Using Social Media to Launch Cyber Attacks

Although this analysis focuses on the various methods of online recruitment and radicalization by terrorist groups, no discussion of Internet terrorist activity would be complete if it failed to mention incidents of previous cyber attacks and the threat of possible cyber attacks to come. Says Weimann, "The online platforms used to promote electronic jihad are also used for operational purposes such as instruction and training, data mining, coordination, and psychological warfare."[39]

A 2008 article in the *Forensic Examiner* discussed another form of cyber warfare favored by terrorist groups. These attacks, known as "denial of service," work by impeding a cyber network's abilities and inundating the system with an enormous number of "pings" to create *message flooding*. The author explains: "A ping is a [data] packet that allows an attacker to determine whether a given system is active on a network. A flood of pings is transmitted to a targeted site. The pings saturate the victim's bandwidth and

---
[35] CIAG, "NETworked Radicalization."
[36] Weimann, "New Terrorism and New Media."
[37] Ibid.
[38] Ibid.
[39] Ibid.

fill up the system's buffer (memory space), causing network performance to deteriorate and the system to hang, crash, or reboot."[40]

To pull off cyber attacks such as these, ISIL and other terrorist groups rely increasingly on technologically savvy younger recruits. With their young recruits' hacking skills, they can penetrate critical infrastructure systems such as electrical grids and the financial sector. For example, in 1997 in Massachusetts, a hacker disabled a computer system that operated the local airport control tower.[41] And in 2000, hackers infiltrated the computer systems controlling the flow of natural gas through Russian pipelines.[42]

This much is certain: if we have seen these attacks before, we will see them again. Terrorist groups are constantly watching the behavior of other groups, learning which kinds of attacks are successful and which to avoid. Thus, they grow ever more sophisticated, with shorter learning curves.

Online Radicalization of Women

As of this writing in 2015, the world is late on the scene in examining the Islamic State's strategies for recruiting and radicalizing women. Terrorist groups have always recruited both males and females, but ISIL has made clear a specific agenda of enlisting women in its cause. In June 2014, ISIL declared itself a caliphate, and since then it has ramped up its efforts to gain both recruits and territory across Iraq and Syria. Its recruitment of women has turned into a massive endeavor that caught the West flat-footed, and we are only beginning to understand the implications. Throughout 2014, scarcely a week passed without news articles about young women in the West leaving home and fleeing to Syria. The stories left us perplexed and infuriated. Our emotional response to the ongoing news has been warranted, certainly, though it has been largely unhelpful. We have paid attention to the pattern of female recruitment, but we have not stopped to question the logic of this particular pattern.

Why ISIL Recruits Women

ISIL understood that to grow its ranks in the long-term and not just in the present, it needed a cadre of women to give birth to the next generation of fighters. Women living in Europe and countries outside the Middle East are among the least likely people to be suspected of Islamist extremism, which makes them prime targets for recruitment. Also, women are not often seen as posing an imminent threat and may be able to travel more freely than men without arousing suspicion. Interestingly, the ISIL campaign to recruit females was carried out primarily by European women who left their home countries, joined ISIL, and relocated to Syria. The recruitment campaign consists of ISIL propaganda, instruction on how women can communicate with ISIL members, and logistics on how Muslim women (of all ages) can travel to Syria and join the cause.[43]

---

[40] Marie Wright, "Technology and Terrorism: How the Internet Facilitates Radicalization," Forensic Examiner, Winter 2008.

[41] Mudawi Mukhtar Elmusharaf, "Cyber Terrorism: The New Kind of Terrorism," Computer Crime Research Center, Apr. 8, 2004, www.crime-research.org/articles/cyber_terrorism_new_kind_terrorism/.

[42] Ibid.

[43] Rita Katz, "From Teenage Colorado Girls to Islamic State Recruits: A Case Study in Radicalization

Researcher Rita Katz reports, "IS women recruiters created dozens of social media accounts, urging women to move to the 'Land of the Caliphate.'"[44] Travel manuals can be found online giving logistics for travel to Syria, where to cross borders undetected, and what to pack for a new life in ISIL. Photos have been posted online showing Western women who have fled their homes and traveled to Syria to join the caliphate. These photos depict young women carrying weapons and surrounded by ammunition. Tweets from women who have joined ISIL include flowery descriptions of a positive environment where they have been embraced in the prophesied land of the new Islamic State.[45]

In February 2014, the United States and its European allies learned that ISIL had formed Umm al-Rayan, "a female brigade, with the purpose of exposing male activists who disguise in women's clothing to avoid detention when stopping at the ISIL checkpoints."[46] According to Islamic custom, men cannot physically touch women on the streets. But the women of Umm al-Rayan can stop anyone dressed in hijab. Each woman receives "a monthly salary of 25,000 Syrian liras (less than $200) and is only allowed to be employed within the brigade."[47] As of this writing, the brigade had not yet taken part in acts of terrorism, but its existence, coupled with ISIL's demonstrated willingness to commit violence and murder against women in the territories it controls, is causing concern.

Why Women Join

It is perhaps puzzling why women might be willing to support a group that oppresses females, or that they would choose to support bombings that kill innocent women and children. But women have indeed shown repeatedly that they are willing to take up arms for extremist organizations and violent causes. They have been on the front lines of battle in extremist groups the world over, including Peru's Shining Path, Chechen rebel groups, and the Sri Lankan Tamil Tigers, and their participation, during the 1970s and 1980s, in terrorist organizations such as the Weather Underground in the United States, and the Red Army Brigades and the Red Army Faction in Europe, is well known.[48] Hence, it should come as no surprise that they are participants in ISIL's particularly brutal extremist violence.

The possibility also exists that despite the terrible oppression of women in many terrorist groups, women join to prove the worth of their gender in the hope of making strides toward women's rights. But while this might cross the minds of some, it is not the force driving their loyalty to the group. Far more commonly, women decide to leave their families, friends, and lives behind and join extremist groups for the same reason as

---

via Social Media," *Insite Blog on Terrorism and Extremism*, Nov. 11, 2014, http://news.siteintelgroup.com/blog/index.php/entry/309-from-teenage-colorado-girls-to-islamic-state-recruits-a-case-study-in-radicalization-via-social-media.

[44] Ibid.

[45] Ibid.

[46] TRAC, "Umm al-Rayan," TRAC report summary, 2014, www.trackingterrorism.org/group/umm-al-rayan.

[47] Ibid.

[48] Nimmi Gowrinathan, "The Women of ISIS: Understanding and Combating Female Extremism," *Foreign Affairs*, Aug. 21, 2014, www.foreignaffairs.com/articles/middle-east/2014-08-21/women-isis.

men: they believe in the cause. If policymakers ever hope to understand the prevalence of female radicalization and recruitment, they must first put the stereotypes to rest.

What, then, is the apparent particular vulnerability of Muslim women to radicalization and recruitment? It may stem from deep feelings of marginalization. These women may be attracted to the idea of belonging to something greater than themselves, and to the elevated status that they expect to achieve by joining ISIL. When women feel they are seen as the "other" in the country where they live, and their religious ideology tells them they have a greater purpose and duty as a Muslim, these are prime conditions for recruitment. Online interactions encourage these feelings because they can foster a sense of community and belonging in a virtual world, which may not exist in the real world. And just as online social media are used to generate fear and spread terror, they are also used to create relationships and feelings of closeness and fitting in with a group.

Fighting Fire with Fire: How to Deter Online Radicalization

The obvious question arises: how to deter this growing threat of recruitment and radicalization using online social media and other new technology? The 2014 analysis by Jerry Mark Long and Alex Wilner, examining al-Qaeda's perspective on social media and the Internet, also suggested a number of practicable ways to diminish the appeal of radical messages online. The authors suggest that the process may begin with depriving extremist groups of their socioreligious appeal. If the very message that the group promotes can be debunked as a false teaching and an obscene misinterpretation of Islam, the very rationale for joining the group vanishes. Stripping away the group's legitimacy will diminish both its community of support and its target pool of recruits. For instance, by challenging al-Qaeda's religious appeal, one may debunk the very way the group legitimizes its acts of terrorism.[49]

An apt example of challenging the jihadists' religious appeal is the campaign to deradicalize inmates in Saudi Arabia's prisons. This program places radicalized individuals with religious scholars who educate them in the teachings of the Quran. According to a PAK Institute report, in 2004 the religious counseling program had some 2,000 prisoners enrolled. By 2007, about 700 had been released upon completing the program.[50] The program has achieved a high success rate through (a) helping individuals learn the true teachings of the Quran and thus avoid buying into the extremist message; (b) rehabilitating former jihadists; and (c) reintegrating them back into their communities.[51] The report attributes much of the program's success to the social support that the prisoners receive after completing the rehabilitation program—for example, assistance in finding employment and living arrangements.

Exploiting the Narrative

As Long and Wilner point out, all extremist groups have a narrative, and it is within the West's power to turn that narrative on its head and use it against them. If the group's manifesto appears questionable, the group will lose sympathizers and, with them, re-

---

[49] Long and Wilner, "Delegitimizing al-Qaida."
[50] Saba Noor and Shagufta Hayat, "Deradicalization: Approaches and Models," PAK Institute for Peace Studies, Apr. 2009, http://san-pips.com/download.php?f=116.pdf.
[51] Ibid.

cruits and revenue. Thus, the West can and must deter potential sympathizers by manipulating the group's message. According to the PAK Institute, "In strategic terms, if we can target a message, we thereby deter/compel those who would have adopted it, along with those who currently employ it. And the degree to which al-Qaida's message loses traction with Arab and Islamic publics is the degree to which deterrence by delegitimization will have succeeded."[52]

Sometimes, the terrorist group itself, through its actions, inadvertently delegitimizes its own message. For example, al-Shabaab lost support after employing cruel tactics, such as suicide bombings that caused the death of many civilians. The group also withheld goods and services from the people of Somalia, causing many civilians to starve.[53] Whenever a terrorist group incurs such a loss of confidence, this creates a golden opportunity for Western news media and intelligence agencies to highlight that group's brutality, thus deterring potential new recruits.

A terrorist group can survive only when it has a following of people who support its cause. History has shown us that without a critical mass of supporters, the group will shrink in size and eventually become irrelevant. Men and women contemplating joining an extremist group will not do so if they feel that the group's belief system is flawed or disingenuous. An examination of past and present terrorist groups shows that when the recruitment pool deems the group's behavior offensive and illegitimate, the group will soon become obsolete.[54]

The International Centre for the Study of Radicalisation and Political Violence (ICSR) has discovered numerous online messages revealing that many of the women married off to jihadi fighters in Syria are mourning husbands who have died fighting for ISIL. And the ICSR has seen several online posts from British women describing the harsh living conditions in Syria, including the lack of hot water for bathing, and the high incidence of illness from the cold.[55] Such online social media posts, painting life within ISIL in a negative light, can be an effective tool in counterradicalization efforts. Intelligence agencies can and should take advantage of social media to routinely monitor what is going on within the ranks of groups such as ISIL. Online social media posts from members themselves provide an excellent opportunity to infiltrate the group.

Understanding Cultural Implications and Shifting the Cultural Identity

It is crucial for policymakers to understand how these groups grow their community of support and portray themselves to the world as a force to be reckoned with. And it is equally important to understand how the extremists' narrative "guides its interpretation of history and of contemporary Western policies"—for example, al-Qaeda's narrative that the actions of the West are monolithic and that they are aimed entirely at expanding

---

[52] Long and Wilner, "Delegitimizing al-Qaida."

[53] Armin Rosen, "How Africa's Most Threatening Terrorist Group Lost Control of Somalia," *Atlantic*, Sept. 21, 2012, www.theatlantic.com/international/archive/2012/09/how-africas-most-threatening-terrorist-group-lost-control-of-somalia/262655/.

[54] Long and Wilner, "Delegitimizing al-Qaida."

[55] Mark Townsend, "How a Team of Social Media Experts Is Able to Keep Track of the UK Jihadis," *Guardian*, Jan. 17, 2015, www.theguardian.com/world/2015/jan/17/social-media-british-jihadists-islamic-state-facebook-twitter.

Western policies and advancing the West's economic stance in the world.[56] This understanding is essential to deterring the threat that extremist groups pose online.

The CIAG report mentioned earlier discusses another key factor in deterring radicalization: understanding the link between how a group thinks and how it acts, based on its cultural identity. Not so long ago, geography played a much greater role in determining an individual's worldview.[57] In today's virtual world, this is often no longer the case.

"Whether contained within a country of origin or within ethnic or immigrant communities, spread of large group identities was only as effective as the limited transportation possibilities at the time. . . . It may be that the Internet is transforming large group identity formation from a lateral, physical process to a metastatic, technological process. Previous boundaries have little relevance."[58]

In short, a successful counterterrorist message must include these elements:

- Be easily accessible.
- Be able to reach the masses.
- Reach out to women.
- Be as globally pervasive and ubiquitous in countering the terrorist message as ISIL is in promoting it.

A surprising example of a counterterrorist message that had a significant impact online originated with the hacker group known as Anonymous. In early 2015, in response to the January 2015 terrorist attack on the satirical Parisian newspaper *Charlie Hebdo*, Anonymous hacked into a large number of ISIL Twitter and Facebook accounts, shutting them down and also sending this message: "You [ISIL] will be treated like a virus, and we are the cure."[59] Hacker attacks such as this against ISIL and other terrorist groups will not prevent the spread of extremism online. But they certainly can present a strong counter message, and a reminder to the online community that illicit power structures do not own the space unimpeded, but are as vulnerable as any other entity that depends on cyberspace to further its objectives.

How to Deter Women from Online Radicalization and Recruitment

As Badran points out, even in patriarchal societies, women often have a solid and powerful network of connections to their surrounding community. This allows them to exert influence in their neighborhoods, towns, villages, and universities. This connection to community is vitally important when studying female recruitment efforts by jihadist groups because, across most cultures, women connect with other local women. Women can provide a powerful voice that should not be discounted in the effort to prevent other women from becoming radicalized.[60]

Counterterrorism officials need to recognize that women's roles within a community can be a key source of intelligence about existing radicalization in their neighborhoods.

---

[56] Long and Wilner, "Delegitimizing al-Qaida."
[57] CIAG, "NETworked Radicalization."
[58] Ibid.
[59] Chris Perez, "Anonymous Attacks ISIS Supporters Online," *New York Post*, Feb. 10, 2015, http://nypost.com/2015/02/10/hacking-group-anonymous-attacks-isis-supporters-online/.
[60] Badran, "Women and Radicalization."

Women are often quite savvy about what goes on in their community. Despite their status as second-class citizens in many cultures, women are predominantly the ones raising the next generation. Their children will grow up, perhaps to become extremists, but perhaps instead to be the first in their family to graduate from college.[61]

The Organization for Security and Cooperation in Europe (OSCE) has consistently stressed the importance of reaching out to women in communities vulnerable to extremism. Stressing best practices in security and development, the OSCE emphasizes gender inclusion as a cornerstone of effective security sector reform, security force development, and community stabilization and resilience:

"Women can have special potential in countering VERLT (violent extremism and radicalization that lead to terrorism). The involvement of women as policy shapers, educators, community members and activists is essential to address the conditions conducive to terrorism and effectively prevent terrorism. Women can provide crucial feedback on the current counter-terrorism efforts of the international community and can point out when preventive policies and practices are having counterproductive impacts on their communities."[62]

The ideal target audience for helping with efforts to deradicalize women is also the same target audience that jihadi groups hope to recruit: modern, young, second generation, perhaps studying abroad. Terrorist groups understand, however, that targeting women from this particular demographic is risky because they may already believe in and support gender equality and, in fact, are *not* prone to radicalization. For feminist transnational networking organizations, this same demographic comprises a whole new generation of Muslim women who have grown up exposed to new Islam-grounded gender-egalitarian ideas. When properly empowered, these women can reach out to other women and play a vital role in deterring their online radicalization and recruitment.[63]

It is not enough to focus on the straightforward technical solution of removing or blocking radical material online. This strategy will not stop terrorist groups from gaining recruits. Indeed, a purely technical solution will probably not be cost effective and may, in fact, work in the terrorists' favor. The independent research organization ICSR, which concluded to help policymakers and practitioners find more intelligent solutions in dealing with radicalization and political violence, concluded that a successful strategy to dissuade the recruitment of women must include the following:

- Dissuade people from posting extremist messages online, by removing such websites and prosecuting the owners. This will send a strong message that those involved in online extremism are within the reach of the law and will be penalized.

- Help online communities take matters into their own hands and vet all material posted to their group pages. They can do this through creation of an *Internet users panel* that would provide a way for members of the online community to report complaints.

---

[61] OSCE, "Women and Terrorist Radicalization Final Report," Mar. 13, 2012, www.osce.org/atu/99919?download=true.
[62] Ibid.
[63] Badran, "Women and Radicalization."

- Diminish the terrorist message by having a larger "counter message" online.[64] For example, develop more websites (largely for women) that spread an antiviolence message of Islam as a peaceful religion that supports women's rights.

- Encourage the publication of progressive messages online, thus discouraging extremist propaganda. Efforts to counter online radicalization of women through social media must include using precisely this technology, not shying away from online social media methods but using those same sites to counter radicalization messages.[65]

Conclusion

Illicit power is not new. Illicit power structures' use of the Internet is not new. Illicit power structures' use of social media and video game technology, to the extent that ISIL is using them, *is* new. And it is dangerous. ISIL's radicalization and recruitment efforts via social media and cyber technology have been very successful in gaining sympathizers worldwide. Terrorists and other illicit organizations understand that the young generation spends a huge amount of its time online, and they target that population for this reason. They realize how vitally important it is to have the young generation in their ranks.

These young people have tremendous cyber capabilities, which can help the terrorists not only radicalize and recruit new blood but also conduct cyber attacks on their enemies. Attacks on electrical systems, transportation systems, and the financial sector can be devastating. We have seen such attacks in the past, and we will continue to see this method of warfare in the future. Extremist groups know that they must adapt and change to survive on the changing landscape of war. They will continue to use the Internet to maintain their current methods and will also try to eclipse those methods with new, more effective ones.

Illicit power structures will continue to use the Internet and new technology—to spread their extremist message, to execute attacks, and to recruit women to their cause. The importance of women within these groups worldwide must not be discounted. If counterterrorism officials can work with other women and focus on changing the hearts and minds of females within these groups, perhaps these mothers at home will begin to teach their sons and daughters to fight with their words and counter the extremists' distortion of the Quran, instead of taking up arms.

The situation is by no means hopeless, nor is it a one-way street. In 2012, the NGO Invisible Children launched its KONY 2012 campaign as "an experiment." "Could an online video [take on an illicit power structure by making its leader,] an obscure war criminal, famous? And if he was famous, would the world work together to stop him?"[66] According to Invisible Children's website, "the experiment yielded the fastest growing viral video of all time. The KONY 2012 film reached 100 million views in 6 days, and 3.7

---

[64] ICSR, "Countering Online Radicalisation: A Strategy for Action," Jan. 28, 2009, http://icsr.info/projects/the-challenge-of-online-radicalisation/.

[65] Ibid.

[66] Invisible Children, "Kony 2012," 2012, http://invisiblechildren.com/kony-2012/.

million people pledged their support for efforts to arrest Joseph Kony."[67] Few outsiders dispute Invisible Children's numbers, and even though the campaign failed to lead to Kony's capture, it riveted world attention and began the conversation on how we can use the Internet and social media to counter extreme cases of impunity when the nation state where it resides is either unwilling or unable to take action.

Three years later, following the massacres in Paris of *Charlie Hebdo* leaders and staff, online counter messaging triumphed with the launch of the *Charlie Hebdo* app. This app was approved by Apple's App Store and made available for users to download less than two hours after developers contacted the Apple CEO. This is highly unusual in that it typically takes a week and a half for an app to get approval and become available for users.[68] The app allowed the user to support the Charlie Hebdo cause of free speech, subscribe to the magazine, and also acquire the issue that was published after the attack.[69] By downloading this app, the user simultaneously supports the Charlie Hebdo creed: "Because a pencil will always be better than barbarity . . . because freedom is a universal right," and demonstrates how technology and social media can be turned against those who would use them to kill and enslave others.[70] Or, as Weimann points out, "Online social media is not only a potent way to promote terrorism, but also a necessary tool in preventing it. . . . This is the emerging challenge for the West: to regain the cyber territory it has long ceded to extremists."[71]

---

[67] Ibid.

[68] Alyssa Newcomb, "Charlie Hebdo Launches App Version of Latest Issue," *ABC News*, Jan. 20, 2015, http://abcnews.go.com/Technology/charlie-hebdo-launches-app-version-latest-issue/print?id=28345203.

[69] *Charlie Hebdo*, "Charlie Hebdo," app, Apple iTunes Preview, Jan. 2015, https://itunes.apple.com/us/app/charlie-hebdo./id957966299?mt=8.

[70] Ibid.

[71] Weimann, "Social Media's Appeal to Terrorists."

14. Make It Matter: Ten Rules for Institutional Development that Works

Mark Kroeker

The preceding chapters show how illicit actors function as primary roadblocks in the path to peace. Illicit power structures emerge and are energized in the vacuum left by the chaos of war, civil upheaval, subregional disorder, and the attendant destabilization. First to go during conflict are the legitimate power bases that arise from the legal framework of the state. Constitutions, laws, and codes of criminal procedure, even vehicular codes, fall by the wayside as the channels of power disbursement are upended and the relevant, legally appointed leaders abandon their posts. Courts are looted, and judicial officers withdraw in fear as tribally and self-appointed power grabbers dictate the law. Police institutions are taken over by militias that impersonate the actual police, usurping the legitimate power of the state to enrich themselves and their criminal organizations. Prisons become the locus for illegal detention at best, and torture and extrajudicial executions at worst.

Interim "leaders" exploit the chaos and uncertainty of transition to further their own illegitimate financial objectives. These leaders see prisons as a lucrative franchise for holding captives for ransom by family and friends. Few, if any, of these inmates are actually charged with a substantive offense under the law. And meanwhile, the truly culpable enjoy the impunity brought about by power deals amid the chaos. These are the illicit networks that institutional development can best address.

Peace Agreements: Where Institutional Reform Begins

The typical postconflict path to peace often includes agreements sometimes referred to as "comprehensive peace accords." These are hammered out by high-level negotiators, diplomats, or emissaries whose main focus is to find common ground between warring parties, stop the killing, and establish a "peace to keep." In some instances, there may be language that refers obliquely to reforming or restructuring the police. But unfortunately, peace accords, treaties, and other agreements rarely contain language specifically addressing institutional development through reconstituted justice systems, including courts, police, and prisons. The result is that the path to peace is distracted or derailed by the illicit actors that a carefully designed rule of law institutional development plan should have addressed.

But regardless of where the efforts begin, ample historical evidence has shown that launching or rebuilding the institutions that make up the justice and rule of law sectors is the principal way to invade the vacuum, overtake and overpower the illicit structures, and pave the road to peace. Without this enormous effort, the political landscape is abandoned to a modern tragedy that consigns the lives of ordinary citizens to a sad and sorry state.

The foremost positive examples of this process are seen when the leaders of power networks operating outside the law (and often in concert with the military, malicious leaders, or interim or ad hoc political leaders) give way to a legitimate return to the rule of law. This can enable an acceleration of the peace process. And it creates a window during which peacekeepers and transitional or nascent governments can address further national enhancements such as education, industry, investment, and, most significantly, amelioration of the lives of ordinary citizens. The whole constitutes a sea change in the political ecosystem and brings a reversal to the otherwise downward spiral of lawlessness, conflict, and chaos.

When Institutional Reform Works, and Why

Bosnia and Herzegovina: Ensuring Equitable Representation and Preventing Illicit Capture of State Institutions

It is always useful to start with a look at what works, before criticizing what does not. One realistic and relatively successful example emerges from postwar Bosnia and Herzegovina. In 1995, the Dayton Peace Accords and other agreements included reforming, restructuring, certifying, and rebuilding police agencies of the two political entities: the Federation of Bosnia-Herzegovina and the Republika Srpska. Developing a Federation-based police academy at Suhodol, with young recruits including Muslims, Serbs, and Croats, was a major struggle and was aggressively opposed by powerful leaders who profited from the chaos of Bosnia's civil war. This was a war that had claimed the lives of more than 200,000 citizens and laid waste the functioning societies, along with their legal systems and structures. And during the reconstruction that followed, factional leaders from the conflict mounted considerable opposition to opening the police academy.

With much fanfare, the academy opened, but when it did, these factional leaders insisted on a segregated approach to housing. In practice, this would have required that the police recruits from each faction be housed on separate floors, thus defeating the purpose behind having a unified national police institution. So factional opposition was rejected, and the academy was launched, with three beds to a dormitory room, assigned by alphabetical order of last name. The only segregation was male/female. This provided random integration, often resulting in a dorm room occupied by a Serb, a Croat, and a Muslim recruit. Proving naysayers wrong, the recruits got along peacefully and amiably. Desegregation sent a powerful signal that police agencies were to be formed along professional lines that included ethnic, religious, and geographic diversity.

Institutionalizing this element provided momentum for the institution-building process and marginalized the loud-talking political leaders who profited from the postwar chaos. But the decision came not without cost. One sad outcome was that a car bombing subsequently took the life of a Croat minister of interior who had supported the police training development program. He had also fought local mafias representing the Croat subculture that formed during the war. Building institutional capacity by establishing a fully diversified national police academy had unsettled the power bases at the heart of the roadblocks to the peace process. And although the minister paid with his life, the institution survived.

Liberia: Restoring Control and Legitimacy within the Rule of Law

Another historical example can be found in Liberia, whose 14-year civil war had claimed some 240,000 lives. During that time, in the vacuum of institutional influence, illicit profiteers formed overlapping power zones. Liberia's constitution, penal code, and code of civil procedure were put on the shelf, and lawlessness prevailed. The result was that Liberia became a major transshipment hub for weapons, drugs, and persons, in a market operated by criminal networks that had profited from the chaos of war.

In late 2003, immediately after the signing of the Accra Peace Accords, none of Liberia's three principal civil war factions—Liberians United for Reconciliation and Democracy (LURD), the Movement for Democracy in Liberia, or the Government of Liberia—could claim legal legitimacy, since each operated clearly outside any law other than the orders issued by rapacious military despots. Actual power emanated from fighters who later emerged as self-proclaimed leaders and power brokers. These criminals carved up the spoils of war with unbridled violence.

The violence was exemplified prominently on Bushrod Island, just across from an infamous bridge that provided entry into Monrovia from the country's outer reaches. The bridge had been the scene of intense wartime fighting, with enormous loss of life. The Bushrod Island Police Station, at the heart of what had been the area's main commercial zone, had been looted, and in late 2003, both the bridge and the station stood in a state of severe infrastructural disarray. The station had also served as a strategic location for LURD's criminal activities. The group saw it as a profit center and occupied it with pseudo "police"—LURD fighters who had no legal authorization to perform police functions. Operating like a street gang, LURD fighters established a zone of fear on the densely populated island, taking into custody anyone they saw as able to fulfill a profit need. The pseudo police would hold the detainees until families or friends arrived and paid ransom money for their release. They performed gangland-style extortion or "protection" for a fee, using power and influence based on the wartime legacy of fear and violence, rather than abiding by Liberia's constitutional law.

At the urging of the UN Police Commissioner, the UN Mission in Liberia Police designated the Bushrod Island Police Station as a "Model Police Station Quick Impact Project." The UN Police very publicly kicked out the LURD fighters, rebuilt the station with a modest amount of UN Quick Impact Project funds, and equipped it for colocated UN Police who were operating alongside carefully selected legitimate Liberian leaders and officers from the Liberian National Police ranks. The illicit power structures that had been engaging in extrajudicial imprisonment, kidnapping for ransom, extortion, and torture were replaced through a focused demonstration of legitimate institution building. This sent a strong signal to the much relieved Bushrod Island communities that the police were going to be functioning once more and that they were there to serve the community, not ravage it.

Justice Reform

While these two examples illustrate police-related legitimacy, many similar examples pertain to courts and prisons. During the chaos of war and transition, all rule of law institutions tend to fall to illicit power, which disguises itself with institutional titles and trappings. Postwar or failed-state conditions exacerbate the problem through the state's inability to exercise legitimate government control, and increase the profit potential for criminal organizations that thrive on the resulting chaos.

One of the principal needs of human society is justice, and illicit actors know this and exploit it. Because of this, institutions that provide justice are strategic pivot points toward stabilization and make attractive targets for the illicit actors that dominate during chaos. To illustrate, we can look at the successful transitions that took place in Latin America in the 1980s and 1990s. Countries such as Colombia and Panama, for example, succeeded because peace was accompanied by wholesale transformation of the justice and security systems writ large, along with modifications to the legal framework that underpinned the transitions. Such examples show that in undertaking institutional reform, illicit power structures that dominate rule of law systems are high-value targets. And their successful dismantling produces the greatest results in a classic path to peace.

Three Ingredients for Success

To successfully deploy institutional development as a tool for disrupting illicit power, three principal elements are needed.

Early Architecture

First, there must be a governance framework that enables systemic, rule of law-based institutional reform. The governance architecture must be accounted for *during* the peace negotiations, not after. It should be included in the articles of peace accords, mirrored in and empowered by UN Security Council resolutions, and codified in memorandums of understanding. Language in these documents must include provisions that can authorize and guide necessary legislative reform as well as the reform of courts, prisons, and police institutions.

Too often, agreements simply refer to reforming and restructuring the police without providing any guidance on the form and structure that should result. The effect is that deliberate shaping of the critical institutions and human capital that operate in the postwar geopolitical environment occurs only much later, and often by default. When these vital aspects are addressed late, rather than in early planning, the golden hour is lost. The low-hanging fruit—reforms that are easier to implement while a strong international advisory and security presence is in place and in charge—becomes progressively harder to reach as a new or restored government consolidates its control. Angels, not devils, are in the details when these are drafted early on, with long-range goals in mind.

An excellent example of the degree of detail that can be included in guiding documents and agreements is seen in Plan Colombia.[1] This codification of both national policy and the peace agreement it furthered established a clear framework for reform of the entire Colombian security system. And it provided the road map for the peace process with the active insurgent groups, for Colombian development, and for the international assistance that supported both peace and development in the years ahead.

Similarly, the Bonn Agreement (Afghanistan), along with its accompanying compacts, agreements, and declarations, established a framework for interim governance and long-term development following the U.S.-led invasion of Afghanistan in 2002. Although imperfectly executed, the agreement did contain relatively precise guidelines for restoration of the institutions and functions necessary to restore and strengthen the rule of law. And it informed both domestic Afghan and international assistance during the decade that followed.[2]

Authentic Partnerships

Second, there must be a clear systemic pathway for forming partnerships between international donors and domestic actors within the host nation. These partnerships must be formed with a common vision, mission, goals, and strategic approach. Most importantly, the international community must be willing to accept unity of effort, together with a systemic ideology that the host nation itself sees as legitimate and culturally acceptable. International actors must form authentic partnerships with stakeholders and must be willing to subordinate their narrower preferences in the interest of a larger common objective. Nongovernmental organizations and local players are not exempt from this principle. Even though they may not have been party to the agreements that facilitated more formal, international intervention, they, too, must accept a common view if their work is to add value. They should work toward common goals based on internationally accepted rule of law ideals that include human rights considerations and accepted norms of police; courtroom; and prison doctrine, standards, policies, and procedures. Where gaps exist in domestic systems, international treaties and conventions should be assessed for their applicability. Particularly in the field of prison administration, which tends to be fraught with abuse, serious penal reform should be considered wherever the existing standards are unacceptable under international humanitarian and human rights law.

In the absence of a solid domestic legal framework, interveners can use model codes as a starting point for engagement. A good one is the Penal Law and Code of Criminal Procedure, reproduced through an enormous effort as the Model Codes for Post-conflict Criminal Justice Project.[3] Development of the Model Codes was an initiative of the

---

[1] "Plan Colombia: Plan for Peace, Prosperity, and the Strengthening of the State," USIP Peace Agreements Digital Collection, May 15, 2000, www.usip.org/sites/default/files/file/resources/collections/peace_agreements/plan_colombia_101999.pdf.

[2] UN Security Council, "Agreement on Provisional Arrangements in Afghanistan Pending the Re-establishment of Permanent Government Institutions," Dec. 5, 2001, http://peacemaker.un.org/sites/peacemaker.un.org/files/AF_011205_AgreementProvisionalArrangementsinAfghanistan%28en%29.pdf.

[3] Colette Rausch and Vivienne O'Connor, eds., *Model Codes for Post-conflict Criminal Justice*, vol. 1 (Washington, DC: USIP Press, 2007).

United States Institute of Peace (USIP), the Irish Centre for Human Rights, The UN High Commissioner for Human Rights, and the UN Office for Drugs and Crime. The goal was not to create a prescriptive framework for all places at all times, but to provide something useful that could fill the gap during the postconflict stabilization and reconstruction period in those states that did not have a legitimate or acceptable basis of criminal law in place. The Model Codes could be used by interveners as an interim measure, and provide the starting point for legislative action by the host nation.

Much has also been said about respect for cultural differences, and the need for "local buy-in" for any institutional development effort. But although it is indeed essential that the international community be sensitive to culture, and although local ownership is key to sustainability, not all the criticism is valid. Illicit actors are particularly skilled at arguing why an international standard of accountability or inclusiveness, for example, should not be applied. The fact remains that at least some of the clamor supporting these so-called cultural mores, norms, and values is just a thinly veiled attempt to retain a corrupt status quo or shore up an illicit power base. This is often seen in the so-called cultural reluctance to adapt to international norms for civilian oversight, professional policing, judicial power, anticorruption, and refusal to protect the rights of minority and vulnerable populations.

These are important issues for effective institution building, and resistance needs to be carefully examined. History provides ample evidence that illicit actors will use every tool, including the guise of national norms and procedures and need for "buy-in," to disrupt the building of institutions that ultimately threaten their selfish interest. Introducing internationally respected codes, such as those still available as legacy work of the international law and human rights partners, can go a long way to overcoming pseudo cultural resistance.

Accurate Measurement of Results

Third, results must be rigorously measured. Clearly, certain intangibles will never be measurable. But the significant measurements can be entrusted to established tools available to assess the forward motion, or lack of it, in institutional reform. The common navigational distress produced by countless news "events," superficial accounts, shallow papers, and distorted anecdotes ginned up amid the abundance of hysteria and through social media all work together to form a distorted view of the state of play in the local environment. When reformers get too caught up in the news cycle, they tend to become reactive in their development activities, and any chance for consistency and sustained progress will be lost.

One excellent measurement tool that has been widely used in Liberia, Haiti, and South Sudan is the Rule of Law Indicators (ROLIC).[4] A product of the UN Department of Peacekeeping Operations and the UN High Commissioner for Human Rights, ROLIC produces navigational coordinates of a national rule of law state of play. Using a template consisting of available administrative data, stakeholder surveys, and expert findings, ROLIC reports on the multiplicity of competences within each of the three

---

[4] United Nations, "UN Rule of Law Indicators: Implementation Guide and Project Tools," 2011, www.un.org/en/events/peacekeepersday/2011/publications/un_rule_of_law_indicators.pdf.

principal security-sector institutions that first come to mind when we think of the rule of law: courts, police, and prisons.

While not conclusive, the ROLIC produces indicators of progress, in and easy-to-digest, linear fashion. As multiyear reports become available over time, trends emerge and create opportunities to engage with development partners and to identify weaknesses in the system. ROLIC data and the framework itself help illuminate weaknesses in the security system and in accompanying institutional development programs, which are particularly vulnerable to illicit power. ROLIC is not infallible, but it can produce longitudinal measurement for players and facilitate greater transparency for development processes. It can also serve the most fragile communities very well. Weak points and corrupt institutions are more clearly visible, and salutary efforts are more effectively identified in a way that clarifies rather than obfuscates.

Other measurement tools can and should also be deployed to measure institutional growth, and there is a growing body of learning on metrics and on measures of effectiveness for quality assurance. Unfortunately, often the wrong things are measured (for example, the numbers of police in a police organization, the numbers of prisoners in the prison system, and the numbers of cases being disposed of in the judicial system). While these are fairly easy to measure, the measurements do not take into account the levels of competence, for example, among the police being graduated from accelerated police training programs designed to produce numbers. Similarly, the reason for prison overcrowding is often a direct failure of the country's court system rather than an increase in police effectiveness. Often, a prison system's effectiveness or capacity is paired with indicators intended to measure the courts' capacity to dispense justice fairly and equitably. While it is useful and essential to look at how justice functions as a system, the subsystems within it must be carefully examined within their own peculiar set of authorities and functions.

Recommendations for Effective Postconflict Institutional Development

While it may seem that any effort to build strength into broken systems—especially justice systems in fragile or postconflict states—would be welcome, the truth is that we in the international community often worsen rather than improve the situation. Funds are wasted, sincere host nation-generated reform initiatives are neutralized, and we contribute to the very dysfunction that is at the heart of corruption. When this occurs, illicit power structures are energized rather than dismantled.

No one would argue against the notion that weak systems constitute a fertile hotbed for the emergence of illicit power. What we rarely admit, however, is that we ourselves can and do make things worse. Below is a checklist of recommendations that should help accelerate institutional development in the justice and rule of law sectors.

1. *Seize the golden hour.* While delays are inevitable in the absence of agreements and other broad-based resolutions, be ever aware that the earliest interventions are usually the most effective.

2. *Build the capacity of institutions by taking on entire justice and rule of law systems.* Transformation of broken systems must be done not in parts but in the whole.

3. *Measure the right things.* There is a tendency to measure the most easily identified and visible aspects of institutional reform, such as recruitment numbers and demographics, graduation rates, arrest and conviction rates, and improvements in physical infrastructure, while overlooking the trickier ones and the intangibles that make for solid institutions. Issues such as improvements in institutional culture, the relationship between police and prosecutors, professionalism within the legal and law enforcement sectors, public awareness and understanding of rights and remedies, and judicial willingness to enforce them are all indicators of institutional development progress. These are difficult to quantify and can be measured accurately only over time, but they also represent sustainable progress toward overcoming the biggest hurdles to credible, capable, and enduring institutions and systems.

4. *Do not let donors contribute to division or discord.* The term "donor interference" has emerged as stark reminder that when donors are not coordinated in a commonly accepted architecture, the result is delay at best and dysfunction at worst. Coherence across the donor community should be considered a metric for programmatic performance.

5. *Wrest ownership of postconflict justice institutions away from self-serving politicians.* Bringing justice systems under oversight of legally constituted leaders is essential. But in fragile environments, there is a tendency toward extralegal grabbing of institutions. Police need to be focused on serving communities, courts need to be unfettered by political interference, and prisons need to be driven by international standards of human rights and the need to ensure that inmates serve only their lawfully prescribed sentences.

6. *Avoid military oversight of the police, and if it is necessary in the near term, end it as quickly as possible.* During and following internal and subregional conflicts, police become nothing more than tools of militias that have entirely different doctrines, procedures, and institutional cultures. After war, there is a tendency to conflate all types of security forces and, in particular, to merge former military combatants with the police. This, too, is a mistake. Community policing, for example—a powerful tool in the fight to control illicit power—is typically in direct conflict with military purpose and doctrine.

7. *Never underestimate the importance of the prison system.* There is probably no clearer indicator of postwar chaos than overcrowded prisons in a state of system disarray, with prisoners dubiously incarcerated in inhumane conditions and with no term end in sight. This prison dysfunction contributes to impunity of those few—especially those linked to illicit power—who have successfully bought their freedom.

8. *Remember that results are incremental, generational, and rarely immediate.* Those who look to nominally functioning justice systems or hastily stood-up police forces as an exit strategy need to look longer range, toward the strength of the institutions, not to numbers and individual personalities. Where institutions are being restructured from scratch, building an accountable, enduring institutional culture takes time, effort, and adaptation as challenges and opportunities emerge. Where institutions are being reconstituted following failure or co-optation, it is critical to recognize that often, the institutional damage occurred over a generation or more. The repair cannot be seen as requiring anything less.

9. *Demystify the language of our papers and memorandums.* The reluctance toward robust and swift transformational leadership often stems from misinterpretation of vague language in the papers, agreements, or mandates.

10. *Join hands instead of pointing fingers.* In postconflict environments, there is a tendency toward disparaging other players and stakeholders. Governmental and nongovernmental players, as well as contractors and local players, would do well to expend their energy in forming authentic partnerships based on common alignments, rather than in criticism.

Conclusion

Institutional development is a powerful tool in the fight against illicit power. Done right, it produces indigenous security forces that can protect the population within the framework of the law, protect and ensure the fair and equitable administration of justice, address grievances that were underlying drivers of conflict, and control and isolate those elements of the population who remain committed to conflict and undermine legitimate government. But if done hastily or incorrectly, institutional development enables illicit power brokers, political patronage networks, criminal enterprises, and other bad actors to consolidate power in the immediate postconflict chaos and disorganization. It leads to the capture of state resources and institutions for personal rather than public gain. And finally, it leads to a resurgence of the very grievances that led to instability and conflict in the first place. For these and many other reasons, it is essential that we get institutional development right.

15. The Hitchhiker's Guide to Intelligence-Led Policing

Clifford Aims

The term "intelligence-led policing" (ILP) has been in use at least since the 1990s, and its origins trace back even further, to the early 1970s. The term shows up everywhere in plans, policies, and procedures for transition from foreign intervention to domestic operations in enforcing the rule of law. And yet, most planners and policymakers have little practical understanding of what ILP means, what it looks like when it works, and what it takes to build sustainable, civilian-led ILP capability within the host nation's security and justice system. Intelligence-led policing is critical to a state's ability to check the rise of illicit power and control illicit organizations and activities where they already exist. Thus, a solid understanding of the principles behind illicit power, and of the capacity required to conduct it, is key. This chapter intends to fill our own ILP knowledge gaps so that planners and implementers alike will understand its impact and ask the right questions when assessing need, capacity, and risk.

Defining ILP

Like other strategies and methods used widely in modern law enforcement, such as "community-oriented policing," "problem-oriented policing," or "broken windows theory of policing," ILP defies any simple single definition. To understand ILP, at a minimum we must be aware of who is using the term, what context the method is being applied to, *why* it is being applied, and what the desired outcomes are for its use. Without such a frame of reference, ILP is largely a meaningless concept with no logical application. This is a particularly acute issue when related to security sector reform (SSR), since poorly defined efforts to introduce ILP will have hit-and-miss results at best. At worst, the unintended consequences will exacerbate existing problems and even create new ones. Such an approach violates the golden rule of SSR: "first, do no harm."

The term "intelligence" alone is charged with misunderstanding, raising apprehensions by its mere association with police or law enforcement activities. Thus, it helps to be familiar with some of the more commonly used definitions of ILP, which can help clarify what it is *not*.

- "Intelligence-led policing is the application of criminal intelligence analysis as an objective decisionmaking tool in order to facilitate crime reduction and prevention through effective policing strategies and external partnership projects drawn from an evidential base."[1]

- "Intelligence-led policing is a collaborative philosophy that starts with information, gathered at all levels of the organization, that is analyzed to create useful intelligence and an improved understanding of the operational environment. This will assist leadership in making the best possible decisions with respect to crime control strategies, allocation of resources, and tactical operations."[2]

---

[1] Jerry H. Ratcliffe, "Intelligence-Led Policing," Australian Institute of Criminology, Trends and Issues in Crime and Criminal Justice Series, no. 248, Apr. 2003, 3, www.aic.gov.au/media_library/publications/tandi/ti248.pdf.

[2] New Jersey State Police, "Practical Guide to Intelligence-Led Policing," Manhattan Institute for Policy

- "The collection and analysis of information related to crime and conditions that contribute to crime, resulting in an actionable intelligence product intended to aid law enforcement in developing tactical responses to threats and/or strategic planning related to emerging or changing threats."[3]

- "Intelligence-led policing is defined as *the collection and analysis of information to produce an intelligence end product designed to inform law enforcement decision making at both the tactical and strategic levels.*"[4]

One of the shortest definitions, from the Netherlands, simply states, "Intelligence-led policing is the use of analyzed information by decision makers to decide on police resources."[5]

A Tailored Approach

Anyone involved in security sector reform or capacity building within domestic security institutions, or faced with the challenge of containing illicit activities and organizations, should understand the potential benefits of ILP strategy and process, and the complex situations where it will likely be applied. Very briefly, ILP is a policing strategy that relies on a well-defined plan of collecting data related to criminal activity and then applying a rigorous method of analysis to clearly identify the type(s) of crime being committed and any discernable patterns, with the ultimate goal of determining how such criminal activity can be anticipated, prevented, or disrupted. This information is then used to guide and inform decision making on the tactical and operational levels for making a response. ILP can be effective in dealing with problems as diverse as bicycle thefts on a college campus, and human and contraband trafficking by multibillion-dollar transnational criminal syndicates. There are many case studies on the successful application of ILP from all around the world, and even a cursory review of this information should make one critical central fact evident:

There is no one "right" way to do ILP. The strategy must be tailored to meet the unique needs of each situation, and the methods must be legally acceptable and practical for the stakeholders' use.

The implications for international personnel engaged in SSR and other capacity building or security assistance are enormous. It is easy to fall prey to the seductive logic that "ILP worked there, so it will work here." But such thinking can unduly weight the decision to import a particular solution. A far more useful strategy involves working through a process to help the recipients of SSR assistance objectively identify the problems and determine how those problems can best be addressed *within the realities and*

Research, Sept. 2006, www.njsp.org/divorg/invest/pdf/njsp_ilpguide_010907.pdf.

[3] David L. Carter, *Law Enforcement Intelligence: A Guide for State, Local, and Tribal Law Enforcement*, 2nd ed. (East Lansing, MI: Michigan State Univ., 2004), www.cops.usdoj.gov/pdf/e09042536.pdf.

[4] U.S. Justice Dept., "The National Criminal Intelligence Sharing Plan," Oct. 2003, https://it.ojp.gov/documents/ncisp/National_Criminal_Intelligence_Sharing_Plan.pdf.

[5] Mariëlle den Hengst and Jan ter Mors, "Community of Intelligence: The Secret behind Intelligence-Led Policing," Delft Univ. of Technology, 2012, http://pubs.iids.org/index.php/publications/show/1879.

limitations of the indigenous system. To use a grossly simplified medical analogy, we can liken ILP to surgery. The surgery that worked great for patient X may not have the same results for patient Y, because patient X had appendicitis, whereas patient Y's similar symptoms stemmed from a bad gallbladder. That is, both surgery and ILP are effective, but only if designed to address carefully identified and analyzed problem sets.

Another aggravating factor is the basic human predilection to operate in an environment that is familiar. Experience has shown repeatedly that international SSR efforts often try to import practices that the implementers know and are comfortable with, whether or not they happen to be well suited to the problems at hand.

There can also be confusion regarding the roles of SSR implementers. ILP is an inherently domestic law enforcement operation and must be undertaken only by those who are authorized to conduct such operations by the domestic law of the country where they are carried out. Except in certain highly specialized and unusual situations, international staff engaged in SSR are not charged with directly carrying out law enforcement functions on behalf of the host nation. Therefore, it is critical for anyone reading this article to remember that it is not your role to "do" ILP. The various indigenous entities you are working with are the ones lawfully empowered to enforce the law. And you are there to help them determine whether ILP is an appropriate strategy and how it can best be applied. In other words, your responsibility is to help your local counterparts be better able to arrest the right people, for the right reasons.

At the same time, remember that while ILP is a proven strategy and tool, it is not a magic solution to any problem, and it cannot be applied in a vacuum. Even the best-designed ILP program will be rendered ineffective if the other sectors of the host nation's justice system are compromised or institutionally weak. Simply put, making more arrests does not, in and of itself, result in greater stability, security, or public safety. There must also be comparable institutional capacity to prosecute, adjudge, and incarcerate offenders.

Designing Assistance for ILP

Unlike many other security sector assistance-related development programs, designing a program to help a host nation adopt ILP does not begin with the development strategy—far from it. Since ILP is a type of law enforcement operation, it is likely that ILP, or an equivalent operational technique, already exists, has been previously informed by legislative, policy, or judicial decisions, and merely needs to be refined or restored. Without a comprehensive review and thorough understanding of the existing justice system, we cannot know whether it makes sense or is even legal to introduce a new ILP concept or methodology. At best, you may very well come up recommending a solution for a problem that does not exist. At worst, you could initiate a strategy that contravenes host nation law or custom, thereby undermining the effectiveness and legitimacy of the entire effort.

Assessing the Current Justice System

Fortunately, a number of handbooks and tools are available that can guide the assessment of a justice system. Among these are the OECD's SSR handbook, "The INL Guide to Justice Sector Assistance," and the DCAF backgrounder on police reform.[6]

It is important to note that these and other related resources do not have the answers. But they do explain the critical *questions* that must be asked in order to shape and inform any SSR effort and, particularly, to help determine what form of ILP will work in the particular context. At the same time, there is no substitute for having experienced SSR professionals assist with the assessment process. This is not a role for amateurs or generalists. It is a highly technical field in any system, and every effort should be made to bring in the right expertise, even if only temporarily, to help give the program design process a solid foundation. Having expert help at the start of the SSR process greatly improves your ability to start a program in the right direction and reduces the likelihood of having to make major changes in the future.

ILP is not a cure-all for law enforcement failure, however. Implementing ILP can have a number of unexpected negative consequences. Some of the more obvious include putting a strain on the courts and the prison system through increased numbers of arrests, and pushing criminal activity out of more effectively policed areas into areas less prepared to deal with it. Less intuitively, an ILP program that is effective against organized crime can result in political and social backlash against the police because the patronage system of targeted illicit power structures becomes threatened, affecting the livelihoods of citizens not directly engaged in criminal activity (e.g., shopkeepers, bar and restaurant owners). In extreme cases, the targeted criminal elements may be so intertwined in the political establishment that their arrest could create instability within the government. None of this should be a disincentive to adopting ILP. But it is vitally important to consider the second- and third-order effects that could impede overall progress in stabilizing an area and bringing about needed reforms. The implementer must recognize them when they occur, and be prepared to make programming adjustments to mitigate the risks.

Preparation and Training

ILP operations can significantly affect the law enforcement institutions themselves. This is particularly true when ILP is part of a transition from military to civilian control in a postconflict environment, or from a less investigation-oriented policing approach. For example, ILP will very likely require specialized training for police *and* prosecutors, who must learn how such operations can be technically managed and properly applied within the existing legal system. The roles and responsibilities between police and prosecutors may shift, particularly in a civil law jurisdiction, and both need to understand

---

[6] OECD, *OECD DAC Handbook on Security System Reform: Supporting Security and Justice*, Feb. 2008, www.oecd.org/governance/governance-peace/conflictandfragility/oecddachandbookonsecuritysystemreformsupportingsecurityandjustice.htm; U.S. State Dept., "INL Guide to Justice Sector Assistance," 2010, www.state.gov/documents/organization/222048.pdf; DCAF, "Police Reform," Oct. 2009, www.dcaf.ch/Publications/Police-Reform.

clearly what is required in order to bring evidence collected through such investigations to trial. Additional technical and legal training for judges may also be warranted. One critical step is a thorough review of the existing laws and codes of criminal procedure. In some locations, many of the practices commonly used by police in the course of an ILP case may not be allowable, and some degree of legislative reform may be necessary. Where regime change has occurred, there may also be resistance to the use of criminal procedure codes that were enacted by the prior regime. In such cases, the use of interim codes may be required, but this creates an additional training and education burden that will have to be addressed across the full range of the security sector.

Aside from training requirements, adoption of ILP may require institutional reorganization or reform of the police and prosecutorial institutions. Since ILP investigations can be complex and require increased confidentiality, law enforcement and prosecutorial agencies may have to set up special units and divert resources to support the new initiative. This, in turn, can cause institutional resistance to ILP if existing units see their resources or prestige threatened by the change. In some cases, it may be a simple matter of someone feeling marginalized and "left out of the loop." In any case, the unexpected institutional knock-on effects of the transition to ILP can create resistance to its adoption, reduce or limit its effectiveness, and hinder overall SSR efforts.

Using ILP against Terrorism

Just as ILP works against any other form of violent criminal activity, when done well, it is also effective against terrorism. Unfortunately, this is the area most prone to *misuse* of ILP, which can threaten SSR efforts in general. Acts of terrorism such as the 2013 Boston Marathon bombing, the 2008 attacks in Mumbai, and less publicized activities such as kidnappings for ransom, receive near-universal condemnation. When such acts occur, there is wide support for the affected government to respond forcefully in capturing and punishing the perpetrators. In this context, ILP can be tremendously useful, but it can just as easily become a tool of repression, particularly in societies with weak oversight-and-accountability mechanisms or with a history of policing as a tool of the regime rather than as a public service.

Beware the Slippery Slope

Implementers of SSR must always remember that ILP should be viewed as a method for improving the host nation's ability to provide public security and the rule of law. Thus, assistance strategies have to be designed to instill this idea, and the host nation's security services must be held to account when they begin to stray from this central tenet. Also, while ILP is very effective at identifying individuals or organizations likely to carry out criminal acts such as terrorism or crimes against internal security, how the host nation defines "terrorism" or "insurgency," and who it labels as "terrorists" or "insurgents," can become an issue. Overly broad legal or political definitions will subvert the nature of ILP, degrade the public perception of policing's legitimacy, and damage the credibility of wider SSR.

Abuses of ILP are not limited to the pursuit of terrorists and insurgents. There is great potential for ILP to be misused or, more insidiously, used selectively in pursuing regular criminals to the point of detention, but not through the whole of the justice process. Selective application of ILP is most likely to occur when the larger justice system is weak or incomplete—a common situation in areas recovering from conflict or other societal breakdown. In weak systems, criminals quickly identify the capacity gaps and seams and exploit them for illicit gain. Even a government that is largely not corrupt and not reliant on the criminal activity knows full well that high levels of criminality can cause a lack of public confidence in the legitimate institutions. In such cases, law enforcement entities are often under tremendous pressure to reduce the crime rate, but this can come at a cost. The risk that law enforcement will use ILP to conduct extrajudicial detentions and incarceration is very real.

On the positive side, in such a scenario, if donors can act quickly before bad practices are entrenched and illicit relationships formed, most police will be happy to receive training and other assistance that increases their capabilities, and are therefore likely to adopt ILP as quickly as possible. The SSR assistance will likely result in immediate and positive results (e.g., higher arrest rates) that would appear to be a good thing. But again, caution is in order, for these results can mask certain underlying systemic problems where police capacity exceeds that of the other justice institutions. In a weakened justice system, the police will know how able the prosecutors are to deal with suspects. They will also know how effective the independent oversight mechanisms are that regulate police conduct. It can be very tempting for police to use ILP only to the point that they identify a suspect and make an arrest, without doing the full investigation necessary to build a solid case that can be presented in a court of law.

Knowing that the prosecutorial function is weak and oversight mechanisms are ineffectual, police can be inclined to rely on confessions (possibly obtained through less-than-transparent means) rather than on hard evidence as the foundation for a case that is referred for prosecution. The implication for those implementing SSR and wanting to introduce ILP is that such an initiative, without a thorough understanding of the indigenous justice system, may do little more than enable the police to go out and "round up the usual suspects." As mentioned earlier, ILP can lead to improved *results,* such as higher arrest rates, but not advance (and, indeed, even hinder) achievement of overall *outcomes,* such as establishing a strong justice system that promotes stability and the rule of law.

Ultimately, the important thing to remember is that intelligence-led policing is more than just the insertion of "intelligence" into particular types of policing operations. Effective ILP engages the whole of the justice sector, within the host nation's domestic legal framework. It can often require highly specialized and technical capabilities, but it is just as often about low-tech community engagement and personal interaction. In a shifting environment, ILP is an essential tool in the fight to control illicit power. But it is equally prone to abuse. And it can produce unintended consequences for weak or nascent governance and justice systems if it is not carefully overseen and calibrated to match the capacity of the other government institutions that its outcomes depend on. As governance takes hold, however, ILP, when done right, lends strength and legitimacy to nascent governance institutions and ultimately preserves the monopoly of force for licit actors, rather than ceding it to illicit power.

TIMOR-LESTE
National capital
District seat
Town, village
Airport
International boundary
District boundary
Road
Savu Sea
Timor Sea
Strait of Wetar
Wetar Island
Atauro Island
Lirah Island
Kisar Island
Jaco Island
INDONESIA
Dili
Liquica
Maubara
Baucau
Manatuto
Lospalos
Viqueque
Same
Ainaro
Suai
Maliana
Aileu
Gleno
Pante Macassar
LIQUICA
DILI
AILEU
MANATUTO
BAUCAU
LAUTEM
VIQUEQUE
MANUFAHI
AINARO
ERMERA
BOBONARO
COVA LIMA
OECUSSI
0 10 20 30 km
0 5 10 15 20 mi
The boundaries and names shown and the designations used on this map do not imply official endorsement or acceptance by the United Nations.

16. Security Sector Reconstruction in Post-Conflict: The Lessons from Timor-Leste

Deniz Kocak

As history makes clear, illicit power arises wherever there is a vacuum of security, justice, and accountable governance. In response, over the past 15 years, the international community has increasingly begun to apply security sector reform (SSR) activities in postconflict countries. Usually done under UN auspices, SSR is an effort to reestablish the rule of law and mitigate the instability and lawlessness caused by dysfunction in the security sector following political transition or conflict. SSR has been recognized as a crucial element of peacebuilding operations worldwide.[1] Timor-Leste was long regarded as an ideal setting for UN-led peacebuilding. After managing to establish a local police force and security governance institutions, it was held up as an example of successful externally led security sector transformation.

But in 2006, violent clashes between security forces cast doubt on the stability of the newly established security system, thus drawing into question one of the mainstream policies of international approaches to security governance in postconflict settings. In an effort to address the emerging problems, the UN mission's mandate was amended to help rebuild the dysfunctional security sector. But by that time, five years into the intervention, the mission's ability to further influence development was openly challenged by growing local resistance to external interference in the sensitive domain of security. Simultaneously, the mission faced assertive local approaches to security governance, and Timorese demands for local ownership.

This chapter looks at these questions: What approach to SSR was applied in Timor-Leste before and after the 2006 security crisis? Which aspects were successfully implemented? Which ones failed, and why? Using the Timor-Leste experience as an example, this analysis will argue that standardized "tool kit" SSR implementations, insufficient attention to the local context, and inherent local dynamics that did not mesh with external donors' policy agendas did not lead to a stable security situation or to a legitimate, civilian-controlled security sector as the Western security governance paradigm advocates. Rather, externally imposed, misaligned policies led to unintended negative consequences when local actors selectively adopted parts of the external SSR agenda while neglecting other parts. This outcome of partial resistance and adaptation by local actors is a critical point for further research on the possibilities and limits of SSR in postconflict countries.

An earlier version of this chapter was prepared for the annual meeting of the International Studies Association in Montreal, March 2011, and published as a working paper. Deniz Kocak, "Security Sector Reconstruction in a Post-Conflict Country: Lessons from Timor-Leste," SFB-Governance Working Paper Series no. 61, Oct. 2013. Deepest thanks to Fairlie Chappuis for her valuable comments on the manuscript.

[1] Irene Bernabéu, "Laying the Foundations of Democracy? Reconsidering Security Sector Reform under UN Auspices in Kosovo," *Security Dialogue 38*, no. 1 (2007): 71-92; Ursula C. Schroeder and Fairlie Chappuis, "New Perspectives on Security Sector Reform: The Role of Local Agency and Domestic Politics," *International Peacekeeping* 21, no. 2 (2014): 133-48; Fairlie Chappuis and Heiner Hänggi, "Statebuilding through Security Sector Reform," in *Handbook of international statebuilding*, ed. David Chandler and Timothy D. Sisk (London: Routledge, 2013), 168.

The Historical Backdrop

The history of Timor-Leste is characterized by foreign rule and domination. The island of Timor, in the Lesser Sunda Islands, attracted Portuguese and Dutch merchants and missionaries in the sixteenth century. In the following centuries, both colonial powers gradually consolidated their rule on Timor, ultimately formalizing their territorial claims in the Treaty of Lisbon (1859), which designated the western part of Timor as Dutch, and the eastern part as Portuguese.[2] Whereas the Dutch left Timor during the decolonization process and the formation of the Indonesian Republic in 1949, Portugal clung to its colonial possession in Southeast Asia until the Carnation Revolution of 1974. The breakdown of the Salazar/Caetano regime in Portugal eventually led to the disintegration of the Portuguese colonial empire. Soon, newly formed local political parties in Timor-Leste struggled for political control. A coup by the pro-Portuguese party União Democrática Timorense against the left-leaning, socialist party Frente Revolucionária do Timor-Leste Independente (FRETILIN) failed, and FRETILIN emerged victorious.[3] But only a few days after the declaration of independence by FRETILIN in November 1975, pro-Indonesian militias and Indonesian armed forces invaded Timor-Leste, killing several thousand Timorese civilians.

Although the official incorporation of Timor-Leste as the 27th province Timor Timur (East Timor) into the Indonesian Republic was carried out in early 1976, the actual pacification of the new Indonesian province lasted several years. Particularly the armed wing of FRETILIN, the FALINTIL (Forças Armadas da Libertação Nacional de Timor-Leste), resisted Indonesian rule through guerrilla tactics and clandestine networks.[4] Despite international protests, Indonesia maintained its occupation. But over time, several factors led to a change of Indonesia's attitude toward Timor Timur. Changed international political developments at the end of the Cold War, the internationally condemned massacre of several hundred Timorese civilians by Indonesian soldiers in the Santa Cruz cemetery in 1991, and, finally, the financial crisis of 1997 in Southeast Asia led to the fall of the Suharto regime, which led to political reforms (Reformasi) in Indonesia and altered its policy toward Timor Timur as well.

The political turning point came in 1999, when Indonesian President B. J. Habibie granted the people of Timor Timur a referendum on the future political status of the eastern part of Timor within the Indonesian state. After some delays and with the help of the UN Mission in East Timor, the referendum was finally held in August 1999. Although most East Timorese supported autonomy, pro-Indonesian militias incited unrest, wreaking such devastation that the UN Security Council mandated a multinational force to intervene and stabilize the security situation (S/RES/1264). The UN Transitional Administration in East Timor (UNTAET) thus assumed administrative control accord-

---

[2] Andrea Katalin Molnar, *Timor Leste; Politics, History, and Culture* (London: Routledge, 2010), 32; Hans Hägerdal, *Lords of the Land, Lords of the Sea: Conflict and Adaptation in Early Colonial Timor, 1600-1800* (Leiden: KITLV Press, 2012).

[3] Sara Niner, "Martyrs, Heroes and Warriors: The Leadership of East Timor," in *East Timor: Beyond Independence*, ed. Damien Kingsbury and Michael Leach (Clayton, Australia: Monash Univ. Press, 2007).

[4] John G. Taylor, *Indonesia's Forgotten War; The Hidden History of East Timor* (London: Zed Books, 1991); Peter Carey, "East Timor under Indonesian Occupation, 1975-99," in *A Handbook of Terrorism and Insurgency in Southeast Asia*, ed. Andrew T. H. Tan (Cheltenham, UK: Edward Elgar, 2007), 376.

ing to S/RES/1272. Several UN missions continued the stabilization work, though none before 2006 were explicitly mandated to conduct SSR in Timor-Leste.

The East Timor Police Service (PNTL)

According to the UNTAET regulation 2001/22, which was based on S/RES/1272, the East Timor Police Service (later Policía Nacional de Timor-Leste [PNTL]) came into being in August 2001. Training and selection processes for police recruits, conducted by the UN Police (UNPOL), had already begun in early 2000, however. The building up of the local police was to be completed as soon as possible, and for this reason, UNPOL resorted to recruiting Timorese who had served in the Kepolisian Negara Republik Indonesia (POLRI), the national police force during the Indonesian occupation of Timor-Leste.

Despite the Timorese population's concerns about having former POLRI members on the force (because of their history of repressive tactics and collaboration with the former occupiers), the United Nations relied on them and placed them in high positions in the new police corps.[5] Also, these former POLRI members, numbering about 400, underwent fast-track training of only one month before being admitted to the new police service, whereas the vetting and training processes for the other recruits were quite different. Selection of non-POLRI recruits relied exclusively on a Western questionnaire, favoring English-speaking candidates over recruits with knowledge of several local dialects.

Political allegiances played a role in the selection or rejection of recruits as local politicians influenced the selection process.[6] Training for the non-POLRI recruits took three months at the police academy, followed by three to six months of on-the-job training, mentored and accompanied by UNPOL officers. From the beginning, the recruitment procedures conducted by UNPOL departed from recruitment guidelines. Although the special treatment of former Indonesian police officers during the recruitment and training can be explained by the need for experienced officers, the necessary vetting processes were not applied.

Beyond recruitment and training, the logistics and maintenance capacities of the new Timorese police force were poor, even after the UN Mission in East Timor (UNMISET) took control over the PNTL in 2002. UNPOL was overwhelmed with the task of providing security, leaving little time or capacity for simultaneously developing and training the PNTL.[7] Because of the double burden, UNPOL did not manage to strengthen institutional capacity and development or create the necessary self-concept for a constitutionally based police force acting in accordance with the rule of law. Instead, according to Ludovic Hood, the mission was characterized by "inadequate planning and deficient

---

[5] Gordon Peake, "Police Reform and Reconstruction in Timor-Leste: A Difficult Do-over," in *Policing Developing Democracies*, ed. Mercedes S. Hinton and Tim Newburn (New York: Routledge, 2009), 150.

[6] Henri Myrttinen, "Poster Boys No More: Gender and Security Sector Reform in Timor-Leste," DCAF Policy Paper no. 31, 2009, 24, www.iansa-women.org/sites/default/files/newsviews/DCAF_PP31_PosterBoys_2.pdf.

[7] Ibid., 25.

mission design; unimaginative and weak leadership; and negligible Timorese ownership of the process."[8]

In 2005, UNMISET handed control of the PNTL over to the Timorese administration and downsized UNPOL's presence in Timor-Leste. But the inclusion of former POLRI officers had a detrimental effect on the fragile PNTL, and conflicts erupted between former POLRI officers and recruits who had been affiliated with the resistance movement or had suffered from the repression of Indonesian military and police forces during the occupation.

Moreover, the minister of the interior, Rogério Lobato, who had full authority over the Timorese police, took also a leading position in the veterans association AC75. With the help of AC75, Lobato pressured Prime Minister Mari Alkatiri for political concessions in favor of Lobato and his allies.[9] Lobato also exploited his informal links to PNTL officers and used PNTL personnel for his own political and financial purposes.[10]

The rampant politicization of the security sector in Timor-Leste since its reconstruction in 2000-2001 was alarming. While the new Timorese defense forces, the Falintil-Forças de Defesa de Timor Leste (F-FDTL) were filled mainly with former FALINTIL resistance combatants in the first recruitment phase, the PNTL, by contrast, had a high proportion of former POLRI officers in its ranks. Hence, political orientations and affiliations hardened within the two security organizations until there was an insurmountable division between them. Also, conflicts emerged between regional identity and political network affiliations: while the PNTL consisted of "Kaladis," people from the western provinces of Timor-Leste, most of the F-FDTL recruited in the first round consisted of "Firakus," people from the eastern provinces of Timor-Leste.[11] Later recruitment phases, from 2003 to 2005, complicated the conflict patterns as new recruits for the F-FDTL came from the western provinces, while new Firaku recruits joined the ranks of the PNTL. Thus, beyond the overarching institutional rivalry between the PNTL and the F-FDTL, new identity-based rifts also emerged within both organizations.

The situation led to open conflict as inadequate mission definitions for the two rival security organizations triggered several armed clashes between PNTL and F-FDTL members since 2002.[12] Ultimately, the pressure on UNPOL to establish a functioning local police force resulted in an approach based on train-and-equip rather than on developing an effective and democratically controlled police institution. This contributed to the politicization of the PNTL. By disregarding regional and political conflict lines

---

[8] Ludovic Hood, "Missed Opportunities: The United Nations, Police Service and Defence Force Development in Timor-Leste, 1999-2004," in *Managing Insecurity: Field Experiences of Security Sector Reform*, ed. Gordon Peake, Alice Hills, and Eric Scheye (New York: Routledge, 2008), 65.

[9] Edward Rees, "Under Pressure FALINTIL Forças de Defesa de Timor Leste: Three Decades of Defence Force Development in Timor-Leste 1975-2004," DCAF Working Paper no. 139, 2004, 50.

[10] Carolyn Bull, *No Entry Without Strategy. Building the Rule of Law under UN Transitional Administration* (Tokyo: UN Univ. Press, 2008), 186; Sven Gunnar Simonsen, "The Role of East Timor's Security Institutions in National Integration and Disintegration," *Pacific Review* 22, no. 5 (2009): 575-96.

[11] Sven Gunnar Simonsen, "The Authoritarian Temptation in East Timor: Nationbuilding and the Need for Inclusive Governance," *Asian Survey* 46, no. 4 (2006): 575-96.

[12] International Crisis Group (ICG), "Timor-Leste: Time for the UN to Step Back," ICG, Asia Briefing no. 116, 2010, 9; Andrew Goldsmith and Sinclair Dinnen, "Transnational Police Building: Critical Lessons from Timor-Leste and Solomon Islands," *Third World Quarterly* 28, no. 6 (2007): 1091-1109.

in conducting recruitment and vetting, the United Nations recreated existing cleavages and aggravated tensions within the security organizations.

F-FDTL: The New Timorese Defense Force

The United Nations' unwillingness to train the former guerrilla forces and integrate them into the security sector of the new Timorese nation was reflected by FALINTIL's lack of mention in Security Council Resolution 1272. Rather, FALINTIL was perceived as a problematic residue of the 24-year struggle for independence—a view that neglected the popular support and legitimacy that FALINTIL enjoyed, at least in the eastern provinces of Timor-Leste.

FALINTIL members were gathered in cantonment sites in preparation for a disarmament, demobilization, and reintegration (DDR) process that had not been properly thought out. The forces held out in the cantonment sites for more than a year until, in 2000, a plan proposed by King's College was adopted to establish a defense force in Timor-Leste. Without a clear perception of the role this new Timorese defense force would play, the recruitment process nevertheless began in early 2001. UNTAET delegated the process completely to the former commander of FALINTIL, Taur Matan Ruak, and his staff. The recruitment process drew heavy criticism from political observers for disregarding important steps of DDR, such as a thorough screening of recruits for past crimes and for their reliability and loyalty to the constitution. Eventually, the recruitment process was biased toward men from the eastern provinces of Timor-Leste, who were loyal to Ruak and the former rebel leader, Xanana Gusmão. About 650 former FALINTIL members entered the new Timorese defense force in February 2001.[13]

The remaining ex-FALINTIL combatants, who were not considered for service in the F-FDTL or who voluntarily waived military service (about 1,308), were disarmed, demobilized, and reintegrated under the auspices of the FALINTIL Reinsertion Assistance Program (FRAP) of the International Organization for Migration (IOM). The process ended in December 2001 because DDR was not part of UNTAET's mandate.[14] The FRAP was considered successful because most of the participants got training in soft skills, received payments, and returned to civilian life.

But several veterans who, because of their political alignment, were not chosen to be part of the new F-FDTL became disgruntled and formed various illicit, largely clandestine groups and veterans' associations. These groups, including Colimau 2000, Isolados, and AC75, among others, opposed the government and the politicized army command and reportedly had strong links to organized crime.[15] Also, because of the severe eco-

---

[13] Edward Rees, "UN's Failure to Integrate Falintil Veterans May Cause East Timor to Fail," *On Line Opinion*, Sept. 2, 2003, www.onlineopinion.com.au/view.asp?article=666; Henri Myrttinen, "Guerrillas, Gangsters and Contractors: Integrating Former Combatants and Its Impact on SSR and Development in Post-conflict Societies," in *Back to the Roots: Security Sector Reform and Development*, ed. Albrecht Schnabel and Vanessa Farr (Münster, Germany: LIT, 2012), 230.

[14] Myrttinen, "Guerrillas, Gangsters and Contractors," 230.

[15] Timor-Leste Armed Violence Assessment, "Groups, Gangs, and Armed Violence in Timor-Leste," TLAVA Issue Brief no. 2, Apr. 2009, www.timor-leste-violence.org/pdfs/Timor-Leste-Violence-IB2-ENGLISH.pdf; John McCarthy, "FALINTIL Reinsertion Assistance Program (FRAP): Final Evaluation Report," USAID, IOM, June 2002, http://siteresources.worldbank.org/INTLICUS/Resources/388758-1187275938350/4101054-1187277377932/TimorLesteFRAP.pdf.

nomic situation in Timor-Leste, unemployed Timorese youth joined gangs and linked with the clandestine groups. All major political parties, as well as several members of the police and the armed forces, were tied to clandestine groups or youth gangs through patronage networks or family bonds.[16] While the FRETILIN party had ties to the martial arts gang KORKA, the Partido Democrático held strong links to the rival PSHT. The group Colimau 2000 initially cooperated with FRETILIN but eventually allied with Gusmão's CNRT party.[17]

Despite FRAP's officially proclaimed success, its DDR measures were limited because it did not include all actual members or veterans of the FALINTIL. Neglect of the sociopolitical context, as well as a lack of local knowledge about guerrilla structures, led to misinterpretation of the term "veterano." In particular, women who served in the FALINTIL or in supportive equivalent groups were excluded from the DDR process despite their crucial role during the independence struggle.[18]

Training for the newly established F-FDTL was provided mainly by external donors on a bilateral basis. Australia and Portugal, among others, led the F-FDTL training from 2001. The 12 donor nations were grouped in the Office for Defence Force Development, which was nominally subordinate to the special representative of the UN secretary-general but operated with relative autonomy from other UN departments in Timor-Leste.[19] Training consisted of a three-month fast-track program. There was no attempt to establish viable democratic and civilian oversight over the F-FDTL or improve institutional capacity.[20] Therefore, the decisionmaking authority concerning new recruits, the strategic administrative planning, and the task of defining the F-FDTL's actual role lay in the hands of the Timorese military command and its affiliated political network.

The United Nations did not play a meaningful role in DDR of the FALINTIL, nor did it significantly support the establishment and training of the F-FDTL. With DDR delegated to the IOM, the United Nations failed to monitor and lead the vetting and recruitment processes for the F-FDTL. Moreover, both the United Nations and bilateral donors neglected the formation of robust and effective institutions, such as a democratically controlled planning staff, which led to the inexorable politicization of the Timor-Leste armed forces. As a result, new recruits from the western provinces of Timor-Leste, who were linked to another political network, clashed with senior officers and members of the rank and file from the first recruitment phase.[21]

Development of the Timorese armed forces was fully relinquished to local actors. Although there was indeed local ownership from the start, the emergence of the F-FDTL in domestic security matters indicated a development outside the Western paradigm of

---

[16] Matthew B. Arnold, "'Who Is My Friend, Who Is My Enemy?' Youth and Statebuilding in Timor-Leste," *International Peacekeeping* 16, no. 3 (2009): 386; ICG, "Timor-Leste: Stability at What Cost?" ICG, Asia Report no. 246, May 18, 2013, 20, www.crisisgroup.org/en/publication-type/media-releases/2013/asia/timor-leste-stability-at-what-cost.aspx.

[17] James Scambary, "Anatomy of a Conflict: The 2006-2007 Communal Violence in East Timor," *Conflict, Security and Development* 9, no. 2 (2009): 271, 286; Henri Myrttinen, "Up in Smoke; Impoverishment and Instability in Post-Independence Timor-Leste," KEPA Working Paper no. 11, 2007, 9.

[18] Myrttinen, "Poster Boys No More," 10; Myrttinen, "Guerrillas, Gangsters and Contractors," 231, 238.

[19] Ludovic Hood, "Security Sector Reform in East Timor, 1999-2004," *International Peacekeeping* 13, no. 1 (2006): 71.

[20] Peake, "Police Reform and Reconstruction," 151.

[21] Myrttinen, "Poster Boys No More," 22.

security governance. And the absence of a clearly defined role for the F-FDTL, as well as its rivalry with the UN-subsidized PNTL, made it an unpredictable force within a fragile new nation.

Assessment of the UNTAET Effort

Main points of SSR are the establishment of viable security governance institutions that can guarantee democratic civilian oversight over national security actors, improvement of security provision by security actors, development of strong local ownership of the SSR process, and sustained reforms of the security and justice sector.[22] But UNTAET focused its SSR efforts on training and equipping the local police, meanwhile neglecting the work of building effective institutions. Therefore, democratic civilian control over the security sector in Timor-Leste never happened.

Rees rightly emphasizes the United Nations' indifference and to, and denial of, the problems within the F-FDTL as main reasons for the rampant politicization of the Timorese military. Leaving the task of overseeing the buildup and formation of the F-FDTL completely to the former resistance fighters' high command and to politicized Timorese officials made problems of democratic oversight inevitable.[23] The screening-and-recruitment process of former guerrilla fighters into the Timorese military was conducted improperly, and former conflict parties were allowed to establish partisan power bases within the Timorese armed forces. This laid a fragile foundation for the security sector.

An obvious weakness of the UN mission in Timor-Leste was the lack of strategic planning before and during the reformation and rebuilding of the security sector. Because UNTAET and UNMISET had no SSR policy or strategy, they failed to implement democratic oversight over the security organizations, thus missing the opportunity to incorporate trusted and competent Timorese politicians into the SSR process.[24] Therefore, the necessary local "ownership" for a sustainable, accountable, and effective security sector was of limited suitability from the beginning. Also, according to a UN security adviser, cooperative planning for SSR-DDR, as well as the division of labor between the United Nations and the bilateral donors, was undertaken without a concept and lacked substance.[25] Differing practices in programming and funding were the result.[26]

Analysts explain the United Nations' SSR failure in Timor-Leste as a consequence of overburdening the mission's capabilities. Because there was no local police service in place after the Australian-led intervention of 1999-2000, the United Nations was responsible for training new local recruits and providing police services at the same time.[27]

---

[22] Organisation for Economic Co-operation and Development, *OECD-DAC Handbook on Security System Reform: Supporting Security and Justice* (Paris: OECD, 2007), 21.

[23] Rees, "UN's Failure to Integrate."

[24] Hood, "Missed Opportunities," 63.

[25] Author interview with UN security adviser, July 2012, Dili.

[26] Beth K. Greener, *The New International Policing* (Basingstoke, UK: Palgrave Macmillan, 2009), 48.

[27] The International Force for East Timor was a multinational non-UN peacekeeping task force, organized and led by Australia in accordance with UN resolutions to address the humanitarian and security crisis in East Timor from 1999 until the arrival of UN peacekeepers.

A further problem was UNPOL's own heterogeneous composition. Most of the deployed international police officers had no experience instructing police recruits and also lacked the language skills to be effective. As a result, UNPOL officers relied on their individual policing experience in their home countries and tried to communicate them to the PNTL recruits. Since this ad hoc approach lacked any form of standardization, UNPOL's police training created confusion, rather than a coherent understanding of professional police practice, among the local recruits.[28]

Also, neglecting the informal structures in Timor-Leste—including the FALINTIL resistance movement—and hesitating to incorporate them into the UN-led reconstruction of the security sector produced a serious rift between FALINTIL veterans and the United Nations.[29] While the United Nations excluded Timorese officials from the official reform process of the security sector, infighting between political camps went on behind the scenes. The security actors' informal allegiances to various groups politicized and destabilized the already fragile security sector.[30]

The Collapse of 2006, and Its Aftermath

In 2005, the United Nations, perceiving overall stability, reduced its presence in Timor-Leste. Thus, the UN office there focused mainly on advisory and training tasks for the Timorese administration and the PNTL.

But Timor-Leste faced a collapse of its security sector in April 2006, when clashes began between PNTL and F-FDTL members, as well as associated youth gangs and organized groups of armed civilians. The incumbent administration could not establish order, and at the end of May 2006, the Australian-led Operation ASTUTE, of the International Stabilization Force, intervened to end the violence. In August 2006, the UN Security Council mandated the UN Integrated Mission in Timor-Leste (UNMIT), according to S/RES/1704, to rebuild and reform the institutions of the security sector, conduct and supervise rehabilitation of the PNTL, and provide security with UNPOL. UNMIT's tasks were specified and amended by a supplemental agreement, which focused on consolidation and reform of the security sector, especially the PNTL.

Again UNPOL deployed about 1,600 police officers from more than 40 countries to maintain security and train the local police force. The core of the PNTL training was the Reform, Restructuring and Rebuilding (RRR) program, which encompassed the registration of officers, vetting, additional training and mentoring, and the concluding certification.[31] The vetting process aimed first and foremost at identifying and rejecting police officers who had committed crimes or other unlawful acts during the clashes in 2006. But various suspected officers were freed without charge because of unsubstantiated claims and missing evidence. This resulted, on the one hand, from the UNPOL officers' inexperience in criminal investigation and their lack of communication skills and, on the other hand, from benevolent acquittals of the suspected officers (since Timorese govern-

---

[28] Author interview with international security adviser, July 2012, Dili.
[29] Peake, "Police Reform and Reconstruction," 148.
[30] Author interview with international policy adviser, June 2011, Dili.
[31] Bu V. E. Wilson, "To 2012 and Beyond: International Assistance to Police and Security Sector Development in Timor-Leste," *Asian Politics and Policy* 4, no. 1 (2012): 79.

ment officials, for political reasons, were not interested in prosecuting PNTL officers).[32] The start of the registration and the vetting procedure itself was slow because of insufficient UNPOL resources. Once again, UNPOL had a mixed mandate of policing Timor-Leste while also training and supervising the PNTL. Moreover, the mentoring phase of six months was reduced to eight weeks. Ultimately, about 3,110 officers were screened and certified in December 2007.[33]

Unlike in previous missions, this time Timorese officials were involved in the vetting process and included in the certification of PNTL officers. Therefore, the success of the RRR program depended heavily on good cooperation between UNMIT and the Timorese administration. But cooperation may not have been as great as UNMIT believed. The RRR program eventually failed, and one Timorese state official, in a 2011 interview with the author, blamed Timorese officials who, he said, did not like the program and therefore obstructed the process.[34] In fact, relations were troubled by miscommunication and contradictory perceptions concerning the handling of police officers suspected of misconduct.

The mission's chances for success were also weakened by high turnover rates, lack of professional expertise, a heterogeneous police body, and insufficient cultural and contextual knowledge within UNPOL. In this regard, UNMIT was repeating the mistakes that UNTAET and UNMISET had made during the earlier UN missions in Timor-Leste.[35] So while the vetting process after the security crisis in 2006 should have constituted a new beginning for the Timorese police, it did not. The ineffectiveness of the vetting itself, as well as the unwillingness of local officials to charge suspect officers with misconduct or dereliction, undermined this important measure to reform the security sector in Timor-Leste.[36]

Strategically, UNMIT should have been on far more solid ground than the previous UNTAET mission had been. This initial phase of UNMIT's activities was accompanied by the deployment of the UN Standing Police Capacity (SPC) and the Security Sector Support Unit (SSSU).[37] The SSSU had the task of monitoring and reviewing the security sector reform process and, if applicable, proposing modifications for further programs. The SPC served as a tool to strengthen relations between UN organizations and the Timorese administration, as well as maintain supervision of the PNTL's reform process.[38] But despite the United Nations' positive self-assessments of the first-ever deployment of an SSSU,[39] the evidence suggests that the unit had only very limited impact in the first

---

[32] Nicolas Lemay-Hébert, "UNPOL and Police Reform in Timor-Leste: Accomplishments and Setbacks," *International Peacekeeping* 16, no. 3 (2009): 397-400.

[33] ICG, "Timor-Leste: Security Sector Reform," ICG Asia Report no. 143, 2008, 5-8.

[34] Author interview with Timorese state official, July 2011, Dili.

[35] Lemay-Hébert, "UNPOL and Police Reform in Timor-Leste: Accomplishments and Setbacks," 399; Peake, "Police Reform and Reconstruction," 154-57.

[36] Ursula C. Schroeder, Fairlie Chappuis, and Deniz Kocak, "Security Sector Reform from a Policy Transfer Perspective: A Comparative Study of International Interventions in the Palestinian Territories, Liberia and Timor-Leste," *Journal of Intervention and Statebuilding* 7, no. 3 (2013): 393.

[37] Elisabeth Lothe and Gordon Peake, "Addressing Symptoms but Not Causes: Stabilisation and Humanitarian Action in Timor-Leste," *Disasters* 34, no. S3 (2010): 436.

[38] UNMIT, "Security Sector Review in Timor-Leste," June 2008, http://unmit.unmissions.org/Portals/UNMIT/SSR/Project%20document%20for%20SSR%20signed%2013June2008.pdf.

[39] Wilson, "To 2012 and Beyond," 82.

years because most of its staff lacked the necessary contextual knowledge and language skills to be accepted by the Timorese as a serious partner.[40]

Despite the still fragile nature of the PNTL as an institution, repeated reports about unprofessional performance, and PNTL officers' links to politicians as well as to criminal organizations, UNMIT began handing authority back to the PNTL, district by district, starting in 2009.[41] UNPOL assessed its progress using a catalogue of requirements, including that the district police be able to provide adequate security, that 80 percent of officers within the district pass the certification process, and that the district police force be sufficiently equipped and able to operate. ICG, however, criticizes the lack of standardized indicators to determine whether a district fulfilled these requirements.[42] And Della-Giacoma notes that the handover process was merely an exercise in "ticking boxes" by the assessment teams, rather than a thorough evaluation.[43] The final handover of executive control of the PNTL, for the whole of Timor-Leste, eventually took place in March 2011.[44]

Impunity as Political Strategy?

Even after the United Nations' renewed attempts since 2007 to reform the Timorese police, the institution remained fragile. Because of dysfunctional internal oversight, Timorese police officers' involvement in criminal activities such as human trafficking, prostitution, and gambling was not adequately prosecuted.[45] And when criminal investigations were made against members of the Timorese police, they eventually came to naught because of corruption and bribery of the investigators. According to a UN official, the involvement of middle-ranking and even senior police in criminal activities posed a serious impediment to meaningful and thorough police reforms by the United Nations.[46]

The withdrawal of the Indonesian military and administration from Timor-Leste in 1999 also meant a serious "brain drain" since most of the public servants during the occupation were Indonesians. As a consequence, the Timorese justice sector lacked qualified personnel to fill the vacancies.[47] Moreover, the Timorese judges and attorneys, quickly appointed by the UN Transitional Administration, had only limited practical knowledge or experience to apply to their tasks. A lack of resources and financial means, combined with the ambiguous legal framework of the transitional period, severely limited the Timorese justice sector's effectiveness.[48]

---

[40] Author interview with Timorese state official, July 2011, Dili.

[41] Greener, *The New International Policing*, 52.

[42] ICG, "Handing Back Responsibility to Timor-Leste's Police," ICG Asia Report no. 180, 2009, 11.

[43] Jim Della-Giacoma, "Police Building in Timor-Leste: Mission Impossible?" GRIPS Discussion Paper 10-04, 2010, 9, www3.grips.ac.jp/~pinc/data/10-04.pdf.

[44] ICG, "Timor-Leste: Stability at What Cost?" 20.

[45] Bu V. E. Wilson and Nélson de Sousa C. Belo, "The UNPOL to PNTL 'Handover' 2009: What Exactly Is Being Handed Over?" SSRC Conflict Prevention and Peace Forum, 2009, 13, https://fundasaunmahein.files.wordpress.com/2009/11/cppf_briefing_paper_-_unpol_to_pntl_handover_2009_final-1.pdf.

[46] Author interview with UN official, Aug. 2012, Dili.

[47] Bull, *No Entry without Strategy*, 190.

[48] Tanja Hohe, "Justice without Judiciary in East Timor," *Conflict, Security and Development* 3, no. 3 (2003): 335-38.

Since independence, the courts' effectiveness has been hampered not only by limited funds and inexperienced staff but also by the reconciliation strategy carried out by the Timorese political leadership. This proclaimed "national reconciliation campaign" by the former president and later acting prime minister, Xanana Gusmão, has been used to exempt prominent former independence fighters, and their affiliates, who were involved in arson, assault, and other criminal activities during the violent clashes in April-May 2006, from legal punishment.[49]

The Centre for International Governance Innovation (CIGI) has dubbed the judicial-political developments in Timor-Leste a "culture of impunity."[50] And indeed, the courts have exonerated several high-ranking politicians whom the United Nations Independent Special Commission of Inquiry for Timor-Leste suspected of instigating and forwarding the outburst of violence between members of the police, the military, and affiliated armed civilian groups in 2006.[51] Despite having been proven guilty, then-Prime Minister Mari Alkatiri, then-military commander Taur Matan Ruak, then-Minister of the Interior Rogério Lobato, and then-Minister of Defense Roque Rodrigues, among others, never faced prison time.[52] Lobato, one of the few who were sentenced to imprisonment, could leave the country with his family before serving the custodial sentence.[53] And only four members of the Timorese armed forces were sentenced to 10 years in jail and ordered to financially compensate the families of murdered police officers.

Despite the conviction, all four F-FDTL members received their regular pay and were not dismissed from the military. And the stipulated financial compensations were never paid to the victims' families.[54] Likewise, thanks to the flawed reconciliation policy, several members of the group who planned and conducted assassination attempts against the Timorese president and prime minister in 2008 faced minor jail terms and were later pardoned by presidential decree.[55] While prominent and well-networked persons benefited from the reconciliation scheme, ordinary Timorese citizens have suffered from draconian application of the law for even minor criminal offenses. It is this application of double standards, as one senior UN official points out, that leads to the Timorese's strong discontent and mistrust toward their government.[56]

---

[49] Gusmão left office in the fall of 2014. See Laura Grenfell, "Promoting the Rule of Law in Timor-Leste," *Conflict, Security and Development* 9, no. 2 (2009): 220; Eva Ottendörfer, "Contesting International Norms of Transitional Justice: The Case of Timor Leste," *International Journal of Conflict and Violence* 7, no. 1 (2013): 23-35.

[50] CIGI, "Security Sector Reform Monitor: Timor-Leste no. 2," May 2010, 17, www.cigionline.org/sites/default/files/ssrm_timor_leste_v2.pdf.

[51] United Nations, "Report of the United Nations Independent Special Commission of Inquiry for Timor-Leste," Oct. 2006, www.ohchr.org/Documents/Countries/COITimorLeste.pdf.

[52] Geoffrey C. Gunn, "Re-enter the United Nations: A Role for the Peacebuilding Commission in East Timor?" *Lusotopie* 15, no. 2 (2008): 211; Selver B. Sahin, "Building the State in Timor-Leste," *Asian Survey* 47, no. 2 (2007): 263.

[53] Gordon Peake, "Rogerio Lobato: From Inmate to President?" Interpreter, Feb. 15, 2012, www.lowyinterpreter.org/post/2012/02/15/Rogerio-Lobato-From-inmate-to-president.aspx.

[54] United Nations, "Facing the Future: Periodic Report on Human Rights Developments in Timor-Leste: 1 July 2009 to 30 June 2010," UN OHCHR, 2010, 11, www.ohchr.org/Documents/Countries/PeriodicReportHRDevelopmentsJuly09-June10.pdf.

[55] Bu V. E. Wilson, "The Exception Becomes the Norm in Timor-Leste: The Draft National Security Laws and the Continuing Role of the Joint Command," CIGJ Issues Paper no. 11, 2009, 6; CIGI, "Security Sector Reform Monitor," 17.

[56] Author interview with senior UN official, June 2011, Dili.

Timorese Self-Determination in Local Security Governance

Although the F-FDTL was one of the main initiators of the violent clashes in 2006, the United Nations failed once again to subject the F-FDTL to a robust vetting process. Thus, even after 2006, the F-FDTL was trained and equipped on a bilateral basis by Australia, Portugal, and other external donors, without a meaningful disciplinary review. The Timorese administration announced new recruitment rounds for 2006 and 2009, in accordance with the 2007 national strategy paper "Force 2020."[57] According to the chief commander of F-FDTL, the "Force 2020" paper was intended to put forward alternatives to imported Western concepts of security planning and to assert Timor-Leste's self-determination in matters of national security.[58]

This increasing tendency toward proclaiming Timorese self-determination was connected to mounting Timorese criticism of the United Nations' handling of the security sector review and, indeed, to growing criticism of the Western concept of SSR in Timor-Leste since 2007. The points of criticism were the insufficient police training given by UNPOL, the questionable professionalism and lack of expertise of its personnel, the ad hoc nature of the SSR programs themselves, and, finally, the United Nations' continual refusal to provide training for the F-FDTL.[59] Further indicators of the Timorese administration's increasing rejection of external SSR practices were its reluctance to prosecute suspect PNTL officers and its open objection to the vetting procedures. (Instead of prosecuting, Timorese officials argued that judicial prosecution would only fan resentment and conflict within society.)[60]

At the same time, the Timorese administration and security organizations were feeling a rise in self-confidence vis-à-vis the United Nations.[61] In two articles published online, Secretary of State for Defense Pinto came out openly against persistent UN paternalism concerning the application of security models in Timor-Leste. Pinto asserted that Timor-Leste maintains international sovereignty as a state and therefore wanted to pursue its own strategy of security management. On the establishment of the Joint Command in 2008 and its proven effectiveness, Pinto argued that Timor-Leste could respond adequately to security issues by pursuing a "Timorese way."[62] As a senior PNTL officer pointed out, the police still need to improve and, therefore, must rely on external support, but they do not wish to be patronized or controlled, and insist on self-determination in the overall process of transforming the police as an institution.[63]

A chance for the Timorese to convincingly assert their autonomy in security matters occurred in 2008. Responding to assassination attempts by rebels against the president and the prime minister in February, the Council of Ministers declared a state of emer-

---

[57] Myrttinen, "Poster Boys No More," 30; ICG, "Timor-Leste: Time for the UN to Step Back," 10.

[58] Simonsen, "The Role of East Timor's Security Institutions," 590.

[59] Júlio Tomás Pinto, "Reforming the Security Sector: Facing Challenges, Achieving Progress in Timor-Leste," *Tempo Semanal*, Aug. 18, 2009, http://temposemanaltimor.blogspot.com/2009/08/ssr-in-timor-leste.html; ICG, "Timor-Leste: Time for the UN to Step Back," 26.

[60] Lemay-Hébert, "UNPOL and Police Reform in Timor-Leste," 400.

[61] Myrttinen, "Poster Boys No More," 29.

[62] Pinto, "Reforming the Security Sector"; Júlio Tomás Pinto, "UNMIT Mission: Development or Destruction?" *Tempo Semanal*, June 7, 2011, http://temposemanaltimor.blogspot.com/2011/06/unmit-mission-development-or.html.

[63] Author interview with senior PNTL officer, July 2011, Dili.

gency and the formation of a Joint F-FDTL and PNTL Command. Although the PNTL was still ostensibly operating under UNPOL authority, the Joint Command brought it under the F-FDTL. This meant a loss of legitimacy for UNPOL, and the relationship between local security forces and UNPOL worsened.[64] Despite the relative success of the Joint Command and the surrender of the remaining rebels, the state of emergency was retained until May 2008 to enable additional Joint Command operations against alleged illegal weapons arsenals.[65] Following a return to normality, the administration announced the establishment of a new Centre for Integrated Management of Crisis, which would be incorporated into the structures of the Ministry of Defense and would become active in cases of internal security crisis or natural catastrophe.[66]

The roles and responsibilities of the PNTL and the F-FDTL within the Timorese security system remain unclear, and it should be pointed out that the Joint Command's inauguration only added to this vagueness. On several occasions in 2009 and 2011, PNTL and F-FDTL forces were combined in "ninja operations" against organized crime and clandestine gangs.[67] While the joint missions could ease friction and rivalry between the security organizations, command structures and the division of labor for further Joint Command operations are still unclear and could add to tensions between the PNTL and F-FDTL.[68]

According to Wilson, the merger of the security organizations was not only a declaration of Timorese sovereignty but could also be interpreted as a blow against the Western model of security governance.[69] Moreover, it seems that the Timorese had begun to learn from the United Nations' negative example: from 2006 onward, various actors from the administration and the security institutions tried to evade and mitigate the impact of UN-driven SSR programs. Notably, the merger of the PNTL and the F-FDTL during 2008 and during other joint missions blurred the boundaries between external and internal security organizations even more.

Conclusions and Recommendations

Conclusions

Security sector reform and disarmament, demobilization, and reintegration are the main international focus for the security sector in postconflict countries. Recognizing that a dysfunctional security sector poses a threat to the population, SSR aims at removing or transforming dysfunctional elements, and it advocates for effective and legitimate, democratic civilian control of security sector organizations. DDR, on the other hand, is an important step in pacifying former combatants and reintegrating them into civilian life.

---

[64] Yoshino Funaki, "The UN and Security Sector Reform in Timor-Leste: A Widening Credibility Gap," Center on International Cooperation, May 2009, https://fundasaunmahein.files.wordpress.com/2009/07/funaki-timor-ssr-final.pdf.

[65] Wilson, "The Exception Becomes the Norm in Timor-Leste," 37; CIGI, "Security Sector Reform Monitor."

[66] Wilson, "The Exception Becomes the Norm in Timor-Leste," 7.

[67] CIGI, "Security Sector Reform Monitor."

[68] ICG, "Timor-Leste: Time for the UN to Step Back," 10.

[69] Wilson, "The Exception Becomes the Norm in Timor-Leste," 12.

The case of Timor-Leste demonstrates the difficulties that attend selection of local partners by international donors. Integrating key local actors into the DDR and SSR processes is crucial. Since the country's future security sector depends on institutionalization and on ownership by local actors, an integrated approach in both the planning and the implementation stages of DDR and SSR is vital.[70]

Neglecting the local sociopolitical and historical context can lead to an irrelevant SSR program that does not fit the realities of the country and, at worst, alienates local security actors. It is important to recognize that key local actors will likely play to their strategic strengths and can influence the SSR process massively. Moreover, particular local actors may have divergent interests concerning SSR and, therefore, may manipulate the SSR implementation process to serve their interests, by placing their political followers in the new security forces or security-related political organs.[71]

Missing the short time frame of transitional peace to conduct DDR and SSR in a tense environment may lead former combatants to pick up arms again and reignite conflict.[72] At precisely this stage, it is important for international donors to prevent the emergence of disgruntled former combatants who may turn into "spoilers," and of actors outside the state security institution who could sabotage further SSR processes and pose a threat to the monopoly on the use of force in the fragile postconflict state.[73] Forms of opposition may range from covert resistance to open defiance. "Inside spoilers" use a strategy of stealth: officially cooperating with external powers, they eventually undermine programs from behind the scenes.[74] Noncooperation, noncompliance, and selective cooperation are other ways that local actors can impede policy implementation or discriminate between externally imposed programs.[75]

An intervention by external actors always has a massive impact on the social structure of a country and turns the external actor into just another power broker among others.[76] That is, the external actor is by no means nonpartisan, because he makes a clear statement by choosing a certain group of local actors to cooperate with.[77] But by choos-

---

[70] Mark Sedra, "Towards Second Generation Security Sector Reform," in *The Future of Security Sector Reform*, ed. Mark Sedra (Waterloo, Ont., Canada: CIGI, 2010), 105; Alan Bryden and Heiner Hänggi, "Reforming and Reconstructing the Security Sector," in *Security Governance in Post-conflict Peacebuilding*, ed. Alan Bryden and Heiner Hänggi (Münster, Germany: LIT, 2005), 37.

[71] Alan Bryden and Vincenza Scherrer, "The DDR-SSR Nexus: Concepts and Policies," in *Disarmament, Demobilization and Reintegration and Security Sector Reform*, ed. Alan Bryden and Vincenza Scherrer (Münster, Germany: LIT, 2012), 12.

[72] Hugo de Vries and Erwin van Veen, "Living Apart Together? On the Difficult Linkage between DDR and SSR in Post-conflict Environments," Clingendael, Oct. 1, 2010; Sean McFate, "The Link between DDR and SSR in Conflict-Affected Countries," Washington, DC: USIP Special Report no. 238, 2010, 2.

[73] Stephen John Stedman, "Spoiler Problems in Peace Processes," *International Security* 22, no. 2 (1997): 6.

[74] Ibid., 7.

[75] Roger Mac Ginty, "Hybrid Peace: The Interaction between Top-Down and Bottom-Up Peace," *Security Dialogue* 41, no. 4 (2010): 403.

[76] Robert Egnell, "The Organised Hypocrisy of International State-building," *Conflict, Security and Development* 10, no. 4 (2010): 480.

[77] McFate, "The Link between DDR and SSR," 6; Madeline England, "Security Sector Reform in Stabilization Environments: A Note on Current Practice," in *Security Sector Reform: Thematic Literature Review on Best Practices and Lessons Learned*, ed. Madeline England and Alix Boucher (Washington, DC: Stimson Center, 2009), 17.

ing some local stakeholders over others, international actors may create potential local adversaries who could obstruct the future peacebuilding process.[78]

Ultimately, the United Nations and associated international donors failed to appreciate these realities. Rather, UN security sector management, combined with the additional bilateral involvement in security affairs, led to unintended consequences—a recurring topic in the area of external security governance.[79] And as Schroeder points out, these unintended consequences often conflict with a policy's original intent.[80] In the Timorese case, the recruitment of partisan former guerrilla fighters into the newly founded Timorese military, as well as UNPOL's unobservant recruitment of former POLRI officers into the Timorese civilian police, laid the foundation of a politicized and troubled Timorese security sector. Organizing and training the PNTL as quickly as possible led to blatant flaws in the recruitment and vetting process. As former POLRI officers received special treatment in training and promotion, the foundation was laid for future friction and resentment within the police force. Added to this, the excessive demands of a mixed mandate that required internationals to be both security providers and security development specialists, combined with the desire for quick results, led to a hollow and vulnerable police institution.

Implementing sustainable security sector reform and security governance in a postconflict setting requires personnel who have training expertise and can serve as role models for new recruits. But the UN police contingent comprised individuals from more than 40 nations, which meant great heterogeneity concerning the level of professionalism and perception of policing.

Standardization of police contingents' professional skills in a postconflict scenario is crucial, and development missions require consistent behavior patterns among their personnel if they are to train, advise, and assist nascent security forces effectively. UNPOL failed to attain the necessary level of cohesion.[81] And as Greener points out,[82] after a week-long preparation course for Timor-Leste, most of the UNPOL officers were still not sufficiently trained or skilled in instructing recruits or developing the police sector of a war-torn country.

Moreover, UNPOL's ability to deal with Timorese customs, conventions, and languages was inadequate to the task. Also, high turnover rates of UNPOL officers, and lack of knowledge transfer within the officer corps prevented UNPOL from gaining experience and achieving its goals.[83]

The reconstruction of the PNTL, and the United Nations' neglect of the FALINTIL were not based on careful consideration at the strategic level but instead rested on ad hoc decisions. According to Rees, the UN decision not to monitor and manage the FALINT-

---

[78] Marie-Joelle Zahar, "SRSG Mediation in Civil Wars: Revisiting the 'Spoiler' Debate," *Global Governance* 16, no. 2 (2010): 226-69.

[79] Christopher Daase and Cornelius Friesendorf, "Introduction: Security Governance and the Problem of Unintended Consequences," in *Rethinking Security Governance: The Problem of Unintended Consequences*, ed. Christopher Daase and Cornelius Friesendorf (London: Routledge, 2010), 8.

[80] Ursula C. Schroeder, "Unintended Consequences of International Security Assistance: Doing More Harm than Good?" in Daase and Friesendorf, eds., *Rethinking Security Governance*, 88.

[81] Peake, "Police Reform and Reconstruction," 148; Myrttinnen, "Poster Boys No More."

[82] Greener, *The New International Policing*, 49.

[83] Peake, "Police Reform and Reconstruction," 150.

IL's transformation into a national defense force was a fundamental mistake, which impeded further development of the Timorese security sector for the coming years.[84] Giving the FALINTIL high command too much leeway in forming a defense force out of former rebels, without monitoring the recruitment process and without demanding the development of legitimate, effective control-and-oversight institutions behind the politicized high command, was a grave mistake.

The UN mission after 2006 did not differ much from the pre-2006 missions. Problems concerning UNPOL's composition and capabilities remained essentially the same. Once again local political and cultural contexts were not considered.[85] UNPOL struggled again with the mixed mandate of providing security while training and supervising the PNTL. As a result, the institutions of the PNTL and the F-FDTL are still fragile, and a clear-cut role allocation for the security organizations has yet to materialize. Local ownership by the Timorese administration has changed since 2006. The formation of the Joint Command and the exclusion of UNPOL during these operations indicated Timor-Leste's clear shift from client to self-determined actor. But the "strong government ownership for reform," as Funaki puts it, led security sector governance to develop in ways contrary to the donors' expectations.[86]

It is striking that the United Nations did not learn from the lessons of either its previous missions or the experiences of UNTAET and UNMISET after 2006. Rather, it maintained a flawed PNTL training program that had failed to produce professionalism or operational effectiveness, it repeatedly conducted ineffective vetting that enabled rather than prevented factionalism and politicization of the force, and it neglected PNTL institutional development in favor of a short-term, train-and-equip mentality. Moreover, it continued to ignore political developments within the F-FDTL that later hampered coherence and accountability across the whole of the security sector.[87] As Peake notes, UNMIT was the second chance for the United Nations to implement—or at least lay the foundations for—comprehensive SSR in Timor-Leste.[88] The rebuilding of security forces after 1999 enabled the United Nations to develop and implement second-generation SSR, and the violent clashes of 2006 allowed for a further review of the security sector. And yet, the United Nations did not succeed in either case.

Recommendations

It must be stressed that SSR does not treat *only* national security agencies in its reform attempts. SSR is, first and foremost, a political task, designed to transform all the formal and informal institutions, and their dependent elements, that affect national development and political stability.[89]

---

[84] Rees, "UN's Failure to Integrate."

[85] Greener, *The New International Policing*, 51.

[86] Funaki, "The UN and Security Sector Reform in Timor-Leste," 13.

[87] Myrttinen, "Poster Boys No More," 33; Wilson, "To 2012 and Beyond," 74.

[88] Peake, "Police Reform and Reconstruction," 142.

[89] See Peter Albrecht, Finn Stepputat, and Louise Andersen, "Security Sector Reform, the European Way," in Sedra, ed., *The Future of Security Sector Reform*, 84; Albrecht Schnabel, "The Security-Development Discourse and the Role of SSR as a Development Instrument," in Schnabel and Farr, eds., Back to the Roots, 30.

- Training-based reform of the local security actors by external donors is an important step in preparing the security agencies to maintain the state's monopoly on the use of force. But training and equipping the local security actors does not, by itself, guarantee an effective and accountable security sector. The security sector encompasses the transformation of all relevant institutions, such as the ministries of interior, defense, and finance, but also the national parliament and parliamentary committees that deal with matters of local security. Thus, a holistic approach to SSR is necessary to ensure reliable, nonpartisan, democratically controlled security forces.[90]

- The case of Timor-Leste highlights the problem of external donors neglecting the political institutions. As a consequence, dysfunctional ministries and bypassed formal hierarchies gave way to patterns of corruption, abuse of office, and the establishment of illicit practices within the local police and military. To prevent misuse of this important sector by illicit local interests, a thorough reform of local political institutions, with clear benchmarks, is essential.

- Because SSR is, above all else, a political endeavor, an in-depth knowledge of the local context is crucial to being able to evaluate local political developments and react accordingly. Understanding the local historical and sociopolitical context will help external actors choose local partners to cooperate with in the long view. In Timor-Leste, however, the United Nations' uncritical inclusion of former police officers who served in the Indonesian police led to the Timorese population's massive loss of trust in the newly founded Timorese police from the start.

- Contemporary analysts increasingly advocate for a closer interaction, or synergy, between SSR and DDR because successful SSR depends heavily on an effective and timely DDR process. Best-practice models for the sequencing and scope of application in combining SSR and DDR have not been established in the policy discourse yet, which may be explained by the widely divergent individual contexts of the countries in question.[91] But evidence suggests that in the case of Timor-Leste, necessary coordination between the external donors in executing SSR and DDR was insufficient. Moreover, the United Nations neglected important steps of DDR, such as obligatory screening and vetting procedures for former Indonesian policemen.

- To avoid producing a highly politicized and partisan local security actor, external actors must apply *all* the standard operational steps of DDR in any given mission. And the external actors must resist excessive politically motivated pressure from local actors on the recruitment process. A short temporal horizon, which generally weighs on every operation in a foreign territory, can also put an ill-advised premium on hasty execution of SSR- and DDR-relevant processes. Although a

---

[90] See Heiner Hänggi, "Conceptualising Security Sector Reform and Reconstruction," in *Reform and Reconstruction of the Security Sector*, ed. Alan Bryden and Heiner Hänggi (Münster, Germany: LIT, 2004).

[91] See Bryden and Scherrer, "The DDR-SSR Nexus," 10, 22.

fast-track processing of SSR and DDR may be easier for the executing external agencies, these shortcuts will undermine the local security sector in the long run. Therefore, SSR and DDR processes should be planned and thoroughly carried out according to elaborated and custom-tailored guidelines, for these measures are the foundation for the future security sector.

It is clear that today, Timor-Leste has taken control of its own reform agenda, but whether it can manage its complex political and security environment remains to be seen. For the UN missions, the lessons from the SSR efforts of a decade may be summed up in the cliché "haste makes waste." For the Timorese, the international missions offered an opportunity to lay a foundation for accountable security sector performance, overseen by a security sector governance structure that was adaptive and responsive to the needs of the state and its population. But in the rush to get military and policing boots on the ground, opportunities to create a culture of public service, accountability to the law, and professionalism may have been lost. Fortunately for the Timorese, the story of SSR in Timor-Leste is still being written, but for the international community, the lessons should be painfully clear.

17. A Granular Approach to Combating Corruption and Illicit Power Structures

Scott N. Carlson

In postconflict and fragile states, corruption is always a core challenge to stability. The impact of corruption on efforts to establish rule of law and to create or restore stable economic markets is clear and well documented.[1] Simply stated, corruption is the abuse of public authority for private gain. While the definition may be clear, the appropriate international responses to prevent and counter corruption are anything but. Meanwhile, illicit power structures continue to employ corruption and benefit from the lack of coherent measures to combat it.

This chapter recommends some principles and practices that support a "granular" approach to countering corruption. It is through the capture, preservation, and organization of a government's information about its own transactions that the "grains" to form a factual foundation for government operations may be established. Discrepancies between these facts, or grains, and the representations of government personnel—as well as those of illicit power structures—can then serve as a basis for investigation and corrective action. With the appropriate legal and administrative measures in place, remedial actions can include recovery of misappropriated assets and, ultimately, prosecution of culpable parties.

The "granular" approach proposed here is not a comprehensive strategy for combating corruption and illicit power structures. But the approach does marshal foundational elements that form a necessary and proper pillar of any effective countercorruption strategy. In postconflict and fragile states, the patient state is in triage, and its very existence is imperiled. Policymakers and practitioners must be able to categorize and sequence the appropriate remedies and respond in a way that stabilizes the patient state. Interventions that protect core functions can enable the patient state to live another day. The granular approach emphasizes securing these basic building blocks first before expecting the damaged state to show a full recovery.

Why Top-Down Anticorruption Programming May Not Be the Best Strategy

Historically, many anticorruption efforts have aimed at high-level changes in governing strategies and structures, such as constitutional reforms that establish an ombudsman, anticorruption commission, or similar office. And while these top-down structural changes can be useful in and of themselves, they do not necessarily result in meaningful change that benefits the larger population. Experience shows that international attention and expectations commonly focus on establishing such institutions to prosecute high-level abuses, but the underlying policies, procedures, and capabilities to empower the institutions for such politically charged tasks are frequently lacking.

Illicit power structures recognize such weaknesses and, as "rational" economic actors, respond accordingly. Ironically, the very institutions and individuals that the in-

---

[1] See, for example, Alix J. Boucher, William J. Durch, Margaret Midyette, Sarah Rose, and Jason Terry, "Mapping and Fighting Corruption in War-Torn States," Stimson Center, Mar. 2007, www.stimson.org/books-reports/mapping-and-fighting-corruption-in-war-torn-states/.

ternational community has publicly designated as a primary threat to the success of criminal networks then become the focal point for the networks' corruption efforts.

By contrast, anticorruption efforts targeting *grassroots* change are dispersed, have a lower profile, and are less likely to provoke elaborate criminal engagement. Illicit power structures tend to staff their lower-level corruption efforts with unsophisticated "bagmen." While targeting a select group of high-level institutions and individuals is logistically manageable, few criminal organizations have demonstrated an ability to train and equip their bagmen to function in a complex, coordinated manner. And yet, the international community continues to emphasize high-level reforms while often neglecting the opportunities available at the grassroots level.

The reasons for failure of high-level reform efforts are complicated, but common themes emerge:

- First, in most postconflict and fragile states, patronage and nepotism are endemic. National identity is commonly weak because of the internal customs and pressures to identify oneself with subgroups. Cessation of hostilities offers warring parties from different ethnic, tribal, religious, political, and linguistic subgroups an opportunity to regroup and secure their economic and social interests. Given the varying degrees of distrust for those outside one's subgroup, it is generally difficult to staff a high-level, central anticorruption institution with individuals who can withstand the pressure to favor "their" people.

- Second, top-down government anticorruption activities are very often viewed as political payback from those now in power. Of course, a government official may be corrupt even when his or her exposure or prosecution *is* primarily political payback, but this is generally not what matters most to the common citizen. Elites trading charges of "grand" corruption can easily be viewed as a continuation of business as usual, where different groups, including illicit networks, argue over the spoils associated with being in power.[2] "Petty" corruption, on the other hand, involves tertiary, or low-level, government functions that interact directly with the citizenry at large. Not surprisingly, the average person is frequently more concerned about these tertiary skims and scams than with those perpetrated on a grand scale. The personal palace built with embezzled donor funds may prompt headlines and general indignation, but the illicit tax of corruption on basic subsistence issues is of much more immediate concern—indeed, sometimes a matter of life and death—for the average citizen in a postconflict or fragile state.

- Third, even in the best cases, the analysis, conclusions, and charges that these high-level anticorruption institutions produce rarely show a high degree of professionalism. To be fair to the professionals tasked with making these cases, the administrative capacity and recordkeeping in postconflict or fragile states is

---

[2] Transparency International made the distinction between "grand" and "petty" corruption. Grand corruption focuses on large-scale bribes and corrupt practices, whereas petty corruption focuses on smaller payments, or "favors," that average citizens and businesses may pay to a civil servant for a government service.

generally weak to nonexistent. Thus, investigations of grand corruption may be unable to gather sufficient objective evidence, and to the extent that legal analysis even exists, linkage of evidence with specific elements of a criminal, corrupt practice will likely be correspondingly weak. Embassies, citizens, and international organizations may express concern with poor performance, but they often struggle to identify solutions to the problem.

The bottom line is this: if the appropriate data management structures are missing, the necessary government data needed to properly challenge corrupt practices will not be captured, stored, and made available to interested parties. Of course, the manual assembly and maintenance of these data files is laborious and time consuming. Nevertheless, the need for proper records is an organizing principle of proper public administration. Illicit power structures all over the world have demonstrated a capacity to thrive in the absence of good government recordkeeping. Moreover, they have done so with impunity.

Proper Public Recordkeeping: A Necessary Step in Combating Petty Corruption

The leaders of criminal organizations commonly show off their power and influence with ostentatious displays of wealth. The difference between legitimate business profits and the houses and riches that corrupt elites flaunt are obvious to all involved. In fact, the relative social status of an illicit power structure may profit from the state's inability to counter such impunity, because citizens can only conclude that illicit power will not be penalized. Even worse, citizens may begin to view the illicit power structures as de facto authorities and valid employers, in control of the very keys to their survival.

But history reveals that with good government data management, even the most powerful and notorious gangsters and their criminal networks may be brought to justice. Well before the advent of computer technology, countries conducted sophisticated manual data analysis, identified discrepancies, and investigated and prosecuted the perpetrators of illegal activity. In the United States, the halls of the Internal Revenue Service (IRS) have been adorned with posters reminding employees of their role in fighting illicit power structures during the Great Depression. In the 1930s, the IRS was on the forefront of fighting organized crime, placing the notorious gangster Al Capone behind bars where others had failed. And today, experts focus increasingly on documenting the details of government internal operations. The advent of affordable, robust computer information systems has made it possible to track vast amounts of data, using small units of trained staff. Developed and developing countries alike are embracing this new capacity to root out fraud and corruption. As an example, the U.S. IRS identified fraud in which 655 bogus refund requests were traced to a single address in Lithuania.[3]

Therefore, an emerging intervention strategy is to start with measures countering petty corruption through the systematic capture, organization, and disclosure of government data, while also enabling ordinary citizens' complaints about government

---

[3] *Economist*, "SIRF's [stolen identity refund fraud] Up," Nov. 30-Dec. 6, 2013, 27-28, www.economist.com/news/united-states/21590912-cleverer-use-data-and-investigative-collaboration-can-help-cut-fraud-sirfs-up.

mismanagement. When information is systematically stored and disclosed, the citizens themselves are empowered to serve as a watchdog for abuses within the government bureaucracy. A civil servant who improperly alters a citizen's data will leave electronic fingerprints that are virtually impossible to cover and easy for the citizen to identify and document.

Why emphasize petty corruption?

Illicit power structures that enjoy high-level government complicity are unlikely to focus on such "small beer," at least at the early stages. And while elimination of all petty corruption may not be readily feasible, a significant reduction in its prevalence in government services can produce dramatic changes for the average citizen and businessperson. These changes foster greater confidence in government. Also, they are an investment in the success of government institutions, motivating citizens to oppose corrupt practices actively.

The capacity for mobilizing grassroots campaigns is large. As one method of public shaming, and demonstrating the grassroots appetite to engage, the Indian-based 5th Pillar movement is handing out fake rupees to corrupt officials asking for bribes. More than 2.5 million have been handed out thus far.[4] An increasingly wired global citizenry is actively seeking out new tools to combat illicit conduct by government officials and their patrons.

Identifying and addressing these "grains" that consist of venal public-private transactions—even though they occur in a society devastated by graft—permits corruption to be countered in manageable units amenable to programmatic solutions. This granular approach is being employed in a number of efforts to promote accountability and transparency.

An example of the increasing popularity of this type of approach is the Open Government Partnership, a multilateral initiative that calls on members to make a shared commitment to publicizing government data (the Open Government Declaration), create national action plans in collaboration with citizens, and monitor and evaluate implementation. Founded in 2011 in eight countries, it has more than quintupled in size. Moreover, it has demonstrated its utility in the poorest of postconflict states. In Sierra Leone, for example, OGP efforts linked government and civil society organizations in a feedback loop that developed a National Action Plan in record time, demonstrating the citizenry's appetite to engage in the fight against corruption when the government engages them.[5]

Successfully executed, this approach undermines the ability of illicit power structures to capture government, because it empowers citizens to own and use their data for government's effective management. The more citizens who step up, the harder it becomes for criminal organizations to find space for their corrupt transactions.

---

[4] *Economist*, "Small Change," Dec. 7-13, 2013, 63, www.economist.com/news/international/21591198-increasingly-popular-weapon-fight-against-corruption-fake-money-small-change.

[5] Marcella Samba-Sesay, "Open-Government Partnership Process in Sierra Leone: Engaging in Mutually Respectful Manner and Finding Common Ground to Actualise the Reforms We Need," *Open Government Partnership*, June 24, 2015, www.opengovpartnership.org/blog/marcella-samba-sesay/2015/06/24/open-government-partnership-process-sierra-leone-engaging.

A Basis in International Law and Development Best Practices

Moreover, this granular approach is grounded in international law. The United Nations Convention against Corruption (UNCAC), which came into force in 2005, suggests exactly such a focus on civil servants and the public.[6] The UNCAC now has 168 signatory nations and constitutes one of the most widely accepted statements of binding legal principles employed to fight corruption.

The international law of the UN Convention has been embraced and applied in practice. Also, in 2005, the United States Agency for International Development (USAID) proposed useful working definitions of "transparency" and "accountability," with the publication of their TAPEE (transparency, accountability, prevention, enforcement, and education) Framework.[7] Regrettably, this practical approach, based on years of Europe and Eurasia experience, has not received the global attention it deserves. Nonetheless, the TAPEE Framework guidance helps unpack the UNCAC principles and demonstrates their practical application. It is a lesson that should be learned from the region, not lost to the rest of the world.

Making Government Information Publicly Available and Meaningful

The TAPEE Framework emphasizes that without access to adequate information, the relevant stakeholders cannot evaluate whether public-sector decision making is in compliance with law and does not constitute an abuse of authority for private gain. Where such "information asymmetries" are present (i.e., where transparency is lacking), institutional mechanisms to combat corruption may well be underutilized or ineffective. To counter information asymmetries, TAPEE recommends programmatic interventions that will promote this essential access to relevant information.[8]

In today's world, access to information is based on information technology, but its ultimate form is flexible. The key requirement is that it be systematic, consistent, and verifiable. As described below, the advent of e-government technology is a quiet revolution that is meeting this need and sweeping the globe. What once was a massive investment is now imminently affordable. Moreover, there is an emerging cadre of international professionals capable of bringing customized solutions to even the remotest locations.

Establishing the Rules: Transparency by Law

Even though transparency and accountability may be anchored in international law, effective mechanisms typically need a domestic legal framework that emphasizes transparency of government information. Recent studies stress that "sunshine is the best disinfectant." So legal structures and mechanisms that make public access to information

---

[6] UNODC, "United Nations Convention against Corruption," G.A. Res. 58/4 of 31 Oct. 2003, www.unodc.org/unodc/en/treaties/CAC/.

[7] USAID, *TAPEE: An Analytical Framework for Combating Corruption and Promoting Integrity in the Europe and Eurasia Region*, Aug. 2005, http://pdf.usaid.gov/pdf_docs/pnadd630.pdf.

[8] Ibid., 18.

the default, not the exception, should be the rule. The Millennium Challenge Corporation, for example, a relatively new U.S. government aid provider, has made fighting corruption a core principle guiding its efforts to promote sustainable development.[9]

Ideally, this framework should be based in constitutional provisions that guarantee a citizen's access to information, and it should be supplemented with implementing legislation, regulations, and policies.[10] These types of foundational legal reforms are commonly part and parcel of the aforementioned programming designed to attack grand corruption, but they are equally important to granular approaches that focus on the contact points between public employees, and citizens and businesses. The essential thing to a granular approach is to make sure that these reforms actually guarantee citizens' access to information.

The importance of focusing on the adequacy of a legal system at the ground level is not new. In 1989, Hernando de Soto's research concluded, "Informals suffer not only from their illegality but also from the absence of a legal system that guarantees and promotes their economic efficiency—in other words, of good law."[11] Shifting people from informal status—in which they live outside the formal economy and legal system—to becoming full-fledged members of the formal economy, governed by law, is first and foremost a matter of making the law and its systems available in practice. Where successful, this transition creates stakeholders in good government, increases individual control of economic decisions, and reinforces rule of law.

Applying the Rules: Transparency Fundamentals in Practice

Transparency implies that information is readily available. Publication and distribution of government policies, procedures, fees, regulations, and other information are at the foundation of what the TAPEE Framework refers to as "substantive transparency."[12] If relevant stakeholders are not aware of the basic substance of public administration, not only will they be unable to evaluate whether corruption is present, but they will also be poorly positioned even to pose the right questions. Even with legal access to government data, the issue remains of how this information relates to the actual process of administration.

If relevant stakeholders are not involved in the process of public administration, public-sector management may well be an opaque machine, which is inherently more subject to improper influence. To counter this tendency, TAPEE suggests that anticorruption programs incorporate measures to increase "procedural transparency."[13] Such measures involve interactions between stakeholders—such as public hearings on new legislation, question-and-answer sessions with lead public-sector officials, and creation

---

[9] Millennium Challenge Corporation (MCC), "Building Public Integrity through Positive Incentives: MCC's Role in the Fight against Corruption," *working paper*, 2006, 11, https://assets.mcc.gov/reports/mcc-workingpaper-corruption.pdf.

[10] See UNCAC, Article 10, which obligates states to make transparency in public administration a reality in law, regulatory framework, and practice.

[11] Hernando de Soto, *The Other Path: The Economic Answer to Terrorism* (New York: Basic Books, 1989), 158.

[12] USAID, TAPEE, 18-19, fn. 12.

[13] Ibid., 19-20.

of independent administrative bodies that receive and review alleged violations of law — incorporating protections from political influence. In each of these scenarios, access to basic information about government operations is a prerequisite.

Both types of transparency depend on a system of government recordkeeping that is robust, secure, and verifiable, as well as accessible to appropriate parties. This type of recordkeeping should capture the financial transactions of government offices and objectively document the affirmations, representations, and involvement of all parties to a particular transaction. To establish such a system requires attention to detail in the design, maintenance, integration, and auditing of public information. But once established, this information provides a bulwark against improper influences. Multiple levels of internal verification frustrate standard attempts to manipulate data from external or internal sources, by automating data exchange and creating an audit trail of inappropriate tampering with files.

In 2006, the MCC Threshold Program in Albania supported an integrated e-government platform, linking business registration, tax, and public procurement. The theory of change was that each of these government functions had historically been abused by rent-seeking civil servants and that corruption could be visibly curtailed if the discretion of the civil servant was constrained using e-government technology. The legal framework was established, and the e-government systems were developed and deployed. The systems were established with automated data cross-checks and audit trails, dramatically limiting civil servant discretion. The results were dramatic. For instance, within two years of the project's launch, the time it took to register a business went from 47 days to one, and the percentage of persons who reported having paid a bribe went from about one in five to zero.[14] Less than 10 years earlier, the Republic of Albania had erupted in civil war, requiring international intervention. The progress is a vivid illustration of the positive change that can arise from the ashes of conflict.

Capitalizing on Transparency to Secure Accountability

Illicit power structures depend on the government's inability to keep track of its own business and prevent manipulation of records. Without secure and systematic records, the lowest-level civil servant enjoys the "discretion" to strike whatever deal she or he can with bagmen as well as with the citizenry at large. Particularly in a postconflict or fragile state, a civil servant with a low wage may well conclude that the private benefits of a bribe or other corrupt practice outweigh the risks of any official sanctions associated with corruption. For instance, in many such jurisdictions, new customs agents will enter the job with the understanding that they will be "caught" and fired, but if they first accumulate enough bribes to buy a house, the benefits outweigh the risks. But if a government system captures information that cannot be readily tampered with, adequate records supporting prosecution become a credible threat. This changes the cost end of the equation from merely having to find a new job to potential fines and imprisonment. This dynamic is present regardless of the ministry or government department (e.g., tax, customs, procurement), highlighting the core necessity for organized, secure

---

[14] MCC and USAID, "Strengthening Governance in Albania: Support to Albania's Millennium Challenge Account Threshold; Agreement Final Report," Sept. 2008, Annex A, http://pdf.usaid.gov/pdf_docs/pdacm504.pdf.

recordkeeping. The availability of information technology makes accurate, affordable data capture, retrieval, remote secure storage, and verification accessible through e-government solutions.

Unfortunately, international donors who are steeped in old technology may be slow to suggest, develop, and fund modern e-government solutions. Western e-government legacy systems frequently rely on distributed information technology solutions that require multiple expensive hardware components and licenses to maintain. In today's world, Web-based systems can be developed, deployed, and maintained at a fraction of the time and cost required even a decade ago. For example, the East African Court of Justice, based in Arusha, Tanzania, recently designed and installed a state-of-the-art case management and video recording system, which links to courts in all members of the East African Community. The project took less than a year and cost less than one million dollars. The past decade is replete with examples of donors paying millions of dollars for court case management systems that took years of development simply to reach a pilot stage.

Transparency: Putting Horizontal and Vertical Accountability into Practice

While there are instances of paper-based systems that supported transparency and accountability, their vulnerability to manipulation and destruction—particularly in a post-conflict or fragile state—are manifest. Consequently, the effective use of e-government technology to capture information is an important first step to countering these concerns. But the storage, management, and verification for such records is equally important. There must also be an e-government solution for accessing the information in the records to compare it against the performance of government institutions, the employees who staff them, and the inputs of the general public.

For example, suppose that a citizen can self-assess and file tax payments online. And suppose that a proper e-government system provides confirmation of receipt and associates the payment with a particular tax liability at a particular time. If a tax assessor challenges the payment of the tax, the citizen has a system that operates outside the discretion of the assessor verifying the citizen's payment. No longer is the payment of tax a face-to-face transaction wherein the civil servant can seek a side payment to register a tax as paid, or subject the citizen to a shakedown for it later. The tax assessor is blocked from tampering with the record, and while legitimate disputes may still arise, the resolution is anchored in established fact. No longer is the citizen powerless against the whims of a civil servant's unfettered discretion. This correspondence between e-government data and actions is the essence of accountability. The TAPEE Framework describes two types of accountability: *horizontal* and *vertical*. The former involves checks and balances between government units, and the latter addresses oversight that citizens and societal actors exercise over government units.[15] Both are crucial to the equation of verifying accurate government recordkeeping and using it to control government operations.

Taken together, these two types of accountability define an environment where safeguards are in place to measure civil servants' compliance with their defined roles and re-

---

[15] Ibid., 22.

sponsibilities and to ensure that they perform their duties in a way that benefits citizens and businesses. For example, due to the Albania project described above, a procurement officer can now exchange e-government data with a fellow civil servant in the business registry to verify that a business registration is in good standing (horizontal). Likewise, Albanian citizens can now confirm that their tax submissions to a civil servant have been credited to the appropriate accounts (vertical). Should any of these e-government verifications reveal a discrepancy, those involved are informed, an audit trail of the interactions is preserved, and corrective action is not merely possible but probable. These granular improvements of the public record provide an increasingly firm factual foundation for all subsequent government actions.

The Challenges, Limitations, and Promise of the Granular Approach

Through the use of e-government information technology, granular transparency and accountability are both feasible and affordable in today's world. Countries with legacy computer infrastructure may resist, and legacy companies will complain when their revenues are lost, but institutional change is by definition difficult. When it involves reduction in a civil servant's scope of work, resistance should be expected. But where infrastructure is most feeble, creating an e-government network of transparency and accountability may, ironically, be easier given the recent advances in affordable information technology and the lack of high-level sophistication in subverting it.[16]

With e-government, horizontal and vertical accountability can be achieved and maintained. What gets reported is verified within the government and without. In that way, government units have integrated workflows that force collaboration toward common goals. Likewise, interaction between civil servants and the public is reduced, further limiting opportunities for corruption. Examples of one-stop shops for business, procurement, and tax registration are increasingly common, eliminating discretion and limiting rent-seeking behavior across multiple government offices. Integrated e-government systems allow public procurement agencies to verify whether bidders are properly registered businesses and current in meeting their tax obligations, while simultaneously ensuring that citizens get credit for compliance. Moreover, implementing such integration through an e-government platform allows the entire process to be controlled, overseen, and exported to existing central oversight bodies and to the general public with secure, verified data. These types of e-government systems deprive illicit power structures of their freedom to operate, and promote rule of law at the same time.

Of course, illicit power structures are not static. They evolve to respond to changes in circumstances and seek new pathways for corruption. Moreover, the granular approach cannot immediately bring a halt to grand corruption. When the Kabul Bank scandal exploded in Afghanistan, President Karzai dug in and refused to address the pervasive high-level corruption, and the Western donor community blinked, backing down for

---

[16] The promise of e-government is not new. A decade ago, the World Bank, in the comments on its draft anticorruption strategy, received explicit feedback supporting e-government. See World Bank, "Feedback from Initial Consultations on Strengthening Bank Group Engagement on Governance and Anticorruption," Aug. 30, 2006, 4, www.improvinggovernance.be/upload/documents/governancenote.pdf.

geopolitical reasons. Granular programming might have provided better documentation of the crimes perpetrated, and average investors might have had more warning to protect themselves, but it seems unlikely that granular interventions could have stopped this defrauding of the international community and the Afghan people by this illicit power structure. Perhaps nothing could have.

But it is crucially important to ask what might have been accomplished in Afghanistan if the anticorruption programming dollars focused on grand corruption had been directed to granular engagements. At the very least, a number of government systems and services could have been established that brought Afghans into the formal, licit economy. In turn, their investment in the formal and legal economy would have constricted the space for illicit activities, created licit norms and benefits, and, ultimately, modeled examples of tangible rule of law. Over time, these core elements build and solidify capacity and political will to tackle grand corruption. In fact, there may not be any rapid fixes to grand corruption.

Conclusion

In postconflict and fragile states, it is admittedly difficult *not* to focus on programs that target grand corruption, because all too frequently, lingering conflict and fragility are linked to an irresponsible set of elites who have literally robbed their citizenry. At the same time, the grim reality is that these elites typically still control substantial assets, political power, and international influence. History shows that they may well be those most likely to avoid responsibility for their actions in the early years of transition. If they do not lead an illicit power structure, someone in their subgroup will likely emerge to do so. Attempts to prosecute these corrupt influences without good data and political will only contribute to circular public discourse based on hearsay, promote cynicism, and consume resources. Moreover, the international community has dramatically demonstrated that in the face of grand corruption, geopolitical considerations trump justice and the rule of law.

Rather than set target objectives that even the most sophisticated state may struggle to achieve, a more modest approach is to emphasize combating petty corruption that erodes the ability to engage in normal life and build a functioning state. Restoring state legitimacy and legal protections for the average citizen charts a course away from conflict and fragility, which in turn fosters political will for attacking corruption. Today, e-government systems are serving citizens around the world and providing the infrastructure to properly manage their affairs with protection from corrupt interference. Indeed, the process is a gradual one, but every time a citizen is empowered to participate fully in the formal economy, the constituency for rule of law grows. When that investment in a legal state is rewarded with objective data that can protect citizens from abuses and punish the perpetrators, the citizen becomes an agent of change. Eventually, these individual citizens and the grains of formality they represent can form a beachhead against a corrupt government, removing the informality and instability that illicit power depends on.

Conclusion: What Should We Have Learned by Now? Enduring Lessons from Thirty Years of Conflict and Transition

Michelle A. Hughes

Ten years ago, when Michael Miklaucic and I began studying the impact of power structures on conflict and transition, we started with the proposition that formal power is only one dimension of state building and stabilization. We believed that governance capacity and legitimacy stem from a complex interplay of formal, informal, and illicit power. How governments come to grips with each is a powerful indicator of whether a nation state can function in partnership with its population and within the rule of law.

We also believed that none of this was new. Power struggles are part of human history—recurring themes in literature and the historical record as far back as recorded time. So why, we asked ourselves, do we as a nation and we as an international community struggle to understand the power dynamics in modern conflict and transition? And why, in our collective response, do we get it so wrong? How do we get our responses right?

It should come as no surprise, then, that one of the enduring insights from the *Impunity* case studies is that we continually fail to learn the lessons of our own experience. We identify lessons correctly, but we do not act on them. We reinforce failed approaches. We replicate success only (it seems) when we have no other choice.

Impunity: Countering Illicit Power in War and Transition was an attempt to codify lessons learned from both failure and success. Throughout this project's more than 10 years of collaborative study, we allowed ourselves to become students of history while, at the same time, looking to the future. We studied the history, not of war per se but of transition—and, importantly, of *containment*. In the effort to anticipate and check the problem of illicit power, we tried to deconstruct issues that cross geographic and cultural boundaries. And then, looking deeper at the enablers of illicit power and at the means of confronting them, we hoped to bridge the gap between information that seems promising even though we do not know quite what to do with it, and information that, if applied, could help civilian and military leaders, strategists, planners, and implementers confront illicit power. Thus, the question becomes, how do we turn unstructured data from multiple sources into useful predictive analysis? And because we are implementers ourselves, we always kept the frontline operator in mind.

So what does "success" in confronting illicit power look like? We know it is not Iraq, even though, when the last U.S. combat forces left the country in 2011, our government leaders were saying it was, and many people genuinely believed it. Is it Sri Lanka, which, having militarily defeated the Liberation Tigers of Tamil Eelam (LTTE), now struggles to restore governmental accountability, legitimacy, and acceptance within the international community? Is it Colombia, undeniably transformed but still seeking a final, enforceable accommodation with the Revolutionary Armed Forces of Colombia (FARC) after nearly a century of conflict and instability? Is it Sierra Leone, improved but still marked by deep social injustice and the resource curse, or is it Liberia, whose supposedly reformed political culture remains riddled by patronage and corruption? Or is it maybe something else?

Our case studies—both those we documented and those we did not—indicate clearly that the metric for success is more about process than about an end state. In both mature and developing countries, success in confronting the problem of illicit power is about containment rather than destruction. An Italian carabiniere officer colleague, speaking about the Puglian mafia, used to tell me: "Even when we are most successful, they never go away. We are just able to limit their activities to certain areas so they do not get in the way of legitimate business and so their violence does not hurt the community." Is this success? And if so, how do we get there? Both our research and experience indicate that the key is to anticipate the problem of illicit power *before* it becomes critical—to prevent it from becoming so influential that it can no longer be contained—and to prevent it from undermining accountable governance, security, and the rule of law. Ignoring the risk of illicit power, while hoping for the best and curbing our own political will and commitment to countering it, has always been and will ever be a disastrous course of (in)action.

The Assumption of Power by Elizabeth I: Setting Conditions for Success

Many historians consider Elizabeth I one of England's greatest monarchs ever.[1] But from the moment she ascended the throne in 1588, at age 25, she was forced to contend with the problem of power—internal and external, licit and illicit. Hers was a political transition that followed decades of war and internal strife, political instability, and near economic collapse. The English throne, one of the weakest in Europe, ruled a bankrupt nation torn by religious discord and threatened by the great powers France and Spain. Its borders were insecure, particularly its border with Scotland, where Mary, Queen of Scots, contended that she had a greater right than Elizabeth to rule England. In fact, Elizabeth's right to succession was questioned openly at all levels of English society, and most Europeans considered her illegitimate. Her political power base was tenuous, and her sex handicapped her ability to build strong political support. Even Elizabeth's most ardent supporters believed her position dangerous and uncertain.

Fortunately for Elizabeth, however, she was a realist and did not underestimate the gravity of her situation. She knew that such a momentous transition of power, even under the best circumstances, raised possibilities of counterclaims and rebellion. Everyone, from the highest nobles and ministers of state to the lowest public officials, would vie for patronage and influence. Ambitions had already been stoked in aspiring courtiers. Religious leaders, both Catholic and Protestant, were poised to invoke the power of the pulpit to secure their sectarian interests. Understanding the risks, Elizabeth and her advisers were prepared. Immediately upon her accession, her new secretary of state, William Cecil, executed a "memorial" consisting of essential tasks that would lay the foundation for long-term security, stability, and reconciliation. They included the following:

- A formal proclamation announced that Elizabeth was sovereign ruler. The proclamation was to be distributed immediately to all "places and sheriffs" and put into print. Doing so would ensure that the people knew who their legitimate ruler was, and who it was not.

---

[1] The following discussion is based on historical events as described in Jane Dunn, *Elizabeth and Mary: Cousins, Rivals, Queens* (New York: Knopf, 2004), 12-67.

- The Tower of London was put into the hands of "trusty persons" and made ready to receive Elizabeth should she need the safety of its defenses while she settled her officers and council. Thus, the sovereign's personal safety and security were ensured during the time required to consolidate the new government administration.

- Elizabeth's name was "written in" with of the keepers of castles and forts throughout the land. This step established her authority over the nation's defenses and helped seal its borders.

- All ports were temporarily closed, and particular care was taken with those closest to France and Scotland, both of which proffered claimants to the English throne. The closures were necessary to further control ports of entry and protect England from external threats. The action also helped control the outward flow of persons of interest and much-needed economic resources.

- No money could be taken out of the country without the queen's express permission. This bought time for the new regime both to assess and to stabilize England's dire economic situation.

- New justices of the peace and sheriffs were to be appointed in each county. This would ensure that law enforcement authority was in the hands of those loyal to the regime, who would act within the scope of Elizabeth's sovereign directives.

- Preachers were "to be dissuaded, in the short term," from touching on anything doctrinally controversial in their sermons. Elizabeth knew that long-term stability would require resolution of the religious strife that had torn the country apart, but she would need time to reach a settlement. Stirring up emotions during transition would be counterproductive and complicate eventual negotiations with religious leaders.

- A final directive concerned the "preacher at Paul's Cross," a notorious firebrand known for rabble-rousing sermons against the establishment, instructing "that no occasion be given to him, to stir any dispute concerning governance of the realm." In other words, the regime meant to limit the opportunities for extremists to get their message out.

The "memorial" tasks succeeded in their aim. Elizabeth managed the extraordinary risks and achieved a peaceful assumption of power that set conditions for success, in both tone and substance. Defying expectations, she reigned for 55 years, until her death from natural causes. When she died, another peaceful transition of power occurred—to her nephew, James I of Scotland. Elizabeth's reign was less than perfect, and the struggles were real. But she eventually achieved political and religious reconciliation, successfully defended England against its enemies abroad, and presided over a period of unprecedented economic growth and intellectual and creative achievement. Historians have often characterized Elizabeth's reign as a golden age of progress.

For anyone not a student of English history, it is easy to overlook how fragile her reign was in the beginning and simply to attribute her success to her remarkable leadership. But she was also a discerning leader, and the immediate steps she took to mitigate the risk from the myriad power struggles in play when she took control had a major role in her ability to rule.

It would be easy to dismiss lessons from sixteenth-century England as irrelevant, but they are far from it. Indeed, they are universal. When we review the 1999 Lomé Peace Agreement, which settled the civil war in Sierra Leone, it is striking how many of the provisions in the (eventually) successful peace accord parallel the tasks that Elizabeth's secretary of state set forth.[2] The Lomé Agreement is surprisingly comprehensive. It is also heavy-handed in a way that clearly anticipates the threats from illicit power structures across the spectrum of politics, security, economics, development, governmental authority, and protection of sovereign territory.[3] And its specific provisions for process and procedure, oversight, accountability, transparency, reconciliation, and public information mirror the tasks in William Cecil's Elizabethan "memorial." As the Sierra Leone case study (Chapter 8) points out, this particular agreement was not the country's first. The Abidjan Agreement, which preceded it, failed because it was less anticipatory and less inclusive. The Lomé Agreement succeeded largely because it was the first agreement to realistically address both known and expected risks from the various power structures that would have a stake in the success or failure of a new Sierra Leonean regime.

Insights from the Case Studies

Taking a historical perspective, what are the enduring, generalizable insights from the *Impunity* case studies? With the amount of material we had to work with, codifying the countless lessons would itself have been a book, so rather than try to capture each lesson here, we refer the reader to the case studies themselves. But one of the central insights is this: when looking at the problem of illicit power, we often fail to ask the right questions or identify the most important issues and indicators. Therefore, we saw that it would be useful to highlight the broad categories for consideration. Within each category are a number of crosscutting observations. The intent here is to provide, rather than a checklist, a guide that facilitates (a) adaptive thinking and critical application of the lessons and insights to future mission analysis; (b) strategic and operational planning; (c) policymaking; (d) programming; (e) and tools, tactics, and procedures (TTP).

We are vulnerable, of course, to the criticism that most of the insights are strategic and policy oriented. We "get" that for a military commander or development specialist on the ground, such macro insights may seem inapplicable. But this is seldom the case. Illicit power is a strategic problem, but it manifests at every level, permeating neighborhoods, houses of worship, businesses and bazaars, police checkpoints, and community resource centers. The bad actors we encounter locally are small cogs in a greater system of systems. Thus, until we understand the larger issues that drive their actions, it is almost impossible to respond appropriately, limit their activities, and effect real change. Each case study in this book—and the countless others we did not include—contains examples of how early local actions and decisions by either the host-nation government or international interveners allowed a nascent or weak illicit power structure to emerge and coalesce. Subsequent action, reaction, and inaction enabled the power structure to

---

[2] *Peace Agreement between the Government of Sierra Leone and the RUF* (Lomé Peace Agreement), UN Peacemaker, July 7, 1999, http://peacemaker.un.org/sierraleone-lome-agreement99.
[3] Ibid.

grow to the point of intractability. This truly is a problem where tactics must be employed with deliberate strategic purpose and understanding.

The insights fall into three main categories: the operational environment, accountability and the rule of law, and institution building and security sector reform. Also, we need to understand the relevance of peace agreements and accords.

The Operational Environment

Before we can address the problem effectively, we have to understand the operational environment. In "It Takes a Thief to Catch a Thief," we devote an entire chapter to unraveling complex geographic, sociocultural, and economic settings, while also providing analytical frameworks to better understand them. This is not just a problem for outside interveners. We saw in Sri Lanka, for example, how the government consistently misread the LTTE, even though the LTTE was a wholly indigenous insurgency. And at each stage of the Sierra Leone conflict, the government struggled to comprehend the nature of the Revolutionary United Front (RUF). Its responses failed accordingly.

"Operational environment" refers to a broader landscape than the immediate environment of the illicit structure. This is because licit and illicit networks alike rely on the same *licit* mechanisms for success. Thus, illicit activities are almost always intertwined with licit ones. When we do not understand the licit environment, not only do we fail to detect anomalies, but our responses tend to be "either or," as if there were always the option for a clean solution.

Phil Williams and Dan Bisbee, in their study of the Jaish al-Mahdi in Iraq, conclude: "[I]n the chaos and anarchy that followed the toppling of the Baathist regime, the line between licit and illicit power was blurred—an ambiguity never fully appreciated by the United States. This set in motion a series of missteps reflecting a profound lack of understanding of Iraqi traditions and politics, a failure to realize that common sectarian identity was no guarantee of harmony, and a sense of bewilderment when U.S. forces were not universally treated as liberators rather than occupiers."

Context is key, and long-term analysis is critical. Illicit power structures are organic. They emerge and submerge in response to opportunities and threats. They evolve. They, adapt. They form alliances. They break according to their own factional interests. They are parasitic, feeding off licit networks and organizations. They exhibit tendencies and preferences in the ways they operate. Motivational factors influence their methods, and as our studies show, motivations also change over time.

Successful illicit power structures are never static. Thus, understanding them requires a long-term perspective that captures the context in which they arise and thrive, and uncovers trends over time. Colombia and Sri Lanka present two striking examples of illicit power structures that evolved and adapted over the course of decades, as did their government foes. Both the FARC in Colombia and the LTTE in Sri Lanka began as creed-based power structures that arose out of perceptions of injustice and need, and both continually adapted in response to government action and inaction. Both took advantage of public outrage when government security forces either failed to protect the population or overreacted with unwarranted escalation of force and wanton human rights violations.

In these two cases, both governments also adapted. When they did so positively and proactively, they created opportunities to diminish the illicit power structures' strength. In each case, the ways in which each side—the government and the illicit power structure—reacted to the other are instructive. For the Sri Lankan government, the LTTE's use of the cease-fires of 2002 and 2004 to refine its strategy and eliminate moderate Tamil political rivals in Tamil-controlled provinces should have come as no surprise—if not in 2002, then certainly in 2004, between the periods that contributing author Thomas Marks refers to as "Eelam III and Eelam IV." Why should the government have expected this outcome? Because, in 1991, the LTTE had used the failed Indian peacekeeping mission as an opportunity to consolidate militarily and politically, which then enabled it to step its operations up to the next level of violence. The Colombian government had similar precedence for its interaction with the FARC. During past cease-fires and negotiations, the FARC rarely demonstrated genuine commitment. Instead, as Carlos Ospina points out, the FARC each time took advantage of the situation to reorganize, expand, and train its cadre, as well as to expand its involvement in the drug cycle. In retrospect, the FARC's objective was never to reach an agreement, but to buy time for a general offensive. Studying the patterns should enable the Colombian government to be more predictive and to adapt its security operations and negotiating strategy accordingly.

Nuanced, long-term, contextual understanding is not easy to achieve, however. Indeed, many of our current practices and procedures discourage it. In Chapter 10, we argue that truly understanding the operational environment requires three levels of analysis: continual assessment of risks and vulnerabilities of the host government and its sociopolitical environment, in-depth study of the emergent illicit power structure itself, and an intimate understanding of the illicit power structure's enabling environment. This requires time, resources, expertise, and collaborative relationships seldom readily available to decision makers. Spotting trends also requires long-term intellectual commitment—which is not compatible with our modern tendency, in difficult, security-challenged environments, toward short-term deployments and remote monitoring and engagement. The tendency is quick-in, quick-out technology-based approaches. So for diplomatic, military, law enforcement, and intelligence personnel, achieving any continuity in understanding a nontechnical, organic problem has grown increasingly difficult.

Finally, contextual understanding requires that leaders and decision makers be willing to listen and learn. In-depth knowledge and nuanced understanding do not come to us through PowerPoint. They demand commitment by those responsible for designing and executing strategies to counter illicit power. Such understanding also requires willingness to engage in highly detailed examination, and intellectually rigorous speculation, all while tolerating a high degree of uncertainty.

Mischaracterizing an illicit power structure increases the risk that our response will fail. As we approached the various research projects that eventually led to this book, a debate ensued about what could or could not be classified as an illicit power structure. But that debate obscures the more important insight: whether an organization meets a set of criteria making it an "illicit power structure" matters less than what *type* of organization it is. The typology matters—very much, in fact. When we incorrectly bin an illicit power structure as a terrorist organization instead of an insurgency, a militia instead of a criminal gang, or terrorists instead of protesters, we handicap our response and limit our options. Surprisingly, however, if we do not understand the operational environment correctly, we are less likely to correctly characterize the illicit power structures within it.

When we get this wrong, the most obvious problem is authority, particularly if the question arises whether law of armed conflict (LOAC) or domestic criminal law applies. Even when an international or internal armed conflict exists, thereby triggering LOAC, rules of engagement should be quite different when approaching a criminal versus an armed belligerent. Because procedure can matter greatly in a criminal case, the TTP for interdiction will be distinct. Classification raises jurisdictional questions that can affect the ability of elements within host-nation security forces to make detentions or arrests. In Afghanistan, for example, arrest authority over crimes classified as "internal security" crimes belonged to the Afghan National Directorate for Security rather than to the Afghan National Police (ANP). Thus, how the International Security Assistance Force (ISAF) coalition chose to classify the activities of bad actors made a difference in which agency it partnered with (depending on whether the goal was effective prosecution or battlefield detention). In the Joumaa Web cases, once banks and other organizations were classified a certain way, U.S. and international law enforcement agencies could invoke specific sanctions and law enforcement cooperative agreements to attack their networks. The classification had to be accurate, however, or sanctions would not hold.

International mandates also must accurately match the situation; thus, when understanding or classification is inaccurate, both the mandates themselves and national caveats imposed by troop-contributing nations do not support stabilization. Haiti provides a striking example of the problem. The United Nations was, as David Beer points out in his case study on the gangs of Cité Soleil, unprepared for the level and nature of the violence. The UN mandate did not support executive policing, which was required to counter the threat from a policing perspective. It also did not empower UN military units to support policing efforts in what had become full-blown guerrilla-type urban warfare. Instead, formed police units were deployed to maintain civil order and were unprepared and ill-equipped to confront the gangs.[4] Faced with mounting violence, contributing nations began to impose even more stringent national caveats than the mandate allowed, further constraining the forces' ability to confront the illicit power structures that they were certain to encounter. By incorrectly defining the threat from illicit power, the UN mission found itself in an untenable position. Meanwhile, the gangs increased in strength and capacity, and the mission's reputation was in tatters.

Getting the classification right is important also because the narrative must fit the circumstances on the ground. The Sri Lankan government experienced this both at the beginning of the insurgency and at its conclusion. In the beginning, reacting to a per-

---

[4] "A Formed Police Unit (FPU) is a team of 140 police officers, which is deployed as a group, who undertake crowd control, protect UN staff and material and escort UN personnel when they must visit insecure regions of a mission area. . . . FPUs are rapidly deployable, well equipped and trained to act as a cohesive body capable of responding to a wide range of contingencies. They are self-sufficient, able to operate in "high-risk" environments and are deployed to accomplish policing duties such as crowd control rather than to respond to military threats." UN Police, "Sustainable Peace through Justice and Security," 2010, www.un.org/en/peacekeeping/sites/police/units.shtml.

ceived terrorist threat, the government overmilitarized its response to student protests and random acts of violence, thus helping the LTTE mobilize public support and transforming what had been a Tamil protest movement into a full-blown insurgency. Conversely, by the end of the conflict, the government's characterization of the LTTE as terrorists brought international condemnation when it applied a scorched-earth, total-war strategy to end what was by then a civil war.

It is important to note that we may deliberately classify an illicit power structure as one thing even while fully aware that it is quite another. This is usually done for political reasons or to limit the scope of engagement or the type of tactics that can be employed. In 1990, for example, the U.S. Country Team in Colombia was ordered not to use the term "narco-terrorism" when referring to the FARC. The FARC was an "insurgency" and nothing else. Even though everyone working counternarcotics knew that the FARC was beginning to team with the drug cartels, the U.S. Congress had limited foreign assistance to the "war on drugs"—U.S. policy was not to get involved in Colombian counterinsurgency operations. An important further reason for this classification decision was to maintain a consistent public message about why the U.S. military was even deployed in Colombia in the first place.[5] In such cases, understanding the effect of the deliberate mischaracterization becomes an essential part of understanding the operational environment.

Planning assumptions about efforts to "follow the money," "shut down their transportation, weapons, and supply networks," etc., are seldom informed by a realistic understanding of what such an objective actually means, how long it takes, and who is involved. Part of the operational environment that we examined included enablers. In particular, we wanted to look at the role of financial networks, arms traffickers, social media, and other elements of what, in Chapter 10, we call *the enabling environment*. One of the lessons is that these enablers themselves are complex networks and systems. Therefore, attacking such networks requires several things: long-term commitment; international legal and regulatory systems that can be exploited to control illicit use of licit systems; strong interagency coordination and information sharing; coherent, focused diplomatic support; jurisdictions and venues where legal cases can be fairly tried and adjudicated; high-end, specialized investigative and intelligence collection capabilities; and, ultimately, the ability to enforce sentences, sanctions, and remedies. This is a complex set of factors. Even mature legal systems in stable countries struggle to pull these strands together. Both licit and illicit actors are implicated, and protecting the licit constrains our ability to pursue the illicit.

Planning horizons and expectations regarding results must be realistic. The Joumaa Web case study on "following the money" is instructive. Among its many lessons, it teaches that a complex financial-crimes case will likely take years to develop and will continue in the court system for yet more years (not months) on top of that. By the time

---

[5] During 1990–91, I led the U.S. military counternarcotics tactical analysis team embedded in the U.S. embassy in Bogotá, Colombia. Upon my arrival in Bogotá, the chief of mission and the U.S. defense attaché instructed me not to use the term "narco-terrorism" in any analytical reports and not to support any activities that would create "targeting ambiguity." If the FARC or either of the other two insurgent groups active in Colombia at the time were identified as narcotics traffickers, the risk was too great that the U.S. military might then target them for direct action, thereby violating U.S. law and policy.

that criminal cases and civil actions conclude, the illicit power structure that relied on the network will have found alternative means to finance its operations. "Cutting off the flow of weapons and materiel" is similarly difficult. As the Odessa Network case study illustrates, arms brokers often operate under the protection or imprimatur of powerful nations. Attacking those networks means going after their political protection. For reasons of sovereignty and diplomacy, this may not be a viable option. Often, the most realistic alternative is to create barriers that will increase the cost of doing business and eliminate the profit motive.

The prohibition against "blood diamonds" from Africa paints a realistic picture of what a successful campaign against enablers can achieve. While no reasonable person would argue that exploitation and corruption in the diamond mining sector has been completely overcome, it has now been limited to the point that bad actors and illicit power structures can no longer use it with impunity to finance their wars and motivate their soldiers, as the RUF and Charles Taylor did in Sierra Leone and Liberia, respectively.

The difficulties presented by constraints and planning assumptions do not mean that we should not go after enablers. Obviously, limiting enablers is part of the tool kit for countering existential threats and known illicit power structures. But we must also go after them for greater long-term, strategic purposes. Left unchecked, the illicit use of banking, finance, commerce, transportation, communications, and armaments systems will eventually make those systems useless for legitimate purposes. Again, we must be realistic about what success looks like. "Success" may mean nothing more than that our willingness to go after enablers signals our partners that we are serious and expect them to be, too. The strategy is as much about mitigating and messaging as about cutting off ways and means. We also must clearly understand that near-term tactical victories against enablers may have no substantive long-term impact. The use of social media for radicalization and recruitment is an apt example. Maeghin Alarid points out in her study that no sooner is an ISIS Twitter account disabled than another one appears. The point is always this: *never* concede the platform to impunity. The message is that illicit use of a licit system is not okay.

Accountability and the Rule of Law

Accountability and the rule of law are not secondary to security. Indeed, any impediment to them is a primary security threat.

It is easy to dismiss rule of law and good governance as the province of development and put them on the back burner as secondary to the immediate security threat. But the absence of rule of law creates an environment where bad actors can thrive, and encourages competition between government security forces and illicit power structures, for the monopoly of force. Left unchecked, illicit actors can perpetuate drivers of conflict, undermine governmental legitimacy, and establish control over the population, resources, and essential government services. Illicit capture of state institutions follows. Agreements and accords designed to restore peace and stability become meaningless. The rule of law is not, as Afghanistan's President Karzai declared in the early days of his regime, a "luxury" that can be addressed later, after security has been restored. By then it is usually too late.

Failure to recognize the risks posed by organized crime, corruption, and warlordism leads us to empower those who will later be the greatest threats to stability. In the chapter "Traffickers and Truckers," Gretchen Peters warns, "Never dismiss organized crime and corruption as 'secondary issues.'" Precisely. Crime and corruption remain some of the most enduring challenges to viable peace. In almost all the cases in this volume, corruption created grievances, siphoned critical state resources from the economy, and led to a failure of legitimacy for the governments, institutions, and officials involved. In stability operations, intervening government officials, whether military, law enforcement, diplomatic, or developmental, need to be alert to, and intolerant of, corruption and criminal behavior by their local counterparts. They need to support the emergence of healthy state institutions, rather than look the other way when those considered allies are involved in illicit activities.

Afghanistan illustrates perfectly how a foreign intervention ensures its own defeat when it turns a blind eye to local corruption and illegal trafficking. Accommodating corruption costs the international community more in the long run because fragile states remain aid dependent and fail to evolve into stable, self-sustaining nations that can become durable partners. The Odessa Network exemplifies what can happen during extreme political transition. Sudden privatization without sufficient regulatory controls enabled former Soviet power brokers to obtain a monopoly on shipping and port services in Ukraine and to use them as licit cover for illicit arms trafficking. At the time that privatization occurred in the early 1990s, Western democracies took a laissez-faire approach to the newly emerging economies, assuming that market forces would ensure a smooth transition. They failed to recognize that the institutional controls taken for granted in the West did not exist in those countries emerging from Communism.

Our own failure of accountability and oversight is often the single biggest enabler of illicit power. Forsberg and Sullivan, in their study of criminal patronage networks in Afghanistan, put it so well, it bears repeating here:

> *In insecure states with underdeveloped institutions and weak rule of law, any massive infusion of international resources to build local capacity, if disbursed with inadequate oversight, is likely to be accompanied by a surge in corruption and organized crime.[6] International forces and their interagency counterparts conducting counterinsurgency or stability operations must anticipate this development and be prepared, in the earliest stages of their mission, to put in place mechanisms to mitigate and monitor the problem . . . At the same time, expectations for transparency and accountability should be articulated to officials in the supported government.[7]*

Afghanistan and Iraq provide stark and—thanks to highly publicized disclosures by the special inspector general for Iraq reconstruction and the special inspector general for Afghanistan reconstruction (SIGAR)—detailed examples of how the international community consistently fails to hold itself and the assistance it provides to acceptable

---

[6] See, for example, UNDP, "Fighting Corruption in Post-Conflict and Recovery Situations," June 9, 2010, www.undp.org/content/undp/en/home/librarypage/democratic-governance/anti-corruption/fighting-corruption-in-post-conflict---recovery-situations.html; Fredrik Galtung and Martin Tisné, "A New Approach to Postwar Reconstruction," *Journal of Democracy* 20, no. 4 (2009).

[7] See chapter 1 (citations omitted here).

standards of accountability. SIGAR's criticism of development assistance and coalition accountability is well known. Less widely understood in the United States is the degree to which the UN Mission in Afghanistan (UNAMA) contributed to the culture of impunity by its own lack of oversight. In 2012, following years of criticism, UNAMA became mired in scandal when it was revealed that the officials managing the Law and Order Trust Fund Account (LOTFA)—established to manage the disbursement of international funds to pay for the Afghan National Police (ANP)—were involved in procurement fraud and collusion with their Afghan counterparts. LOTFA officials knowingly paid the salaries of police officers who did not exist, and created high-paying positions for well-connected Afghans.[8] Despite removal of the LOTFA country director, the problem of oversight in the troubled program continued. In late 2013, UN and ISAF officials were again confronted with evidence of LOTFA mismanagement and unaccountability when audits revealed that tens of thousands of salaries continued to be paid to "ghost soldiers." In part because of concerns about distracting the Ministry of Interior from the need to increase policing capacity in advance of the 2014 elections, a decision was made to defer resolution of the problem.[9] Later SIGAR efforts to spotlight accountability were rebuffed, and ISAF made SIGAR's investigation more difficult by overclassifying critical information on Afghan National Security Forces' (ANSF) strength, attrition, equipment, personnel sustainment, infrastructure, and training, as well as on anticorruption initiatives at the Ministry of Defense and Ministry of Interior.[10]

While the LOTFA scandal may have been only the latest high-profile example of external interveners' inability or unwillingness to police themselves, the experience in Afghanistan is hardly unique. In the 1990s, the UN Interim Administration in Kosovo (UNMIK) was also riddled with scandal. Not only did UNMIK facilitate widespread graft and corruption, but security forces under UNMIK authority *fueled* the sex trafficking trade, empowering Kosovar and Albanian mafias, many of which continue to operate with impunity throughout the region. In some instances, these illicit organizations succeeded in consolidating political and operational control over law enforcement agencies, security forces, and local governments.[11] In 2007, when veteran UN diplomat James Wasserstrom tried to expose the corruption within UNMIK, he was immediately fired and detained by UN police, who ransacked his apartment, searched his car, and put his picture on a "wanted" poster.[12] Seven years later, a UN dispute tribunal finally

---

[8] Dion Nissenbaum, "UN Staff Suspended in Afghan Fraud Probe," *Wall Street Journal*, June 17, 2012, www.wsj.com/articles/SB10001424052702303379204577472461229403068; Katrin Park, "A 200 Million USD Scandal in Afghanistan Shows UN Mismanagement," RAWA News, Oct. 15, 2014, www.rawa.org/temp/runews/2014/10/15/a-200-million-corruption-scandal-in-afghanistan-shows-un-mismanagement.html.

[9] Author interview with a senior (director-level) UNAMA official who participated in the decision briefs on whether to cut off LOTFA disbursements until an accurate count of ANP personnel could be established and verified, Kabul, Jan. 2014.

[10] SIGAR, "Quarterly Report to the United States Congress," Jan. 30, 2015, www.sigar.mil/pdf/quarterlyreports/2015-01-30qr.pdf.

[11] UN Peacekeeping, "Human Trafficking and United Nations Peacekeeping DPKO Policy Paper," Mar. 2004, www.un.org/womenwatch/news/documents/DPKOHumanTraffickingPolicy03-2004.pdf. Additional insights derived from author's observations and interviews while conducting security sector reform fieldwork in cooperation with the U.S. Mission to Albania during Apr.-Oct. 2009.

[12] Author interviews with James Wasserstrom in Kabul, 2011, and Washington, DC, 2014. See also Julian

cleared Wasserstrom and compensated him for his losses. But just as UNAMA has done with LOTFA, UN leadership in the Wasserstrom case stonewalled the investigations and denied any institutional culpability.[13]

With the stakes so high, why are we reluctant to impose rigorous standards on ourselves and our host-nation partners? For several reasons. Some donors would rather operate with less transparency. There are political implications, particularly when only one of several donors may be principally at fault. Do we "name and shame" the offenders, or do we look the other way to maintain the appearance of harmony and cooperation? Lack of security is another impediment to oversight. Insecure environments make for high staff turnover and limited ability to put "eyes on" development projects. Overseeing the disbursement of aid becomes especially challenging, so opportunities for corruption and mismanagement abound.

Also, effective oversight and accountability are difficult in purely practical terms. In 2010–11, NATO Training Mission-Afghanistan (NTM-A) tried to conduct a complete personnel and vehicle inventory of the Afghan National Security Forces. The labor-intensive effort dominated staff attention and resources, within both NTM-A and the Afghan Ministries of Interior and Defense, for more than five months. Transportation assets had to be dedicated to moving human resources and logistics personnel to every corner of the country, and battlespace owners had to assist with security and life support. As accountability dominated the partnering agenda, advise-and-assist relationships between coalition members and their Afghan partners became strained. Because this happened during the height of the "surge," understandable tension developed between NTM-A and coalition ground force commanders, who often saw the inventory as a distraction that they could ill afford while conducting ongoing security operations. In the end, the audit was only a snapshot in time, and U.S. and coalition personnel openly questioned whether the effort had been worth the cost. For the military coalition, which was preparing to draw down its presence as Afghan forces took over the security lead, the oversight process was neither replicable nor sustainable. Thus, "What's the point?" was a widespread sentiment. For the Afghans, the only disincentive for failure of accountability was loss of face. Equipment would continue to flow. No one was willing to limit distribution of fuel or spare parts, and coalition-led training would go on uninterrupted. The political will to enforce accountability by cutting off the Afghans' international support, at that juncture, was nonexistent.[14]

Obviously, this level of accountability needed to come about much sooner than the final year of the surge. Retrospectively applying the *Impunity* case studies to Afghanistan, the teaching point is that accountability procedures, incentives, and disincentives should have been established in the immediate wake of the Bonn Agreement in 2001 and maintained consistently throughout the duration of ISAF's time there. But even at the late date of 2011, there were positive effects. The inventories demonstrated a pro-

---

Borger, "UN Tribunal Finds Ethics Office Failed to Protect Whistleblower," *Guardian*, June 27, 2012, www.theguardian.com/world/2012/jun/27/un-tribunal-whistleblower-james-wasserstrom. Wasserstrom later took a position within SIGAR.

[13] Borger, "UN Tribunal."

[14] Author's personal observations while senior rule of law adviser to the policing development mission within NTM-A (2010-11).

cess that, though perhaps not replicable for us, could work for the Afghans. It exposed, within both ministries, important capability gaps that were not fully appreciated before the inventory, and it put accountability at the forefront of ANSF development discussions. As the Ministry of Interior began to take control of its own strategic vision later in the year, accountability as a goal and a process figured prominently in the ministry's revised National Police Strategy, its National Police Plan, and the succeeding minister of interior's official vision for the future of the ANP.[15] The audit caused NTM-A to reconsider how it was measuring anticorruption efforts, and it made some (although not nearly enough) adjustments to how it assessed capabilities milestones. In some instances, NTM-A recovered improperly expropriated vehicles. Even a year later, some Afghan commanders recalled the inventory and used it as a teaching point in their efforts to develop junior officers.[16]

Ultimately, our own failure of accountability creates a crisis of perception, which sees interveners more as part of the problem than as the solution. It opens us to criticism that we do not take accountability seriously. In postconflict stabilization and reconstruction, our failure of accountability undermines security assistance. It perpetuates linkages to organized crime and supports the corruption of licit power structures. It provides fodder for opposition elements, obstructionists, and negative media campaigns that undermine legitimacy—both ours and the host-nation partners'. Unaccountability feeds social vulnerability and undercuts the licit economy. It is an assault on the rule of law, creating conditions for failure rather than for success.

When law enforcement is viewed as a luxury, things fall apart, and illicit power structures are the first to fill the vacuum. It is convenient to think of security as, first and foremost, a military problem. And under the laws of war and occupation, citizens' security is indeed a military problem. But illicit power is also a law enforcement problem, and when there is insufficient capacity to enforce the law, illicit power structures will inevitably fill the void. They do so in two primary ways. The most obvious is by taking advantage of gaps in law enforcement to break the law. Thus, we saw illicit power structures such as the RUF in Sierra Leone, or rebel groups in Liberia, move across borders with impunity, illegally transporting weapons, diamonds, and people for profit and military advantage. We saw how the Pashtun trucking networks in Pakistan and Afghanistan evolved from transporting licit goods illicitly for profit, to transporting illicit commodities such as weapons, materiel, and heroin to facilitate violence and insurgency against the ISAF coalition and the nascent Afghan government. The failure of law enforcement to competently and humanely control political unrest helped enable the rise of the LTTE in Sri Lanka and the Moro Islamic Liberation Front in the Philippines, and the consolidation of Haiti's Cité Soleil gangs. Indeed, civil disorder and inadequate border management are capacity gaps that come into play in every case study in this book.

The other problem in the law enforcement void is that illicit power structures not only exploit the void, but also replace it to further their own ends. It should be axiomatic

---

[15] See, for example, Afghan Ministry of Interior Affairs, "Ten-Year Vision for the Afghan National Police: 1392-1402," 2014, http://moi.gov.af/en/page/5718/5729.

[16] Ibid; author's further observations and interviews during fieldwork in Kabul, Helmand, Balkh, Laghman, and Wardak provinces (Oct.-Nov. 2012), and Kabul (Jan. 2014).

that communities do not function without rules to control behavior, and some form of enforcement. Where legitimate rule of law is absent, the strong rule. Illicit power brokers establish their own set of rules, which the population must obey if it is to survive. The impact on the population can be devastating, as current events in the Middle East, where ISIL has taken control of Iraqi and Syrian territory, illustrate in stark relief.

The impact on legitimate governance and stability is equally devastating. By the time Plan Colombia was initiated in 1999, almost 80 percent of Colombian sovereign territory was outside government control and in the hands of the FARC or the major drug cartels. Restoring government control over the population, resources, and institutions required billions of dollars in international assistance, a complete overhaul of the Colombian justice system, and a massive investment in security operations and security sector reform. Colombia's very existence as a democracy was at stake because illicit power structures owned the monopoly of force and were using it to protect their activities and impose their will on the population. Although the Colombian military plays a critical role in internal security, it was strengthening governance and reforming the criminal justice system that turned the tide.

A capable, credible, accountable military is a powerful tool for stabilization. But ultimately, respect for national sovereignty, and compliance with the law rely on policing. As external security providers, the police should be viewed as secondary or subordinate to the military. But under most constitutions and most legal systems, police are the principal tool for internal security, the face of the government, and the primary instrument for enforcing the rule of law. To counter illicit power requires positive law enforcement engagement with local authorities and the people; strong relationships between police, prosecutors, judges, prison administrators, and other actors in the criminal justice system; and, if applicable, effective processes to coordinate and deconflict military and policing operations without relegating police to a supporting role. Law enforcement, if done right, respects and protects positive local customs and culture and sends the message that the government is in charge. It is not a luxury. It is essential.

Grievances that go unresolved create opportunities and incentives for illicit power structures to emerge, and enable them to build an enduring base of support. When assessing options to control illicit power, the tendency is to focus on the criminal justice sector. But the wider problem of dispute resolution; respect for basic equality, human rights, and property; and civil enforcement is strategically just as important to containment. Illicit power structures tend to arise out of popular frustration over inequality, and a sense that the government is unresponsive to the needs of vulnerable or minority populations. Power brokers who can address local concerns gain wide legitimacy and support, even where the population decries their methods. Restoration of the rule of law, in a way that addresses both governmental and nongovernmental impunity and protects fundamental fairness, reduces vulnerabilities that allow illicit power structures to emerge. It creates confidence in the government and invests the population in the success of legitimate governance structures, institutions, and initiatives. Lieutenant General H. R. McMaster points out in his foreword to this volume:

> *People fight today for the same fundamental reasons the Greek historian Thucydides identified nearly 2,500 years ago: fear, honor, and interest. . . . Crafting effective strategies to address the challenge of*

weak states must begin with an understanding of the factors that drive violence, weaken state authority, and strengthen illicit actors and power structures. Terrorist, insurgent, and criminal networks exploit fear and anger over injustice, portraying themselves as patrons or protectors of a community in competition with others for power, resources, or survival.

In Chapter 7, on the Moro Islamic Liberation Front in the Philippines, Joseph Franco writes, "The overarching driver of conflict in Mindanao all throughout its recorded history was the maintenance of specific economic rights. Discourses of religious and ethnic strife were the *effect* rather than the cause of conflict." Phil Williams tells us that in Iraq, "the Sadrist movement emerged as, and remains, the advocate for a large, young, and hugely disadvantaged sector of the Iraqi population, whose grievances and concerns must be met for Iraq to have any hope of long-term stability." Countering illicit power requires that we confront the challenge of injustice and shore up state authority to address the emotional, human dimension of conflict. This becomes even more critical in times of extreme political transition, when the distribution of power is still uncertain and the population is assessing where to place its support.

Institution Building and Security Sector Reform

Security sector reform and institution building, done right over time, can have a positive effect in countering impunity.

The Organisation for Economic Co-operation and Development (OECD) invested several years codifying best practices in security sector reform, leading to the 2007 publication of the "OECD DAC Handbook on Security System Reform: Supporting Security and Justice."[17] Since then, it has become fashionable, in the aftermath of failed SSR interventions in places such as Yemen and South Sudan, to bash the OECD's conclusions about the importance of a comprehensive approach as grand, unrealistic, and ultimately ineffective. But is the approach faulty, or is the problem that it has been executed imperfectly, with unrealistic expectations over too short a time? Several of this volume's contributing authors point to the limitations of technical capacity building approaches, and rightly so. But excessive focus on technical capacity misses the fact that where the focus has been on institution building, real progress occurs.

Sustainable institution building requires political will, backed by sustained assistance over time. Modern history is rife with examples where overemphasis on speedily training and equipping security forces, without security sector governance to manage and oversee them, has failed. Yemen, Iraq, and Syria are only the latest to hit the news cycle. But what about the success stories? Why do not they not get more attention? Among the case studies in *Impunity*, Sierra Leone, Liberia, Colombia, Haiti, and the Philippines are examples where a sustained institution building approach yielded progress. But in all cases, progress lasted only as long as the focus remained on the institutions rather than on the operating forces themselves. Timor-Leste, on the other hand, is a case study in good intentions poorly executed, without regard for political reality. Deniz Kocak con-

---

[17] OECD, "Security System Reform and Governance," May 31, 2005, www.oecd-ilibrary.org/development/security-system-reform-and-governance_9789264007888-en.

cludes, "It is striking that the United Nations did not learn from the lessons of either its previous missions or the experiences of UNTAET and UNMISET after 2006. Rather, it maintained a flawed PNTL [Timor-Leste National Police] training program that had failed to produce professionalism or operational effectiveness, it repeatedly conducted ineffective vetting that enabled rather than prevented factionalism and politicization of the force, and it neglected PNTL institutional development in favor of a short-term, train-and-equip mentality."

Colombia is often cited as a success story because the comprehensive security sector reform that took place under Plan Colombia did, in fact, reverse decades of failed train-and-equip efforts. The international community's ongoing commitment to reform in Sierra Leone formed the basis of the OECD effort and continues to this day. Liberia struggles, but the commitment to whole-of-government reform has at least given it a chance.

In Eastern Europe, Ukraine, threatened by Russian incursions and fearing for its future autonomy, is reforming itself faster than the international community can respond.[18] And the Republic of Georgia continues to progress even though it no longer enjoys significant donor attention.[19] The U.S. State Department's Security Governance Initiative in Africa represents a deliberate approach to security sector governance that places a greater premium on institutional capacity building and less emphasis on training and equipping. And it is beginning to get results in a handful of the countries that were selected to participate.[20] Even in Afghanistan, after NTM-A recalibrated its capacity building efforts for the police in 2010 to focus more heavily on ministerial development and professionalization, there was a marked increase in public trust and confidence in the ANP.[21]

In all cases, certain political common denominators support success. National ownership and leadership of the reform agenda is essential, and the host nation has to want its institutions to succeed. Leaders, institutions, and the population all must believe that they have significant interests at stake should reforms fail. As international donors, we must be committed to an assistance strategy that is long term and comprehensive, is culturally, contextually, and politically appropriate, and reflects the host nation's vision for itself, rather than our vision for it.

Countering illicit power requires special skills and capacity across the justice system, and capacity development must address that need. Often overlooked is the fact that confronting illicit power structures, their networks, and the networks that enable them is a highly specialized undertaking even in mature systems. All the illicit power structures in our

---

[18] Author interviews with U.S. State Department officials, Washington, DC, Sept. 2015.

[19] Ibid.

[20] Ibid. See also White House, "Fact Sheet: Security Governance Initiative," Aug. 6, 2014, www.whitehouse.gov/the-press-office/2014/08/06/fact-sheet-security-governance-initiative.

[21] Asia Foundation, "Afghanistan in 2014: A Survey of the Afghan People," Nov. 18, 2014, 42, http://asiafoundation.org/publications/pdf/1425. Notably, 2012, the one year during 2010-14 when public perceptions of the ANP declined, coincided with a period when NTM-A reduced its emphasis on ministerial development and professionalization in order to refocus on its "core" mission of rapid training and equipping of "boots on the ground." Author interviews with senior officials in NTM-A, ISAF, and the European Police Development Mission, Kabul, Oct.-Nov. 2012.

case studies had extensive geographic, functional, and political reach. Power structures such as the Jaish al-Mahdi in Iraq, Pashtun trucking networks in Afghanistan and Pakistan, and Charles Taylor's National Patriotic Front of Liberia were regional players. The Odessa Network and the Joumaa Web routinely operated on a global scale. As their power grew, they established relationships, created alliances, and expanded operations to the point where they transcended borders. Countering such complex networks poses challenges across the entire end-to-end process of any criminal justice system.

Viewed as a functional continuum, the capacity required to confront impunity begins with the ability to detect and prevent crime. At one end of the detection spectrum is community policing and community vigilance. The other end of the detection spectrum includes cross-border and cross-domain intelligence collection, analysis, and dissemination. Once detected, illicit power structures must be interdicted and investigated. This involves another set of highly specialized skills, attributes, and institutions. One vital niche capability, for example, is the ability to understand and audit highly complex financial transactions. This is not a job for amateurs. Chapter 10 discusses the challenge the Kenyan government faced when it tried to conduct small-boat maritime operations to interdict Somali extremists who were entering the country through its major seaports. The investigative ability to conduct maritime forensics simply did not exist in the Kenyan police or prosecution services. Thus, there was no justice endgame for the interdiction operations.

The challenge of capacity continues through to prosecution, adjudication, and appeal. Along the way, there must be a prison system capable of containing and controlling those who have been brought into the system (arrested) as well as those merely under suspicion. In Chapter 14, Mark Kroeker emphasizes the importance of prison systems to effective security sector reform. He notes that prisons are usually overlooked and underresourced, with disastrous results. What criminal law practitioners know intuitively is that this entire sequence has to work as a horizontally integrated process. A breakdown at any point along the continuum usually leads to total failure. But Kroeker also explains, "While it is useful and essential to look at how justice functions as a system, the subsystems within it must be carefully examined within their own peculiar set of authorities and functions."

The Drug Enforcement Agency (DEA), more than any other agency, has long recognized the need for a systemic approach—but one focused strictly on its mission of countering the flow of illicit drugs. Its formula for creating vetted units teamed with specially trained investigators and prosecutors, and establishing protected, specialized courts to handle drug cases under criminal codes designed to confront the problem of drug trafficking, provides one model. The DEA comes under criticism that its approach is not generalizable. But if we look at other successful strategies for confronting specialized threats, such as financial crimes, tax fraud, and grand corruption, that exist in developed nations with mature legal systems, we quickly see that the approach is not the weak link. What makes it impracticable is our own reluctance to apply the concentrated resources and diplomatic force necessary to make these systems work.

Effective institution building requires uncompromising accountability, and enforcement of both discipline and professional standards. We have discussed the consequences of our unaccountability as interveners, and emphasized the corrosive effect of corruption on a host nation's ability to prevent and confront the rise of illicit power. And in security sector performance, the negative impact is magnified. Accountability, transparency, and adherence to the rule of law must be front and center in approaching the challenge of security sector reform. And yet, our pattern, over and over again, is to sideline accountability as a problem for the lawyers, the inspector general, or Internal Affairs.

Successful, sustainable, self-reliant capacity building has three main components: (1) inculcation of a culture of accountability and public service; (2) inoculation of the system, to reduce or eliminate opportunities for impunity; and (3) strong discipline and enforcement mechanisms. The three must go hand in hand, and they are not easy to implement.

The first requires vision. What is the big idea behind the capacity that is being built, and how is that vision communicated? This is more than a messaging strategy, although the message is certainly an essential part of the whole. Laying out a vision, and a set of standards to support it, is key.

To this end, vetting, done with a high degree of transparency and public participation, is a valuable tool. In Chapter 4, Will Reno discusses the problem of impunity as part of Liberia's political culture. It is therefore notable that success may be emerging in the current Armed Forces of Liberia. During the postconflict reconstitution of the country's security forces, two different approaches were applied: one for the army and one for the Liberian National Police. The army (and the civil service within the Ministry of Defense) benefited from a gold-standard vetting process that required strict compliance with newly articulated qualification standards, and a lengthy process of public notice and comment for every recruit.[22] The result, after two years of painstaking development, was that the first graduation of military recruits became a national event. The Liberians were proud of their new force, and the soldiers were proud of themselves and the fresh start they represented.

Police and Ministry of Interior development followed a less rigorous, more ad hoc "check the box" approach, so that by the time the first army cohort was being fielded, the Liberian National Police force was already falling into disrepute.[23] Some would argue that haste was necessary so that policing could be restored and the international community could shift the burden of security to the Liberians, but the actual effect was that international civilian police units were forced to maintain a more active, sustained presence than their mandates anticipated; otherwise, law enforcement throughout the country would suffer.[24] The Liberian army, although it took longer to establish, has proved to be the more capable, credible force. Because the memory of its past abuses and atrocities is still raw, it fights an uphill battle in the court of public opinion. But leaders continue to enforce the standards that were put in place during the reconstitution. Only a few years after it was reformed, the army is increasingly finding its place as a credible tool of national and regional security, and emergency response.[25]

---

[22] Author interviews with Colonel (ret.) Thomas Dempsey, former U.S. defense attaché to Liberia, and, later, member of the assistance team that designed and conducted the Liberian vetting program.

[23] Author interviews with Liberian officials and members of the UN Mission in Liberia, attendance at town hall meetings, and observations while conducting security sector reform fieldwork, Monrovia and Lomo County, Liberia, 2007.

[24] Ibid.

[25] Helene Cooper, "Amid Ebola Crisis, Liberian Army Sees Its Chance at Rebranding," *New York Times*,

Inoculating the security system against impunity requires attention to the bureaucratic processes and mechanisms through which the institution's business is conducted. The e-government initiatives that Scott Carlson describes in Chapter 17 as part of a "granular approach" to combating corruption are apt examples of how rudimentary technology can be used to increase transparency, public access, and oversight, making systems and institutions less vulnerable. Another useful basic management technique is to provide actual position descriptions for government jobs so that merit-based hiring and promotion has a proper foundation. Internal and external audits and reviews; asset disclosure requirements; whistle-blower protection; fraud, waste, and abuse hotlines; ombudsmen; and inspections are all ways to increase accountability and decrease opportunities for abuse.

Finally, strengthening discipline and enforcement requires, first and foremost, that enforceable standards be clearly established. It was shocking that in Afghanistan, the ANP did not have an official, signed code of conduct until the summer of 2011—nine and a half years after international efforts to rebuild the National Police had begun. This oversight is a testament to the overwhelming focus on numbers—"boots on the ground"—in the recruitment and training of the ANP. Even after the minister of interior signed the police code of conduct, few in the ANP or the international community recognized the code's existence or significance. Dissemination was spotty at best, and how it should actually be enforced had not been thought through and was not being taught anywhere in the leader development training at the time.[26]

Part of the problem with strengthening discipline and enforcement is the commonly held view that it revolves around legal prosecutions rather than administrative discipline. International advisers often assume that the administrative measures taken in functioning organizations to correct misbehavior or enforce job performance standards already exist or will emerge as the organization matures. But it is important to remember that Western expectations and experience reflect a very democratic view of security as a public service. In countries where the main role of security services (and their colleagues throughout the justice sector) was to protect the regime rather than the population, the idea that the population has a right to demand that public duties be performed for its own benefit represents a sea change in attitude and institutional culture.

Ultimately, the key to developing successful accountability within the security sector is to make it a primary effort that is mainstreamed throughout every aspect of capacity building. Accountability, oversight, transparency, and anticorruption cannot be secondary to training and equipping. TTP must be vetted through the lens of accountability and enforcement. What standard will be applied? At which levels? How will the standard be taught, put into operation, and enforced? What internal processes must be put into place? What are the external controls? What does governance look like for each function, mission, and organization? What is the role of the public and civil society? Of the media? How can the public and civil society be educated about their roles? How will public participation be ensured? These are not incidental activities. They are primary, and in the fight against the rise of illicit power, they are essential to successful capacity building.

Oct. 11, 2014, www.nytimes.com/2014/10/12/world/africa/amid-ebola-crisis-liberian-army-sees-its-chance-at-rebranding.html.

[26] Author's firsthand experience as an NTM-A representative to the UNAMA-led advisory team that assisted in developing the Code of Conduct for the ANP, Kabul, 2011.

Inclusion matters – a lot. Deniz Kocak's excellent analysis of security sector reform in Timor-Leste contains many important lessons, and one of the most important is how critical it is that the security sector represent the population as a whole. As Kocak points out, SSR is, above all else, a political task. In Timor-Leste, as elsewhere, we see the harm done when local power brokers capture security forces and use them to empower their own ethnic, tribal, religious, familial, or political constituencies. The result is loss of trust and legitimacy.

Inclusion also matters in effectiveness. Security institutions that are not representative of the population cannot engage effectively with that population to prevent and detect the emergence of illicit networks. Successful intelligence-led policing, as Cliff Aims explains in Chapter 15, begins with the ability to identify the problems and determine how those problems can best be addressed *within the realities and limitations* of the indigenous system. Police must be able to interact with the whole population, detect anomalies, and develop reliable sources of information and evidence. To do this effectively, they must be objective and free from political, gender, ethnic, religious, or familial bias.

And prosecutors and judges must share the same objectivity. If members of a diverse citizenry are to come forward with their information and their concerns, they must first feel confident that the system represents and values them. Where vulnerable populations have been victimized by bad actors, it is particularly important that they be represented. Security and governance institutions must be accessible to both men and women, and accommodating to their trauma and their sensitivities. Confronting illicit power and impunity requires courage and confidence, which will be present only if security institutions truly reflect the population they serve.

Illicit Power and the Peace Process

With all the focus on the existential challenge that illicit power structures present on the ground, it is important not to overlook the critical role that peace agreements and processes play in our ability to contain illicit power. Viewed from the other side of the glass, what impact do illicit power structures have on efforts to reach viable peace?

The presence of illicit power adds layers of complexity to any peace agreement or process. Looking across the case studies, we can conclude that peace agreements have a chance of success if, as Gretchen Peters affirms, they take the illicit economy into account and address negotiable interests, however distasteful that may be. Peace agreements that gloss over difficult issues of power sharing, distribution of resources and authority, reintegration, and reconciliation are problematic from inception and may even be doomed to fail. Not only do they wish away the most intractable points of contention, but they also fail to provide a realistic framework for reform. Experience teaches that capacity building cannot succeed unless it addresses politics, power, and factionalism. An effective peace agreement must provide the structure around which these issues can be resolved.

While working on *Impunity*, we had the opportunity to teach the material as part of a graduate course on security and development in complex operations. So we incorporated a discussion on the importance of peace agreements to development that counters illicit power. The students were mostly current or former Army Civil Affairs officers. Almost all had served in either Afghanistan or Iraq, typically as part of a stabilization or reconstruction effort. We polled those who had served in Afghanistan, asking them whether they had ever read the Bonn Agreement, which codified provisional arrangements for the reestablishment of Afghan governance and security.[27] Many of our students had deployed in positions where they were advising the Afghans on governance and the rule of law, yet *none* had read the Bonn Agreement. Only four were even aware of its existence.

Even after we introduced the Bonn Agreement, the students were initially skeptical of its relevance, since they had mostly been working at the subnational level in provinces, districts, and villages. But they later realized that the Bonn Agreement had had a direct impact on their mission. They noted how highly aspirational it seemed. And they were surprised at how much the United Nations' role depended on decisions of an Afghan Interim Authority consisting only of individuals who were personally present in Bonn. They remarked that most of the illicit power structures they later encountered in the battlespace were not parties to the agreement, and they wondered how the agreement could have such force given the scale of the unrepresented interests. They found the proposed legal framework ambiguous. Indeed, those who had worked on rule of law-related issues while deployed thought the legal framework in the Bonn Agreement inconsistent with what they had been trying to apply in the field. And in light of their experience dealing with local power brokers, everyone believed that reliance on the Afghans' ability to achieve consensus on complex issues of future governmental authority was misplaced or, as one student put it, "a joke." Ultimately, most felt that it would probably have been a good idea to read it before they tried to accomplish in the field things that did not exactly track with the agreement reached in Bonn.

The students' analysis was not far off. As the authors of Chapter 1, "Criminal Patronage Networks and the Struggle to Rebuild the Afghan State" point out, the 2001 political settlement in Afghanistan neither accurately nor adequately addressed the risks presented by the various competing power structures that would be present during the reconstruction. This is especially ironic given Afghanistan's turbulent history of competition for power. Western expectations that the agreement represented a process to build a democratic, sovereign Afghanistan were not shared by the Afghans, who viewed it as a grand bargain for distribution of power between the most dominant (and present for the negotiations) ethnic and political factions. Instead of constraining the various power brokers, the Bonn Agreement actually created ambiguity and political space for illicit and factional leaders to assume power through unchecked patronage, violence, and coercion. Or, as the authors state, "Afghanistan's political settlement protected and, at times, empowered the country's CPNs [criminal patronage networks] and rendered ineffective many of the coalition's governance and development efforts."

The contrast between the Bonn Agreement and the Lomé Agreement in Sierra Leone[28] is striking. The Lomé Agreement is far less aspirational and much more pragmatic.

---

[27] UN Dept. of Political Affairs, "Agreement on Provisional Arrangements in Afghanistan Pending the Re-establishment of Permanent Government Institutions (Bonn Agreement)," 2001, http://peacemaker.un.org/afghanistan-bonnagreement2001.

[28] *Peace Agreement between the Government of Sierra Leone and the RUF* (Lomé Peace Agreement), UN Peacemaker, July 7, 1999, http://peacemaker.un.org/sierraleone-lome-agreement99.

It contains specific guidelines for how all former combatants should be treated upon cessation of hostilities, and how they were to be reintegrated into society and, in some cases, government security forces. All the belligerents who had negotiable interests were parties to the agreement, and the disposition of the intransigents was anticipated and addressed in considerable detail. The RUF's future political role was specified, as were the tasks that each party must complete to make it succeed. All tasks were framed within the structure of existing Sierra Leonean law and administrative governance. Major reforms would not begin until stability had been achieved. Oversight, transparency, accountability, enforceability, and public information were recognized as crosscutting concerns and were covered throughout. Nearly every major provision contained an agreement on how it would be promulgated and guaranteed. The authority of the United Nations, regional forces, and other interveners was clear, unambiguous, and heavy-handed. Withdrawal of foreign forces would be conditions based.

The Lomé Agreement was not immediately successful and might have been abandoned if the international community and the parties believed in it less. The main problem was that the RUF was divided between those with negotiable interests and those without. The internal schism caused the cease-fire to unravel, and violence ensued. But the international community held on to its commitment to the agreement, and increased troop strength so that capacity matched worst-case requirements. Reconcilable elements of the state security forces who had gone rogue recommitted themselves to the process, and the resurgent threat from the RUF was defeated.

The mission was saved, and the agreement was saved, which turned out to be a good thing. The Lomé Agreement had, in fact, addressed quite well the vulnerabilities and risks presented by the myriad competing illicit power structures and political factions. Once security was restored, the agreement provided a competent, realistic framework for conflict resolution, reintegration, reconciliation, resource management, and security sector reform. It was, in the end, a peace agreement that correctly anticipated the threat and fit the operational environment of illicit power in Sierra Leone and surrounding states.

As we conclude the study of *Impunity*, Colombia is back at the negotiating table with the FARC, with breakthroughs being announced almost weekly. A final deal will supposedly be worked out by March 2016, and the parties to the agreement—the Colombian government and the FARC—have six months to work out the final details of implementation. Key points of agreement include justice and reparations, rural development and land reform, and ending the FARC's involvement in the drug trade. The FARC's future as a political party, as well as its disarmament, demobilization, and reintegration, is not yet resolved, but both sides agree on the need to agree. Impunity during the conflict is also being addressed, and the processes to assign accountability and ensure just punishment for perpetrators of human rights violations are to apply to both sides. It remains to be seen whether Colombia will finally be able to resolve its decades-long war with the FARC, and the FARC its grievances with the Colombian government. But it is encouraging that—in this round of peace negotiations, at least—central issues of power, impunity, and the connection with the illicit drug trade are all being addressed. Peace in Colombia may finally have a chance.

Conclusion

We said that from the beginning we allowed ourselves to be students of history, and we have tried to do just that. In the end, our most enduring insight may be that the fight against impunity is all about governance—good, capable, credible, accountable governance. As Michael Miklaucic stated in the introduction:

> *Regardless of the idiosyncrasies of any particular illicit organization or network, the fundamental issue at stake is accountability. States are, or should be, accountable both to their citizens and to the international community of states. Illicit power structures, organizations, and networks are accountable only to themselves. They have no commitment to the broader public good beyond their parochial interests. And to the extent that they succeed in carving out operating space within a polity, they erode that polity's legitimacy by creating accountability-free zones, or zones of impunity.*

So what does success look like? Here is what the cases studies suggest: Success is more about a process than about a product. It is the demonstrated ability of a government and its people to work together toward a culture of accountability. It is the political will, within and among nation-states and international actors, to shine a light on impunity—to root it out wherever it is found, to call it what it is, and to commit the necessary resources to contain, transform, or destroy it. It is the ongoing education and development that enables legitimate institutions to strengthen and reform from within. It is vigilance in enforcing of the rule of law, even when this is not convenient, quick, or profitable. Success is also marked by the willingness to continue to engage—to return again and again to the negotiating table, even when those on the other side are belligerent and distasteful. Success requires a sincere, ongoing effort to resolve grievances, improve government services, and demonstrate through action and not mere words that impunity will not be tolerated. Success comes from comprehensive, long-term commitment, not from an on-the-fly injection of resources and training. Successful approaches are generational. They span decades, and they are comprehensive. If defeating impunity sounds difficult, it is because it is. But as our research illustrates, the impact of unchecked illicit power is far worse.

CONTRIBUTORS

Clifford Aims retired in 1996 from a 20-year law enforcement career in the United States and worked in a variety of peacekeeping operations and Security Sector Reform (SSR) missions in several regions of the world until 2004. From 2005 through 2009, he was part of the Interagency Team in the Experimentation Directorate at the U.S. Joint Forces Command in Suffolk, Virginia. There, he worked extensively with SSR counterparts from around the world to build on lessons learned from peacekeeping and postconflict stabilization missions and to develop the most current set of best practices for dealing with instability and illicit power structures. Since 2009, Aims has been leading SSR efforts for the U.S. Department of State.

Maeghin Alarid is the lead policy analyst at the USAF Institute for National Security Studies at the U.S. Air Force Academy in Colorado Springs, where she conducts research and analysis in the areas of arms control, deterrence, and strategic stability. She previously held a position at USNORTHCOM at Peterson AFB, as the joint training systems specialist. Alarid also spent seven years at the Defense Threat Reduction Agency, where she was an exercise planner at the Defense Nuclear Weapons School at Kirtland AFB in Albuquerque, New Mexico. There, she instructed on radiological terrorism and developed the school's first course on female suicide bombers. Alarid has a BA in ethnology and Portuguese from the University of New Mexico, an MA in international security and homeland defense from the University of Denver, and a graduate certificate in terrorism analysis from the University of Maryland.

David Beer is director of the Ottawa Bureau of MediaBadger (www.mediabadger.com) and serves as international policing adviser at the Pearson Centre (www.pearsoncentre.org). He served 35 years with the Royal Canadian Mounted Police in a variety of operational and command roles. He has a unique depth and breadth of international policing experience. At the time of retirement, he served as director general of international policing, with responsibility for international operations, peacekeeping operations, Interpol Canada, and international policy development. Beer deployed to conflict and postconflict areas, representing governments and international agencies. This experience included Iraq, Central African Republic, Liberia, Democratic Republic of Congo, and an extended time in Haiti, including bilateral capacity building dating back to the 1990s. In 2004-5, Beer commanded the UN Police Mission in Haiti. In 2007, he was elected to the executive of the International Association of Chiefs of Police, serving as vice president and chair of the International Division until 2011 (www.theiacp.org). He holds a BA in Sociology and Anthropology, Carleton University (Ottawa) and an MA in political science and international relations, University of Windsor (Ontario).

Dan Bisbee is a PhD student at the University of Pittsburgh's Graduate School for Public and International Affairs, researching warfare and governance challenges in urban terrain. Dan served with the U.S. Army and State Department on the Baghdad Provincial Reconstruction Team during multiple tours in Iraq from 2005 to 2008, developing counterinsurgency strategy, implementing reconstruction plans, and conducting diplomacy with Baghdad's municipal leadership. He has an MA in world history from the University of Pittsburgh and an MA in transatlantic security policy from the University of North Carolina at Chapel Hill. Bisbee is currently conducting research on his dissertation, "Metropolitan Battlefields: Urban Combat and City Politics."

Lieutenant General (Ret.) Tej Pratap Singh Brar was commissioned into the elite First Maratha Light Infantry in 1966 and fought in the 1971 war that resulted in the creation of Bangladesh. Later, commanding his own Maratha battalion in Jaffna Peninsula, Sri Lanka, as a part of the Indian Peace Keeping Force, he was awarded the Yudh Seva Medal (YSM) for distinguished service in an operational context. Later assignments included Indian Army liaison officer to the British Infantry School (Warminster) and commander of the important XVI Corps in Jammu & Kashmir theater, during which he was wounded in a July 22, 2003, terrorist suicide attack. For his exceptional service, he also has been awarded the PVSM (Param Vishisht Seva Medal). He completed his service as commandant of the Defence Services Staff College at Wellington, Tamil Nadu.

Scott N. Carlson has over two decades of experience developing, implementing, and evaluating complex international legal assistance programs in conflict, postconflict, and other transitional environments. For the U.S. Department of State, he deployed to Camp Phoenix, Afghanistan, in 2011, where he served as technical director for the Interagency Rule of Law Office, planning, coordinating, and synchronizing ROL efforts across the country with the United Nations, NATO, EU, World Bank, USAID, and DFID. Most notably, he oversaw the first interagency evaluation of the Provincial Justice Centers, for which he received a Meritorious Honor Award. At Main State, Carlson provided expert assistance on international ROL and anticorruption programming for the Bureau of International Narcotics and Law Enforcement (INL) and the Office of the Coordinator for Reconstruction and Stabilization, serving as an adviser in various countries, including Cambodia, Haiti, Lebanon, and Thailand. He also served as a delegate to international working groups, such as the UN Access to Justice Working Group. Before INL, he supplied expertise to leading international development organizations. For the United Nations Department of Peacekeeping Operations, he developed the *Primer for Justice Components in Multidimensional Peace Operations*. For the Millennium Challenge Corporation, Carlson designed and implemented an anticorruption program in Albania, which boosted transparency and accountability across three institutions, using an integrated e-government platform to reduce discretion of civil servants working in tax, public procurement, and business registration. He graduated with honors in law from Georgetown University, LLM in international and comparative law, and the University of Georgia, JD. Carlson is fluent in Albanian, proficient in French, and learning Croatian. Currently, he is principal at New-Rule LLC.

Carl Forsberg is a PhD student in the History Department at the University of Texas at Austin, where he is also a Donald D. Harrington graduate fellow and a Clements Center doctoral fellow. His doctoral research focuses on U.S. attempts to build an alliance system in the Middle East from the Second World War to the present. During 2011-12, Forsberg served in Kabul, Afghanistan, as an adviser to the commander of Combined Joint Interagency Task Force *Shafafiyat*, the countercorruption coordinating body for the International Security Assistance Force. Previously, he spent two years as an analyst at the Institute for the Study of War in Washington, DC, where he published reports on counterinsurgency strategy, Afghan politics, and Afghanistan's corruption problem. His research led to invitations to testify before Congress and to join a team conducting research in Afghanistan for General David Petraeus. He holds a BA in history from Yale College.

Joseph Franco is an associate research fellow at the Centre of Excellence for National Security (CENS), specializing in radicalization studies, counterterrorism, and counterinsurgency. Franco was awarded an MS in international relations from the S. Rajaratnam School of International Studies at Nanyang Technological University, Singapore, through an ASEAN graduate scholarship. He previously worked for the chief of staff, Armed Forces of the Philippines (AFP) and the deputy chief of staff for operations (J3), AFP. His portfolio with the AFP covered research on internal conflict, peacekeeping operations, defense procurement, Asia-Pacific security, and special operations forces. He was involved in revising the AFP National Military Strategy and other AFP-wide policies. Franco was the lead writer of the *AFP Peace and Development Team Manual,* a novel, community-based approach to counterinsurgency, which remains in use among all AFP units involved in internal security operations. At CENS, Franco is a prolific writer of commentaries and articles on internal conflict and terrorism in the southern Philippines and maritime Southeast Asia. His work has been featured in such periodicals such as the *Straits Times,* the *Jakarta Post,* and the *Nation*. Franco also shared his research findings with various audiences in major industry events such as Global Security Asia and in Singapore-based institutions such as the Civil Defence Academy. He is also frequently interviewed as a resource for international media outlets such as *TIME, Channel News Asia,* and Deutsche Welle. Currently, Franco is working on research projects on target displacement theory and terrorist networks in Southeast Asia.

Michelle Hughes is a lawyer, educator, writer, and consultant. Her work focuses principally on building capacity for multinational, interagency, civil-military, and public-private cooperation to build and strengthen the rule of law, resolve violent conflict, and enable sustainable peace. She is the founder, president, and CEO of VALRAC Innovation LLC, a company dedicated to preparing the next generation to restore and strengthen the rule of law at home and abroad. She is also a senior fellow with the National Defense University Joint Forces Staff College, the senior PROLAW development adviser for Loyola University Chicago School of Law, and a fellow with the Center for Advanced Defense Studies. Hughes was a senior executive in the U.S. Department of Defense, where she was the only designated "highly qualified expert" for rule of law and security sector reform. She has field experience in 12 conflict countries and deployed multiple times to Afghanistan, where her role was to advise senior military commanders on how to connect security force development to governance and justice. Originally from Buffalo, NY, Hughes graduated from the University of Florida in 1979 with a BA in English and was commissioned as a second lieutenant in the United States Army. As a military intelligence officer, she was one of the first women to serve in the elite Eighty-Second Airborne Division, and the only woman officer ever assigned to the Army's parachute demonstration team, the Golden Knights. Hughes graduated at the top of her class from Regent University School of Law in 1996 and practiced complex civil litigation, prosecution, and criminal defense with the firm of Williams Mullen. She is admitted to the Bar in Virginia and the District of Columbia.

David E. A. Johnson is executive director of C4ADS, a Washington, DC-based nonpartisan, nonprofit organization dedicated to improving global security. C4ADS won the Google Chairman's 2014 New Digital Age Grant. Johnson is a graduate of the United States Military Academy at West Point, the Command and General Staff Course, and the

Joint Defense College (War College) at the École Militaire in Paris and holds a master's degree in the history of strategy from la Sorbonne. His career as a decorated Special Forces combat veteran and Army strategist with service on six continents and multiple overseas contingency operations is outlined in the Congressional Record. Before coming to C4ADS, he worked for Intel Corporation as director of digital security products in the Software and Solutions Group. He speaks French, Russian, and Arabic.

Deniz Kocak is currently a temporary lecturer in political science and a research affiliate with the Berlin Graduate School for Transnational Studies at the Free University Berlin. Previously, he worked as a research associate with the Special Research Unit 700 "Governance in Areas of Limited Statehood," Free University Berlin, on security sector reform and security governance transfers. He studied, worked, or conducted research at Humboldt University of Berlin, the University of Potsdam, Free University Berlin, Chulalongkorn University Bangkok, and Singapore Management University. While his doctoral thesis dealt with police reform and institutional change, his overall research interests include security governance, security sector reform, and civil-military relations in non-OECD countries, with a focus on Southeast Asia.

Mark Kroeker is senior partner of Kroeker Partners LLC, a company that offers a wide range of global services, including justice, rule of law, and security sector development efforts, with a heavy focus on postconflict and fragile environments. He is temporarily assigned as assistant secretary-general in the role of deputy special representative of the secretary-General for rule of law, ad interim, in the UN Mission in Liberia, (UNMIL). Following 32 years of service in the Los Angeles Police Department, where he rose to the rank of deputy chief of police, he served as deputy police commissioner of the UN Mission in Bosnia and Herzegovina. He later became police chief of Portland, Oregon, where he served for almost four years. In 2003, he was appointed the first police commissioner for UNMIL, and later he was appointed police adviser and director of the Police Division in the UN Department of Peacekeeping Operations in New York. There, he oversaw UN police operations in all peacekeeping missions. For five years, he served as vice president for global intelligence, threat analysis, and crisis management for the Walt Disney Company. Before launching Kroeker Partners in 2015, he served for three years as senior vice president for justice and rule of law at PAE, a global government contracting company. He was a member of the UN Model Criminal Codes Committee, the UN Working Group on the Protection of Civilians, the International Policing Advisory Committee, and the American-Israeli-Palestinian Anti-incitement Committee. In 1988, he founded the World Children's Transplant Fund, and he continues to serve as the board's chairman. He has a BS degree from California State University at Los Angeles, and an MS in international public administration from the University of Southern California.

Thomas A. Marks is head of department, War and Conflict Studies, College of International Security Affairs (CISA) of the National Defense University, in Washington, DC. He completed his doctoral work in his home state of Hawaii, where, for 14 years, he was chair and professor of social science at Academy of the Pacific, a private high school. While there, for more than two decades he was a highly successful cross-country and track coach at all levels of competition. He was a leading authority on terrorism and insurgency (especially Maoist) when asked to join CISA as a consequence of 9/11. He has written hundreds of publications, several dozen of them on Sri Lanka. His instructional positions in the counterterrorism field include the Oppenheimer Chair of Warfighting Strategy at the Marine Corps University (Quantico), and longtime adjunct professorships at the Air Force Special Operations School (Hurlburt Field) and the intelligence community's Sherman Kent School (Washington, DC). A former army and U.S. government officer, he subsequently worked as an independent contractor for, among others, Control Risks of London. For a decade during the heyday of *Soldier of Fortune,* Marks was the magazine's chief foreign correspondent.

John Robert McBrien is a consultant specializing in economic sanctions and interdisciplinary countermeasures against transnational threat networks. An authority on the employment of U.S. economic sanctions programs, he had a seminal role in the conception, design, and development of targeted sanctions against nonstate foreign adversaries. His initiatives have been a key factor in the evolution of economic sanctions as major instruments of national security policy. At retirement, he was the associate director for global targeting in the U.S. Treasury Department's Office of Foreign Assets Control (OFAC) and responsible for the Specially Designated Nationals programs directed against the operatives and networks of sanctioned countries, regimes, and nonstate foreign adversaries. McBrien retired in December 2012 after 42 years of government service. Most of his career, including 25 years with OFAC, was with the Treasury Department, where he handled a broad spectrum of cross-cutting national security, intelligence, and law enforcement issues. He was involved in the U.S. counterterrorism program from 1972 forward and in counternarcotics programs since 1985. His career in Washington began as an attorney in the U.S. Justice Department's Organized Crime and Racketeering Section. He also played a role in developing President Obama's National Strategy on Transnational Organized Crime and the Executive Order against transnational criminal organizations. A graduate of the National War College, he also served as a visiting scholar at the Center for Strategic and International Studies. A member of the bar, He holds a JD from St. Louis University School of Law, and a BA from Southern Illinois University Edwardsville.

Michael Miklaucic is the director of research, information, and publications at the Center for Complex Operations (CCO) at National Defense University, and the editor of *PRISM,* the journal of CCO. Before this assignment, he served in various positions at the U.S. Agency for International Development (USAID) and the Department of State, including as USAID representative on the Civilian Response Corps Interagency Task Force, as the senior program officer in the USAID Office of Democracy and Governance, and as rule of law specialist in the Center for Democracy and Governance. In 2002-3, he served as the Department of State deputy for war crimes issues. In that position, he was responsible for U.S. relations with the International Criminal Tribunal for Rwanda, the Special Court for Sierra Leone, war crimes issues and negotiations in East Timor and Cambodia, and early implementation of the Sudan Peace Act. His university education was at the University of California, the London School of Economics, and the School for Advanced International Studies. He is adjunct professor of U.S. foreign policy at American University, and of conflict and development at George Mason University. He sits on several academic and professional advisory boards.

Carlos Ospina is a retired general and former commander of the Colombian Armed Forces, serving as Colombia's top military officer until his retirement in 2006. At that point, he became the chief of defense chair and professor of national security affairs at the Center for Hemispheric Defense Studies at the National Defense University, in Washington, DC. Since 2014, he has served as a distinguished professor of practice at the College of International Security Affairs (CISA). Commissioned from the Colombian Military Academy in 1967, he rose steadily, commanding at every level, including the Second Mobile Brigade (BRIM), Fourth Brigade (Medellín), Fourth Division, and the Colombian Army. He emerged as one of Colombia's most decorated combat soldiers, having been wounded in action while commanding Fourth Brigade. As a major, he was decorated when a small jungle tracking unit he commanded, less than a squad, assaulted more than a hundred M19 guerrillas who had landed on the southwestern coast of Colombia—an action that led to their complete neutralization once reinforcements arrived. Simultaneously, General Ospina played a key role in the crafting and implementation of the approach that turned the tide of the counterinsurgency and restored the strategic initiative, which Colombia still holds. Together with Generals Tapias and Mora, he is credited with reversing what was seen as a hopeless situation and doing so in a manner that won the respect of his soldiers and his country.

Gretchen Peters conducts research and analysis work to help governments understand and disrupt transnational organized crime. She has supported efforts in Kenya, Mozambique, and Gabon to fight elephant and rhino poaching, focusing on disrupting wildlife trafficking networks at the transport and financial levels. She also cochairs the OECD Task Force on Wildlife and Environmental Crime. Her prior work focused on the links between narcotics and terrorism in Pakistan, supporting the deputy assistant secretary of defense for counternarcotics and global threats, U.S. Central Command's J2 (Intel), and Special Operations Command. Peters is considered a leading authority on D-Company and other Pakistani drug-terror syndicates. She is the author of *Seeds of Terror*, a groundbreaking book that traced the role the opium trade has played in three decades of conflict in Afghanistan. She spent five years researching the book, which *Barron's* magazine called "a well-written, well-documented and exemplary work of journalism." In the past, she worked as a foreign correspondent and investigative reporter, covering Pakistan and Afghanistan for more than a decade, first for the Associated Press and later for *ABC News*. She has published editorials in the *New York Times*, the *Washington Post*, and *Foreign Policy* and has also reported from Mexico, Cambodia, China, Taiwan, Burma, Vietnam, Laos, Thailand, Kosovo, India, and Egypt.

Ismail Rashid is a professor of history at Vassar College, where he has been teaching since 1998. He grew up in Freetown, Sierra Leone. He received his BA honors in classics and history from the University of Ghana, his MA in race relations from Wilfrid Laurier University, Canada, and his PhD in African history from McGill University. His primary teaching interests are modern African history, enslavement, resistance and emancipation, and the African diaspora. His research interests include subaltern resistance against colonialism, and social and military conflicts in contemporary Africa. Tecent publications include "Epidemics and Resistance in Sierra Leone during the First World War," *Canadian Journal of African Studies* 45 (2011): 415-39; "Decolonization and Popular Contestation in Sierra Leone: The Peasant War of 1955-1956," *Afrika Zamani* 17

(2009): 115-44; *West Africa's Security Challenges,* coedited with Adekeye Adebajo (Lynne Rienner, 2004); "Religious Militancy and Violence in West Africa: A Study of Islam in Sierra Leone" (cowritten with Kevin O'Brien), in *Militancy and Violence in West Africa: Religion, Politics and Radicalization,* ed. James Gow, Funmi Olonisakin, Ernst Dijxhorn, (London: Routledge, 2013); and *The Paradoxes of History and Memory in Postcolonial Sierra Leone,* coedited with Sylvia Ojukutu-Macauley (Lexington, 2013, forthcoming).

William Reno is a professor of political science and director of the Program of African Studies at Northwestern University. He is the author of *Corruption and State Politics in Sierra Leone* (Cambridge, 1995), *Warlord Politics and African States* (Lynne Rienner, 1998), and *Warfare in Independent Africa* (Cambridge, 2011) as well as numerous other publications regarding on-the-ground politics of conflict in sub-Saharan Africa. His recent work focuses on the determinants of the organizational strategies of nonstate armed groups, with particular focus on how leaders of these groups manage the challenges of operating in socially fragmented environments of collapsed states. Research for this project has included extensive field investigations in Somalia and elsewhere in Africa to investigate the organizational strategies of these groups. His other projects include a study of the changes in civil-military relations in African states, such as how governments include their militaries in new economic development strategies and how participation in regional peacekeeping forces reshapes these forces. Another study focuses on African approaches to counterinsurgency operations.

Tim Sullivan is a joint MA/MBA candidate at the Yale School of Management and Yale's Jackson Institute for Global Affairs, where he focuses on entrepreneurship and institution building in developing contexts. Before his graduate studies, Sullivan served for over a year in Kabul as an adviser to the commander of Combined Joint Interagency Task Force *Shafafiyat* ("transparency" in Dari and Pashto), which was responsible for the countercorruption efforts of NATO's International Security Assistance Force. Previously, Sullivan was a research fellow in the Foreign and Defense Policy Studies Department at the American Enterprise Institute in Washington, DC. Most recently, he led the business development team of Praescient Analytics, a national security-focused analytic technology services start-up in Alexandria, Virginia. He is a member of the Center for a New American Security Next Generation National Security Leaders program and splits his time between New York and New Haven.

Phil Williams is Wesley W. Posvar professor and director of the Matthew B. Ridgway Center for International Security Studies, at the University of Pittsburgh. Williams has published extensively in the field of international security. He has written books on crisis management, the U.S. Senate and troops in Europe, and superpower détente and has edited volumes on Russian organized crime, trafficking in women, and combating transnational crime. During 2001-2, he was a visiting scientist at CERT, where he worked on cyber crime and on infrastructure protection. During 2007-9, he was visiting professor at the Strategic Studies Institute, U.S. Army War College, where he published two monographs: *From the New Middle Ages to a New Dark Age: The Decline of the State and U.S. Strategy* (2008) and *Criminals, Militias and Insurgents: Organized Crime in Iraq* (2009). Williams has worked extensively on transnational criminal and terrorist networks; terrorist finances; and, most recently, the rise of drug trafficking violence in Mexico, Nigerian organized crime, and long-term trends and their impact on organized crime. He has recently coedited a volume for the Strategic Studies Institute on malevolence in cyberspace and is currently working on the crisis of governance in the northern triangle of Central America, and a cowritten monograph on military contingencies in megacities.